The Indian Space Programme

India's incredible journey from the Third
World towards the First

Gurbir Singh

Gurbir Singh/Astrotalkuk Publications
www.astrotalkuk.org
info@astrotalkuk.org

Front and back cover images adapted from ISRO. Front cover GSLV Mk3 launch on 05/06/2017. Back cover image of the Earth taken on 1/12/2013 by Mars Orbiter Mission while in Earth orbit.

Ordering Information:
Quantity sales. Special discounts are available on quantity purchases by corporations, associations, and others. For details, please contact via an email to info@astrotalkuk.org.

The Indian Space Programme /Gurbir Singh —1st edition.
ISBN: 13 9780956933782
ISBN: 10 956933785

Mohinder – An older brother and an unlikely mentor

About the Author

Gurbir Singh is the publisher of www.astrotalkuk.org and the author of *Yuri Gagarin in London and Manchester* published in 2011 to mark the 50th anniversary of humanity's first journey into space. A former college lecturer, he is working in the information security sector. He has a science and an arts degree.

Once keen on flying, Gurbir holds a private pilot's license for the UK, US and Australia. He was one of the 13,000 unsuccessful applicants responding to the 1989 advert "Astronaut wanted. No experience necessary" to become the first British astronaut, for which Helen Sharman was eventually selected and flew on the Soviet space station MIR in 1991.

Born in India, he has been living in the UK since 1966 except for one year in Australia. He is married, with a nine-year-old daughter and lives in Lancashire in England.

◅ ◈ ▻

Contents

List of Figures

List of Tables

Introduction

Someone once said that if a book you want to read does not exist, it is your responsibility to write it. I was looking for a book that would help me understand how the Indian space programme had evolved, the key players, its technical capability, how space-based services are being used by Indian citizens and its role in shaping the nation's destiny as it ploughs headlong into the 21st century. Such a book did not exist. So, I set out to write it and after almost six years, this in your hands is it.

The central theme that runs through these pages is why in just 15 years after independence, India rolled up its sleeves and chose space technology to help transform a developing nation towards a developed nation. To what extent has it succeeded, and how has this capability shaped the perception of India in the 21st century? The start of India's colonial period coincided with the European scientific Renaissance. The first ship from the UK arrived in India as Galileo looked at the night sky through the newly invented telescope. Was the vast maritime traffic and the inevitable people-to-people communication between Europe and India in the 16th and 17th centuries equivalent to the modern high-speed Internet link, bringing the products of the Industrial Age from the mighty cities of Europe to small villages in mostly rural India?

The story of the Indian space programme, as embodied by the Indian Space Research Organisation (ISRO), is long and complex. It cannot be comprehensively told in one telling. In this book, I attempt to explore the Indian space programme in three broad sections historical (chapters 1 to 6), existing capabilities (chapters 7 to 12) and current and new projects (chapters 13 to 17). The historical section looks at how the traditions of science and technology of ancient India have been preserved through India's long and tumultuous history. India's ancient traditions of science can be glimpsed in its modern institutions, and hints of modern scientific methods can be seen in its past.

Mathematics and the application of scientific thinking were present in the civilisation that first arose around the Indus Valley perhaps 8000 years ago.

Aryabhata, Bhaskara and others continued that scientific tradition albeit to satisfy an astrological demand which surprisingly continues to co-exist in India to this day.

At the outset of the colonial period, the British navy provided an information highway between India and the dynamic 16th-century European Renaissance. Scientists, scientific societies and institutions in India benefited to the extent that they served the British motive for profit. Nevertheless, a few gifted individuals did emerge in India, in some cases supported by the colonial power and have left their mark on modern science. Their names and scientific achievements, such as Bhabha Scattering, the Saha Equation, the Boson, Raman Effect, Bhatnagar-Mathur Magnetic Interference Balance and the Chandrasekhar Limit, are recorded in modern textbooks, although readers may not always be aware of their Indian origins.

Before 1799, Tipu Sultan had used rockets as weapons. The first rocket to enter space from India was launched 1963. In between, a now almost-forgotten actor tested rockets in India as a means of transport. In chapter 5, India's Forgotten Rocketeer, I present perhaps the most detailed account of Stephen Smith's contribution to rocketry in the early 20th century. Starting in the 1930s, in numerous experiments, he demonstrated the delivery of mail and small quantities of items useful in an emergency in the challenging mountainous landscape of Sikkim. He was also the first to test the use of rockets to transport living animals, a small hen and a cock. My thanks to Maureen Evers from the Families in British India Society for helping me track down Stephen Smith's now London-based descendants.

The existing capability of the Indian space programme is a moving target. I have attempted to trace the evolution of India's existing infrastructure of launch pads, launch vehicles and spacecraft and the technologies required to succeed in challenging and complex space missions. In painting that picture, I have also sought to capture the contributions of key individuals, such as Vikram Sarabhai, Homi Bhabha, Abdul Kalam, Satish Dhawan, U.R. Rao, and the international collaboration that they fostered to help kick-start ISRO's infrastructure, particularly during the first four decades. India's tradition of international collaboration continues today through active programmes with the UN and BRICS nations, as well as new ventures with Israel, Japan, NASA and European Space Agency. Since 2014, a new dynamic government with a nationalist and an aggressive economic agenda has been positioning itself to use the Indian space programme as an instrument for regional influence.

The current and new projects for which the Indian space programme has acquired global recognition emerged in the mid-1990s. By then, it had accomplished the goals set by its founders. India had developed an indigenous capability to build, launch and operate remote sensing, meteorological and communication satellites. Since the millennium, the Indian space programme has broken new ground offering services from space that include satellite navigation, search and rescue, air traffic control, military communication and science. With missions to the Moon and Mars and a telescope in Earth orbit, India is developing new skills and processes and instilling skills in its growing workforce for leading-edge space technologies. As its economy grows, India's population of 1.3 billion is beginning to take space-based services as a given, just as peoples of the developed nations.

The evolution of the Indian space programme from a standing start in 1963 to its current stunning capabilities is a beacon to developing and developed nations alike. But this journey has not been a smooth one. Missed opportunities, internal politics and international sanctions reined back the pace of development for many years. Today, with only a single launch site and still without a fully operational heavy-lift launch vehicle, the prospects for the Indian space programme are limited. The heavy-lift launch vehicle capability is essential to explore the surface of Mars, venture into the outer solar system, embark on human spaceflight and launch and operate a space station. The single launch site at Sriharikota further limits India's space capability. ISRO achieved just eight launches in 2016, and it is not equipped to deliver all the services the nation demands.

Attracted by the engaging and captivating writings of historians, scientists and academics, my interest in history has grown with age. It emerged from reading books by writers, such as Michael Wood, Ramachandra Guha, Dr Rajinder Singh, Amrita Shah, William Dalrymple and especially Professor Rajesh Kochhar, who explores in detail the development of science and technology in India. This interest in the past took me to several archivists, including Sowmithri Ranganathan (IISc), Oindrila Raychaudhuri (TIFR), Paul Jenkins (India Study Circle), Lokesh Sharma (Sikkim State Archives) and Nalini Pradhan (Commissioner cum Secretary to the Government of Sikkim), who hosted my research visits. Writing this book has been a long journey in time and space. It started in 2012 and took me to Sikkim in the north, Thumba in the south, Mumbai in the west and Sriharikota on the eastern coast of India. In addition to face-to-face meetings, I used the old-fashioned post, as well as modern media (Skype, email and Facebook) to source the information presented in these pages.

ISRO scientists, many who held senior posts at key times, provided me with current and accurate information, which is at the heart of this story. Several ISRO directors made time to talk to me, despite being engaged in demanding active missions at the time. They include M.Y.S. Prasad (Sriharikota), S.K. Shivakumar (ISAC), Dr Mylswamy Annadurai (ISAC), V.S. Hegde (Antrix), R. Raghunath (Byalalu), S. Ramakrishnan (VSSC) and especially Professor U.R. Rao, who found time for me during my multiple trips to the ISRO headquarters in Bangalore. Special thanks to two individuals without whom my visits to ISRO centres would not have been possible, the Director of Publications and Public Relations Mr. Deviprasad Karnik and the erudite scientist/engineer B.R. Guruprasad, who provided generous support and guidance from the start.

Acquiring first-hand testimony of the unique beginnings of the Indian space programme has been especially rewarding, as the number of individuals with such experiences is dwindling. Some of these individuals include Mrinalini Sarabhai, Dr Praful Bhavsar, Professor Roddam Narasimha and Professor Jacques Blamont. Several people helped me understand India's human spaceflight activities, most of which happened during the 1980s and not much progress has been made since. N.C. Bhat and P. Radhakrishnan shared their excitement at being selected to fly abroad the Space Shuttle and their disappointment when their mission, along with many others, was cancelled following the Space Shuttle Challenger accident in January 1986. In a very matter-of-fact way, Ravish Malhotra expressed his disappointment at not having experienced spaceflight despite completing a comprehensive training programme for it. My gratitude to Rakesh Sharma and his wife for hosting my visit to Coonoor, where we spoke at length about the adventure of his spaceflight. More than three decades on, he remains the sole individual with spaceflight experience supported by the Indian government.

I drew on the existing work by accomplished writers within and outside India on various aspects of the Indian space programme. These include R.P. Rajagopalan, P.V. Manoranjan Rao, Asif Siddiqi, Robert S. Anderson, Ajey Lele, George Joseph, R. Aravamudan, Gopal Raj (his book *Reach for the Stars: The Evolution of India's Rocket Programme* is probably one of the most underappreciated works in this field), GR Hathwar, D.A. Maharaj, Yash Pal, J.C. Moltz, Brian Harvey and Abdul Kalam.

I also want to acknowledge the contribution of several online sources that provide surprisingly detailed information about spaceflight statistics and even mission analysis. They often provide references to publications and other reliable sources available publicly and offer this service without charge. Some of these sources of online independent information include Anatoly

Zak, Gunter Krebs, Norbert Brügge, Jonathan McDowell, Narayan Prasad, Patrick Blau and V.K. Thakur. I also benefited from the immense body of work accumulated over time by organisations including the Union of Concerned Scientists, The Secure World Foundation, Nuclear Threat Initiative and national space agencies including NASA, ESA, CNSA and of course ISRO.

As with any writing project, it is the research and the rewriting that takes the most time and is the most productive. During the writing process, I sought reviews on initial drafts. Once complete I sought further feedback through a beta reading phase. The reviewers were mostly accomplished authors. One of the first reviewers was Brian Harvey, who provided constructive and encouraging feedback at a critical early stage. Others included Bert Viz, David Baker, Dr Jason Held, Andrew Thomas, Akash Yalagach, Bruce Gilham, Phil Clarke, Davide Sivolella, Gopal Raj and Shambo Bhattacharjee. The end product has been shaped by the input from several beta readers including Neil Jaworski, Mike Little, Manoranjan Rao, Tejaswi Shinde, Srinath Ravichandran, Bhargava Srinarasi and Ohsin.

If my writing looks ordered and polished, that has only been possible thanks to the astonishing contribution from my eagle-eyed editor Liza Joseph. I thank Bhushan Hadkar for his creative and technical skills for improving the quality of the images presented here.

During my research trips across India, I stayed in various places, but in Bangalore, I was fortunate to enjoy the warm hospitality and family atmosphere provided by my friend Satish, his wife Sumeetha and daughters Smurthi and Swathi (the S family).

Several individuals, including Bert Viz, Professor Praful Bhavsar, Professor U.R. Rao and Jacques Blamont, Dr Aravamudan, Dr G Joseph and others have kindly shared priceless images from their personal archives. Despite my efforts, I have been unable to identify the copyright owners of a few, particularly older, images. I would welcome your assistance in case you are able to identify them.

The development of India's space technology is the central theme that connects all chapters of this book. Each chapter, however, is self-standing and can be read independently of the others. Given the focus on rocket technology, space and science, inevitably, there are many technical terms. These are explained in the text or are supported by several appendices, illustrations and particularly the list of abbreviations.

You will find some element of apparent inconsistency regarding place names. Whereas Madras is now well established as Chennai, Bengaluru as a replacement for Bangalore has been less successful. To further complicate matters, the city names I use depend on the context so you will see both versions of city names in the text. I also noted discrepancies in dates when, for example, launch events in the US or ESA's launch site Kourou in French Guiana are recorded in India. Typically, the dates can be out by a day given the time zones.

I also regularly found significant variation in the published figures for costs of space missions, the mass of satellites, orbital altitude and other quantitative values. I also found variations in the costs of missions and assets. Thus, they too should be considered approximate given rounding errors and fluctuating exchange rates. I have attempted to represent currency in Indian rupees and US dollars for consistency. The use of lakhs and crores in Indian currency is another challenge for those who do not interact with it on a regular basis. I have added an appendix that captures the fundamental terminology of Indian currency.

Finally, the views and opinions presented here are my own. The only objective of this book is the one I outlined in the first paragraph. Despite all the cycles of checking and reviews, I can only assume that some errors remain. Responsibility for these lies with me. In case you spot any, do let me know and I will make the necessary correction in a potential future edition.

Gurbir Singh, Lancashire England, October 2017

◄ ◊ ►

Chapter One
Rise of National Space Programmes

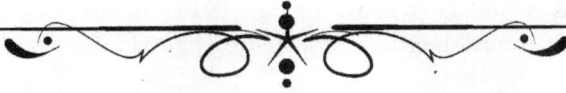

***

Duringthe closing days of World War II, many European cities experienced the terror of death and destruction from rockets. German V1 and V2 rockets landed in many parts of the UK, Belgium and France bringing a new, unfamiliar fear[1]. The first time that British forces experienced this fear of rockets was, however, almost a century and a half earlier in India. On 10 September 1780, Tipu Sultan (1750–1799) of the Kingdom of Mysore in India and his rocket corps inflicted on British forces led by Colonel William Baillie (died 1782) of the British East India Company, what was at the time, their greatest military defeat. The Mughal Empire, which had ruled most of India for over two centuries, was in decline by the 17th century, and European powers were competing to replace it.[2] Starting with small trading posts in the 16th century, the French, Portuguese, Dutch and British had established colonies in India through their respective East India companies.[3] The military campaigns were not only between Europeans and Indian rulers but also between the Europeans in India, especially the British and French. The two embarked on a race to rule the post-Mughal India.

Europeans had been fighting each other in Europe throughout the Middle Ages. By the late 18th century, the UK was in conflict with France on three different continents, America, Africa and India. It was fighting colonists supported by the French in America, Napoleon Bonaparte (1769–1821) in Egypt, and Indian states that had aligned militarily with the French in India. By then, France and Britain had honed their respective navies into formidable military machines. A nation's supremacy at sea was not just a symbol of national pride and wealth, but also a direct measure of its international influence and power, equivalent to a nuclear power status today. For Napoleon, Egypt was a stepping stone to India, where his goal was to disrupt Britain's presence and eventually replace it with France's.[4] In the aftermath of the American War of Independence in 1783, Britain, too, looked east to India to regain its global influence. As in North America, France stood in the way.

Between 1797 and 1798, three brothers arrived in India from Britain. Their collective efforts played a key role in ensuring that Napoleon ambitions failed and that Britain, not France, triumphed in India. Richard Colley Wellesley (1760–1842) arrived as the Governor General of India.[5] His youngest brother, Henry Wellesley (1773–1847), was a career diplomat and officially became his private secretary. Arthur Wellesley (1769–1852) was initially a soldier and later a statesman but found his place in history by defeating Napoleon in the battle of Waterloo in 1815.

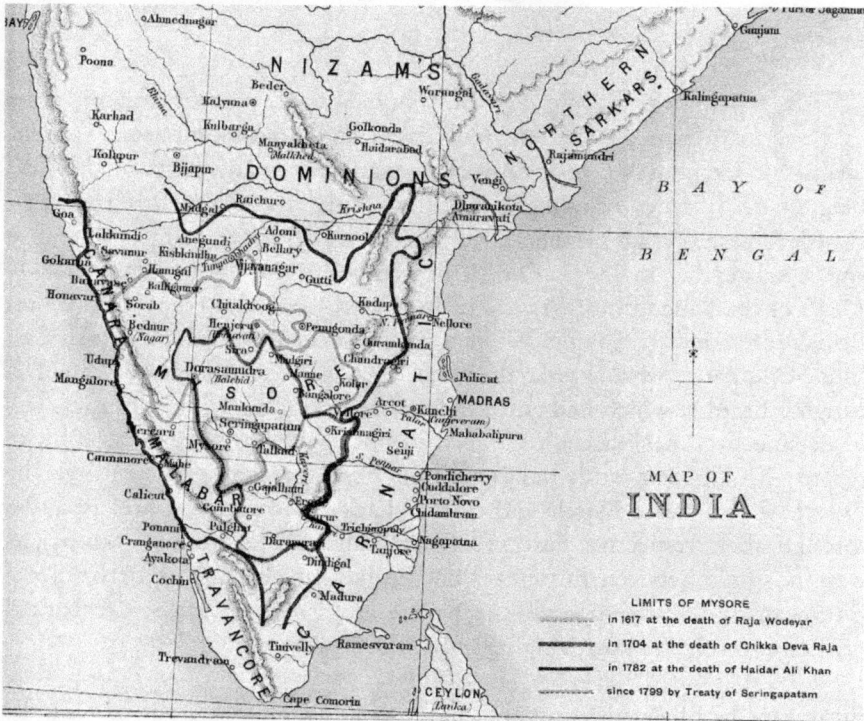

Figure 1-1 History of Mysore 1617-1799. Credit John Bartholomew & Co. 1897

Richard Wellesley introduced and vigorously pursued the concept of Subsidiary Alliance, a protection racket dressed up as foreign policy. Subsidiary Alliance assured Indian rulers of British help against external attack or internal revolt. For this to work, the Indian ruler had to align exclusively with Britain with the British veto preventing any future official relationship with other foreign powers, employ a British official at court, expel all non-British European officials, and permanently base British troops within their territory at the cost of the ruler. Wellesley successfully convinced the rulers of several Indian states, including the Nizam of Hyderabad, the Nawab of Oudh, the Raja of Tanjore and the Rajput States, to take up Subsidiary Alliance. With this single measure, Wellesley successfully removed French influence, secured British military presence inside potentially troublesome princely states, asserted and grew British power and generated huge wealth for the British East India Company. However, not everything went to plan. The ruler of the Kingdom of Mysore would not play ball.

In 1761, an illiterate but gifted military leader Hyder Ali Khan (c.1720–1782) had taken charge as chief minister in the southern Indian city of Mysore. Eventually, he became the ruler of the Kingdom of Mysore, and in his time, Mysore peaked in its military power. Hyder Ali established strong relationships with the French and Dutch outposts in India providing him access to goods and services through the ports they controlled. In the 1790s, his son Tipu Sultan continued the affiliation and established contact with Napoleon when he arrived in Egypt.[6] Given his fierce anti-colonialist stance, shrewd political insight and efficient governance, Hyder Ali, and later his son Tipu Sultan, fought four wars against the British East India Company to prevent what they saw as foreign rule.

Tipu Sultan rejected Wellesley's offer of Subsidiary Alliance. In response, Wellesley offered incentives to the rulers of states neighbouring the Kingdom of Mysore if they helped him acquire it. They would be offered parts of Mysore's territory once it fell to British rule. Tipu Sultan persisted fiercely and continued to oppose the British in Mysore. He resisted breaking ties with the French. Wellesley saw this as a declaration of war, and he proceeded to undermine Tipu Sultan's alliances through manoeuvres (incentivise Tipu's supporters to switch sides) similar to those of Robert Clive's (1725–1774) in 1757 when the latter defeated the Nawab of Bengal in the seminal battle of Plassey. By early 1799 and the Fourth Anglo-Mysore War, Wellesley was ready to finally defeat Tipu Sultan and acquire the troublesome Kingdom of Mysore. On 4 May 1799, Tipu Sultan was killed during the battle of Srirangapatna (known then as Seringapatam) ending Mysore's resistance and the series of four Anglo-Mysore Wars between 1767 and 1799. The British East India Company would have triumphed sooner but for the rockets that Hyder Ali and his son Tipu Sultan deployed with surprising effectiveness.

Tipu's Rockets

The invention and early development of rockets and their use in warfare are generally accepted to have taken place in China. In 1232, gunpowder-based rockets were used by the Chinese City of Kaifeng against a 30,000 strong Mongol force led by the son and successor of Genghis Khan (c. 1162–

1227).[7] Marco Polo (1254–1324), a Venetian traveller and merchant, brought a sample of these Chinese war rockets to Europe from his travels in 1275.[8] Versions of these rockets were deployed in European conflicts in the years that followed but with little or no innovation. It was in Mysore under Hyder Ali that rockets were developed into a sophisticated tactical military force for use as an incendiary and anti-personnel weapon.[9] With his extensive experience in military tactics, Hyder Ali effectively used the newly designed war rockets to preserve his rule and extend the Kingdom of Mysore.

Figure 1-2 Battle of Pollilur. The Ammunition Cart Exploding in the Middle of the Defensive British Square. A Mural in the Summer Palace, Srirangapatna. Credit Author

Mysorian rockets were cylindrical, 30 cm in diameter and 60 cm long, with a case made of metal rather than the traditional design using wood or heavy paper. Attaching a sword or a 3 m long bamboo staff provided stability. With a total weight of around 3 kg, it had a range of about 2 km (1.24 miles). The key innovation was the metal casing, which resulted in higher combustion temperature and pressure, increasing the size of the rocket and its range beyond anything that had existed until then. In the 18th century, the highest quality iron available in the world was being manufactured in India. India had a metallurgic tradition going back to the remarkable bronze statues of the 12th-century Cholan civilisation, which had flourished in the same region where Hyder Ali was developing his war rockets.

Unlike Hyder Ali, Tipu Sultan recorded his military tactics, including the use of rockets. He distributed these tactics in a manual called Fathul

Mujahidin to all his officers.[10] In the Second Anglo-Mysore War, Hyder Ali joined the siege in Arcot, and Tipu led the forces at the battle of Pollilur. On 10 September 1780, more than 7,000 British soldiers were held captive at Tipu's fortress in Srirangapatna.[11] A spectacular explosion of the British arsenal resulting possibly from a rocket impact is one of the images depicted in a mural on the walls of the Daria Daulat Bagh, Tipu's former palace and now a museum.

On 5 April 1799, in one of the early battles of the Fourth Anglo-Mysore War, Richard Wellesley's younger brother Arthur, who later went on to become the Prime Minister of Britain, was almost killed during an ill-prepared night-time encounter with the forces of Tipu Sultan. One vivid account captures the impact on the soldiers experiencing rockets for the first time "Colonel Wellesley, advancing at the height of his regiment, the 33rd, into the tope, was instantly attacked, in the darkness of the night, on every side, by a tremendous fire of musketry and rockets. The men gave way, dispersed and retreated in disorder. Several were killed and twelve grenadiers (these men were all murdered a day or two before the storm) were taken prisoners."[12]

Arthur Wellesley's experience of abject shock and terror was not unique. Many of the troops facing rockets for the first time found the impact to be physically and psychologically profound even though they were battle hardened with the experience of numerous military encounters. Colonel Baillie describes his experience as the rockets approached "The rockets and musketry from 20,000 of the enemy were incessant. No hail could be thicker. Every illumination of blue lights was accompanied by a shower of rockets, some of which entered the head of the column, passing through to the rear, causing death, wounds, and dreadful lacerations from the long bamboos...".[13]

During the last three decades of the 18th century, Hyder Ali and his son, Tipu Sultan, bolstered their military with a contingent of rocket men, initially 1,200 and later strengthened to 5,000 men.[14] Tipu Sultan succeeded his father, Hyder Ali, following his death in 1782, and he was British East India Company's most formidable and tenacious adversary "a fanatical and relentless warrior, he vowed not to mount his elaborate throne until he had vanquished the British."[15] With the help of his French military connections, Tipu imported industrial technology from Europe and also introduced silk production in Mysore with help from China. He continued his father's tradition of developing rocket technology and training his men on how best to deploy rockets to meet a military objective. Mobile structures on wheels capable of launching multiple rockets at a time

5

introduced greater tactical agility. Tipu's rocket men honed their targeting skills by calculating the angle of launch based on the distance to the target and the diameter of the rocket. The metal blades and bamboo shafts spun at high speed like uncontrolled scythes. Sometimes, a rocket would either explode or burn slowly, increasing the unpredictability of its impact.

During the four Anglo-Mysore Wars, state-of-the-art military rocket technology was deployed against the British East India Company. However, Srirangapatna was overrun eventually on 4 May 1799, and Tipu Sultan lay dead. As victorious soldiers of the British East India Company searched the extended area of Srirangapatna, they located 600 launchers, 700 serviceable rockets and 9,000 empty cases.[16] Some of these found their way to London, and another father and son team, both called William Congreve, systematically developed these rockets from Mysore, enhancing their range, size, reliability and thus military effectiveness.[17]

Rockets and Empire

In the second half of the 18th century, Britain's expanding empire made increasing demands on its military infrastructure to support its imperial ambitions around the world. The Royal Arsenal located on the banks of River Thames in south-east London was established in the late 17th century to develop and test ordnance and artillery. Tipu's rockets from Mysore ended up at the Royal Arsenal, and the journey of rocketry development in Europe began.

Initially known as the Woolwich Warren, the Royal Arsenal was developed into a large, sophisticated complex for research and development (R&D), as well as the manufacture of high-grade weapons. It had its own laboratory, foundry, storehouse and a permanent base for army engineers. On 15 May 1778, Colonel William Congreve (1743–1814) was appointed as the Superintendent of Military Machines on a salary of 101 pounds and 25 shillings per year. His remit was to "improve the science and practice of artillery."[18] In April 1783, as the Anglo-Mysore Wars raged in India, William Congreve was appointed as the Deputy Comptroller[19] of the Royal Arsenal, and in 1789 as the Comptroller, to test and develop high-grade gunpowder for Britain.

Congreve introduced a series of innovations to the traditional practices in the Arsenal's production of weapons. Instead of bringing in the private sector as advocated by the prime minister, he introduced new lean working practices, refined quality control, put in place a rigorous testing regime, reduced management layers and increased the number of government sites for gunpowder production in the north and south of London. Using the processes and procedures that emerged from the Industrial Revolution, Colonel Congreve established systems and procedures for the efficient and cost-effective production of military ordnance. He transformed the Royal Arsenal into an economic centralised system of armament production operated by the state's officials under his new vision.[20] Upon his death in 1814, Congreve Senior's role was passed on to his son William Congreve Junior (1772–1828). Both Williams received a baronetcy and a knighthood during their lifetimes and were formally known as Sir William Congreve 1st Baronet and Sir William Congreve 2nd Baronet.

Figure 1-3 Congreve rocket fired at Stonington in August 1814. Credit Stonington Historical Society

Congreve Junior was a colourful character in history. Having studied law, mathematics and chemistry, he went into politics editing a Conservative Party Newspaper, *The Royal Standard* and turned to inventing after a damaging libel action. Living most of his adult life with his mistress and producing two illegitimate sons, he eventually married in 1824. Later in life, he became a businessman, and in 1818, he returned to politics, successfully standing as MP for Plymouth. Accused of fraud, he fled to France, where he died in 1828 bringing an ignominious end to a very creative and productive life. During his lifetime, he recorded 18 patents, two of which were associated with rockets: whale harpoons and flares to light up a battlefield and flares for signalling. His most enduring design, however, is of the unforgeable bank note, which is still in use around the world today.[21]

The new lean environment of the Royal Arsenal was the perfect setting for Congreve Junior to begin enhancing the technical capability of Tipu Sultan's rockets when they arrived in London following Tipu's defeat at Srirangapatna in 1799. With the support of his father, he conducted experiments to investigate their performance. In his 1985 paper, Narasimha Roddam (born 1933) of the National Aeronautical Laboratory and the

Indian Institute of Science (IISc) in Bangalore describes what took place then as "a vigorous research and development programme".[22] Using his understanding of science and chemistry, Congreve Junior systematically analysed the composition of propellants and trajectories using Newton's laws of gravitation and recorded the characteristics of varying sizes of rockets weighing 12, 32 and 42 pounds (6, 16 and 21 kg).[23]

Figure 1-4 Sir William Congreve Second Baronet. Circa 1812. Credit James Lonsdale

He meticulously investigated all aspects of military rocket case design, production methods, cost and practical use in the field. His innovations included new formulations in the manufacture of gun powder, black powder, and warhead design. He developed explosive and incendiary devices with timer fuse, an efficient and fast mechanism for attaching the stabilising stick and a new collapsible light-weight wooden launching frame that could fire volleys of 20 or 50 rockets once every 30 seconds to replace the prevailing heavy-wheeled carriages.[24]

Congreve Junior's commitment to the scientific approach obliged him to meticulously record his data and the conclusions he drew from them. He published three books, the first one in 1810 titled *A Concise Account on the Origin and Progress of the Rocket System*. Perhaps, his most significant contribution was standardised production techniques, a by-product of the

Industrial Revolution. Such a systematic process based on a quantified cost model controlled by a centralised state structure could not have been possible either for Tipu Sultan in India or the 14th century China. Congreve Junior's work was a product of the European scientific Renaissance and industrial methods, modern traditions from which India was remote geographically and conceptually.

To ensure that his work did not remain purely academic, Congreve Junior calculated and recorded the cost of each type of rocket, even itemising the cost of individual items (case, cone, stick, rocket charge, carcass charge, labour and paint). The typical cost of one of the smaller rockets was just over one pound.[25] Through his father, he also had close contact with influential figures in politics, business and the establishment. Further, the British Navy, too, was upbeat and keen on maintaining its superior position post the spectacular success at the Battle of Trafalgar in 1805.

Unlike his father, Congreve Junior had never signed up in the military. However, on that day, under cover of darkness, he participated in a night-time naval attack on the French coastal town of Boulogne. Using his newly designed launching frames, 200 of his new 24-pound rockets were fired in 30 minutes powered by gunpowder of his formulation. Although parts of the town were set on fire, the impact of the rockets was not substantial. However, this being the first instance of the tactical use of rockets in a European war, Congreve Junior could claim a significant strategic advantage.

A year later, with further technological enhancements, over 2,000 Congreve rockets bombarded Copenhagen when the British Navy captured Danish and Norwegian ships in a pre-emptive attack. In the process, "the fifth greatest naval power in Europe was annihilated profoundly."[26] The attack was led by Arthur Wellesley; this time, he would attack with rockets, not defend against them. For this engagement, Congreve also introduced a new boat, a sloop-of-war, designed for launching rockets. The use of rockets from ships brought an additional benefit. Unlike cannons, rockets fired from a ship generated no recoil. Rockets could be fired from smaller ships without the risk of capsizing.

Even though most rockets did not hit their target, when they did, the impact was devastating. One Dane reported that a rocket went through the roof, past three floors and stuck into the side of the building.[27] The might of the British onslaught supported by the new super weapon, evoked a hitherto unknown fear terrifying the Danes into submission. Arthur Wellesley had surrounded Copenhagen in August 1807, and by the end of the first week of September, the Danish fleet surrendered. Initially, the destruction or acquisition of ships was the goal. However, as in India, Wellesley exploited every opportunity to build his personal wealth, reputation and power alongside that of the King and the country. On 21 October 1807, laden with goods looted from Copenhagen's arsenal and 150 ships, almost the entire Danish and Norwegian fleet, Wellesley set sail for the UK.

Figure 1-5 Congreve 32-pounder (15 kg) incendiary rocket. Credit National Air and Space Museum

Through its military victory at Srirangapatna in 1799, the UK had asserted its firm grip on India while diminishing the role of France. Buoyed by the victories in the battles of the Nile, Trafalgar and Copenhagen, the British adopted a tougher position, especially against France and French interests. By the first decade of the 19th century, the British Navy, bolstered by the deployment of its rockets, was the most powerful navy the world had known. There was not much civility or courtesy in the way the Britain wielded that power, which eventually led the US, for the first time in its history, to declare war.

With the Napoleonic Wars being fought in Europe, the British Navy employed an extreme and illegal form of conscription, tantamount to kidnapping or abduction, called impressment to increase its manpower. Young men were coerced into joining the British Navy. Force and violence invariably accompanied this form of recruitment essential for the UK to maintain its naval power and global influence. Many former Britons who had become naturalised Americans were targeted. Impressment was one of the most pressing issues for President James Madison (1751–1836) when the US declared war on Britain, and the fight continued for about two and a half years from 1812 to early 1815.

Many battles during the War of 1812 involved the use of Congreve's rockets. In one, a formidable British flotilla of HMS Ramillies, the frigate HMS Pactolus, HMS Dispatch and HMS Terror arrived off the coast of Stonington on 9 August 1814 and ordered the inhabitants to leave. They refused and fought against the barrage of shells, incendiary and Congreve rockets. The following month, Fort McHenry in Baltimore, close to

Washington, came under a similar barrage. It was witnessed first-hand by a young American lawyer on one of the British ships. He was there to negotiate the release of American civilians who had been captured and was forced to stay aboard the British ships until the end of the battle. As the Sun rose on 14 September, using a telescope, he saw the flag still flying on the Fort and knew instantly that the Fort and hence Baltimore were safe. This unique encounter inspired him to write what later became the US National Anthem. The rocket referred to in the first verse of the anthem had its origin in 18th century India.

O say can you see, by the dawn's early light,
What so proudly we hailed at the twilight's last gleaming,
Whose broad stripes and bright stars through the perilous fight
O'er the ramparts we watch'd were so gallantly streaming?
And the rocket's red glare, the bombs bursting in air,
Gave proof through the night that our flag was still there,
O say does that star-spangled banner yet wave
O'er the land of the free and the home of the brave?

The First Verse of the US National Anthem

Tipu Sultan in India used his rockets to help defend and expand his kingdom. A decade and a half later, refined by Congreve, the same rockets were used by British forces in their fight to assert their dominion over the other side of the world. News of the military use of rockets in the Anglo-Mysore Wars spread across the world in the first few decades of the 19th century. In 1815, Alexander Dmitrievich Zasyadko (1779–1837) developed military rockets in Russia. He built individual rockets and platforms that could be used to launch salvos of six rockets at a time. Five years later, he was appointed the head of the Petersburg Armoury, where he established the first rocket unit of the Russian army. The R7 missile that launched Sputnik into space in 1957 can arguably be traced back to Zasyadko's work inspired by the rocket science in India in the late 18th century.

Invented in China and Mongolia, rockets evolved gradually over centuries with incremental technological innovation in Asia, Europe and ultimately in North America. However, for almost a century following the American War of 1812, it was artillery, not rockets, that saw technological advancements. Then, the early 20th century witnessed an entirely new vision for rockets emerging from gifted writers. For the first time in human history, the confluence of scientific discovery, technological innovation and industrialisation made space exploration and human spaceflight seem more than just a dream.

Founding Fathers of Modern Rocketry

Wernher von Braun (Wernher Magnus Maximilian, Freiherr von Braun, 1912–1977) spearheaded the development of rockets that put the US in space. The day after the launch of Apollo 11 and before its landing on the Moon, he published a piece titled 'Pioneers of a New Age'.[28] While mentioning the contribution of individuals, not just American but across the world and back in history, he also writes about the role of imagination "We sometimes too underestimate the influence of the arts on sciences (and vice versa), particularly in astronautics. It is interesting to note that the three modern rocket pioneers in astronautics ... all had something in common, in addition to their learning and passion for science. They had imaginations that were initially inspired by the fiction of Jules Gabriel Verne (1828–1905), who made space travel sound exciting and even more important technically feasible to young boys with an aptitude for science. The three pioneers he referred to were: Konstantin Eduardovich Tsiolkovskii (or Tsiolkovsky for the West, 1857–1935), Hermann Julius Oberth (1894–1989) and Robert Hutchings Goddard (1882–1945). Between them, they not only imagined the possibilities of space travel but also laid down the theoretical and engineering framework within which it could be realised.

The internal combustion engine was invented in the 19th century triggering an avalanche of inventions in the late 19th and 20th centuries. These inventions transformed agricultural communities into industrial societies, beginning with the north-west of England in the late 18th century. Its march across the world is still in progress. Inspired by the advent of new materials and technologies during the Industrial Revolution, writers conjured up powerful visions of fantastic futures. Gifted illustrators visualised the unbounded creativity of writers and helped produce spectacular images in science fiction publications, such as *Amazing Stories,* first published in 1926. The memory of Jules Verne was still fresh during Tsiolkovskii's lifetime. Accomplished writers, such as H.G. Wells (Herbert George Wells, 1866–1946) and William Olaf Stapledon (1886–1950), penned stories about the possibilities of future technology that remain popular in the 21st century. Arthur Charles Clarke (1917–2008), who first speculated about and then lived through the Space Age, was about to embark on his prodigious writing career. Around 1925 space exploration using rockets came of age in popular culture.

The decade between 1925 and 1935 saw a wave of optimism for the potential of rockets sweep across the globe. Robert Goddard, one of the pioneers mentioned in von Braun's piece on Apollo 11 experienced first-hand through his own work the transition from science fiction to science fact. In 1932, he wrote to novelist H.G. Wells of the "deep impression" that his novel *War of the Worlds* had made when Goddard first read it in 1899. It directly led him to "take up the search for spaceflight."[29] The first half of the 20th century witnessed rocketry groups, amateur and usually unfunded but dedicated, pop up around the world. The German Verein für Raumschiffahrt (Society for Spaceflight, 1927), the American Interplanetary Society (later renamed the American Rocket Society, 1930), the Moscow-based Group for the Study of Reactive Motion (GIRD, 1931) and the British Interplanetary Society (1933) helped to propel imagination into reality.[30]

Figure 1-6 Robert Esnault-Pelterie 1907. Credit San Diego Air and Space Museum

While many of the pioneers in aviation and rocketry are celebrated as national heroes and recognised globally for their achievements, there are many who made key theoretical and technological contributions that cumulatively led to the development of safe and reliable aircraft and rockets in the 21st century but have largely been forgotten. A couple of such early contributors were Robert Albert Charles Esnault-Pelterie (1881–1957) and Ary Abramovich Sternfeld (1905–1980). Robert Esnault-Pelterie invented many elements now commonly found on aircraft, such as the aileron used by aeroplanes when turning, the speed indicator and the joystick in the cockpit for control. He was the fourth Frenchman to gain a pilot's license. As a scientist, engineer and pilot, he published his thoughts on rocketry as early as 1912. He calculated and compared the properties of different propellants and estimated the flight times to the Moon, Venus and Mars.[31] He also proposed the concept of passive temperature control for interplanetary spacecraft.

Passive temperature control relies on surface coatings, a bright polished surface facing the Sun to reflect sunlight and minimise high temperature and a dark matt surface facing the cold blackness of space to minimise low temperature. This concept was implemented in India's first satellites of the 1970s by Professor U.R. Rao (Udupi Ramachandra Rao, 1932-2017) and his team. The significance of Rao's work was acknowledged by Professor Vyacheslav Mikhailovich Kovtunenko (1921-1995), designer of space technology, to have been of "immense value" to the Soviet space programme.[32]

Ary Sternfeld promoted the term 'cosmonautics' in his book *Introduction to Cosmonautics*[33], and long before it became a reality, he had brought the science of space travel into the mainstream, making it a respectable subject of study. In the 1974 edition of his book, he explains: the word "cosmonautics is more correct than astronautics because the definition of science studying motion in interplanetary space should provide a notion of the medium where the motion is assumed to occur (cosmos) but not one of its goals."[34] Though not welcomed initially when first introduced in the USSR in the 1930s, it was firmly established in the Russian language by the time Sputnik was launched in 1957, and today, cosmonautics is an internationally recognised branch of science and technology in Russian-speaking countries. In the US, Europe, India and Japan, the word 'astronautics' continues to dominate. Born in Poland in 1905 to Jewish parents, Sternfeld studied in France and later, at the age of 30, moved permanently to the USSR. Sternfeld survived Stalin's purges and World War II, but he was unemployed and lived a hand-to-mouth existence for many years in the unfulfilled vision of the idealised socialist society that had brought him to the USSR.

National Space Programmes

In 1903, brothers Orville Wright (1871–1948) and Wilbur Wright (1867–1912) realised one of mankind's most enduring dreams by designing, building and successfully flying the world's first heavier-than-air, powered flying machine. Just 66 years later, humans landed on the Moon. Neil Alden Armstrong (1930–2012), the first man on the Moon, was 18 when Orville Wright died. They never met but could have.

Developments in rockets accelerated during World War II and continued with a stunning pace during the Cold War. Each side fearing falling behind the other invested vast amounts of financial resources and political commitment at the expense of societal development. On 4 October 1957, the USSR put the world's first artificial satellite, Sputnik, into space. A dozen years later, the US put two men on the surface of the Moon with Apollo 11. The skill and dedication of hundreds of thousands of individuals, including engineers, technicians, pilots and politicians were essential for the success of Sputnik and Apollo 11. Yet, history remembers individuals for these monumental achievements rather than the complex nation-wide extended teams. For the USSR, it was their chief designer Sergei Pavlovich Korolev (1907–1966). In the US, it was Wernher von Braun; initially German, he became a US citizen in 1955. For India, it was Vikram Ambalal Sarabhai (1919–1971).

Korolev, von Braun and Sarabhai had very little in common. Despite all three being born around the same time, it was a unique journey that led each to play a key role in their respective nation's space programme. Each made a unique contribution to the development and application of space technology to the economic and industrial development of their nation. They were very different individuals, but in one key respect, they were similar. They were building on the profound contributions to rocket science made in the first quarter of the 20th century by the "three modern rocket pioneers"[35], Konstantin Tsiolkovskii, Robert Goddard and Hermann Oberth.

Hermann Oberth was originally from Romania but made Germany his home. In 1922, his PhD thesis on space travel was rejected because it was considered outlandish and unrealistic. Oberth cast aside this criticism and published his thesis *Die Rakete zu den Planetenrumen* (The Rocket into Interplanetary Space) privately the following year. Initially controversial, the book was later a success and went on to inspire not only future rocket scientists but a wider international community of writers and filmmakers, too. It was under Oberth's leadership that the German rocketry society Verein für Raumschiffahrt (Society for Spaceflight) was founded in 1927. Two years later, a young rocket enthusiast Wernher von Braun joined the society. Guided by Oberth, von Braun went on to study liquid-fuel propulsion.[36] Eventually, through the violent, tumultuous events of World War II and the Cold War, von Braun would imagine, design and build the rockets that took men from Earth to the Moon.

Robert Goddard, more an experimenter than a theoretician, was probably the world's first rocket scientist, but he died as World War II ended, the very time that rocket technology started to receive high-level political attention and funding he had sought. On 16 March 1926, during an ostensibly unimpressive flight that lasted 2.5 seconds achieving an altitude of a little over 10 m, Goddard successfully tested the world's first liquid-fueled rocket. Surprisingly, Robert Goddard's huge technological breakthrough of designing, building and flight-testing a liquid-fuel rocket never reached the wider public. Despite his multiple attempts, neither the military nor his own government expressed any interest in the potential of his innovative rocket technology. He filed many patents, including for liquid fuel engines and multi-stage rockets, which National Aeronautics and Space Administration (NASA) purchased in 1960 for $1 million (Rs.4.76 million)[37]

Figure 1-7 Robert Goddard at his launch control shack. Credit NASA

The American press at the time ridiculed Goddard's speculation on rockets from Earth going to the Moon. An editorial in the *New York Times* in January 1920 questioned Goddard's grasp of basic physics, asserting that rockets could not function in the vacuum of space. This criticism attracted a special notoriety that only came to an end when the *New York Times* published a correction on 17 July 1969 as Apollo 11 was on the way to the Moon.[38] Today, launch vehicles around the world use Tsiolkovskii's principle of multi-staged rockets and incorporate liquid-fuel engines as demonstrated by Goddard.

Goddard and Tsiolkovskii never met but received a public introduction in October 1923 via a report in the Soviet newspaper *Izvestia* titled 'Is Utopia Really Possible'.[39] The report covered Oberth's recently published book *The Rocket into Interplanetary Space* and a suggestion within it that current rocket technology was sufficiently advanced for a vehicle to leave Earth and enter space. The news report concluded with reference to a press release from the Smithsonian Institute, in which Goddard speculated on the possibility of a rocket that could impact the Moon and astronomers on Earth being able to observe the resulting explosion. The newspaper report received a surprisingly large coverage, bringing the names of Tsiolkovskii, Oberth and Goddard greater international recognition. One immediate consequence was a disgruntled Tsiolkovskii republishing his 1903 paper, in which two decades earlier, he had published much of what Oberth was publishing now. Reasserting his own work, he complained "Do we always have to get from foreigners what originated in our boundless homeland and died in loneliness from neglect?"[40]

The consensus among modern historians is that Tsiolkovskii , Oberth and Goddard were the founding fathers of rocketry. By the time Vikram Sarabhai established the nascent space programme in India in 1963, only Oberth was alive. At 76, half a century after the mockery at the hands of his PhD examiners, he was present at Cape Canaveral to watch humanity's greatest technological achievement, the launch of Apollo 11 in July 1969.[41] He lived until 1989. Hermann Oberth's life must have been a journey of an extraordinary vision fulfilled.

Scramble for German Rocket Technology

Even before the formal end of World War II, the first signs of the Cold War hostilities began to emerge. In the final stages of the war, as the US army approached Germany from the west, they instigated operation 'Paperclip' with the objective of acquiring German rocket scientists and the associated infrastructure.[42] The USSR was doing the same thing in their approach to Germany from the East. The British joined the scramble for German technology by sending Lieutenant-General Sir Ronald Morce Weeks (1890–1960), Deputy Chief of the Imperial General Staff for this purpose. Weeks considered the acquisition of German equipment as "one of the most vitally important of our immediate post-war aims."[43]

In February 1945, as the defeat of Hitler's (1889–1945) forces became imminent, Major Robert Staver of the US Army arrived in Europe with the mission to seize German technical know-how, equipment and the engineers who had developed it and transport them to the US. Soon after his arrival in Europe, Major Staver experienced the deadly force of a V2 first-hand. His US base in London was caught in a V2 explosion, and he narrowly avoided becoming a victim of his quarry.[44] He succeeded in acquiring not only the key personnel, designs and up to a hundred V2 rockets, but also a wind tunnel. The wind tunnel used for testing the rocket Saturn V was designed by von Braun's colleague Oscar Carl Holderer (1919–2015) and was still in use in 2015.[45]

On 11 April 1945, just ahead of Germany's formal surrender, Nordhausen was liberated by the US forces. Situated on the southern edge of the Harz Mountains in the centre of Germany, Nordhausen today is a picturesque town. During the war, its mountains became the ideal secure location for building weapons of war, including von Braun's V2 rockets. In a frantic operation, the US forces removed trainloads of complete and partially built V2s, component parts and the machinery used to make them from Nordhausen to the Belgian port of Antwerp by 22 May, before onward transport to the US. The haste was necessary because Nordhausen fell in the Soviet, not the American zone of occupation. In February 1945, the Allies had met and agreed at the Yalta Conference that Germany would be divided into four zones of occupation, American, Soviet, British and French. However, when Germany formally surrendered on 8 May 1945, the four allies were scattered across Germany. Nordhausen was formally occupied by the USSR only on 26 May 1945, but by then, the Americans had removed all key V2 assets. The US won the race for German rocket technology, and by doing so, they also achieved the equally significant feat of denying the USSR this substantial advantage. The stage was set for the Space Race to begin.

Korolev, the Chief Designer

Known to the West only as the anonymous chief designer until after his premature death in 1966, Sergei Pavlovich Korolev is widely regarded as the founder of USSR's space programme. Korolev grew up through a period of political unrest and military conflict. Two years before Korolev was born, in 1905, the first of the series of Russian Revolutions had begun. He was not quite yet a teenager when he lived through the 1917 Russian Revolution and the civil war that followed. Korolev was brought up by his mother in Kiev, the Ukrainian capital. He lived through one of the city's most turbulent periods: World War I, Russian Revolution, the Russian Civil War and the Polish-Soviet War. Between 1917 and 1920, Kiev changed hands between the various factions over a dozen times.

Figure 1-8 Sergei Korolev middle of picture transporting his glider to the launch site in October 1929. Credit Natalya Koroleva

Korolev's early education was ad hoc. He made it to college as an uneasy peace gradually replaced the tumult of war. It was an ill-equipped vocational college, and he enrolled for a course on carpentry and roof tiling. In 1913, ten years after the Wright brothers had demonstrated the first powered flight, a five-year-old Korolev saw a man-made flying machine. Sitting on his grandfather's shoulders at a fairground in Ukraine, he saw the plane take off, fly two kilometres and land again. For most watching the plane fly, it would have been a spectacle tantamount to magic. The experience wowed the spectators and perhaps mesmerised Korolev into a future career in aviation. In 1917, his mother remarried, and they moved to Odessa, where he experienced his first flight aboard a seaplane on the Black Sea.

In his late teens, Korolev's interest in flight drew him to gliders. Gliders, then commonly known as sailplanes, were much more popular and practical in the early 20th century than they are today. Korolev honed his basic theory and practical flight skills and progressed to designing and building, as well as flying, gliders. Later in life, with his experience in rockets, he developed a passion for rocket planes well before the jet age. By the 1930s, aircraft engines were becoming more powerful, but he believed that rocket power rather than the combustion engine was the future of aviation.

Korolev had been part of the small team of rocket enthusiast largely self-funded, that in November 1933 led the USSR to launch its first liquid-fuelled rocket. While he was developing those ideas in the mid-1930s, his name appeared on a death list as part of Joseph Stalin's (1878–1953) Great Purge.[46] For unknown reasons, he survived the firing squad. Many of his fellow rocket scientists did not and perished in what came to be known as the Great Terror. Imprisoned in 1938 for 10 years in a harsh Siberian labour camp, he was moved in 1942 to another labour camp where his engineering skills could be put to work to support the Soviet war machine.

As the war ended, Korolev was among the Soviet rocket specialists sent to Germany to assess and acquire for the USSR the technological advances made by the German rocket and aviation research teams. Released on parole, Korolev travelled to Germany in the autumn of 1945. A few weeks later, he met up with his fellow rocket scientist Valentin Petrovich Glushko (1908–1989), who had arrived a few weeks earlier also in pursuit of German rocket technology. The two had worked with each other since the early 1930s, successfully launching rockets using liquid fuel. In Germany, with the Nazis vanquished, they now got a chance to see a V2 in action. In October 1945, British forces organised the launch of three V2s under their control from the northern city of Cuxhaven to evaluate and learn their technical secrets.

Representatives from France, the USSR and US were invited. Only three representatives had been invited from the USSR but five, including Korolev, turned up. Korolev had to pose as a chauffeur to get in. Glushko was one of the three allowed inside the compound while Korolev had to observe from outside the perimeter fence. The sight of a large, powerful guided rocket streaking into the sky must have rekindled the original dreams of spaceflight that had brought Glushko and Korolev together in the 1930s. Their combined forces would pull off the surprise of the century by putting Sputnik, an artificial satellite, in Earth's orbit just twelve years later.

Immediately after the war, most of the German rocket technology ended up in the US, some in USSR and a little in France and Britain. In the USSR, the German missile technology was first replicated and then quickly enhanced with Soviet innovations. A leading German rocket engineer Helmut Grötrupp (1916–1981) had chosen the USSR, as Wernher von Braun had chosen the US. Grötrupp also helped to identify German rocket engineers who, like him, could assist the USSR. In 1946, he wrote a detailed report for the USSR on how they had solved many of the technical problems of the V2. While von Braun helped the Americans to develop their rocket programme, Grötrupp did likewise for the USSR. The starting point in both cases was Nazi Germany's state-of-the-art technological marvel, the V2 missile. With surprising speed, Soviet engineers guided by Korolev developed and transformed the technology of the V2 into their own indigenous design, the R1. By 1953, Korolev was convinced that the German engineers could offer no more assistance, and to keep them at a distance from the new Soviet secret innovations, many including Grötrupp were sent back to Germany.

Korolev was the team leader behind two of humanity's greatest achievements. After the launch of Sputnik, he also headed the team that put the first man in space. The Soviet leader Nikita Khrushchev (1894–1971) designated Korolev's name a state secret. To the outside world, he was known only as the chief designer. Khrushchev's decision to withhold Korolev's identity from the Nobel Prize committee denied Korolev two opportunities of winning the Nobel Prize, one for Sputnik and another for his contribution to Yuri Alekseyevich Gagarin's (1934–1968) flight, and the international recognition that would have followed.[47]

It is remarkable that despite the false accusations, harsh imprisonment and brutal treatment, Korolev went on to dedicate his life's work to the very state that had undermined and unjustly punished him.[48] Today, a street, a town, a museum, a crater on the Moon and another on Mars carry his name. His daughter Natalya Koroleva, who is openly bitter about the treatment that her father had received at the hands of the State, has converted his former home into a museum, a shrine to his memory.[49] It is regularly visited by rocket engineers, historians and space enthusiasts from all around the world. In death, he finally received the respect, and national and international recognition, that had been denied him during his life.

Von Braun and the Moon

Widely recognised as the rocket genius behind the successful American Apollo programme to the Moon, von Braun is one of the most intriguing and controversial characters in the story of mankind's first steps into space. In 1944, Hitler awarded von Braun the Knight's Cross for his contribution in developing the V2 combat ballistic missile.[50] In January 1959, von Braun received the Distinguished Federal Civilian Service Award from President Dwight D. Eisenhower (1890–1969) in the White House. The V2 missiles were introduced in September 1944, and they killed approximately 5,000 people. Built by slave labour, more people were killed in making the V2 than by their military use.[51] Despite the death and destruction caused by the V2s in London during the final months of the War, the London-based

British Interplanetary Society awarded von Braun its first Gold Medal in 1961.[52] The second Gold Medal went to Yuri Gagarin during his visit to London on 11 July 1961.

As Hitler's chief rocket scientist, von Braun developed the world's first rockets with enough power to reach space, and in some accounts, he inadvertently did so as early as 1942.[53] In the final months before his suicide, Hitler ordered the destruction of German resources from which the victors could benefit; this included patriotic skilled Germans. Von Braun spent the last few months of the war evading execution by his own German Army as he doctored an escape out of Germany. Of all the nations, the USSR had suffered the greatest losses in the War von Braun, fearing that "it's the Russians who will take revenge", headed west towards the approaching US forces, along with his team of rocket scientists.[54] His brother Magnus contacted the Anti-Tank Company, 324th Infantry of the US Armed Forces with the words "My name is Magnus von Braun. My brother invented the V2. Please, we want to surrender."[55] In the end, von Braun and members of his team survived the war and, with pretty much all his engineering designs intact, made it safely to America.

Figure 1-9 Wernher von Braun with President Kennedy 16 November 1963.
Credit NASA

At the end of the War, German technological lead in rockets was assessed to be 25 years.[56] Von Braun estimated that he needed a further two years of V2 development to bring the US within its range. Perhaps, it was this that motivated Major Staver to send a remarkable message to the Pentagon. He suggested that "a policy and procedures be established for the evacuation of other German personnel whose future scientific importance outweighs their present war guilt."[57] The programme to exploit German scientists in and for the US was set up as a secret operation called Overcast to be managed by the US military intelligence.[58]

The US placed national interest ahead of justice and concluded that von Braun's unique experience was too valuable to be lost in the gallows. In June 1945, the putative prisoner of war von Braun, along with other members of his team, was loaned by the US to the UK. During the 10 days of his interrogation in a POW camp at Wimbledon in London, von Braun was taken to witness first-hand the site of destruction where one of his V2s had landed only months earlier. The British invited him and his team to abandon the contracts they had signed to work for the Americans and to work for the British rocket programme instead. The UK, after all, was much nearer to their home country, Germany, but he declined.[59]

Initially, some in the US wanted von Braun to be treated like any other potential war criminal. However, the US military decided that von Braun and his team's technical expertise could be useful in the war still raging in the Pacific. Despite being questioned in the US and briefly in London, von Braun was officially able to convince his interrogators of his innocence of any war crimes[60]. The three-way unjust and unconscionable alliance between the US and UK governments and von Braun exploited each other at the expense of the rule of law. Von Braun was never formally tried, and he successfully engineered his future, working on the space programme in the US. In 1955, von Braun became a US citizen. Citizenship was an essential prerequisite for him to have access to and work on secret US military projects. The decision that the US defence programme should benefit from a former Nazi was not welcomed by all Americans. President Eisenhower expressed his disquiet. But in the end, as the Cold War set in, the pursuit of US national interest prevailed.

Von Braun was an "extremely eloquent and self-assured individual" concluded Reginald George Turnill (1915–2013), a BBC journalist who covered the US space programme and interviewed von Braun many times during the Apollo era. During a visit in 1990 to the Dora Concentration Camp in Nordhausen, which was used for the production of V2 rockets, Turnill reflected once again on his assessment of von Braun and concluded that von Braun had "cheated the hangman."[61] Another analysis of von

Braun's ethics concluded "Wernher von Braun's life was one dedicated wholeheartedly to the goal of putting men in space and on the Moon. To achieve this goal, von Braun made many morally questionable decisions. Given the highly emotive context, his decisions and the subsequent reassessments are subjective and are reinterpreted by each generation."[62]

The two architects of the Space Race during the Cold War, Korolev and von Braun, never met. von Braun had left a few months before Korolev arrived in Germany on 8 September 1945. [63] Over the next two decades, both men faced each other in the epic rivalry that came to be known as the Space Race and resulted in the most accelerated period of rocketry development in human history. Despite their profound differences, Korolev and von Braun shared key characteristics. They were technical geniuses, gifted people managers and ambitious men with an almost infinite energy to pursue their goal of spaceflight.

Sarabhai and India's Space Programme

At a time when India had no space scientists or infrastructure and the idea of India having a space programme appeared a fantasy, Vikram Sarabhai embarked on a bold, seemingly impossible mission. Independent India was maturing as a nation when the Space Age arrived. It could either join the Space Race right away or delay and pay the price of catching up later. Although one of the earliest nations to have joined the space club, India entered well after the Space Race had begun. It was on 21 November 1963 that India launched its first rocket into space. Although a scientist, Vikram Sarabhai was not a rocket scientist. He did, however, share with Korolev and von Braun the traits of being a great communicator, skilful manager and highly energetic person. Like Korolev, Vikram Sarabhai grew up through a period of political unrest. In 1930, at the age of 11, he was actively taking part in the Indian independence movement. In 1945, at 26, he was in the UK to complete his PhD that had been interrupted by World War II. His venture into building rockets and exploring space was still over a decade and a half away.

Although there is no prominent record, it is likely that Sarabhai met von Braun during his many trips to the US. With the mission to the Moon successfully completed, von Braun was sent to India in 1973 with another mission: to "explore possible sales of communication satellites and launch facilities. Sales potential for the package was roughly 10 million dollars."[64] By then, however, Sarabhai had died, and Abdul Kalam was assigned to meet him upon arrival in Madras (now Chennai) and subsequently give him a tour of Thumba. When von Braun arrived, India had already established a non-commercial arrangement with NASA for a one-year loan of its communication satellite ATS-6 as part of the Satellite Instructional Television Experiment (SITE) programme. Kalam recalls von Braun's words expressing his passion for rocketry "do not make rocketry your profession... make it your religion."[65] Kalam reflected on the parallels between Sarabhai and von Braun when he remarked "Did I see something of Prof. Vikram Sarabhai in von Braun? It made me happy to think so.

◄ ◇ ►

Chapter Two
From Vedic Astronomy to Modern Observatories

Civilisations throughout human history, both nomadic and settled, have understood the utility of astronomy. It is a prerequisite not only to understanding the concepts of the year and time of the day but also for determining geographical positions on Earth. Referring to the life and death importance of astronomy to marine navigation, Indian historian Rajesh Kochhar asserts that the institutionalisation of modern astronomy was not for the "love of the stars but the fear of death."[66] Throughout its long history, India has had a tradition in science, albeit frequently interwoven with superstition, astrology and mysticism.[67] Preoccupation with understanding the real world was one reason why civilisation took root in India.

Hints of a science-based society in India can be found at the beginning of the Vedic period (a 1000-year period of Indian history beginning in 1500 BC when the oldest Indian scriptures, the Vedas, were composed). The Vedic communities observed and recorded the celestial motions of the Moon, planets and the Sun, and mapped the night sky into 27 nakshatras, hinting at the central role of science at the beginning of the Vedic Period. The tradition of science in India can be traced even further back to the organised societies of the Indus Valley Civilisation, also known as the Harappa Culture that flourished in India and predates the ancient civilisations in Egypt and Iraq).[68] Archaeological evidence of planned cities with streets, drains and tiled flooring suggests that precision measurement instruments and sophisticated building techniques were practised as long ago as 8000 years.

A 7-m high iron pillar stands close to the Qutab Minar in New Delhi, the capital of India. It was forged during the late 4th century BC and weighs over 6 tonnes (6000 kg). Constructed from 99.7% wrought iron, it remains rust free. Stunning statues carved in iron, copper and bronze using the lost-wax process is evidence of advanced metallurgical techniques developed in early India. History provides many examples where a scientific advance is made, forgotten and remade centuries later. The most celebrated example, perhaps, is the Antikythera mechanism discovered in 1901. This clockwork astronomical computer was originally constructed in Greece in about 100 BC. The techniques and technology needed to produce it were rediscovered in Europe around the 15th century.

In the 4th century BC, the astronomy of the Greek's based on mathematics, precision measurements and systematic observations came to India with Alexander the Great's (356 BC–323 BC). This Greek structured approach, however, could not uproot the entrenched culture of astrology that had persisted in India for centuries. Observational astronomy served only to support the existing ancient traditions rather than ushering in a new way of thinking shaped by the scientific method. Over the next few centuries, Indian scientists continued to engage in understanding the natural world through the language of science albeit still steeped in ancient (and non-scientific) Indian rituals and traditions.[69]

Figure 2-1 Jantar Mantar in Jaipur. Credit McKay Savage

In the 6th century AD, long before Nicolaus Copernicus (1473–1543), Aryabhata (476–550) in India had proposed a heliocentric system. He also wrote about trigonometry, algebra and astronomy. Even though most of his writings did not survive, Aryabhata is credited with introducing the concept of zero, calculating the value of pi (π) and establishing the idea of the Earth turning on its axis daily and revolving around the Sun annually; he also provided a new explanation for lunar and solar eclipses. Aryabhata is a celebrated figure in India, so much so that Indian Space Research Organisation (ISRO) in 1975 named its first satellite, Aryabhata.[70]

Another Indian mathematician and astronomer Bhaskara (600-680) formalised and recorded Aryabhata's work. Bhaskara is known for three of his writings, two cover astronomy in verse and one is a commentary on Aryabhata's original work did not survive. Within a century, Bhaskara recorded and progressed Aryabhata's contributions. He took Aryabhata's concept of zero and implemented it as a recognisable circular symbol and established for the first time the positional (units, tens, hundreds, etc.) system familiar today. Until then, numbers were represented by words or allegories rather than numeric characters. Bhaskara developed Aryabhata's work on trigonometry, fractions and lunar and solar eclipses.

Before the telescope was invented, celestial objects were observed with the naked eye and their positions, brightness and patterns in the sky, which changed over time, were recorded. In the 18th century, a Rajput prince and senior general of the Mughal Empire Raja Jai Singh (1688–1743) built a series of large static masonry buildings called Jantar Mantar to function as astronomical instruments, such as the astrolabe that can measure and predict the position of the Sun, Moon and stars. A total of five were built at sprawling sites in Jaipur, Delhi, Varanasi, Mathura and Ujjain, but only two (Jaipur and New Delhi) remain today. Between them, the instruments could measure the time of local noon, longest and shortest days of the year, time of the day and angular positions of stars and the Sun. His goal was to duplicate and improve the star catalogue of 944 stars developed in the 15th century by the Ulugh Beg Observatory, which was situated in modern Mongolia.

While the contributions to science, mathematics and astronomy from the Middle East (Iraq, Syria and Egypt) are well established, those from Central Asia are less familiar. Ulugh Beg Observatory was an advanced centre for astronomy in the 15th century. Despite his larger instruments, Raja Jai Singh was unable to match the accuracy of the measurements recorded by Ulugh Beg two centuries earlier,[71] Now these elegant structures are more a visitor attraction than precision instruments for a scientific investigation.

Colonialism and Renaissance

The European Renaissance (14th–17th centuries) and the Age of Enlightenment (17th–18th centuries) in Europe coincided with the colonial period in India. The British inadvertently provided a conduit for transferring the advances and discoveries of the European Renaissance in arts and science to India. Following the defeat of Tipu Sultan, innovations in rocketry flowed in the opposite direction.

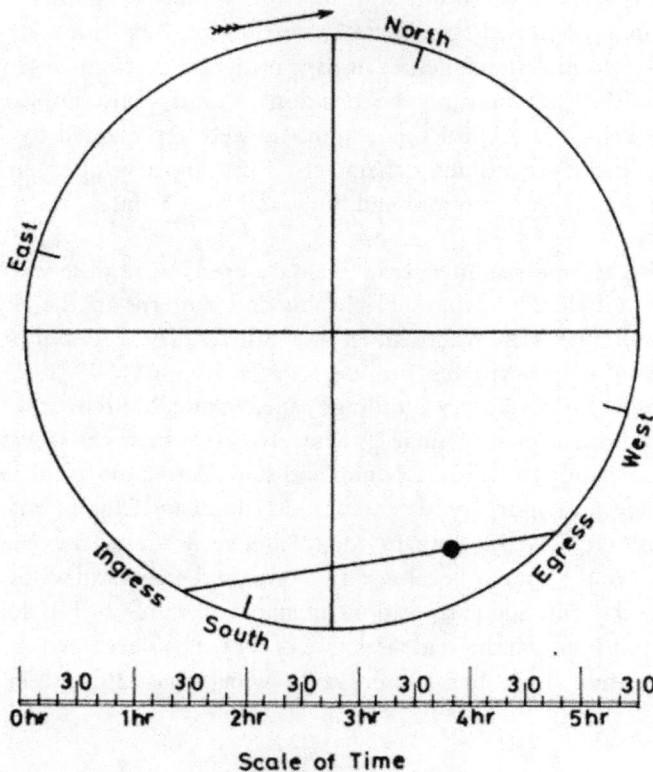

Figure 2-2 Transit of Mercury Recorded by Jeremiah Shakerley from Surat. 3 November 1651. Credit Indian Institute of Astrophysics

The transfer of science between societies has been a constant feature of human civilisation and continues today. The 15th-century invention of the printing press in Germany was timely and facilitated the prompt communication of ideas, inventions and technologies across the disparate lands of the growing Empire. Galileo's observations of the night sky using the newly invented telescope and William Shakespeare's literature from the UK were brought to India by scholars and engineers, who accompanied the soldiers, merchants and traders of the British East India Company. Modern ideas about the utility of science and its potential to shape the economic prospects of an underdeveloped nation were first tested in India during the colonial period.

Galileo observed the night sky using the telescope for the first time in 1609. Four decades later the first astronomical observation using a telescope was made from Surat on India's west coast. Jeremiah Shakerley (c1626–1655), an employee of the British East India Company, recorded the transit[72] of Mercury on 3 November 1651. An experienced observational astronomer, Shakerley had calculated that the transit would not be visible from Europe. It is not known if that was the reason for his arrival in India, but he was probably the second person in history to witness a transit of Mercury. The first was probably Pierre Gassendi (1592–1655), who observed it from Paris in 1631.[73] During the transit, the planet's silhouette would have been seen as a small dot moving slowly across the face of the Sun over a period of about three hours. Shakerley is also credited with making the first telescopic observation of a comet that was visible from India in 1652.[74]

From Pondicherry (now Puducherry) on India's south-east coast, Father Jean Richaud (1633–1693), a French Jesuit priest, used a telescope on 9 December 1689 to observe Alpha Centauri. He discovered it was not one star but two.[75] This was the second 'double star' to be documented. Modern telescopes have revealed a third member in the Alpha Centauri system, Proxima Century, which at 4.2 light years is the nearest star to our solar system. Today, most stars are binary or multiple star systems. The stars of the Alpha Centauri system are the next nearest stars to the Earth after the Sun. Had Alpha Century been visible from northern Europe, this discovery of Earth's celestial neighbour would have been made from Europe and not India.

The British in India recognised that recording accurate observations for time, meteorology and location were essential for a nation that relied on maritime power to administer an empire thousands of miles away. The observation data gathered and processed by the Royal Greenwich Observatory in London established in 1675 underpinned the British Empire's naval prowess. To secure and extend its hold over India, scientific institutions and technologies were introduced, and astronomical observatories were one of the first such institutions of science to be established in India.[76]

The Great Trigonometrical Survey

With Tipu Sultan's defeat, the British East India Company acquired the vast Kingdom of Mysore. It then expanded its control from Madras (now Chennai) on the east coast to Mangalore (now Mangaluru) on the west coast. Keen to quantify his conquest, Madras Governor Lord Clive approved the suggestion that a trigonometrical survey similar to those conducted in France and Britain be repeated in India. The mathematical techniques and the precision measuring instruments used to make astronomical observations could, with some modification, be used to survey the Earth. What came to be known as the GTS of India started on 6 February 1800 and lived up to its name when it was extended to the whole of India on 1 January 1818.

GTS was a monumental scientific undertaking given the vastness and complexity of the terrain. Although established a few years prior to the start of GTS, Madras Observatory found its profile elevated by the GTS as it provided training and organisational support for the survey.[77] The man appointed Surveyor General of India from 1821 to 1827, Colonel Hodgson, was an astronomer. In addition to his surveying obligations, he supervised a series of transit observations made from Calcutta (now Kolkata).[78] The Official Manual of Surveying for India was published in 1851. It was divided into five sections: (i) Geometry and trigonometry, (ii) Surveying instruments, (iii) Surveying, (iv) On native field measurement (khusrah) and (v) Practical astronomy and its application to surveying.[79]

The British East India Company and, after 1857, the British colonial government invested some of their immense wealth in promoting science, purchasing scientific equipment and sponsoring scientists, engineers and surveyors to come to India. Apart from contributing towards astronomical and scientific progress, the instruments of science generated wealth from the discovery of coal in Raniganj, petroleum in Assam and gold in Tipu's former stronghold of Mysore. Railways and ports were developed to facilitate the transport of this wealth out of India. Scientific literacy spread through India following the first GTS and then spread further and faster thanks to the advent of the railways.

Madras Observatory

The Madras Observatory formally came into being on 19 May 1790 when the directors of British East India Company accepted an offer from William

Petrie (1784–1816') to nationalise the observatory he had personally established and operated since 1787 with two 3-inch (7.6 cm) achromatic telescopes, two astronomical clocks and a transit instrument. The first official astronomer of the observatory was John Goldingham (1767–1849), who calculated the time difference between London and Madras to be 5 hours and 21 minutes. This was subsequently used to establish Indian Standard Time of GMT +5:30 that is still used today. The British East India Company was preoccupied with its expansionist ambitions and particularly with defeating the troublesome Tipu Sultan in southern India. Its first significant foothold in India was in Madras (now Chennai), so establishing the first observatory there was a natural outcome. The motivations for establishing the Madras Observatory were to (i) survey the territories it already had, (ii) increase revenue earnings, (iii) ensure the safety of sea passage and (iv) learn about the geography of the country for future expansion.[80]

Figure 2-3 The Madras Observatory 1838. Credit Indian Institute of Astrophysics

The British East India Company recorded its primary aim for the Madras Observatory as "promoting knowledge in astronomy, geography and navigation in India."[81] Following Tipu Sultan's defeat in 1799 and the start of the Great Trigonometrical Survey (GTS) in 1800, the activities of the Madras Observatory accelerated.

The Madras Catalogue

Thomas Glanville Taylor (1804–1848), whose father had been the first assistant at the Royal Greenwich Observatory in London since 1805, arrived at the Madras Observatory in 1830. By 1831, he had installed two astronomical instruments, a 4-foot (121 cm) mural circle and a transit telescope, both having lens 3.25 inches (8.9 cm) in diameter, optically both were substantially powerful instruments. With these instruments, he was able to record precisely the positions of the Sun, Moon, planets and stars.

The transit instrument, a telescope mounted such that it can only move up and down and not sideways, allowed him to measure precisely when stars in the night sky crossed his local meridian (passed directly overhead).[82] The data collected helped to explain the geometry of planetary orbits and through parallax detect the proper motion of stars.[83] With the support of four 'native assistants' over several years, Taylor produced what came to be known as the Madras Catalogue of 11,015 stars, which the Astronomer Royal Sir George Airy (1801–1892) described as the "greatest catalogue of modern times"[84].

Taylor recorded and published his works in many volumes over several years in Madras. In his book, *Astronomical Observations Made at the Honourable East India Company's Observatory at Madras*, he records in detail the challenges of his undertaking. A violent storm on 30 October 1836 blew away the top of his observatory, and torrential rain and high winds further damaged his instruments; it took three months to fully recover from this devastation. This book was the second edition of his fourth volume recording the observations made during 1836–37. In the preface, he explains why the second edition was necessary. The ship carrying the printed copies had sunk with "nearly the whole of the copies of the former edition having been lost in the wreck of the Duke of Northumberland." [85]

Discovery of Helium

In 1814, Joseph Ritter von Fraunhofer (1787–1826) developed the first spectroscope in Germany. A spectroscope was a new powerful analytical tool that could determine the chemicals in a flame just by looking at the light from it. This was true for light from the stars and planets too, but that light was extremely faint. Long exposure photography made it possible to accumulate the faint light from telescopic observations, something that the human eye could not do. Long exposure photographs revealed details that could not have otherwise been seen revealing stars, structures in galaxies and details in planetary atmospheres. In short, astrophotography made the invisible visible. Combining astrophotography and spectroscopy made it possible to investigate the chemistry of celestial objects.

On 18 August 1868, a spectrum of the Sun was recorded in Guntur in

South India during a total eclipse. Many European teams visited India specifically for this event. Often, new scientific discoveries follow the invention of new scientific tools. The introduction of spectroscopy in astronomy held the prospect of something equally profound.[86] The 1868 spectrum of the Sun led to the discovery of helium in the atmosphere of the Sun.

Figure 2-4 Solar Eclipse. Turkey. 29 March 2006. Credit Toni May

Norman Robert Pogson (1829–1891) observed the same eclipse from Masulipatam (Madras was not in the ground-track of the eclipse). Red tape delayed the publication of his spectroscopic observations, thus denying him the credit of discovering helium. Pierre-Jules-César Janssen (1824–1907), who observed from Guntur, published his work with Joseph Norman Lockyer (1836–1920) first. So, Janssen and Lockyer were credited with the discovery instead.[87] Three years later, spectroscopy was used on the much fainter (than the Sun) light from a distant star, star γ Argus. The first recorded spectrum of a star in India was made at the Madras Observatory on the eve of the total solar eclipse of 1871.[88]

The Earth-Sun Distance

Though planetary transits are arguably not as visually spectacular as a total solar eclipse when the Moon completely obscures the Sun, they are unique and have the potential to generate new scientific knowledge. Only Venus and Mercury have their orbits between the Earth and the Sun, and planetary transits occur when one of these comes directly in between the Earth and the Sun and their silhouette appears on the disc of the Sun. The first record of the transit of Venus was made on 24 November 1639 by Jeremiah Horrocks (1618–1641) and William Crabtree (1610–1644) observing in Salford and Preston, about 31 miles (50 km) apart in the north-west of England.

Celestial mechanics determine that pairs of Venus transits separated by eight years repeat every 130 years. Johannes Kepler (1571–1630) had calculated that one would occur in 1631 and accurately predicted that it would not be visible from Europe.[89] The next in the sequence in 1639. Following his observation of the transit in 1639, Horrocks was able to estimate the size of Venus and the Earth-Sun distance to about 50% accuracy. The transit of Venus occurred again in 1761 and 1769, but only the 1761 instance was visible from India. In 1761, the British East India Company was not as well established in India as it would be in the later decades, but instructions for observing the transit from India were published.[90]

The transit was observed from the terraces of Fort St. George in Madras by William Hirst shortly after sunrise on 6 June 1761. Hirst recorded what came to be known as the black drop effect where the sharp circular edge of the silhouette of Venus appeared distorted as it entered and exited (ingress and egress) of the disc of the Sun. In Hirst's words "at the total immersion, the planet, instead of appearing truly circular, resembled more the form of a bergamot pear"[91], an effect similar to the distortion of the circular shape of the Sun as it sets over the horizon. The black drop effect limited the precision of the timing of the transit.

In the 19th century, the first transit of Venus occurred soon after the technique of spectroscopy had demonstrated its potential as a potent tool for science. During the transit, astronomers quantitatively measured, for the first time, the atmospheric properties of another world. A thick atmosphere was detected on Venus, which helped explain the black drop effect, limiting the precision of the timing measurements for the beginning and end of the transit. Observations with modern telescope suggest that the black drop effect was not entirely due to Venus's atmosphere but a result of instrumental effects, Earth's and the Sun's atmosphere.[92] By the late 19th century, astronomers had been observing the heavens with telescopes for over 150 years. Telescopes had improved optically with larger lenses and mirrors; enhanced engineering for gears and drives for telescopes resulted in larger, brighter and clearer images. Developments in clocks also provided increased reliability and precision of astronomical observations.

The "worldwide transit enterprise of 1874 was thoroughly shaped by national political ideologies"[93] converging with technological advances. France, Germany, the USSR and the US organised expeditions with the goal of measuring the Earth-Sun distance. In 1874, astronomers were able to determine an approximate value for the Earth-Sun distance (also known as the astronomical unit), and thereby the scale of the solar system, by measuring with high precision from several geographically distant locations on Earth the time that Venus entered and exited (ingress and egress) the disc of the Sun during the transit. Following the transit of 1874, there was another in 1882, which was not visible from India. Data of both transits collected by astronomers from Italy, France, the USSR, UK, US, Germany and South America were correlated and analysed. The Earth-Sun distance was refined to 92.95 million miles ± 0.19 million miles (149.59 million km ± 0.31 million km).

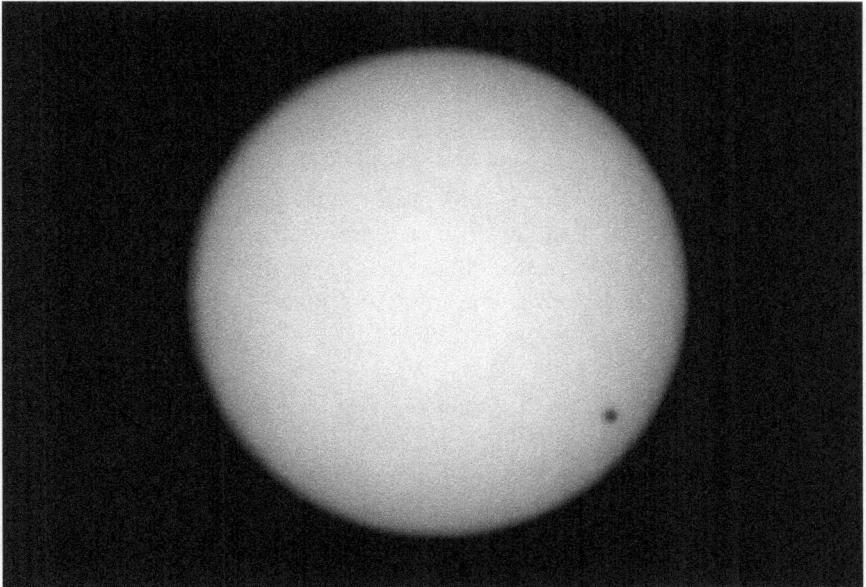

Figure 2-5 Transit of Venus 8th June 2012. Had photography been available the 1761 transit would have looked similar. Credit Author

The Director of the Madras Observatory, Norman Pogson, was taken ill in June 1891 and died a few weeks later from cancer. One of his last actions was to recommend by letter to the Chief Secretary of the India Office that Charles Michie Smith (1854–1922) be his successor. Norman Lockyer did not have high regard for Smith and preferred Kavasji Dadabhai Naegamvala (1857–1938) instead.[94] Naegamvala had set up the Takhtasingji Observatory in Poona (now Pune) with a 20-inch (50.8 cm) telescope, the largest in India at the time. The India Office overruled Lockyer and Pogson's recommendation was selected. The Madras Observatory remained the primary astronomical observatory in India for over a century. Its focus was not just on the routine recording of astronomical observations but also in the pursuit of scientific knowledge. However, attempts to upgrade and develop the instruments and facilities at the Madras Observatory were unfruitful. In 1899, all astronomical activity was moved to the Kodaikanal Observatory, and the Madras Observatory was re-tasked to undertake meteorological observations.[95] The work of astronomical observatories in India was designed to mitigate risks, predominantly of bad weather during long journeys at sea. This would be critical for an Empire with ambitions to grow.

Kodaikanal Observatory

Three solar eclipses (in 1868, 1871 and 1872), three transits of Mercury (5 November 1868, 7 November 1881 and May 1891) and one transit of Venus (9 December 1874), with this series of rare and spectacular celestial phenomena in the late 19th century, the interest in studying the Sun peaked. All of these, by chance, were visible from India. Lunar eclipses (shadow of the Earth on the Moon) can be seen from anywhere on Earth where the Moon is above the horizon at the time of eclipse and can last for around four hours. A total solar eclipse (shadow of the Moon on the Earth) is much rarer, lasts only for a few minutes and is visible from small areas on the Earth.

The Moon is 400 times smaller than the Sun, but because it is 400 times closer to the Earth, they both appear to be about the same size in the sky. This is responsible for the spectacular beauty of a total solar eclipse, the Moon in front of the Sun surrounded by the Sun's atmosphere – the corona. The Moon is much smaller than the Earth and so is its shadow. When it falls on the Earth, the shadow of the Moon has a diameter of only about 93.20 miles (150 km). During a total solar eclipse, this shadow races across the Earth's surface at around 0.62 miles/s (1 km/s) over a ground track approximately 9,320.57 miles (15,000 km) long.[96]Although the total duration of a solar eclipse is over a few hours, totality (that is, the period when the Moon completely obscures the Sun) experienced by a stationary observer on the ground lasts only a few (typically five) minutes.

With most of the celestial events of the late 19th century visible from India, the idea of a solar observatory in India attracted considerable interest and support. Key scientific discoveries, all centred on the Sun, seemed to further validate the intellectual case for building a solar observatory in India, and the Kodaikanal Observatory was sanctioned. The Kodaikanal Observatory initiated the systematic study of solar physics in India.

Figure 2-6 Kodaikanal Observatory. 1908. Credit Unknown Artist

The construction of the Kodaikanal Observatory commenced at the end of April 1895, and it became operational in 1900. The 89-acre site, 1.24 miles (2 km) above sea level, is in southern India on the Western Ghats equidistant from Madurai and Coimbatore. The Observatory was the product of a decade-long struggle by Charles Michie Smith, to build an observatory on a mountaintop. In a letter to the Under Secretary of State for India dated 17 August 1883, the then Astronomer Royal William Henry Mahoney Christie (1845–1922) stated "I am not prepared to endorse Mr Pogson's remarks as to the enormous improvement in defining power necessarily resulting from a great elevation."[97] In the absence of Pogson, Smith defended this argument, and the mountaintop site was eventually approved. Today, all professional astronomical observatories are placed on mountain tops to minimise the atmosphere through which the faint light from distant stars must travel.[98]

The Evershed Effect

Smith was a dedicated astronomer and an industrious administrator without whom Kodaikanal Observatory probably would not have been established. Although he did not make any major scientific contributions, his chief assistant and later successor, John Evershed (1864–1956), did. Evershed arrived at Kodaikanal in 1907. An accomplished amateur astronomer specialising in spectroscopy, he had held the responsibility of Director of the Solar Spectroscopy section of the British Astronomical Association.[99] In 1907, he observed Comet Daniel and, in 1910, Comet Halley using the observatory's 6-inch (15.24 cm) aperture telescope. He recorded spectra and identified nitrogen and carbon in the nuclei of both comets. In the tail of Halley's Comet, he identified carbon monoxide.

Evershed was experienced in observing the Sun photographically and spectroscopically. He had made regular observations of sunspots over many years. On the particularly clear morning of 5 January 1909, Evershed, assisted by his wife Mary, captured from Kodaikanal the spectra two large sunspots. Upon carefully examining the absorption lines, he noted "the spectra revealed a curious twist in the lines crossing the spots which I at once thought must indicate a rotation of gases."[100] He had recorded the evidence of gas emanating from the centre of the sunspots and being pushed radially outwards by magnetic fields associated with the sunspot. This phenomenon had been predicted earlier, but Evershed was the first to capture evidence. The effect has since been known as the Evershed Effect.

The actual value of the pressure pushing the gases was not known until Meghnad Saha (1893–1956) published his work on thermal ionisation in 1920, which helped to quantify the pressure that prevailed on the surface of the Sun.[101] Interestingly, in the following year, Evershed invited Saha to come and work at Kodaikanal, but Saha declined.[102] In 1911, Smith retired, and after a short trip back to Scotland, he returned to India, which he made his final home. He died in 1922 and is buried in India close to the observatory.

Other Observatories

A by-product of the GTS and the astronomical techniques it relied on was a series of astronomical observatories established in India during the 19th century, in Lucknow (1831–49), Trivandrum (1837–52), Dehradun (1878–1925), Calcutta (1879) and Poona (1888–1912).[103] During the colonial period, leaders of all significant scientific institutions in India, including astronomical observatories, were vetted and approved in Britain by organisations, such as the Royal Society, British Astronomical Association and the British Association for the Advancement of Science. In most cases, the funding also came from the Britain, but there were a few exceptions.

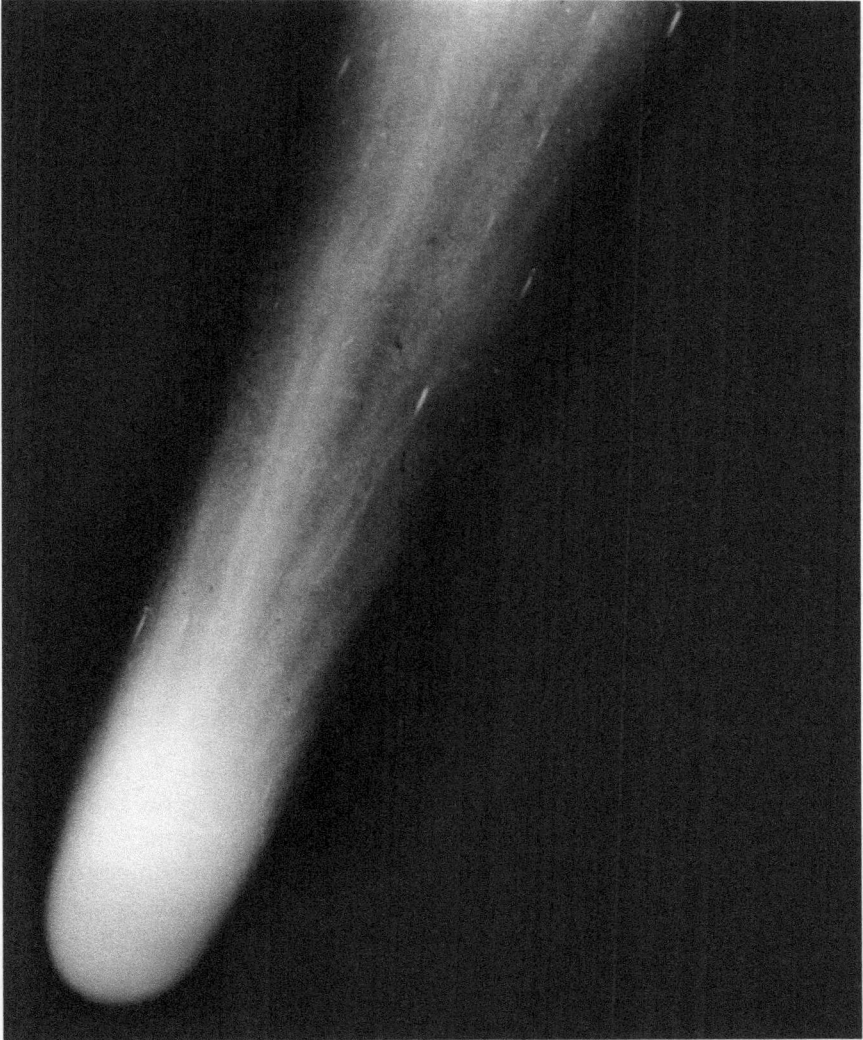

Figure 2-7 Halley's Comet photographed by John Evershed from Kodaikanal 1910. Credit Indian Institute of Astrophysics

In 1831, the Nawab of Oudh set up an observatory in Lucknow, and in keeping with the traditions, he requested a British director to lead it. Major

James Dowling Herbert (1791-1833), a GTS officer, was Lucknow Observatory's first director. The observatory was well equipped, and high-quality observations were made, but none were ever published. When its subsequent director Richard Wilcox (1802–1848) died in 1848, the observatory fell into disuse. Lucknow was the centre of some of the fiercest fighting during the 1857 revolt. When Lieutenant-General James Francis Tennant (1829–1915) of the Bengal Engineers recaptured Lucknow, he discovered the building intact, but the contents had been ransacked.[104]

The motivation for the Trivandrum Observatory came from British scientists. The British monarch provided the funding but with key local support. Swathi Thirunal Rama Vermah (1813–1846), Maharaja of Travancore, "entered warmly into the project" and appointed his astronomer to manage the observatory.[105] Its location was determined in part for the same reason that Thumba would be selected a century later as India's first rocket launch site, the close vicinity to the magnetic equator. Observations taken from a latitude of 8° 30' 35" N were considered likely to yield valuable results owing to its proximity to the equator.[106]

The observatory had several instruments, including a telescope with a 5-inch (12.7 cm) aperture. Although marginally more effective in producing scientific observations due to its location close to the magnetic equator, the Trivandrum Observatory, like the Lucknow Observatory, failed to produce significant astronomical observations during its active period. A decade and a half after its inception, the instruments were in a poor state, and John Broun (1817–1879), the Observatory Director appointed in 1851, changed the role of the observatory. It was re-tasked to record observations in magnetism and meteorology rather than astronomy.

The observation of the solar eclipses of 1868, 1871 and 1872 and the transit of Venus in 1874 helped observational astronomy get a foothold in India.[107] It also supported the case for establishing a more permanent facility in India to monitor the Sun regularly, where it was more reliably accessible than the cloudy British skies. Norman Lockyer convinced the Secretary of State for India, Lord Salisbury (1830–1903), that a telescope already in India for the 1874 transit of Venus should be deployed in a more permanent facility to routinely observe the Sun.

Dehradun Observatory in the foothills of the Himalayas was approved in September 1877, and it started its systematic observations of the Sun in early 1878. As part of the Survey of India established in 1767 (of which later the GTS was a part), photographs of the Sun's disc were taken regularly and sent to the UK on a weekly basis between 1878 and 1925.[108]

The function of the Dehradun Observatory was routine solar observation rather than scientific investigations of the type undertaken at the Kodaikanal Observatory established in 1900.

Figure 2-8 Eugène Lafont (1837–1908). Credit Grentidez

A science enthusiast and Belgian Jesuit priest, Father Eugène Lafont (1837–1908) helped establish an observatory in Calcutta, which survives to this day. Lafont was a professor of science at St. Xavier's College. His rooftop observatory was initially designed to gather meteorological data, and it became operational in 1867. The transit of Venus brought many Europeans to India. One was Professor Pietro Tacchini (1838–1905) of Palermo Observatory in Italy. Jointly, they observed the transit, Lafont optically and Tacchini spectroscopically, from Madhapur north of Calcutta, a site Lafont had identified as suitable.[109] After the transit, Tacchini encouraged Lafont to set up a solar observatory and conduct spectroscopic observations of the Sun to complement his work in Italy.[110]

By 1879, Lafont had raised funds and rebuilt his college rooftop observatory into what became the St. Xavier's College Observatory in Calcutta. It was equipped with a 7-inch (17.78 cm) equatorially mounted telescope housed in a large dome supported by two spectroscopes, one for use with a telescope and the other for visual observations. As part of its role as a teaching facility, the observatory was primarily used to observe sunspots and prominences. As such, it was India's first observatory dedicated to the scientific investigation of the Sun.[111]

Lafont's contribution was more than just his rooftop observatory. He helped nurture the scientific renaissance of the late 19th century in India by holding popular public talks on scientific subjects. Among his students at St. Xavier's College was Jagadish Chandra Bose (1858–1937), who studied physics. Bose demonstrated radio waves in 1895 in Calcutta prior to Marconi's demonstration of transatlantic radio communication in 1901. When controversy arose over who invented the radio, Lafont championed recognition for his former student. Lafont also helped to establish the Indian Association for the Cultivation of Science (IACS), where the Nobel Laureate C.V. Raman conducted his first laboratory experiments. The observatory Lafont set up in 1867 was renovated in 2014 and renamed the Fr. Eugène Lafont Observatory.[112]

Another Indian pioneer in observational astronomy was Gode Venkata Juggarow (1817–1856). He had trained at the Madras Observatory for four years with Taylor. In 1840, he built his private observatory at Daba Gardens in Visakhapatnam on the east coast. He made his observations using a 4.8-inch (12.192 cm) telescope. Juggarow published several scientific papers in the Madras Journal of Literature and Science based on his own observations of Jupiter and Lunar occultations.[113] Following his death in 1856, first, his son-in-law and then his daughter managed the observatory before it was handed over to the Madras government in 1894. Papers published by Juggarow Observatory included measurements of the mass of Jupiter, observations of three solar eclipses (18 August 1868, 12 December 1871 and 17 May 1882), all partials from Daba Gardens and transits of Mercury (5 November 1868, 7 November 1881 and 10 May 1891) and significant work on comets, including Comet Pons-Brooks on 31 January 1884.

The first observatory funded and operated entirely by Indians was the Takhtasingji Observatory founded in Poona in 1882. Initially, Maharaja Takhtasingji of Bhavnagar donated Rs. 5,000 matched by an equivalent sum by the Bombay (now Mumbai) government to establish a spectroscopy laboratory at the University of Bombay (now University of Mumbai). The switch from a laboratory in the University of Bombay to an observatory in the College of Science in Poona was probably driven by an academically gifted Naegamvala, whom Lockyer later wanted to succeed Pogson as the Director of the Madras Observatory. He had strong connections in Bombay and Poona. Naegamvala had secured his Bachelor of Arts in 1878 and Master of Arts in physics and chemistry from Bombay by the age of 21, and then, he served as the professor of astrophysics in the College of Science in Poona.

Having secured the funds in 1882, Naegamvala travelled to Europe and North America to understand the latest technology in astronomical instrumentation and techniques. He visited observatories in Germany, Italy and Spain. He spent time with Norman Lockyer in London, from whom he acquired training in the use of modern equipment and observational techniques, especially spectroscopy and photography.[112] Naegamvala called on the assistance of Astronomer Royal Sir Christie to help him compile a list of equipment for his observatory. In July 1884, Naegamvala visited the premier designer and builder of telescopes, Howard Grub (1844 -1931), in Dublin and placed an order for a 16.5-inch (41.91 cm) reflector telescope. It was the largest telescope in India at the time and remained so for several decades. Having placed the order, he returned to India. Four years later, the Takhtasingji Observatory containing the Bhavnagar telescope was in operation with Naegamvala as its director.

Figure 2-9 Bhavnagar Telescope in Ladakh 1984. Credit Indian Institute Astrophysics

Today, the reconfigured and renovated Bhavnagar telescope is better known for the time it has spent in boxes travelling or in storage than the astronomical observations made using it. During its lifetime, it was returned to Dublin twice for repair and reconfiguration. To mark its centenary in 1984, the Bhavnagar telescope was renovated and relocated to Leh in

Ladakh. After four years of operation, it was packed into boxes and returned to Kodaikanal, where it remains in storage once again. Although only a few in number, these and other observatories helped to foster a tradition of observational astronomy rooted in the scientific method in India for over 200 years. This culture of scientific enquiry is what Homi Jehangir Bhabha (1909–1966) and Vikram Sarabhai drew upon when they initiated the space programme in India in 1962.

Modern Astronomy

In the 21st century, India has a network of observatories working on a variety of international research programmes. Today, most large cities in India promote scientific investigation of the cosmos through planetarium shows, formal and informal educational courses and amateur astronomical societies. For example, the Bangalore Astronomical Society in Bengaluru has one of the largest memberships amongst the many astronomical societies in India with an active programme of observational astronomy and outreach.[115]

Figure 2-10 Devasthal Optical Telescope. September 2015. Credit Aryabhatta Research Institute for Observational Sciences

In the 21st century, India has a network of observatories in India and participates in international scientific research programmes. It has deepened its commitment to the European Centre for Nuclear Research (CERN), which operates the Large Hadron Collider where the Higgs Boson was discovered in 2013, by replacing its current Observer status with that of an Associate Member. The ground-breaking detection of gravitational waves was made in February 2016. The work of several Indian institutions on gravitational waves has been recognised, and it is most likely that India will be confirmed as the third Laser Interferometer Gravitational-Wave Observatory site.[116] In September 2015, ISRO launched Astrosat. Despite being frequently dubbed as the Indian version of the Hubble Space Telescope, Astrosat is not primarily an optical telescope. It carries six sensors that are designed to look away from the Earth. Its primary target is the Sun and deep sky objects that emit high energy radiation in the ultraviolet and X-ray part of the spectrum. At an equatorial orbit of 403.89 miles (650 km), it orbits the Earth 14 times daily. Ten of these 14 orbits pass over India and are used daily to transmit up to 420 gigabits of data to the ground stations below.

India is also committed to the international 30-m telescope project currently in the design phase. It will participate by providing 100 of the 492 smaller mirrors and 3,444 edge sensors that will be used to construct the 30-m telescope. In return, astronomers from India will be allocated time for their experiments on the largest telescope ever built.[117] The largest optical telescope on Indian soil was formally inaugurated by the Prime Minister of India remotely from Belgium on 30 March 2016. The 3.6-m diameter Devasthal Optical Telescope is located at Manora Peak near Nainital in the foothills of the Himalayas. The telescope is managed by the Aryabhatta Research Institute for Observational Sciences (ARIES) and is referred to as the ARIES telescope. Its 3.6-m diameter mirror makes it the largest optical telescope in Asia. It is an international resource that astronomers will use to conduct research on galaxies, stars and magnetic field structures around stars.

◄ ◊ ►

Chapter Three
Emergence of Scientific Institutions

T echnology in the service of humankind - read the vision statement on ISRO's website www.isro.gov.in until the end of 2014.[118, 119] A similar motivation drove Jawaharlal Nehru (1889–1964), Homi Bhabha and Vikram Sarabhai in founding India's space programme, developing the economy of the nation on the anvil of science and technology. This is the vision that has been driving the economies of Western countries and is the epitome of a developed nation. The British Labour Party leader Harold Wilson (1916–1995) in a speech on 1 October 1963 called for a new Britain to be forged in the "'white heat' of a technological and scientific revolution."[120] The European nations and North America developed their economies firmly on engineering, science and technology before, during and especially after the Industrial Revolution. The wealth generated by international trade facilitated by the Industrial Revolution gave rise to a blossoming middle class. Rich, bright men with enquiring minds spent their money and time in the pursuit of their scientific curiosities. The growth of human civilisation throughout history has been made possible by science and the technology it underpins. It is science and the technology it enables that provides infrastructure for civilisation to support ever larger populations to live longer productive and healthy lives.

Innovations based on science and technology in various fields, including medicine, surgical procedures, metallurgy and weaving, can be traced through Indian history. During the colonial period, however, scientific thinking and innovation were not encouraged. European nations had to suppress development within the countries they colonised; to do anything else would undermine the idea of colonisation itself. A nation consumed with superstition and ancient traditions was far easier to rule and administer than one based on rational, informed discourse. It was in the UK's interest to keep millions of Indians scientifically illiterate and let them remain unskilled producers of raw materials and consumers of British exports. Speaking during his 1953 tour of four Australian cities, Patrick Blackett (1897–1974), a Nobel laureate with left-of-centre political views, concluded

"The chief economic interest of England [and other colonial powers] in their respective colonies was to develop them both for markets for the European manufactured goods and as a source of primary products, oil, rubber, tea, coffee and so forth. In general, the industrialisation of the colonial countries was not encouraged, so that they would not compete with the home countries."[121] Blackett was an advisor to the British government during World War II and became Nehru's advisor on science policy following India's independence in 1947.

As the first Prime Minister of independent India, Jawaharlal Nehru irrevocably intertwined India's future development directly with science and technology. Though he had no formal scientific training (he had studied law in Cambridge), Nehru was convinced of the transformational potential of science for developing nations, unlike Gandhi. While in Ahmednagar Fort prison for five months, from April to September 1944, Nehru wrote his book *The Discovery of India*. Although still under colonial rule at the time, an independent India that he had pursued for most of his lifetime was finally becoming imminent. In this book, Nehru put on paper his thoughts and motivations that would in time crystallise his vision of a modern industrialised first-world India.

Despite almost 200 years of colonial rule, the traditions and ideals of India that had developed over the millennia were still present in 1947 and would be the foundation for Nehru's vision of independent, self-sustaining India. He was mindful that the ancient Indian culture and its traditions would need to be adapted where possible, or replaced, to meet the challenges of the 20th century. The India that the British left in 1947 was large and democratic and could not simply return to pre-colonial conditions.

Soon after independence, the Indian government launched Five Year Plans (FYP) to implement its policies for economic development. The vision for the first FYP (1951) was to achieve a "Faster, Sustainable and More Inclusive Growth". The latest is the 12th FYP published in 2011 to cover the period of 2012–2017.[122] FYPs define the national targets for the various government departments (for example, the Department of Space (DOS), Department of Atomic Energy (DAE), Department of Science and Technology (DST)). The DST's 12th FYP envisages that "within the next 20 years, Indian economy would have emerged as a major global economy."[123] This plan is predicated on science being at the heart of strategies driving national development.

Economically prosperous societies owe their prosperity to science, and for that prosperity to persist, investment in science is crucial. Further, a large-scale societal transformation could be accomplished only through scientific institutions. Emerging from two centuries of colonial rule, India was steered by its first independent first government onto the road to economic development through the establishment of scientific institutions. This was not a dramatic shift in direction, but an opportunity for institutions that had existed for many decades to emerge and flourish. While many institutions that could collectively deliver scientific innovations were established around the time of independence, largely at the behest of Nehru, some institutions had much earlier beginnings. Even though many of them ceased to exist long before independence, their collective spirit persisted to inspire future generations. Today, India has institutions for optical and radio astronomy, particle physics, astrophysics and space conducting leading-edge research. It is, however, possible to detect the roots of present Indian institutions in India's past. Even India's burgeoning space programme owes its success to a wide range of scientific capabilities with deep roots that can be traced back into India's history.

A comprehensive assessment of all the scientific institutions that have driven societal development in India is beyond the scope here. However, a sample of Indian scientific institutions founded before and after independence can help illustrate the central role of science as an agent for national economic and social development.

Aligarh Scientific Society

One of the earliest organisations in India established around the Western traditions of science was the Aligarh Scientific Society (ASS). It was set up in 1864 by Sir Syed Ahmad Khan (1817–1898), who had visited Oxford and Cambridge and recognised the fundamental role of science in facilitating the Industrial Revolution and in shaping the British and European societies. Driven by this European vision, the primary objective of the ASS was "not only an attempt in imparting scientific knowledge but also an attempt at social reform through science in India."[124]

The ASS attempted to realise its vision by translating and publishing scientific papers and books, especially rare and valuable oriental works. Communication and discourse is at the heart of the scientific process. At that time, it was almost entirely facilitated by paper-based communication. As Abdus Salam noted almost a century later, isolation from other scientist is the major concern in cultivating scientific knowledge, especially in a developing country.[125] The ASS established a dedicated library with a reading room and subscribed to 44 journals and magazines by 1866. Less than half were in English; the rest were in Arabic, Urdu, Persian and Sanskrit.

Figure 3-1-Sir Syed Ahmad Khan. Credit Unknown

The ASS also communicated through newspapers, magazines and formal presentations at public meetings with the intention of engaging the Indian public. Although it was the vision of a Muslim intellectual and most of its initial members were Muslims, the ASS was an openly secular society from the outset. To emphasise the central role of science and technology, works of religious nature were explicitly prohibited. At the time that the ASS was founded, India was largely an agricultural society with very high illiteracy, and familiarity with science among the general population was all but absent.

The ASS had four key objectives:[126]

- Translating Western literature into local Indian languages
- Popularising and democratising mechanised farming
- Delivering lectures on topics of common interest
- Highlighting the socio-political problems of the country

As the first of its kind, the ASS regarded the role of modern science as more significant for India's future than ancient Indian traditions of ritual and mysticism. Khan was motivated by his first-hand experience of the higher quality of life enjoyed by the inhabitants of the major cities in the West with planned housing, transport infrastructure and mass employment in the manufacturing industry. He was ambitious and wanted Indian societies to be on par with those in the west and understood the critical role of science in that process.

A decade after establishing the ASS, Khan founded the Mohammedan Anglo-Oriental College in 1875. Despite his emphasis on Western science, Khan took his Muslim identity seriously and saw science as a way of securing political power for the Muslim community in India. The Mohammedan Anglo-Oriental College became Aligarh Muslim University in 1920 and is one of the most prominent educational institutions in India today. Despite 109 members attending its first meeting, the ASS experienced a severe decline in membership. It was never patronised by a sufficient number of distinguished individuals. By 1887, its founder had turned 70, and in the absence of an enthusiastic membership, wealthy benefactors, and especially a committed, passionate new leader to take it forward the ASS ceased to exist in 1887.[127]

Indian Association for the Cultivation of Science

A decade after the ASS was founded, another ultimately more successful organisation was established in Calcutta (now Kolkata). IACS founded in 1876 eventually attracted the patronage of brilliant Indian scientists of international repute. Today, it is India's oldest research institute and is actively engaged in pure scientific research. Mahendralal Sircar (1833–1904), the founder of IACS, recognised the potential of science to advance India as a nation. Initially, IACS was financed by the Lieutenant-Governor of Bengal, Sircar's private funds and through public subscription. With additional donations, IACS expanded over time. A laboratory was added in 1892 with the help of the payment received from the Maharaja of Vizianagaram after Sircar had successfully treated him for a rare disease.[128]

Figure 3-2 Original Building of the Indian Association for the Cultivation of Science in Calcutta. Credit IISc Archives

In addition to Sircar, the cumulative efforts of at least three other individuals have been integral in the success of IACS during its initial phase; they were Jagadish Chandra Bose, Acharya Prafulla Chandra Ray (1861–1944) and Asutosh Mookerjee (1864–1924).[129] The scientific achievements of international standard by Bose and C.V. Raman, another early member, also contributed profoundly to the success of IACS. Another key promoter of the IACS during the initial stage was Father Eugène Lafont, the Jesuit priest from Belgium who founded St. Xavier's College Observatory, Calcutta. Lafont was an established scientist of high repute based in Calcutta for over a decade before the IACS was founded.

Although Sircar wanted IACS to promote Indian scientists, he understood that political and financial support from the ruling British authorities was essential. Any Indian institution with hopes of international recognition depended upon British approval. He became a life member of the British Association for the Advancement of Science in 1864. IACS was slow in getting started. It had research facilities and a library, but in its early days, it was more a gentleman's club for social gathering than an active research institute.

Nobel laureate C.V. Raman was one of the first scientists to make active use of IACS's facilities for scientific research. Raman, like Bose, was an experimentalist. Unlike other Indian scientists who had achieved international success in science, Raman was entirely the product of the Indian educational system. He had not studied in the West. Raman was employed in the finance department of the Indian Civil Service and used the laboratory for his scientific research before and after work for a decade (1907–1917) before leaving to join the University of Calcutta. He was awarded the Nobel Prize in Physics in 1930, long after he had left the IACS.

For many decades, the IACS has been and remains today one of India's premier institutes for scientific research. Its broad sweep of subjects has attracted many individuals, including Suri Bhagavantam (1909–1989), Kedareswar Banerjee (1900–1995), L. Srivastava, N. K. Sethi, C. Prosad and Meghnad Saha. Saha served as its leader from 1946. At present, IACS is funded jointly by the DST and the government of West Bengal. With over 200 staff members and 400 research students across 15 different science departments, the IACS conducts fundamental research in chemistry, biology, physics and material sciences, including nano-particles and graphene. It also hosts a supercomputer (CRAY) facility on site to support its research. During the academic year 2013-14, IACS produced almost two dozen doctorates, registered national and international patents and engaged in international collaborative projects.

Astronomical Society of India

The present Astronomical Society of India (ASI) was established at Osmania University in Hyderabad in 1972. It has around 1,000 members, all of whom are professional astronomers or associated with similar disciplines and regularly publish technical work of high scholarly calibre. Although completely unconnected and now almost entirely forgotten, another ASI functioned in India between 1910 and 1920. It was the spectacular display of comet Halley that triggered the creation of this original ASI in 1910.[130]

During its brief existence, the original ASI attracted the patronage of several high-profile astronomers. A leading observer of the Moon and Fellow of the Royal Society, H.G. Tomkins was the main force behind the creation of the

ASI. He was also its first president. Charles Michie Smith, Director of the Kodaikanal Observatory (and later John Evershed who replaced him) led the society's solar section. The vice presidents included the Maharaja Rana Bahadur Bhawani Singh, and C.V. Raman (then just 23 years old) served as its secretary. The ASI held monthly meetings at the Imperial Secretariat, Treasury Buildings in Calcutta; it was then the capital of British India. It conducted much of its work through specialist sections, such as the lunar, meteor and variable star sections. Although membership was restricted initially to the subjects of the Empire, nationals of other countries were later encouraged to join.

Figure 3-3 Seal of the Astronomical Society of India Designed by Member F.C. Scallan in February 1911. Credit Indian Institute of Astrophysics

The ASI also set up a library and exchanged journals with other organisations in India and beyond. They included local astronomical societies in Leeds in the UK, Barcelona in Spain and Turin in Italy; national astronomical societies of Italy and Canada; the Royal Observatory of Scotland; the Vatican Observatory; the Royal Astronomical Society and the British Astronomical Association.[131]

The ASI was rich in the range of topics and the complexity of the debates it hosted. The nature of the papers presented at its regular meetings was specialised and highly technical. John Evershed spoke about his solar observations on the "angular speed of rotation of long enduring prominences". C.V. Raman presented a paper on Astronomical Optics, which he illustrated profusely with mathematical equations.

Some of the papers tackled astronomical themes that are well understood and accepted today but lacked sufficient evidence in the early 20th century. For example, on 11 February 1911, W. A. Lee spoke on the Nebular Hypothesis, the idea that the solar system, the Sun and all the planets condensed out of a giant cloud, a nebula of gas and dust. Lee supported his case by presenting observational evidence "the planets all revolved in one plane, they all revolved in one direction in their orbits, and mostly rotated in the same direction on their individual axes."[132] A wealth of evidence acquired since then from Earth and space-based telescopes confirm this hypothesis. In November 2015, astronomers announced the discovery of a solar system in the making, which was consistent with the Nebular Hypothesis but beyond the Solar System.[133]

A discussion following a paper on meteors on 28 March 1911 presented the possibility that a meteor trail was a "result of the combustion of the meteor."[134] In the early 1940s, the British astronomer Bernard Lovell (1913–2012) detected radio waves that reflected from the trail of ionised molecules left behind as a meteor combusted following its high-speed impact in the upper atmosphere. He confirmed the hypothesis of meteor combustion and went on to advance a new branch of study, radio astronomy. He also developed the world's largest fully steerable, 76-m diameter radio telescope in 1957.[135]

The idea that large planets with correspondingly strong magnetic fields can hold to their atmosphere while a smaller planet with a weak magnetic field cannot is now an accepted and understood scientific concept. For example, Mars has a very tenuous atmosphere, unlike the Earth or Venus. Intriguingly, this idea was raised in the society's correspondence records on 1 May 1914. F.C. Molesworth asserted "We know that the Moon is not heavy enough to retain permanently any gases. Hence, at no stage of its existence could it have had an atmosphere of appreciable density."[136] ASI also debated principles of the scientific method. One of the speakers insisted that it was the "glory of Science that it did change its hypothesis when they were tested and found wanting."[137]

The ASI attracted considerable interest from non-professionals, too. To promulgate the excitement and joy of astronomy, the ASI conducted a series of classes in Calcutta for beginners in astronomy. The classes introduced "rudiments of science and demonstrated how to use star charts". Mr Rackshit, the ASI Director of the meteor section, led those classes, and the first batch was attended by 20 to 30 individuals.[138] Despite the high standards of the scientific contributions from its several members of national and international repute, the ASI did not survive for long. Although financially secure, its scientific contributions did not attract substantial international recognition.

With air and light pollution, a growing city like Calcutta was not a suitable location for an astronomical observatory. However, it was the failure to attract young, talented and industrious members with the ability to conduct and publish original research that led to the demise of ASI within a decade of its inception. Attempts to engage the wider public in astronomy and science is replicated by today's ASI. In 2014, the new ASI established the Public Outreach and Education Committee of the ASI to promote awareness of astronomy and increase public engagement in science.[139]

Council of Scientific and Industrial Research

The hope of nationalists for the independence of India immediately after the end of World War I had been dashed. Two decades later, as the UK fought for its own independence against Nazi Germany, a politically stronger Indian independence movement was determined to end British rule. In preparation, the Council of Scientific and Industrial Research (CSIR) was founded in 1942 to facilitate the R&D essential to realise Nehru's vision of an India built on modern science and technology. Nehru had spent most of World War II in prison and used that time to write. He reflected on the role of science concluding that "the applications of science are inevitable and unavoidable for all countries and peoples today."[140] CSIR today is the largest R&D body in India employing about 21,000 staff members, including research scientists, technical support members and research students in equal measure.

Whereas the IACS and Tata Institute of Fundamental Research founded by Homi Bhabha in 1945, focus on pure research, CISR's mandate has been industrial research. World War II was raging at the time it was created, and this emphasised the critical nature of a national industrial infrastructure, not just for national development but also for defence. The building of the first CSIR laboratory, the National Chemical Laboratory, started on 6 April 1947 in Pune, three months ahead of the formal independence of India. Others followed. The National Physical Laboratory in New Delhi, National Metallurgical Laboratory in Jamshedpur, Central Fuel Research Institute in Dhanbad, Central Glass and Ceramics Research Institute in Calcutta and Central Food Technological Research Institute in Mysore were established in 1950. Since then, the number has grown to over 40 across India employing over 1,000 technical personnel. The laboratories are grouped under five disciplines: (i) physical and Earth sciences, (ii) chemical sciences, (iii) biological sciences, (iv) engineering sciences and (v) information sciences.

Most key posts in the scientific institutions established by Nehru were held by physicists, but he picked Shanti Swaroop Bhatnagar (1894–1955), a professor of chemistry, as the Director-General of CSIR. Bhatnagar was revered as the Father of Research Laboratories and went on to hold the influential post of the Chairman of the University Grants Commission. The CSIR files around 1,000 new patents and publishes about 4,000 scientific papers every year. Its research is not pure research but research that can be directly applied. The diverse disciplines include innovative commercial aircraft, parallel computers and technology for genetic fingerprinting.

Governments of all political colours have consistently supported industrial development since independence. In September 2014, just a few months after coming to power, the Narendra Modi government announced a new initiative called Make in India. The primary objectives included encouraging foreign investment, increasing industrial production, creating employment and training opportunities and generating a larger tax revenue. Government procedures and controls on foreign investment have been relaxed in many areas, including railways, space, automobiles, chemicals, news media, IT, pharmaceuticals, defence, aviation and textiles.

The CSIR has put forward its Broadband Spectrum Confocal Microscope (a microscope using an enhanced optical imaging technique to increase resolution) as a 'humble beginning' to the Make in India initiative. The former ISRO Chairman U.R. Rao, who led the team that built India's first satellite, expressed frustration at the absence of self-sufficiency in non-space sectors in India; the space sector is an exception that has largely succeeded in becoming self-reliant. In November 2015, Rao lamented, "even Brazil is making small aeroplanes. What are we making in defence? We import everything, including Swiss knives. You can't defend a country unless you make it yourself."[141]

Even though CSIR has grown in size and complexity since 1942, its management has not kept pace. It no longer attracts the attention of the highest political office as it once did. On 25 August 1947 "even while riots raged",[142] Nehru ensured he was present for a meeting of the CSIR. The quality and quantity of the scientific work conducted by CSIR today is considered by some to be below its full potential.

Indian Institute of Science

In 1893, Jamsetji Nusserwanji Tata (J.N. Tata, 1839–1904) chanced to meet with Swami Vivekananda (1863–1902) aboard a ship bound to the US from Japan. Vivekananda was probably India's first global celebrity extolling the values and traditions of India (especially Advaita Vedanta and Raja yoga). They discussed establishing steel and other industries and higher education institutions in India for India. This meeting has now become almost a sacred memory in Indian folklore. Tata envisioned an institute of higher education for science that would work for the benefit of India. In the late 19th century, this was a particularly outrageous vision. India, at that time, had little in the way of established universities producing high-quality graduates in numbers anywhere near proportional to its population.

To realise J.N. Tata's vision, Burjorji Padshah (1864–1941), an intelligent and capable manager within the Tata business group, travelled to Europe and America with the task of identifying the model of a university that would best suit India's need. He chose Johns Hopkins University in Baltimore. In 1898, he published a report titled 'Institute of Scientific Research for India', which eventually resulted in the IISc.

It took a decade of surmounting numerous obstacles, but finally, the IISc was established in 1909, five years after the death of J.N. Tata. A century later, the IISc became the first Indian higher educational institution to be listed among the top 100 higher educational organisations worldwide in a highly regarded ranking published in the *British Times Higher Education Supplement*.[143] Today, the IISc offers undergraduate and postgraduate courses and research opportunities in many disciplines, including biology, chemistry, electrical engineering, physics and mathematics.

To get the project moving, J.N. Tata had understood that he needed three key ingredients. He would have to invest his own funds, persuade the colonial government to commit financially and encourage a donation of land from a wealthy Indian princely state. J.N. Tata and Padshah met with the new Viceroy of India, George Curzon (1859–1925), on 31 December 1898, two days after he arrived in India to take up his new post. Their intention was to secure his support for their project. Curzon was not forthcoming. Not only was he not convinced of the viability of the project, but he also saw no need for Indians to acquire higher education, even if they were intellectually capable. After all, working as clerks and administrators in the Indian Civil Services was all that Indians could aspire for. He likened higher education for Indians to "presenting a naked man with a top-hat when what he wants is a pair of trousers."[144]

Tata presented 18 of his privately-owned properties to the government as his side of the bargain to get the colonial government to commit to the project.[145] He did not want his name associated with the project to encourage the widest range of additional sponsors.[146] In 1904, J.N. Tata died. One of his two sons, Dorabji Tata (D.J. Tata, 1859–1932), wrote to the Viceroy informing him of his father's death. In the same letter, he confirmed that the institute project would continue as planned: the government was to provide 30% of all proposed costs as grants, the Royal Society in London was to select professors for the Institute and the new facility was to be a public, not a private, institution.[147]

The difficult and protracted project finally met with some success in 1907. By then, Curzon had been replaced as Viceroy. Tata's sons, Dorabji and Ratan, succeeded in first securing a financial commitment from the government and then 372 acres of land in Bangalore (now Bengaluru) from the Maharaja of Mysore, Krishnaraja Wodeyar IV (1884–1940). After a further two years of complex negotiations, the colonial government formally passed a resolution to establish the Institute. With additional funding from the State of Mysore, the foundation stone for IISc was laid in Bangalore in May 1909, and the formal teaching started two years later.

IISc's first Director, Morris Travers (1872–1961), a chemist, was encouraged to take up the role by his colleague and scientific collaborator William Ramsay (1852–1916). Upon request from the British government to advise on a potential Indian scientific institution, Ramsay had visited India in 1900. He was a celebrated chemist who, working with Lord Rayleigh (1842–1919), had discovered argon, a noble gas, in 1894, an achievement that was recognised with a Nobel Prize in Chemistry in 1904. Subsequently, Ramsay and Travers collaborated and used fractional distillation of liquid air to discover the noble gases neon, krypton and xenon. They were also the first to isolate helium, which had been initially detected from India in a spectroscopic observation of the Sun during the total solar eclipse in 1868. In addition to his skills as an experimental scientist, Travers excelled in designing and making glass instruments, as well as designing and building industrial plants and furnaces. He pioneered a process for producing acetylene gas from calcium carbide. In 1903, using his own apparatus, Travers produced liquid hydrogen (LH2) for the first time during his demonstration in Germany.[148]

When Travers was offered the post of Director of IISc, he was working at the University of Bristol and was not initially keen to take it up "Expatriation and working with people with whom I could not cultivate a bond of sympathy are my main reasons for holding back". Eventually, however, Travers accepted.[149] An increase of salary from his then £ 350 to £ 1800, paid accommodation and a pension probably helped to persuade him. On 2 November 1906, Travers left Marseilles on the steamer *Victoria* to take up his post in Bangalore. He arrived before the building, facilities and organisational structures were complete. As the institute director, those were his initial tasks before students arrived in 1909.

Even though he was able to conduct his research, he found routine tasks a burden. People management, provision of buildings, equipment, teaching and administration, along with developing and fulfilling a vision for the future of the institution, did not hold his interest. One of his first challenges in 1911 was to provide the unique catering facilities to meet the varied dietary needs of students from all over India, vegetarians, non-beef-eating Hindus and non-pork-eating Muslims. His most serious disputes, however, were with Padshah and Tata on the direction the institute should take academically. Travers wanted to introduce humanities, including archaeology. The Tata brothers, who were convinced by the potential of science to develop India, wanted the focus to be on hard sciences, such as bacteriology and tropical diseases.[150] These disagreements with the governing council (similar to what C.V. Raman would experience three decades later) led Travers to terminate his position and return to the UK in 1914. Over time, the IISc added additional subjects electrical engineering in the 1920s, physics in the 1930s and aeronautics in the 1940s.

A year after the outbreak of World War II in Europe, France was occupied by Germany, and Japan had moved to occupy French Indochina. Concerned that Japan could head east to India, the UK invested in a factory to assemble and repair aircraft. It was established by Walchand Hirachand (1882–1953) in Bangalore. To support the demands of war, an aeronautical department was added to the IISc. In 1941, Hindustan Aircraft Limited (now Hindustan Aeronautics Limited) assembled the first aircraft, a Harlow trainer, in India. Today, it produces spacecraft for ISRO. Roddam Narasimha, a student at the IISc in the early 1950s, remembers seeing a Spitfire on the grounds of the IISc, but the landing strip on the IISc grounds used for training had by then been reclaimed.[151]

In April 1939, a German architect Otto Köenigsberger (1908-1999) arrived in Bangalore as the Government Architect of Mysore State. He was the nephew of Max Born (1882–1970), whom C.V. Raman had invited to IISc for a 6-month stint in 1935. Possibly motivated by the imminent war in Europe, IISc went through a phase of expansion, and Köenigsberger was drafted in to help. He designed the departments of aerospace engineering and metallurgy and the dining hall/auditorium in addition to his duties for the State of Mysore.[152]

Figure 3-4 Indian Institute of Science. Credit IISc Archives

The IISc played a key role in the early careers of most Indian scientists who helped shape India's post-independence technical development. Nobel laureate C.V. Raman was the institute's first Indian Director (1932–1938), Homi Bhabha held his first teaching post in India at the IISc from 1939, Vikram Sarabhai conducted his PhD research under C.V. Raman's supervision during World War II and Satish Dhawan (1920–2002) taught there in the 1950s and returned as Director in 1962. The success of the IISc model inspired and fostered other institutions, including the Indian Institutes of Technology, Indian Institutes of Management, the TIFR and the National Institute of Advanced Studies.

When Morris Travers started, the IISc had two departments and 21 students. Today, it has 40 departments with nearly 4,000 students, most of who are engaged in postgraduate or research programmes supported by 500 academic and technical staff. It houses the largest library in India with the annual bill for the subscription of publications amounting to Rs.9 million ($0.13 million). The newer departments deal with topics that preoccupy universities globally, including climate change, nanotechnology, cyber systems, neuroscience, astronomy, aerospace, microbiology, cryogenics and atmospheric and Earth sciences. A *New York Times* survey of more than 5,000 recruiters from 30 countries released on 28 October 2013 listed IISc as 23rd for producing the most employable graduates. The success of IISc is the cumulative product of the effort of many, including Morris Travers, C.V. Raman and Satish Dhawan.[153]

Philanthropy under the Tata name has been reappraised by some in recent times.[154] There were not only alleged high-handed modern business tactics but also shady deals between the Tata family and British businessmen during the opium wars of the mid-1880s from which J.N. Tata would have benefited directly. However, the unquestionable success of IISc is, at least in part, a testament to the vision, determination and tenacity of the Tata family.

Tata Institute of Fundamental Research

Founded by Homi Bhabha and initially set up in Bangalore in 1945, TIFR was the first modern institute in India dedicated to pure scientific research. Set against the institutions in existence in India at the time, TIFR was unique, if somewhat incongruous. It was not obvious in 1945 how the large amounts of money spent on pure research would help a nation in its transition to independence after 200 years of colonisation.

In August 1945, the spectacular demonstration of nuclear power as a weapon of mass destruction in Hiroshima and Nagasaki brought an end to World War II in the Pacific. The potential to use the same nuclear technology for civilian nuclear power was well understood by the late

1930s, in particular by those who had closely worked in the field during the golden age for nuclear physics. The Cavendish and Clarendon Laboratories had pioneered nuclear research in the UK during the 1930s, and Homi Bhabha was at the Cavendish Laboratory in Cambridge during this time. Bhabha's vision for nuclear energy was not limited to the industrialisation of developing nations; he saw a deeper imperative, the preservation of human civilisation itself. At the first International Conference on the Peaceful Uses of Atomic Energy held in August 1955 in Geneva, Bhabha said "For the full industrialisation of underdeveloped countries, for the continuation of our civilisation and its further development, atomic energy is not merely an aid, it is an absolute necessity."[155]

Figure 3-5 Tata Institute of Fundamental Research Mumbai. Credit Author

Bhabha was a gifted intellectual and scientist of international reputation. His family connections with the high society in India, as well as with India's political leadership, brought him into contact with influential players in industry, business, commerce and politics since his childhood. In 1939 Bhabha was on a visit to India from Cambridge. When World War II broke out, he was stranded in India and could not return as planned to Manchester to work with Patrick Blackett.

C.V. Raman recognised Bhabha's potential and hastily created the position of Reader of Theoretical Physics at the IISc, which Bhabha accepted. In 1941, at the annual meeting of the Indian Academy, Raman introduced Bhabha as "the modern Leonardo da Vinci". In the same year, Bhabha was elected as a member of the Royal Society, and 1942, he won the prestigious Adams Prize for his PhD thesis.[156] Bhabha possessed an international reputation unique among Indian scientists and could easily have acquired a prominent position in a Western university. Instead, he chose to stay in India.

Convinced of the need for a "vigorous school of research in nuclear physics"[157] located in India, Homi Bhabha wrote to the Tata Trust on 12 March 1944 seeking sponsorship for what later became the TIFR.[158] This became the cradle for the Indian civil nuclear energy and space programmes. In his letter, Bhabha confided that he had already declined an offer of the Physics Chair at the University of Allahabad and professorship at the IACS. Perhaps, he was swept up by the national fervour of the impending independence, as in the same letter to the Tata Trust, he wrote "it is one's duty to stay in one's own country and build up schools comparable to that other countries are fortunate in possessing". Bhabha also saw the potential for a civil nuclear energy programme in advance of the first use of atomic energy as a weapon. The prospect of nuclear energy ahead of nuclear weapons is one that the British government pursued. Bhabha may have been directly influenced by it. In his 12 March 1944 letter to the Tata Trust, prior to the nuclear explosions in Hiroshima and Nagasaki, he outlines his vision for a nuclear-powered India saying, "when nuclear energy has been successfully applied for power production, say a couple of decades from now, India will not have to look abroad for its experts".

Although the TIFR was conceived and initiated in Bangalore, Bhabha wanted it to be based in Bombay (now Mumbai). The six months in 1935 that Max Born spent at IISc triggered a sequence of events that determined the design and architecture of some of the buildings at IISc and helped shape the initial design for TIFR, too. While at the IISc, Born was asked to recommend a 'non-English' European architect who could help with Mysore's expanding building programme. He suggested his nephew Otto Köenigsberger, who arrived in Bangalore three years later. Born had met Homi Bhabha in Cambridge, where he was teaching after Nazism had forced him to abandon his professorship at Göttingen. Köenigsberger had already visited Mysore in 1933, also fleeing Hitler's Germany.

Bhabha was impressed with Köenigsberger's work at IISc and wanted him to design TIFR in Bombay. In the summer of 1947, Bhabha invited him to Bombay to start the detailed design as the chief architect. Köenigsberger started the work, but fresh legal delays with the TIFR site permissions forced him away, and Bhabha engaged Helmuth Bartsch of Holabird and Root from Chicago instead. TIFR formally came into existence on 1 June 1945. It operated until December of that year from IISc in Bangalore and then moved to Bombay. The custom-built premises it occupies today was formally opened by Prime Minister Nehru on 15 January 1962.

Figure 3-6 Max Born (front row fourth from the right) and Homi Bhabha (left-hand side fourth row) at an Informal Meeting on Nuclear Physics. Institute for Theoretical Physics Copenhagen 1936. Credit IISc Archives

As a nuclear physicist, Bhabha was one of a "handful of scientists in India who understood the fission process and grasped its implications."[159] During his time travelling in Europe and the US, he had also seen first-hand the dependence of developed nations on the availability of reliable electrical power to operate their national infrastructures.[160] For realising his vision of an industrialised self-reliant India, nationwide availability of low cost, dependable electrical power was a pre-requisite. He did not envisage retracing the developmental steps of the West by first building a series of coal-fired power stations but wanted to leapfrog into a future where a large-scale network of nuclear plants would eventually provide all the country's power supply.[161]

In 1960, the Tata Institute of Fundamental Research Automatic Calculator (TIFRAC), India's first computer, was built at TIFR. The computer division of TIFR was co-located within the electronics and instrumentation group in 1955. The division consisted of a handful of physics or electronics graduates, some with master's degrees, supported by students with diplomas in radio engineering. The project to build TIFRAC was headed by Dr R.

Narasimhan (1926–2007), a mathematician who had been working in the US. He was probably the only member of the project team who had "seen a computer, let alone use one."[162] Just as engineers working on India's nuclear reactors had seen and India's space engineers would discover in 1963, the computer engineers had to learn on the job with very limited facilities.

Physically, TIFRAC was about 10 m in length, typical of the size of computers in the 1960s. One of its key innovations was the use of a visual display unit as the output device rather than the traditional teleprinter. TIFRAC used a word length of 40 bits and a memory capacity of 1024 words using the then state-of-the-art, three-dimensional magnetic core memory. Ready built memory was not available, so B.B. Kalia and his team were assigned the task to manually construct the memory matrix consisting of 12,000 magnetic cores by hand.[163] TIFRAC was completed and commissioned on 22 February 1960.[164]

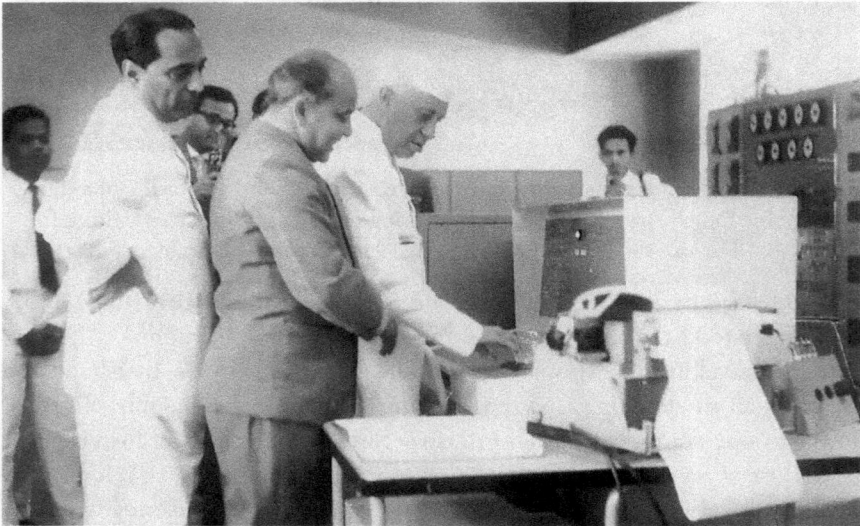

Figure 3-7 India's First Digital Computer TIFRAC. 15 January 1962. Credit TIFR

TIFR was organisationally located in the DAE. The DAE was also the home of the Indian National Committee for Space Research (INCOSPAR) in 1962, which later became ISRO in 1969. The connections between the TIFR and ISRO continue to the present. Many of the instruments and devices used on-board India's communication, science and remote sensing satellites were originally designed and built at TIFR. Five instruments aboard ISRO's first space telescope Astrosat were built by scientists in TIFR.

Before rockets, scientists used high-altitude balloons to collect data on cosmic rays. Bhabha had engaged Osmania University in Hyderabad to launch high-altitude balloons during his time at the IISc. The Balloon Facility, Hyderabad, is now a TIFR entity with an international reputation for building and conducting high-quality research using high-altitude balloons. It was responsible for making the balloon used on 24 October 2014 to set the world record for a human parachute jump from over 42 km (26.10 miles) altitude.[165]

Fundamental research in nuclear physics has been at the heart of TIFR from the outset, but now, the work includes other subjects, such as radio astronomy, molecular biology, semiconductor research and applied mathematics. Its size and organisational structure have also broadened. TIFR has expanded beyond its first purpose-built site at Colaba on the Bombay waterfront to multiple national centres across India, organised under three schools, mathematics, natural sciences and technology and computer science. It has developed a strong tradition of academic research and offers PhD and Master of Science courses, as well as hosts visiting international scholars and researchers. TIFR was formally established through a tripartite agreement between the Government of India, Government of Bombay and the Tata Trust in 1955. Today, it is almost entirely supported financially by the Government of India.

Scientific Temper

Unlike Gandhi, Nehru regarded ancient culture and traditions as a burden that "must be let go".[166] He was convinced that the "methods and approach of science have revolutionised human life more than anything else in the long course of history."[167] In Article 51A of the Constitution of India, which came into effect in 1950, Nehru codified the central role of science in India's future by requiring that every citizen of India shall "develop the scientific temper, humanism and the spirit of inquiry and reform."[168] A year after independence, Gandhi was assassinated. With his views unchallenged, Nehru kept science at the forefront. In addition to his role as prime minister, Nehru kept the key ministries of foreign affairs, atomic energy, natural resources and scientific research with himself. As prime minister, he set about embedding science into his political objectives wherever he could. He established personal contact with the many leading figures in the field of science, Indian and otherwise. Nehru sought as his scientific advisor, someone with high scientific credentials, international connections, experience in advising government at the highest levels and who shared his socialist political ideology. He found Patrick Blackett with extensive experience working for the British Ministry of Defence, as well as the Military Application of Uranium Detonation (MAUD) committee. For Nehru, Blackett's expertise and connections with political leaders in the West were of particular value following India's independence. For Blackett, this would have seemed an opportunity to put into practice his personal left-of-centre principles to shape the world's largest democracy and help it carve out a new independent destiny.

Following the formal declaration of independence in August 1947, Nehru kick-started his plan to modernise India through science. In his haste to drive development through government institutions, such as CSIR, he drew competent personnel away from universities at a critical time. He may have got the running start he wanted but at the expense of an efficient higher education sector. The infrastructure to drive the relationship between a nation's economy, higher education and industry needs to be cultivated over time. It cannot arise spontaneously. As one historian characterised it "government science in general is more government than science."[169]

Today, India enjoys international recognition for its success with information technology and more recently with frugal innovations in its space programme. Although the proportion of India's population in these sectors is tiny, it would appear that Indians do not lack the capability to establish, build and lead large-scale technology programmes. Many household names in the field of technology, including Dell, Google, Microsoft and Adobe, are currently headed by individuals of Indian origin.[170] The image of India and Indians has transformed over the last couple of decades. "The old stereotype of Indians was that of snake charmers and fakirs lying on beds of nails; now it is that every Indian must be a software guru or a computer geek."[171] Nehru would be disappointed by the slow rate of India's progress since independence, but he would be content with its continued commitment to science and technology as a guiding principle in national development.

≺ ◊ ≻

Chapter Four
Science and the Raj

I ndian scientists conducting scientific research in colonial India had to overcome unique challenges defined by the prevailing political and social structures of colonialism. India had its share of talented individuals who laid the groundwork for an environment where Indian scientists could develop science for the benefit of India. However, their names or their achievements are not much known outside India. This is partly a consequence of the suppression and discrimination that accompanied colonialism.[172] While many of the English language accounts of the 20th-century scientists include names, such as Enrico Fermi (1904–1951), Guglielmo Marconi (1874–1937) and Ernest Rutherford (1871–1937), the names of Indian scientists are not well known. During the 20th century, gifted Indians contributed to science, mathematics, biology, physics and literature. Bhabha Scattering, the Saha Equation, the Boson, Raman Effect, Bhatnagar-Mathur Magnetic Interference Balance and the Chandrasekhar Limit are terms found in modern textbooks although readers are not always aware of their Indian origins. In 1913, Rabindranath Tagore (1861–1941) won the Nobel Prize for Literature and C.V. Raman (Chandrasekhara Venkata Raman, 1888–1970) for Physics in 1930.

The Industrial Revolution and the booming economy in the UK had endowed gifted British individuals with substantial wealth to undertake scientific and astronomical research on a scale that, today, could only be contemplated by a university or a research institute.[173] Individuals, such as James Watt (1736–1819), Michael Faraday (1791–1867), Richard Trevithick (1771–1833), Geoffrey de Havilland (1882–1965), Edmund Cartwright (1743–1823) and George Stephenson (1781–1848), had a deep sense of curiosity, technical aptitude and immense personal drive. It is the legacy of their scientific and technological innovation that drives the UK and the Western nations even today.

India produced many men of scientific vision, drive and talent, Raja RamMohan Roy (1772–1833), Syed Ahmad Khan, Mahendralal Sircar and Asutosh Mookerjee, and many used their personal wealth to promote research and science, for example, Maharaja Takhtasingji of Bhavnagar, Jamsetji Tata and Maharaja Krishnachandra Roy (1710–1783) of Krishnanagar. However, in the absence of the opportunities facilitated, in part by the Industrial Revolution, India did not produce individuals with a combination of wealth, vision and talent for science. Indian scientists conducting research in colonial India faced with challenges of colonialism. A couple of such challenges were the low esteem of a subordinate people and the colonisers' discriminatory views prevalent at the time. In the 20th century, the US was astonished when the USSR, a "backward nation of potato farmers",[174] launched Gagarin into space.

For much of the 19th century, formal scientific research was conducted only by the Government of India through organisations, such as the Trigonometric Survey of India (founded in 1818), Geological Survey of India (founded in 1851) and the Meteorological Office (founded in 1864). Despite its name, the Government of India was not working for the interest of the people of India. Neither the British East India Company nor the British government after 1858 initially supported the idea of providing Indians with a higher education in English to the scholarly standards that prevailed in, for example, Oxford and Cambridge. Many British personnel in positions of authority believed that Indians were not intellectually capable of benefiting from advanced education. Others feared that Indians with higher education qualifications would be a potential threat to their position.

It was through the collective effort of many talented individuals over an extended period of time that the groundwork was laid for an environment where Indian scientists could work for the benefit of India. One of the first such individuals responsible for navigating India towards a future shaped by science was Raja Ram Mohan Roy (1772–1833).[175] In Roy's time, science was transforming societies. The European Age of Enlightenment founded on rational thought was shaping the world around him. The Industrial Revolution was transforming small agricultural communities into manufacturing economies with global reach. Inventions, such as the typewriter, electricity, photography, Morse code, railways, large ocean-going ships, canals linking distant cities and electric telegraph providing instant communication across oceans, dramatically changed the way people lived.

Roy recognised the urgent need for India to reform from within to be part of this new future. He spoke up against the caste system and ingrained traditions based on superstition and rituals inherited from scriptures. He is credited with social reforms, such as ending the Indian traditions of Sati, child marriage and polygamy, and promoting education for women and remarriage for widows. Roy was probably the first Indian intellectual to meet with and directly influence the powerful elites of Europe. He left India on the steamer Albion in November 1830 arriving in Liverpool four months later. He was mobbed by workers at public meetings in the great northern industrial cities of Liverpool and Manchester.[176] He met the British King in 1831 and the King of France in October 1832. His visit to Britain coincided with the Reform Bill, popularly known as the Great Reform Act. He was present in London when the bill was eventually passed by the British Parliament in 1832. It was called the Great Reform Act because the reforms it introduced were great. The bill made bribery and corruption very difficult and allowed more of the population to participate in the democratic process leading to a fairer society.[177] He experienced first-hand the social and economic benefits that nations accrued from cultivating science and technology.

During his time in Europe, he would have seen steam engines, cotton mills, mechanised production, new shift working patterns and their collective capacity to improve the quality of life for ordinary people. Roy would never return to India. During a visit to the southern English city of Bristol, Roy became ill and died on 27 September 1833. His biographer later wrote that Roy "stands in history as the living bridge over which India marches from her unmeasured past to her incalculable future. He was the arch which spanned the gulf that yawned between ancient caste and modern humanity, between superstition and science."[178] Though Roy died prematurely, his vision and achievements inspired those who came after him.

Another early visionary who contributed profoundly towards establishing a framework for scientific and technical education in India was Sir Asutosh Mookerjee, mathematician, physicist and an industrious administrator. He helped establish the Bengal Technical Institute in 1906, the Calcutta Mathematical Society in 1908 and the College of Science, Calcutta University, in 1914. The first postgraduate courses in science in 1913 were the product of his efforts. Mookerjee was highly influential and nurtured the careers of many Indian scientists who made a mark in science in the first half of the 20th century, including Srinivas Ramanujan (1887–1920), C.V. Raman, J.C. Bose, S.N. Bose (1894–1974) and Meghnad Saha.

By the late 19th century, the ripples of the European scientific revolution were reaching all parts of the globe. Rapid communication, made possible through the telegraph, telephone and radio communication, was as transformative in its time as the Internet is today. While the primary objective of The British East India Company, the biggest company ever to have existed, was to make a profit for its shareholders, it realised that a few thousand Englishmen could not govern a nation of millions. So, it engaged Indians locally and introduced the methods and products of modern technology. Britain ended up becoming the conduit for India to absorb the methods and techniques of scientific investigation. Gifted and curious Indians engaged and helped embed the principles and the role of science in India and established a vision for India's future development based on science.

Colonial power was not always suppressive. Outside the domain of politics and military, Indians with exceptional talent were encouraged by the British and European academia "if there was an Indian who was competent they would definitely support and help him."[179] For example, Lord Rayleigh supported Jagadish Chandra Bose during the 1880s, Ernest Rutherford guided C.V. Raman since the 1920s and nominated him for Nobel Prize in 1930 and Albert Einstein's (1879–1955) participation cemented international recognition for Satyendra Nath Bose's contributions. Without Godfrey Hardy (1847–1947) and John Littlewood (1885–1977) nurturing Ramanujan's career in Cambridge between 1914 and 1919, the world would have lost out on one of India's greatest mathematical genius.

Jagadish Chandra Bose

Jagadish Chandra Bose was one of the first Indian scientists to conduct pioneering experimental research in science that attracted worldwide recognition. He was born in Bengal in 1858, the same year that the control of India moved from the East India Company to the British government. Bose wanted to study for the highly sought after Indian Civil Service exams, but his father guided him to "rule nobody but himself ... [and] become a scholar, not an administrator."[180]

Bose attended St. Xavier's College in Calcutta (now Kolkata), where he studied under Father Eugène Lafont, a key promoter of IACS during its early days and the founder of St. Xavier's College Observatory. Perhaps, it was Lafont's enthusiasm for experimental science that drew Bose initially to science. After graduation, Bose initially went to study medicine at the University of London in 1880 and then moved to Cambridge two years later to study natural sciences. It was there that he first came in contact with Lord Rayleigh and developed a strong relationship that prevailed for many years after Bose left Cambridge. With a Bachelor of Arts from the University of London and a Bachelor of Science from Cambridge, Bose returned to India in 1885 as a professor of physics at Presidency College in Calcutta, as recommended by Lord Rayleigh. As the first Indian to hold a senior position as a professor, he was offered half the salary of a British professor. He signalled his indignation by accepting the role but not the salary. It took three years of service as an unpaid professor for authorities to relent and pay a salary equal to that of his British peers, backdating it to his start date.[181] Bose replaced the traditionally spoken lectures with practical experiments, and his teaching was innovative, effective and very popular. Bose's hands-on teaching style was probably the influence of his physics teacher, Father Eugène Lafont and in turn possibly passed on to his student, C.V Raman. The annual report of IACS for 1886 states that Bose conducted seven practical demonstrations on electricity and magnetism.[182].

Figure 4-1 Jagadish Chandra Bose at the Royal Society in London. Credit Wikimedia Commons

Bose is known for his contribution to the development of radio. He designed and built equipment to generate, transmit and receive radio waves. Bose first demonstrated the potential for radio communication in 1894 when he triggered the explosion of a small sample of gunpowder using radio waves. This 'action at a distance' in the absence of a physical mediator must have appeared like magic for most of those who witnessed it. In the history of telecommunication, 12 December 1901 was a watershed moment. On that day, Guglielmo Marconi (1874–1937) transmitted a man-made radio signal, three dots representing the letter 'S', from southern England across the Atlantic to his colleagues in Newfoundland 2,174.8 miles (3,500 km) away.[183] Marconi's success, however, relied on something that is argued to have been invented by Bose, an electrical component called a coherer.[184][185]

A coherer was a key component in the development of early radio receivers. When first used in 1891 in Paris, it was a small tube filled with iron filings with an electrical connection (an electrode) at each end. In the presence of radio waves, the iron filings lined up or 'cohered' and made a connection between the two electrodes completing a circuit and triggering the detection of radio waves. However, once the iron filings cohered, they remained cohered until the tube was manually shaken to allow the iron filings to fall away from the electrodes and re-enable the coherer's ability to detect radio waves once more. Bose designed and built a coherer that would recover automatically and continuously without manual intervention, making possible the first ever man-made radio transmission.

On 27 April 1899, his paper titled 'On a self-recovering coherer and study of cohering action of different materials' was read by Lord Rayleigh at the Royal Society in London, and this coherer was publicly described as Bose's invention.[186] However, just prior to the publication of his paper on 27 April 1899, Bose had been encouraged by a multi-millionaire proprietor of a telegraph company not to go public with details of his invention but to patent it for money. Bose went out of his way to declare that he had no interest in the notion of personal wealth. He was content with solving an intellectual challenge that addressed a real-world problem.

In a personal letter on 17 May 1901 to his friend Rabindranath Tagore (1861–1941), another Bengali polymath, who in 1913 achieved international success with a Nobel Prize in Literature, Bose wrote "If I once get sucked into this terrible trap, there won't be any escape! See, the research that I have been dedicated to doing is above commercial profits."[187] Seven months later, Marconi filed the patent for the self-recovering coherer under his own name. The self-recovering coherer turned out to have been a significant stepping stone in the development of radio communication. Bose was openly credited at the time, but many argue that his contribution has not been fully recognised even today. In 1904, Bose was persuaded to file a patent for his galena-based coherer.[188] His use of galena (a lead-rich mineral) in his coherer was the first practical use of a semiconductor in electronics. The concepts of a conductor (materials through which electricity can pass, for example, copper) and non-conductor (materials through which electricity cannot pass, for example, wood) were well understood before Bose's time. However, at the time, semi-conductors were new. Today, all modern computer systems, computer memory and microprocessors that are made using microscopic integrated circuits, rely on semiconductors. In addition to the coherer, he developed a number of specialised devices familiar to radio engineers today, such as waveguides, horn antennas, dielectric lenses, interferometers, couplers and absorbers.[189]

Bose was also the first person to identify and work with microwaves, which today are used for radar, telecommunication and domestic appliances. Michael Faraday's work in the early 19th century revealed light as the radiation of vibrating electric and magnetic fields travelling at the speed of light. Visible light was only a subset of all possible frequencies. The different frequencies are grouped into radio, microwave, light, infrared, all the way up to gamma-rays; collectively, they make up the electromagnetic spectrum. Low-frequency radio waves can have a wavelength of thousands of kilometres and are used underwater by submarines for communication, while high-frequency gamma-rays at a tiny wavelength, the size of an atom, are used to detect some of the most violent events in the universe, like supernovae.

The wavelength of the radio waves that Marconi used for his historic transmission in 1901 was initially considered to be around 366 m, but in a recorded lecture in the 1930s, Marconi said it was 1,800 m.[190] Bose experimented with short wavelengths (a few millimetres), which are now categorised as microwaves. Bose's interest was not restricted to experimental science. In 1896, he published *Niruddesher Kahini* (The Story of the Missing One), a science fiction work that earned him the epithet 'father of science fiction' in India. But Bose was not the first in India to explore the realm of science fiction. Jagadananda Roy (1869–1933), another writer of Bengali origin, had published *Shukra Bhraman* (Travels to Venus) in 1879. Roy imagined large, hairy ape-like inhabitants of Venus a decade before H.G. Wells described his slender, big-eyed, intelligent Martians in *The War of the Worlds*. Bose was knighted in 1917, and he became a Fellow of the Royal Society three years later. As a polymath, his research interests were broad and included archaeology, botany, physics and biology. Many of his experiments straddled multiple disciplines and dealt with questions in physics and biology at the same time. He demonstrated the effects of electromagnetic waves on animate and inanimate matter.

Bose was unique among his scientific contemporaries. Despite his systematic pursuit of knowledge guided by the scientific method, in 1911, he distinguished the Western reductionist approach from his Eastern approach, which attempted to comprehend and combine the multiplicity of phenomena available for observation. Perhaps influenced by ancient Indian spirituality and the belief in 'cosmic unity', he wrote to his friend Rabindranath Tagore on 30 August 1901 saying "there is a great gap between the living and the non-living. I was experimenting on the responses of plants to make a connection between the two. Just now I got the results; Same, Same, all are the Same."[191]

Srinivasa Ramanujan

Srinivasa Ramanujan (1887–1920) was an extraordinary mathematician, who lived a short but extremely productive life. Considered by some as a savant and a social misfit, he was recognised as India's greatest mathematician by Subrahmanyan Chandrasekhar (1910–1995), winner of the Nobel Prize in Physics in 1983. Ramanujan produced exceptionally original work in pure mathematics that confounded many of his peers at the time and continues to inspire mathematicians today. His theorems are being applied in the 21st century in areas as varied as polymer chemistry, crystallography, computing and medicine.

Ramanujan was born into a family of average means in the southern state of Tamil Nadu. His short life was marked by regular periods of illness. It started at the age of two with a smallpox infection. Before Ramanujan was seven years old, he had lost his next three siblings to smallpox within months after birth.[192] Largely self-taught, he excelled in mathematics but showed no aptitude for or interest in any other subject.

Ramanujan completed his formal education at the age of 18 succeeding only in mathematics. This failure in every subject except mathematics and the resulting loss of his scholarship triggered a brief disappearance in 1907.[193] He made his first appearance in a newspaper as the subject of a missing person report. He turned up safe a few days later.[194] In 1909, he married a 10-year-old girl as arranged by his mother although they did not live together until three years later. Motivated by his failure to get into a university and facing the responsibilities of a married man, he found work as a clerk at the Madras Port Trust on a salary of Rs.30 per month (£20 per annum).[195] Confident of his unique mathematical skills, Ramanujan set out to find a sponsor to support him to continue to study his beloved mathematics. He solicited several individuals. In 1913, it began to pay off when he established contact with Gilbert Walker (1868–1958). Walker from Rochdale in England but was then based in Shimla as the Head of the Meteorological Department of India.

Figure 4-2 Srinivasa Ramanujan (1887–1920). Credit Professor Richard Askey

He was a former lecturer at Trinity College, Cambridge. Although they did not meet, Walker inspected Ramanujan's notebooks personally during a visit to Madras (now Chennai) on 25 February 1913. In the absence of formal mathematical training, Ramanujan's approach was idiosyncratic, unconventional and difficult for others to follow. Walker noted that Ramanujan's work was "lacking in the precision and completeness necessary for establishing the universal validity of the results."[196] Nevertheless, impressed by what he saw, Walker contacted Madras University on the following day recommending that "it would be justified in enabling S. Ramanujan for a few years at least to spend the whole of his time on mathematics without any anxiety as to his livelihood."[197] In response, Madras University offered Ramanujan, who had no formal university entry qualifications, a 2-year scholarship of Rs.75 per month (£60 per annum) as the University's first research scholar.[198]

Walker made one other recommendation without which Ramanujan's contribution in mathematics would have been lost to the world. Walker asked Madras University to contact his former colleague, G.H. Hardy, a Fellow of Trinity College, Cambridge, and "assure Mr Hardy of their interest in him."[199] Hardy was a leading mathematician of his time. By chance, Ramanujan had already written to Hardy six weeks prior to Walker's visit, asking Hardy "to go through the enclosed papers. Being poor, if you are convinced that there is anything of value I would like to have my theorems published.'[200] Examining Ramanujan's papers, Hardy's conclusions were similar to Walker's. He saw in Ramanujan's work evidence of intuition, induction and at times mingled argument, hunches and guesses. His written arguments were not coherent and almost always lacked mathematical proof. Nevertheless, Hardy recognised (as Walker had done) Ramanujan's underlying mathematical gift and invited Ramanujan to Cambridge. Glowing recommendations from an authoritative individual, such as Hardy, based in an elite institution like Cambridge prompted Madras University to offer Ramanujan a scholarship of £250 per year plus £100 for passage.

Once he received assurances that he would not need to pay his expenses or undertake any further examination upon arrival in Britain, that his level of English would suffice and that he could remain a vegetarian, Ramanujan agreed to go. He left Madras on 17 March 1914 on the S.S. Nevasa and arrived in London three weeks later. His period of stay in the Britain almost identically matched the duration of World War I. Research at Cambridge had continued through the War years although at a subdued pace. The number of students had gone down from 700 to 150 by November 1915. Ramanujan wrote to his mother in India on 11 September 1915 to alleviate her concerns about his safety stating that there was no war in the England but in neighbouring countries, "as far from me as Rangoon is from Madras."[201]

In his five years at Cambridge, Ramanujan published 21 papers (some jointly with other authors) containing theorems on leading mathematical themes, including asymptotic formulae, infinite series, definite integrals, summation of series, modular equations, Riemann zeta function, analytic number theory, combinatorial analysis, partitions and modular functions. One paper for the Journal of the London Mathematical Society published in 1915 consisted of 62 pages with 269 equations. Despite his unconventional mathematical approach and the peculiarities of his spoken and written English, the brilliance of his original thought shone through. An official report on Ramanujan's work for the registrar of Cambridge University stated "India has produced many talented mathematicians in recent years, a number of who have come to Cambridge and attained high academic distinction. They will be the first to recognise that Mr Ramanujan's work is of a different category."[202]

A well-known anecdote that illustrates Ramanujan's mathematical genius is a conversation between Hardy and Ramanujan. Hardy recalls "I remember once going to see him when he was ill at Putney. I had ridden in taxi cab number 1729 and remarked that the number seemed to me rather a dull one and that I hoped it was not an unfavourable omen. "No," he replied, "it is a very interesting number; it is the smallest number expressible as the sum of two cubes in two different ways."[203] The sum of 1 cubed and 12 cubed is 1729 and so is the sum of 9 cubed and 10 cubed ($1729 = 1^3 + 12^3 = 9^3 + 10^3$). The sums of the cubes of any other pairs of numbers will be higher than 1729, never lower. During his time in the UK, Hardy coached Ramanujan but with utmost care. Ramanujan exhibited an innate mathematical ability, and he had been largely self-taught. It is interesting to speculate: had he gone through formal training, would it have extinguished his genius?

Two years after arrival, Ramanujan was offered a Trinity College Fellowship, along with a salary of £ 250 per year. Ramanujan was now free to pursue his work in mathematics completely devoid of any conditions, including any obligation to teach. He seemed to have it all, international recognition, a significant body of published work and the security of tenure at a leading university. However, the ill health that had dogged him since childhood in India returned. His isolation from his family and wife probably triggered his depression. His last two years in the UK were spent in and out of nursing homes and sanatoriums in Cambridge, Wells and in the Derbyshire town of Matlock, where he complained about the poor quality of care. Later, he spent time nearer Cambridge at Fitzroy House and a hospital in Putney in London.[204] The combination of illnesses and isolation cut short his time in Cambridge and eventually his life.

Ramanujan had been ill for a long period, fever, dysentery, diarrhoea and abdominal discomfort. At a particularly low point in 1918, Ramanujan attempted suicide by throwing himself in front of a train at a London underground station. The attempt failed through the action of a guardsman operating a switch that brought the train to a stop a few feet away.[205] He was briefly arrested (suicide was illegal at the time), but with Hardy's intervention, all charges were dropped. This story was recalled by Hardy in 1936 during an evening dinner at Trinity College, where astrophysicist Subrahmanyan Chandrasekhar was completing his PhD. It was through this account retold by Chandrasekhar during a lecture in Delhi that India first came to learn about this dark episode of Ramanujan's story. In bringing that story to India, Chandrasekhar was himself saddened and depressed by the misrepresentation that followed in the press. Chandrasekhar was accused of defaming Ramanujan and attempting to enhance his own reputation. Ramanujan had been his role model and had profoundly influenced Chandrasekhar's career choices.

Ramanujan's illness persisted throughout his time in the England. He recovered for short periods, one even long enough for him to return to Madras. Ramanujan left the UK on 27 February 1919 aboard the S.S. Nagoya and arrived weak and emaciated in Madras a month later. Upon his return to India, Ramanujan received a national hero's welcome.[206] He was held in the highest esteem for his intellectual prowess, maybe, second only to Nobel laureate Rabindranath Tagore. Back in India, Ramanujan gained a second wind. He spent it almost entirely on writing about several complex concepts in pure mathematics.

A year after his return he died. He had confided in his wife that he regretted not taking her to Cambridge with him. Had he done so, his diet would have been healthier, and the companionship would have sustained him emotionally. The cause of his illness was never known for certain, but food poisoning, vitamin deficiency, gastric ulcer, hepatic amoebiasis and tuberculosis were all potential candidates as the cause of death. A tuberculosis specialist who had treated Ramanujan immediately upon his arrival in India was recorded in the 1930s as saying that Ramanujan could and should have been saved had he not been misdiagnosed early on.[207]

Much of Ramanujan's work was published posthumously with Hardy's assistance, but not all, especially what he had produced in his final year in India. In the spring of 1976, Professor George Andrews (born 1938) discovered what came to be known as *Ramanujan's Lost Notebook* in Trinity College Library.[208] Seen for the first time in 50 years by someone who was capable of understanding their true value, Ramanujan's work once again captured the imagination of the international mathematical community. In the *Lost Notebook*, Ramanujan contributed to several branches of mathematics, including elementary mathematics, number theory, infinite series, integrals, hypergeometric functions, q-series, continued fractions, theta functions and class invariants. There is some speculation that Ramanujan was working on some new over-arching theory, which his premature death prevented him from completing.[209] Although he died almost a century ago, he continues to inspire young mathematicians and scientists in the 21st century.[210] Modern science is using his mathematics to solve problems, such as the superstring theory and complex molecular systems. Physicists are using his work on the mock modular function to understand how black holes evolve.[211]

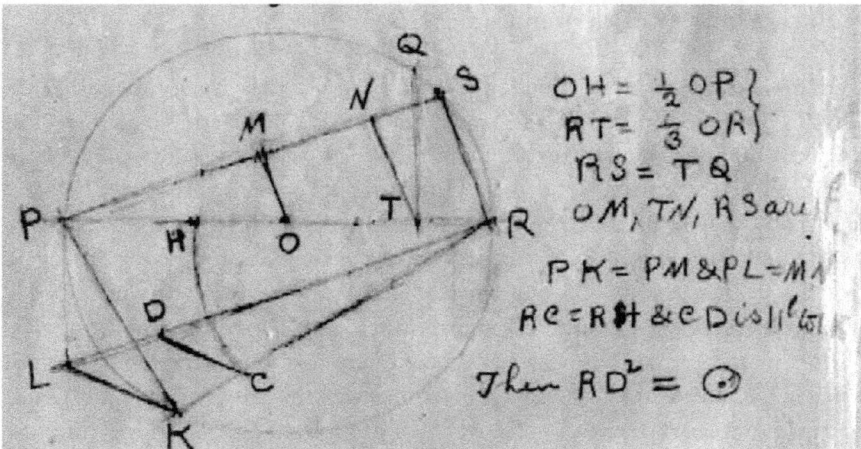

Figure 4-3 Sample from Ramanujan's Lost Notebook. Credit University of Madras

Ramanujan was aware of his special abilities and unique achievements. On his deathbed, he told his wife that his name will be remembered for 100 years and she will always have money.[212] Hardy, himself an accomplished mathematician, rated Ramanujan as the best. Rating mathematicians by pure talent on a scale from 0 to 100, Hardy gave himself a score of 25 and Ramanujan 100. Speaking in 1936, Hardy asserted "I did not invent him but did discover him ... what a treasure I had found."[213]

C.V. Raman

Chandrasekhara Venkata Raman was a gifted self-taught experimental scientist. By the age of 42, he had already achieved the three key accolades that most scientists covet. He was elected a Fellow of the Royal Society in

1924,[214] knighted by the British government in 1929 and won the Nobel Prize in Physics in 1930. What makes his achievements even more remarkable is that he had to overcome the hurdles of being a subject of the British Empire. Unlike other Indian scientists of international repute, including those who made a significant contribution to the Indian space programme, Raman was entirely the product of the Indian education system. His industrious pursuit of scientific research took him into varied branches of science, including acoustics, spectroscopy, optics, astronomy, crystallography and radio. In addition to the many awards he received within India, he accepted others from the US, the USSR and France.[215] In his lifetime, he published 465 scientific papers, two-thirds of which are under his sole authorship, and founded two institutions, Indian Academy of Sciences and the Raman Research Institute. Both continue to play a central role in scientific research in India today.

Raman had accurately predicted that he would win the Nobel Prize in Physics years before he did and was described as a "supreme egotist" by his own biographer and nephew Professor Sivaraj Ramaseshan, an accomplished scientist and an IISc Director. Once when asked why he wore a turban, which had become his hallmark especially during his trips to Europe, Raman quipped "Oh, if I did not wear one, my head will swell. You all praise me so much, and I need the turban to contain my ego."[216] He was cantankerous, impetuous, quarrelsome and self-centred, but he was selfless in seeking out and fostering new talent and in putting the quest for knowledge ahead of the quest for personal wealth. He cherished his child-like wonder for scientific discovery. Throughout his working life, he exhibited complex and conflicting attributes, not unusual for an intelligent, industrious and dedicated man of science. Raman was born to a Tamil-speaking family near Tiruchirapalli in the southern state of Madras in British India. He was the second of ultimately eight siblings, five brothers and three sisters. His father played the violin and, as a teacher of mathematics and physics, maintained a collection of science books that probably nurtured Raman's curiosity of the natural world. He breezed through most of his early education academically ahead of his contemporaries by a few years. He matriculated at the age of 11 and with a scholarship attended the Presidency College in Madras, completing his Bachelor of Arts at the age of 15, followed by a Master's degree three years later. By 1907, while still at the Presidency College in Madras, Raman wrote and published his first research paper, even though the college had no remit, facilities or tradition of research.[217]

Despite his interest in science, Raman followed his father's guidance and successfully completed the competitive Financial Civil Service examination, which secured him a post in the Finance Department in Calcutta as Assistant Accountant-General at a lucrative starting salary of Rs.400 per month. In 1907, at the age of 18, he married Lokasundari Ammaul, and both moved to Calcutta, the capital of British India at the time. A devout vegetarian, he ate simple food usually cooked by his wife. The Indian Association for the Cultivation of Science (IACS) had a research laboratory that no one was actively using and, as luck would have it, it was located around the corner from the house that Raman rented on his arrival in Calcutta.[218] Raman engaged with the IACS conducting research in its laboratory before and after work as Assistant Accountant-General. His work took him away from the IACS, first to Rangoon in 1909 and then to Nagpur in 1910 (from where he observed the lunar eclipse using a 3-inch 7.62 cm telescope.[219] At both places, he converted part of his home into a laboratory so that the disruption to his experimental research was mitigated.

Figure 4-4 C.V. Raman at the IISc. Credit IISc Archives

On returning to Calcutta in 1911, he came across the Astronomical Society of India, which was founded the previous year. He joined as a member on 27 February 1912 and later served in a variety of roles including that of the secretary, librarian and director of the meteor section. He presented his own research papers on topics that included Astronomical Optics, Saturn in a Small Telescope and Eclipse of the Moon; he also wrote spectroscopic notes and a paper on the diffraction phenomena observed in the testing of optical surfaces.[220] Raman's work in astronomy has not attracted as much attention as his work in other fields. It did, however, play a significant role in bringing him into contact for the first time with scientists in prominent posts. He met with John Evershed, Director of the Kodaikanal Observatory, and Gilbert Walker, Director General of Observatories, in India and a wider influential audience, notably the Royal Society in the UK, through the scholarly publications that carried his work in Europe.

In 1921, a congress of the universities of the Empire was held at Oxford in the UK to help coordinate the increasing activities within universities across the Empire. Raman was a member of the delegation that travelled to the UK from the University of Calcutta. While in the UK, Raman extended his personal network through meetings with scientists whose work he had read and was surprised to meet many who had read his.[221] He established contact with three Nobel laureates during that trip, J.J. Thomson (1846–1940, Nobel Prize winner in 1906 for the discovery of the electron), William H. Bragg (1862–1942, Nobel Prize winner along with his son Lawrence for understanding the structures of crystals using X-rays) and Ernest Rutherford (Nobel Prize winner in 1908 for his work on radioactivity).

On his return journey to India aboard the S.S. Narkunda, Raman conducted an experiment on why the colour of the sea was blue. It was accepted wisdom that the blue colour of the sea was due to the reflection of the blue sky. By excluding the reflection and looking directly into the sea, Raman observed that the colour of the sea "far from being impoverished by suppression of sky reflection, was wonderfully improved thereby." He concluded that the blueness of the sea was due to diffraction effect as light passed through the water. He wrote a two-page paper for publication in Nature before his ship docked in Bombay on 26 September 1921.[222]

The discovery that led to Raman's winning the Nobel prize was made on 28 February 1928. In 1927, Arthur H. Compton (1892–1962) was awarded the Nobel Prize in Physics for discovering Compton scattering, a change (increase or decrease) in the energy of an X-ray caused by passing close to an electron in an atom. Raman set out to explore whether a similar effect could be seen with light instead of X-rays. He found it. Raman discovered what came to be known as the Raman Effect (also called Raman scattering). He did so with the assistance of K.S. Krishnan (1898–1961) who conducted multiple experiments using a combination of bright sunlight, an 8-inch (20.32 cm) lens, a mercury arc lamp, a direct vision spectroscope and polarisation filters). The scattering was faint and the cause was not identical to that for Compton scattering, but it was real. The announcement appeared in the Associated Press of India on the following day under the heading 'New Theory of Radiation. Prof. Raman's Discovery'.

Raman's ego, self-confidence, arrogance and hubris were on par with his scientific brilliance. Self-nomination for the Nobel Prize is prohibited and soliciting nomination is considered to be unbecoming. Further, the nominations for the Nobel Prize are kept secret and published only 50 years after the award. Raman missed the 1929 window. However, in a letter dated 6 December 1929, he wrote to Niels Bohr (1885–1962) saying "I feel sure that if you give your influential support, the Nobel Committee for Physics may recommend that the award for 1930 may go to India for the first time."[223] Raman did not know at the time that Bohr had already nominated him for the 1929 award and repeated his nomination for 1930. Raman received 10 nominations, including those from Ernest Rutherford and Charles T.R. Wilson.[224]

Perhaps, it was through such private communication that Raman had secured sufficient confidence to book two tickets on a steam ship from India to Stockholm for November 1930, five months prior to the formal announcement that he was the recipient of the 1930 Nobel Prize in Physics. During his acceptance speech, he referred to Nehru indirectly stating that he had received a congratulatory telegram from his "dearest friend who was now in jail". Raman noted that he had to accept the prize under the British Union flag and lamented "I realised that my poor country did not even have a flag of her own."[225] In acknowledgement of his new-found international status, Raman was selected as the first Indian Director of the IISc in 1933 with the remit for the "promotion of advance instruction on original investigations in all branches of knowledge and their utilisation for the benefit of India."[226]

The first three directors of IISc were all British. Raman was selected as director from a list of 20 other applicants from all over the Empire, including Vancouver, Johannesburg, Colombo, London, Manchester and Bristol, as well as two from within India.[227] He took up his role as Director on 6 October 1932 at a salary of Rs. 3,000 per month, which included accommodation on IISc grounds. Following the remarkable success of his 1930 Nobel Prize, Raman could have taken a senior role in a research institute in the US or Europe, but he chose to stay in India.

In his new role as the Director of IISc, Raman set up a new physics department where he conducted most of his research in optics, crystallography and acoustics. He reorganised other departments and helped develop workshops that produced scientific instruments and components rather than relying on Western imports. He also established the Indian Academy of Sciences with the intention of promoting pure and applied research and documenting the research in formal scholarly papers for national and international dissemination. It also served as a source of authoritative scientific advice for the government. Six years before he arrived at the IISc, he had started a *Bulletin of the Indian Association* that published results of original research. Today, the bulletin has grown into a full science journal, the *Indian Journal of Physics*.

As director, he experienced pressure against his proposed reforms at IISc. His management and organisational skills honed during his time as the Assistant Accountant-General were strong, probably too strong but he had weaknesses too. His approach threatened some who held well-paid influential posts at the IISc. Even his admirers acknowledge his weakness in developing effective professional relationships. Over time a small but strong opposition evolved determined to oust Raman as the IISc director. To establish a centre of excellence in scientific research at the IISc and raise IISc's international status, Raman wanted to persuade European scientists fleeing Nazi Germany and the War to make India their new home.[228] With this end in mind, he attempted to secure a full-time post at IISc for Max Born, a German physicist and mathematician who went on to win the Nobel Prize in Physics in 1954.

By the autumn of 1935, Born's post in Cambridge was coming to an end, and as chance would have it, the Nazi Party revoked his German citizenship rendering him stateless. This offer at the IISc was timely and suited Born. Born came with his wife Heidi, and both settled well in Bangalore. Raman sought to make Born's temporary post as Reader into a permanent post as Professor of mathematical physics at the IISc. A humiliating incident at the formal vote of the IISc Council to ratify Born's appointment to the IISc led to Raman terminating his own role own as Director of IISc. The political manoeuvring to oust Raman had succeeded. It was the end product of "a group of antagonists that had been working for a year and a half to disgrace Raman."[229] On 1 June 1937, he wrote to the Chairman of the Council saying that "having considered all the circumstances, I feel it would be best that I offer to terminate my contract of service with the Institute as its Director." Raman was expected to remain the Director until 1948 but continued instead as a professor.

There is no direct evidence of Raman playing a critical role in creating or nurturing the space programme in India. He declared in his characteristic forthright style that it was "nothing but sheer raving lunacy" to "shoot men into space and make them walk there."[230] Speaking in 1966, he concluded that "it is militarism, very thinly disguised" and that the spaceflight was the "most sinister aspect of progress of science in the last 60 years."[231] These comments, though they sound harsh and idiosyncratic, were made in the context of a lesser known space race between military and civilian factions in the US and USSR. The US Air Force sought approval for two overtly military manned space projects, the Dyna-Soar spaceplane and the Manned Orbiting Laboratory. Both were cancelled, one in 1963 and the other in 1969.

Figure 4-5 Raman in Europe. Credit IISc Archives

However, during the period of World War II, two individuals came under Raman's influence who had a very direct impact on India's space programme, Vikram Sarabhai, who completed most of his PhD at the IISc, and Homi Bhabha, who taught at the institute, first as a reader and then as a professor. Perhaps, inspired by Raman, both went on to dedicate their professional lives to scientific development in India. Had Raman managed to attract scientists and theoreticians from Europe, the tradition of high-class research in India today would probably have been established sooner. With deeper roots in science, perhaps the technological progress that India is making today could have been made a decade or two earlier.

Satyendra Nath Bose

Science has provided answers to some fundamental questions, such as how old the universe is, how big it is and what it is made from. Physicists classify everything that exists in the universe either as fermions or bosons. Fermions make up all the matter in the universe. Four bosons mediate the four known forces, gravity, strong nuclear force, weak nuclear force and electromagnetic force, between them.[232] The term 'boson' was coined by Paul Dirac (1902–1984), a British theoretical physicist, acknowledging the work of Bengali mathematician and physicist Satyendra Nath Bose (1894–

1974).[233]

Bose was born and educated in Calcutta (now Kolkata), the epicentre of the scientific renaissance in India during the early 20th century. He was a polyglot, who was comfortable communicating in French, German, Italian, Sanskrit, Bengali, English and Hindi. He developed an interest in experimental physics, the subject in vogue at the time. Optics, spectroscopy and wireless technologies were emerging just as his formal education was coming to an end. During his career, Bose designed and built equipment for X-ray (crystallography, diffraction and spectroscopy), optical spectroscopy and wireless technology,[234] but his unique theoretical work in mathematics and quantum physics remains his greatest contribution.

The regular supply of fresh scientific literature coming into India dried up in 1915 as World War I engulfed Europe. This turned out to be a blessing in disguise for Bose. A chance contact with an Austrian instructor P.J. Brühl, working at the Bengal Engineering College, got Bose reading books on mathematics, magnetism and physics in German. He borrowed from Brühl's personal collection several books, including Max Planck's (1858–1947) *Theories der Wärmestrahlung* (Theory of Heat Radiation) and Max von Laue's (1879–1960) *Das Relativitätsprinzip* (Principles of Relativity). This chance encounter with the German language and the leading-edge work in physics and mathematics being done in Germany eventually put Bose in touch with Albert Einstein that would eventually elevate Bose to international recognition.

In 1918, Bose, jointly with Meghnad Saha, wrote his first research paper titled 'The influence of the finite volume of molecules on the equation of state'. The paper dealt with the properties and behaviour of actual (but complex) gases instead of idealised (but simple) gases. It was published in the *Philosophical Magazine* in London, the same publication where C.V. Raman had published his first scientific paper a decade earlier. In late 1923, Bose submitted another paper to the *Philosophical Magazine*, but it was rejected by the referees.[235] In the rejected paper, Bose was covering new ground in quantum theory, which itself was so new and revolutionary that not many people, probably including the referees, understood its underlying concepts.

In 1900, Max Planck had introduced the concept of quanta where the energy of electromagnetic radiation existed only in predefined fixed quantities (quanta) rather than any value from within a continuum. His work introduced a new fundamental constant, Planck's constant, and ushered in the quantum theory (also known as quantum mechanics) to explain the phenomena at the subatomic level that classical physics could not. During the early 20th century, physicists had still not resolved the bigger question of what light was made of, waves or particles. In the paper that was rejected by the *Philosophical Magazine*, Bose had developed Planck's work using a new form of statistics, in which he introduced properties of waves, as well as quanta, to establish a quantum theory of 'ideal' gases.

Respected Sir,

I have ventured to send you the accompanying article for your perusal and opinion. I am anxious to know what you think of it. You will see that I have tried to deduce the coefficient $8\pi v^2/c^3$ in Planck's Law independent of the classical electrodynamics, only assuming that the ultimate elementary regions in the phase-space have the content h^3. I do not know sufficient German to translate the paper. If you think the paper worth publication, I shall be grateful if you arrange for its publication in Zeitschrift für Physik. Though a complete stranger to you, I do not feel any hesitation in making such a request. Because we are all your pupils though profiting only by your teachings through your writings. I do not know whether you still remember that somebody from Calcutta asked your permission to translate your papers on Relativity in English. You acceded to the request. The book has since been published. I was the one who translated your paper on Generalised Relativity.

Yours faithfully,

S.N. Bose

Figure 4-6 Transcript of S.N. Bose's 1924 Letter to Albert Einstein. Credit S.N. Bose

Confident of the quality of his work, Bose wanted his rejected paper published in a highly respected German publication, *Zeitschrift für Physik*. Although he had a good understanding of the German language, Bose was not sufficiently competent to write a full scientific paper in German. In 1924, Bose wrote to Einstein, by then a Nobel laureate, asking him to forward Bose's paper for publication to *Zeitschrift für Physik* if he considered it worthy, a remarkably audacious act for a young unknown researcher in a corner of the British Empire[236] This was not the first interaction Bose had had with Einstein. In 1918, Bose had contacted Einstein requesting consent to translate from German to English Einstein's jointly authored book *The Principle of Relativity*, a collection of writings, including Einstein's seminal paper, 'The General Theory of Relativity', first published in 1915. Einstein had agreed, and the first English translation of Einstein's 'General Theory of Relativity' was published in India by the University of Calcutta in 1920.[237] In 1924, four years after completing the translation of Einstein's work from German to English, Bose wrote to Einstein asking him to return the favour.

Einstein was impressed with Bose's paper 'Planck's Law and Light Quantum Hypothesis'. He translated it into German and submitted it for publication, adding "In my opinion, Bose's derivation of the Planck formula signifies an important advance." He informed Bose with a postcard dated 2 July that he had submitted it for publication. This one postcard helped Bose secure a two-year research trip to Europe with a generous "stipend, a separation allowance for the family, with sumptuous travel allowance with round-trip fare", and ultimately crystallised his place in the history of science.[238] Bose departed from Bombay in September and arrived in Paris on 18 October 1924. He had never travelled out of India before. During the following two years in France and Germany, Bose worked and made personal contact with leading figures in science, many of whom would achieve the ultimate recognition as Nobel laureates. In France, he had the opportunity to communicate in French and get hands-on experience of X-ray spectroscopy at the laboratory of Maurice de Broglie (1875–1960) and of radioactivity with Madame Curie (1867–1934) at her Radium Institute in Paris.

Einstein extended Bose' original paper and they jointly predicted a new state of matter that when cooled to almost absolute zero (-273K) produce this new state of matter called Bose-Einstein Condensate (BEC). BEC exhibits two unique properties, superfluidity, where the flow is devoid of friction, and superconductivity, where there is no electrical resistance. BEC was a product of the statistics that Bose had introduced, but the technology to create it did not exist in his time. While predicted in 1924–25, it was created for the first time in 1995 by scientists in the US. Today, BEC is used to help understand dark matter, one of the key topics that preoccupy present-day cosmologists.[239]

In October 1925, Bose arrived in Berlin, where he gave and attended lectures, meetings and seminars. Bose held seminars in German on low-temperature statistics, which concluded with detailed and lengthy question and answer sessions. During his time in Berlin, Bose conducted experiments on the refracting index of X-rays at the Kaiser Wilhelm Institute. In his letter dated 9 May 1926, Herman Mark (1895–1992) states "his most valuable quality which makes him of inestimable value to a collaborator is his deep and clear insight into the fundamentals of science."[240]

He also met several luminaries from Germany and beyond, including Max

Born, Theodore von Kármán (1881–1963), Max Planck, Leo Szilard (1898–1964), Fritz Haber (1868–1934), Werner Heisenberg (1901–1976), Hans Geiger (1882–1945), Wolfgang Pauli (1900–1958), Herman Mark, Lise Meitner (1878–1968) and Albert Einstein. In an interview recorded in 1974, Herman Mark recalled a jovial and friendly Bose.[241] During the many social gatherings, Bose even sang in German. Herman also recalled the time Bose and Einstein spent together walking in the grounds of the Institute in Dahlem. Although they spent much time together, they did not undertake any further joint work, and nothing was formally published as a result. As Bose's time in Europe was coming to an end, Herman Mark took Bose to Vienna for a few days, where he was invited to give a seminar, and he made personal contact with Erwin Schrödinger (1887–1961), Hans Thirring (1888–1976) and Ludwig Flamm (1885–1964).[242]

On his return to India in 1926, Bose accepted the role of the professor of physics at Dacca University (now University of Dhaka). Over the next two decades, starting from scratch, he developed laboratories and libraries and nurtured a tradition of experimental research. He did not limit the activities of his department to areas of his personal interest. He encouraged disparate specialisms, which eventually included spectroscopy, magnetic properties of matter, wireless and crystallography. Laboratory equipment (including a demountable X-ray unit, differential thermal analyser and fully automatic scanning spectrophotometers for thermoluminescence studies) was not imported from Europe but made in-house using local material by students under Bose's guidance.

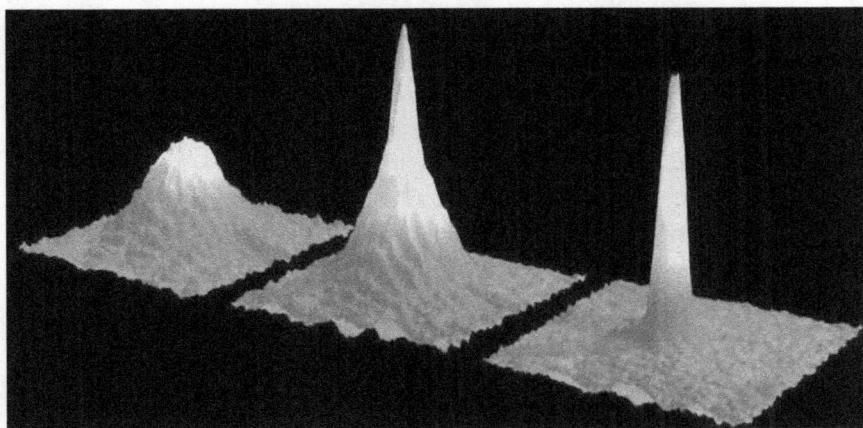

Figure 4-7 Bose-Einstein Condensate. Credit Massachusetts Institute of Technology

During his lifetime, Bose published 24 papers. Following his return from

Europe in 1926, he devoted most of his time to building his department and published only 17 papers. When Madame Jacqueline Eisenmann (1904–1998), a scientist Bose met during his first visit to Paris in 1924 and again in 1951, asked Bose why he had not published more, he said "he had spent a great deal of time in preparing experimental research work for his pupils in Dacca".[243] Bose was dedicated to his students. Madame Eisenmann recalled in an interview in July 1973 that Bose "had no ambition for himself, ... a very modest man who had an extraordinary heart."[244]

For a man with a tranquil temperament, Bose harboured strong anti-colonial tendencies. He had experienced tumultuous events in Bengal throughout his life. In 1905, while he was still in school, Bengal was partitioned. Later in 1943, Bengal suffered a severe famine with a loss of life on a catastrophic scale. After India gained independence from the Britain in 1947, Bengal experienced extreme violence during partition into West Bengal and East Pakistan, and this was followed by the Bangladesh Liberation War in 1971. During his student days, he was a member of several secret societies and attended meetings in the pursuit of Indian independence. His visit to Europe was not just motivated by his science but his politics, too.[245] Some accounts claim that Bose out of pride chose not to visit the Britain prior to India's achieving its independence. Bose, however, had written to Ernest Rutherford and William Bragg during his 1924 trip to Europe but neither could host him at that time.[246]

Bose's work was recognised, somewhat belatedly, when he was elected as a Fellow of the Royal Society in 1958. Subrahmanyan Chandrasekhar, the Indian astrophysicist (later an American national), who personally knew many of the key Indian scientists, including Homi Bhabha, Meghnad Saha, C.V. Raman and J.C. Bose, said "from human point of view, [S.N. Bose] was the best of them all. He was very generous, gentle, easy-going and not particularly caring about the glamorous aspects of science." Rabindranath Tagore, perhaps the first intellectual from India to receive widespread recognition in the West, published a collection of essays on science, *Visva Parichay—An Introduction to the Universe*, and dedicated it to Bose.

Homi Jehangir Bhabha

Homi Bhabha lived through perhaps one of the most intensive periods of fundamental discovery that physics has ever seen. Many of the heavyweights in physics, Albert Einstein, Patrick Blackett, C.V. Raman, Niels Bohr, Wolfgang Pauli, Enrico Fermi, Werner Heisenberg, Paul Dirac, Erwin

Schrödinger and John Cockcroft (1897–1967) were his contemporaries, teachers or colleagues and a few his friends. Bhabha came from a successful family of industrialists synonymous with India. Dorabji Tata was his uncle, a name that continues to be associated with Indian commerce and industry in the 21st century. His father had envisaged that Bhabha would study engineering and join the extended family firm Tata Steel as a metallurgist, but that is not how things turned out. He excelled in his education, contributed unique research and developed a reputation as a high-calibre scientist in a premier European scientific institution. In the two decades that followed India's independence, Bhabha was instrumental in establishing India's first pure research institute, civil nuclear energy programme, nuclear weapons programme and space programme.

In 1924, at the age of 15, Bhabha passed the Cambridge entrance exam in India with distinction. Since the minimum age for Cambridge was 18, he spent the next three years studying Art and Science before arriving at Gonville and Caius College, Cambridge, in 1927, to study mechanical engineering.[247] Europe at this time was a hothouse for scientific discoveries in physics. By chance, much of it was happening at the Cavendish Laboratory in Cambridge, exactly where Bhabha was studying. Intelligent, ambitious and inquisitive, it was almost inevitable that he would be hooked by physics. In 1928, he wrote to his father expressing his interest in physics saying "I seriously say to you that business or a job as an engineer is not the thing for me. It is totally foreign to my nature and radically opposed to my temperament and opinions. Physics is my line ... I am burning with a desire to do physics". His father relented but insisted that he complete his engineering degree first before moving to physics. He earned a degree in mechanical engineering, and in 1932, another in mathematics. Both first class.

In the next three years, he completed his PhD, won additional scholarships and split his time between writing and publishing scientific papers and working with noted physicists in Europe and the US. The year 1932 was a particularly productive one for Cambridge, and Bhabha was there to experience it first-hand. In February 1932, James Chadwick (1891–1974) discovered the neutron (one of the two elementary particles inside the nucleus of the atom), and in April, John Cockcroft and Ernest Walton (1903–1995) demonstrated the idea of transmutation, changing one element into another. They changed large lithium atoms into smaller helium atoms by firing high-speed protons at a sample of lithium. While in Europe, Bhabha travelled to meet Enrico Fermi in Rome, Niels Bohr in Copenhagen and Wolfgang Pauli in Zurich. While in Zurich in 1933, with assistance from Pauli, Bhabha published his first paper in German 'Zur

Absorption der Höhenstrahlung' in *Zeitschrift für Physik* on the absorption of cosmic rays.[248] Cosmic rays are a mixture of high-speed particles (mostly protons) and high-energy radiation (X-rays and gamma-rays) from outside the solar system. After travelling for potentially billions of years through the vacuum of space to Earth, cosmic rays interact with the molecules in the Earth's upper atmosphere. Bhabha investigated these interactions for his PhD in nuclear physics, which he completed in 1935.

In 1897, physicist J.J. Thomson (1856–1940) discovered the first subatomic particle, the electron. An indivisible bit of matter had been contemplated since the beginning of human civilisation. A new category of particles called antiparticles was discovered in 1932 by Carl Anderson (1905–1991).[249] Anderson's discovery of a particle identical to the electron but with a positive charge ushered in the era of antimatter. Nuclear physicists had to invent new terms and mechanisms to understand and calculate interactions associated with antiparticles.

Figure 4-8 Homi Bhabha (second from the left) in Cambridge. Credit. Tata Institute of Fundamental Research

The positron had been predicted in 1933 that "the positron appeared as one of a pair of positive and negative electrons produced when a gamma-ray was converted into matter."[250] This idea that energy can be transformed into matter was established by Einstein's famous equation $E=mc^2$. If matter can be turned into energy, then energy, too, can be turned into matter. This is what nuclear scientists were seeing in the interactions between nuclear particles. Three years after antimatter was discovered, Bhabha published a paper describing what happens when matter and antimatter particles meet. He described mathematically the interaction between an electron and a positron moving at very high speeds. When they meet, they mutually destroy each other (annihilate) producing a gamma-ray (mass changed to energy). Bhabha went on to describe the scenario in reverse. When high energy particles collide, the interaction produces gamma-rays. Gamma-rays decay producing an electron and positron pairs spontaneously (energy changed to mass). This is Bhabha Scattering, and his work is used in modern particle accelerators to calibrate positron beams.[251] Before the age of 30, Bhabha had established his name in the field of cosmic rays and within the community of physicists responsible for the fundamental discoveries made during what turned out to be the golden age of physics.

In 1937, Bhabha attended a weekend conference in Manchester, the UK, where Werner Heisenberg and Patrick Blackett were also in attendance. Heisenberg had already been awarded a Nobel Prize, and Blackett was still a decade away from his. Despite Blackett's senior position, the 28-year-old Bhabha confidently argued his assertion that only a new unknown particle, heavier than an electron, can explain the penetrating attributes of cosmic rays observed in experiments. Blackett insisted that quantum theory would fail at higher energies associated with heavier particles. Subsequently, Bhabha was proved right. Bhabha was confident, bright and self-assured. Although neither of them knew it then, Bhabha would engage Blackett to help him embed science in rebuilding independent India. Heisenberg and Bhabha would have cooperated further, but the tumult of the impending war changed the destinies of nations, as well as individuals.

Bhabha had developed an international reputation and could have easily found a post heading a department of physics in many of the leading universities in the West. In one instance in 1939, he was interviewed but not selected for a teaching post in Liverpool University because the work would be "drudgery to a man like Bhabha who was a most exceptional man."[252] This was the conclusion of the interviewer James Chadwick, a Nobel Laureate who had discovered the neutron in 1932. Perhaps, this was a genuine assessment of his intellectual ability or rejection concealed in compliment. However, Bhabha successfully turned to the Royal Society for a grant to work with Blackett in his laboratory in Manchester.[253] Before taking up that role, Bhabha returned to India for a holiday. A few months later, World War II broke out in Europe, preventing his return. Instead of returning to Europe, Bhabha joined IISc in Bangalore as Special Reader working alongside C.V. Raman in 1940. Following his nomination as a Fellow of the Royal Society in March 1941, arguably his highest international recognition, he became a full professor at the IISc in 1942.[254]

Despite the uncertainty of war, one thing was clear; Indian independence was imminent. As a new nation, independent India would need to be built up from scratch. On gaining independence, India would offer the unique opportunity to build in science and technology into the fabric of a national economy from the outset. Bhabha had spent time in Europe and understood the key role of science and technology in industrialised societies. India, at the time, was devoid of scientific institutions of international repute and lacked the infrastructure of higher education and a culture of science within the population at large. Bhabha contemplated his future and how he could use science to shape the destiny of a new emerging nation.

Figure 4-9 Indian Institute of Science prior to the reorganisation in 1948. From Altor Homi Bhabha Registrar A.G. Pai Director J.C. Ghosh J. Taylor C.V. Raman. Credit IISc Archives

A prerequisite for successful leaders of large national programmes is direct access to the highest political office. As an example, Sergei Korolev, the chief designer responsible for Sputnik in 1957 and Yuri Gagarin's space mission four years later, had direct access to the Soviet leader Nikita Khrushchev. In the US, Wernher von Braun had the ear of the US President John F. Kennedy after he had committed the US to land a man on the Moon. To achieve his objective of transforming India into an economically developed nation through industrialisation, Bhabha too required personal access to India's highest political office. For Bhabha, such access was easily achieved, not least because Bhabha and Nehru had close personal ties for many years, but also because Nehru shared the vision of India built on science and technology. In Bhabha, Nehru saw someone who could help deliver the vision. India approached independence as the World War II approached an end in Europe. Having founded the TIFR in 1945, Bhabha acquired perhaps the highest profile of any scientist in India. For a national programme of development based on industry, nationwide power infrastructure was a fundamental prerequisite. Bhabha, a physicist, looking into the future, not the past, believed that nuclear power, not coal, was the route for India.

Figure 4-10 Sketch of C.V. Raman by Homi Bhabha. 1945. Credit IISc Archives

Even before India was independent and Nehru became its Prime Minister, Bhabha had started the preliminary work on India's nuclear energy programme. It was clear to him that international collaboration was going to be the key ingredient to fast-tracking India's ambitions of national development. Bhabha visited Sir John Cockcroft, the Director of the Atomic Energy Research Establishment in the UK, and W. Bennet Lewis (1908–1987) in Canada. In June 1947, he made his first request to the National Research Council of Canada for a tonne of crude uranium oxide so that Indian scientists could start to experiment. Although India had no natural deposits of uranium, it did have a large supply of thorium. With an eye on India as a future source of thorium, Canada supplied the uranium ore, and neither the UK nor the US objected.[255]

One of the most influential of Bhabha's contacts was Patrick Blackett. Apart from his long-standing professional contact, Blackett had high-level experience as an advisor to the British government before, during and after the War in all aspects of science and technology. Importantly, Blackett was motivated by socialist ideals and was sympathetic to the needs of the developing world.[256] In a letter dated 9 January 1948, Bhabha wrote to Blackett saying that he was very "anxious to see you before you leave Manchester for the US". The purpose of the meeting was to "explore the possibility of cooperation in atomic energy research between your country and mine."[257] By the end of 1948, India was independent, Nehru its Prime Minister and the Atomic Energy Commission (AEC) was in place. The AEC was established in August 1948, primarily under the guidance Bhabha had provided to Nehru, and with it, he had given Bhabha the nod to start the journey to set India on a path of self-sufficiency in energy.

As the 1950s gave way to the 1960s, technological successes, such as Sputnik and Gagarin, raised the appetite for political and financial commitment to science and technology, not just in the US and the USSR, but in Europe, too. It is conceivable that Bhabha's strong connections with Europe triggered his vision for an Indian space programme. Fearing being left behind once more, in 1962, Bhabha stated "We are on the same ground floor as the Western nations. They are leading us only by about 4 or 5 years .... Hence during 10–20 years, we must be able to equal them."[258]   In August 1962, as the Chairman of India's leading scientific body, the DAE, Bhabha proposed that India should pursue space research, too. The Physical Research Laboratory (PRL) was identified as the appropriate centre for it, and INCOSPAR (later ISRO) was established.[259] Bhabha selected Vikram Sarabhai, a cosmic ray physicist, to lead it. With great haste, a year later, India had constructed a launch facility in Thumba and launched its first rocket into space. INCOSPAR operated under the AEC and until transferred to the DOS in 1969. Bhabha built two institutions, TIFR and the Atomic Energy Establishment (now the Bhabha Atomic Research Centre (BARC)), Trombay. The objective for both was to help develop a prosperous and self-sufficient India. TIFR has been Bhabha's legacy for ISRO; it still produces state-of-the-art instruments used onboard ISRO's. He wanted India to be militarily strong, too. In 1958, he told Blackett that he hoped to develop nuclear weapons.[260] Bhabha, however, could not persuade Nehru for approval to develop nuclear weapons. Nehru instead that the nuclear programme was limited to the production of energy civilian use.

The US and the former USSR had developed nuclear weapons as a national military programme from the outset. Other nations ostensibly started with civil nuclear energy projects that later evolved into nuclear weapons programmes. Today, nine nations are believed to have a national nuclear weapons programme, the US, the USSR, China, France, the UK, India, Pakistan, North Korea and Israel. The successful nuclear test by China in October 1964 finally forced the then new prime minister, Lal Bahadur Shastri (1904–1966), Nehru's successor, to give Bhabha the go-ahead. Bhabha simultaneously held the three most important positions in the Indian atomic energy programme. He was the Chairman of the AEC, Secretary for the DAE and Director of the Atomic Energy Establishment in Trombay, while he continued to be the Director at TIFR, too. Consequently, his sudden death in 1966 was traumatic for the higher echelons of the Indian government. Who could step into his shoes? In January 1966, his routine Air India flight had crashed into the Alps as it descended towards Geneva.[261] India's future in scientific research, atomic energy, as well as its space programme suddenly found itself in a crisis. Just three years after its first rocket launch, the fledgeling Indian space programme had to absorb the sudden and profound shock of losing its founder and champion, Homi Bhabha. Bhabha's death, however, introduced a dramatic change to Vikram Sarabhai's career. Although not the first choice, Sarabhai succeeded Bhabha as the Chairman of the AEC, while continuing to lead the space programme.[262]

The Nehru-Bhabha partnership, with key contributions from many others during the first two decades after independence, moved India dramatically towards a developed country. That pace of development suffered a sudden shock with the demise of Nehru in 1964 and of Bhabha two years later. Despite Bhabha's enormous contribution, his achievements will always be dwarfed by the scale of the tasks demanded by nation-building.

◁ ◊ ▷

Chapter Five
India's Forgotten Rocketeer

O n 25 May 2012, an American private company SpaceX, used a rocket to transport supplies from the surface of the Earth to the International Space Station in low Earth orbit (LEO). However, this was not the first time that rocket power had been used as a transport mechanism by a non-governmental agency. Over half a century earlier, on 10 April 1935 at 15:35, rocket power was used for the first time in India to transport materials.

Figure 5-1 Stephen H. Smith. Credit Superior Galleries

Rockets had already been used to transport mail, but in this instance, a disparate collection of everyday items, including toothbrush, cigarettes, a spoon and a handkerchief were transported across River Ranikhola from Saramsa (now Saramsa) to Rey in Sikkim. The man behind the flight was Stephen Hector Taylor-Smith (1891–1951), usually abbreviated to Stephen Smith, a name that is not well known today either within or outside India. Like all early rocket pioneers around the world, his endeavour was largely self-funded and on a private small-scale. Smith received encouragement and active participation from the King of Sikkim, His Highness Sir Tashi Namgyal (1893–1963), along with several British representatives of the Indian Civil Service. On 6 June the same year, another of Smith's rockets transported a half-kilogram payload about a kilometre across the River Roopnarayan in West Bengal. The launch was a symbolic demonstration of the delivery of materials required in the immediate aftermath of an emergency, like the devastating earthquake in Baluchistan weeks before this flight. The consignment included rolls of bandages, lint, iodine and aspirin. The flight demonstrated the simplicity and astonishing speed with which rockets could be used for rapid transport in times of urgent need, especially in a difficult mountainous terrain.

Figure 5-2 Livestock transport by rocket 29 June 1935. Credit Superior Galleries

Around three weeks later, in a record-making flight on 29 June 1935, Smith transported livestock, a small cock and a hen, about a kilometre across the

Damodar River, demonstrating that living beings were capable of sustaining the immense acceleration generated during a rocket flight. His rocket did not use parachutes and remarkably the passengers, christened Adam and Eve, survived the hard landing. Acknowledging his own surprise, Smith recorded in his diary that "Sheer luck with a capital L, the wind and the soft sand helped me. These were in my opinion the greatest factors to the birds being alive."[263] Both birds were thriving 18 months later when Smith visited them at his patron Sir David Ezra's private zoo in Calcutta.

Smith's achievements were recorded in the international press and specialist magazines on rockets and spaceflight.[264] A global revolution was taking place in aerial transport just when Smith was starting out on his adult life. By the early 1920s, when pigeon mail was still in routine use, India saw the introduction of aerial transport of cargo, regular airmail and scheduled passenger flights.

BUCKINGHAM PALACE

The Private Secretary is commanded by the King to thank Mr.Stephen H.Smith for the letter which he has sent to His Majesty by the First Flight India to Great Britain.

16th April, 1929

Figure 5-3 Letter from the King to Stephen H. Smith Marking the First Airmail Flight between Britain and India. Credit Eric Winter

Large countries, like India, with varied topography, benefit from air transport more than others. In the same week that Smith turned 20, the world's first airmail flight had taken off from Allahabad on 18 February 1911. It would have made international headlines and would have inspired Smith, who was in school in Asansol 600 km from Allahabad, to take an interest in tracking and recording the developments in airmail. Just as steam engine locomotives had inspired enthusiasts in the past and spaceflight does today.

In the 1920s, aeroplanes joined airships in flying long-haul routes crossing seas, oceans and continents. Initially, a series of tentative and exploratory short hops explored routes between Europe, Middle East and Asia; they were pioneered by Imperial Airways to help administer the growing empire.[265] These routes were later used for regularly scheduled passenger flights, commercial cargo and military transport, all of which would eventually become commonplace in many countries. The first airmail service from England to India was flown by Imperial Airways. The flight left Croydon airport on 30 March 1929 and arrived in Karachi on 6 April. On the return flight the following day, Smith sent a letter to the King at Buckingham Palace in London. The King's Private Secretary responded on 19 April on behalf of the King acknowledging his letter.

India was not alone in experimenting with rocket-assisted transport. In Europe, similar experiments were first conducted in the late 1920s. German engineers and adventurers experimented with rocket-powered cars, motorbikes and planes. The German industrialist Fritz von Opel (1899–1971), along with Max Valier (1895–1930), promoted rocket-powered cars through public demonstrations on open roads and on rail tracks. In 1930, Max Valier was killed when a rocket engine exploded, and he became perhaps the first victim of rocketry. On 11 June 1928, Fritz Stamer (1897–1969) successfully demonstrated in Germany that rockets could be used to power a glider in flight.[266]

Air and Rocket Mail

The world's first rocket mail experiment was conducted by an Austrian, Friedrich Schmiedl (1902–1994).[267] Having completed six test flights over the preceding three years, on 2 February 1931, he fired his mail-carrying rocket from the summit of Schöckl Mountain to Radegund in the valley below in south-eastern Austria. His work was largely unrecognised during his time.

Schmiedl's interest was in launching rockets from an unusual platform. Inspired by Nobel laureate Victor Hess's (1883–1964) work in the early 1920s, Schmiedl developed what he called a Strato balloon. Hess used to conduct his balloon flights during the day and night at great personal risk, given the limitations of the early 20th-century balloon technology. Unlike

Hess, Schmiedl did not go up in his balloons. Instead, he would launch a balloon with an attached rocket. To track the balloon, he incorporated a magnetised rod that maintained the position of an aluminium flag, which reflected the Sun towards the Earth-based observer. Once the balloon was at 15 km, the rocket would automatically launch. After Schmiedl's successful experiments in Austria in 1931, rocket mail experiments were repeated and developed further in many countries around the world, including Australia, Austria, Belgium, Cuba, Denmark, France, Germany, Holland, Italy, Yugoslavia, Luxembourg, Spain, Switzerland and the US. [268] However, Smith was the only one launching rockets in India. He experimented with rocket launches from ship to shore, shore to ship, at night time, across rough terrain and rivers.

Rocket mail was not just the domain of enthusiastic amateurs but accomplished rocket engineers, too. Robert Goddard, whom history remembers as the founding father of rocketry, had pioneered the use of LOX and kerosene to power rockets. Oxygen is liquid at -183°C, and this was a daunting engineering challenge that Goddard surmounted in 1926. On 2 July 1936, he conducted the first international rocket mail experiments by sending 1,072 covers from Texas, USA, to Taaulipas in Mexico.[269] Before the day was out, Goddard crossed the border and repeated the experiment from Mexico.

Figure 5-4 Regulus I Missile Fired from USS Barbero. 8 June 1959. Credit Smithsonian Postal Museum

The most advanced demonstration of rocket mail came out of a collaborative experiment between the US Department of Defence and the US Post Office Department.[270] It was more missile mail than rocket mail. A modified Regulus surface-to-surface training missile launched from the US Navy submarine USS Barbero (SSG-317) off the coast of Virginia delivered

3,000 commemorative letters to Florida. The missile landed on a runway at the Naval Auxiliary Air Station in Mayport, Florida, 22 minutes after launch under automatic control. One of the 3,000 letters were then delivered by hand to the US President Eisenhower in Washington. This was the first and the only time that missiles were used for mail transport. On the strength of this single but impeccably successful experiment, the US Postmaster General Arthur Summerfield (1899–1972) declared that "Before man reaches the moon, mail will be delivered within hours from New York to California, to Britain, to India, or Australia by guided missiles. We stand on the threshold of rocket mail."[271] This was in 1959. The space race was on, but Kennedy's declaration to go to the Moon was still two years away. Perhaps, at the time of the experiment, the Postmaster's vision was entirely realistic.

Figure 5-5 Covers flown to the Moon on Apollo 15 July 1971. Credit NASA

The ultimate example of rocket mail was the mail that went to the Moon. Between 1968 and 1972, 24 men went to the Moon on nine manned missions of the American Apollo programme, half of them walking on its surface. Each astronaut was permitted to take a limited quantity of personal items (2.5 kg aboard the command module and 0.25 kg in the lunar module). Declaring the contents of these Personal Preference Kits was at the discretion of the astronaut. These kits included signed covers, which could then be sold to collectors once the Apollo missions were over. One such cover flown on Apollo 15 was sold for $10,350 (Rs. 435,243) in 2007.[272] Not all covers were flown. Some remained with family members on Earth as a form of insurance. The astronauts received standard salaries paid to pilots during the space programme, and all spaceflights were high risk. In the event of an astronaut's death, these signed covers left with their families on Earth could generate some financial compensation.

A few months after the return of Apollo 15, some of its flown covers came onto the market while the Apollo programme was still in progress. This was contrary to the arrangements made by the Apollo 15 Commander. The covers were to be sold only after the Apollo programme ended. In the wake of the resulting adverse publicity, NASA brought an abrupt stop to the practice. Subsequently, Al Worden, the Apollo 15 Command Module Pilot, was deemed responsible, and he became the only Apollo astronaut ever to be sacked by NASA. He came under suspicion because he was the only crew member who was also a stamp collector at the time.[273] Worden describes in detail this episode and the eventual vindication in his 2011 biography Falling to Earth.[274]

Rocket Mail and World War II

During the Spanish Civil War in 1936, General Franco's rocket squadron deployed rockets containing thousands of sheets of propaganda delivering his messages written in Spanish and Arabic to his enemies in Madrid, the legionnaires and moors. The rockets exploded over their targets distributing the leaflets directly to the soldiers below with messages like "to prolong resistance is the sacrifice of life needlessly". In an aerophilately magazine The Aero Field (No. 6, 1957), Dr Max Kronstein[275] writes "The introducing of the rocket as a propaganda tool in the Spanish campaign was reported in British newspapers during the war. And the Daily Telegraph of London reported on 12 February 1938, that even at that time both sides used for this purpose a rocket with a range of a mile and a half, which scattered 1000 pamphlets at a time."[276] Having served in the medical corps during World War I, Smith would have understood the potential grave consequences of war on mainland India. Just as the aeroplane became an instrument of war, Smith would have been aware of the potential utility of rockets for war. All of Smith's rockets were small and were not suitable for military use. There is no evidence that the military expressed any interest in Smith's rockets. Rather, according to Jal Cooper, a friend and fellow philatelist of Smith, the "British military authorities showed much annoyance and dis-favour at Mr Smith continuing his rocket experiments". During the War, Smith continued with testing his rockets in the suburbs of Calcutta where he lived.[277]

Within a few days after the start of World War II, Smith launched what he called his first 'war rocket'. Despite the name, these war rockets had no military use other than the potential, as General Franco had shown, to deliver propaganda. His first two war rockets, rocket number 207 and 208, were fired in the suburbs of Calcutta on 19 September 1939 carrying 20 covers each. They were called war rockets only because they were launched after the War had been declared. These were the very early stages of the War. It is unclear if Smith was publicizing to the military the potential of his work should the conflict arrive in mainland India. On 25 July 1940, Smith launched two rockets, number 230 and 231, from Park Street, Calcutta, probably in the park very close to his home on Elliot Road. Each carried a brownie camera with an intention to take aerial photos. Both were on the whole unsuccessful and no pictures were recorded. The first flight reached an altitude of 500 m and a range of 100 m, while the second achieved an altitude of 70 m over 500 m.[278]

In February 1936, Smith had joined the British Interplanetary Society (BIS) that had been founded in Liverpool in England three years earlier. He was probably the BIS's first member from India. The BIS was one of the several societies around the world established in the 1930s to promote the development of rocket technology and its potential application for space travel. Most similar societies around the world, including the American Interplanetary Society (later the American Rocket Society), were incorporated into their respective national defence programmes and thus ceased to exist as independent bodies. The BIS, however, was not and it continues to operate independently in the UK today. Through the BIS's monthly journal, which had an international reach, Smith would have been aware of the technological developments in rocketry worldwide, at least until 1939 when the BIS's formal activities were suspended for the duration of the War.

Initially, Smith's rockets were supplied by the Oriental Fireworks Company based in Calcutta (now Kolkata) and James Pain & Son established in London. Later, he developed his own rockets using leather, cardboard and tin. Smith's grasp of the science of rocket technology appears to have been basic. Although Sputnik and the Space Age were still a quarter of a century away, more technologically sophisticated rocket experiments were being conducted elsewhere. Perhaps, inspired by some of the writings in the literature he had access to from his membership of international organisations, like the BIS, Smith introduced design changes to his rockets. By adding control surfaces, wings and rudder, he developed what he called a Boomerang Rocket with which he experienced some success. A boomerang rocket could fly out, drop a payload and then return to the area of launch. A report in the Star of India on 25 April 1938 describes rocket number 163 as having "stubby double-finned rudder and peculiar sharp pointed wings". The report of a test launch says, "it first shot forward following a straight course and then slowly turned returning towards the firing point."[279] Smith fired two rockets, rocket 255 and 256, on 1 July 1941, at Hastings Maidan in Calcutta. He called them Parachute Rockets, but neither had a parachute. Instead, each carried 10 signed covers with a blue and white label depicting an image of a descending parachute.

Figure 5-6 King of Sikkim Igniting One of Smith's Rockets. April 1934. Credit Stephen H. Smith

Not much technical data about his experiments survive, and less was recorded and published once the War started. Smith does not appear to have had any technical training or experience. His diary shows a systematic approach to recording each rocket launch, but the data is more qualitative than quantitative reflecting his limited skills in engineering and science, as well as his lack of instruments for measuring wind speed, distance and altitude. Smith's final rocket tests were conducted in 1944. He experimented with alternative forms of propulsion. On 31 October, he launched a rocket using compressed air that he recorded as "compressed air projectile". It carried 28 signed covers, but the record has no details on the design and technology of how the compressed air was used. On 4 December 1944, he launched two rockets. One used compressed air and the other he called gas-propelled projectile. Once again, no details are available on which gases or how they were used.[280]

Between 1934 and 1944, Smith conducted nearly 300 rocket experiments. Just about every rocket carried covers signed by Smith himself and others. Even those rockets that carried parcels, livestock and cameras also had covers. There is no comprehensive record of how many covers were taken in each of his rockets and not all attempts were successful. The tally of his first 10 flights indicates that 1,313 items were flown. As his experience grew, so did the capacity and reliability of his rockets, and the number of covers flown also increased. In the first week of February 1937, Smith attended the All India Boy Scouts Jamboree in New Delhi and launched 19 rockets over three days, some in the presence of high-ranking officials, including the Viceroy and Lord and Lady Baden-Powell. The total number of covers carried during the Jamboree was 6,358.

Spectacular failures in the presence of the public and media undermined the immediate prospects of rocket mail. German businessman Gerhard Zucker (1908–1985) was popularising rocket mail in Europe as Smith was in India. Zucker, who was simultaneously considered a pioneer, showman and a charlatan, conducted a fateful rocket experiment on a beach in Western Scotland on 31 July 1934.[281] He claimed that his rocket could cover nearly 400 km travelling at 3,000 km/h, but the rocket exploded at launch in the presence of a large contingent of the press. Smith in India also experienced failure in the presence of the Calcutta press during the launch of three rockets on 28 February 1935.[282] The first rocket launched and exploded (or burst as Smith records in his diary) in mid-air. The other two, both larger, also failed. The first just managed to leave the launch rack before falling to the ground and the second burnt itself out on the launch rack.

A combination of rocket technology's unreliability and the increasing prevalence of airmail prevented rocket mail from getting a firm foothold. As commercial airline services grew around the world, the delivery of mail by air became convenient and cost effective. In April 1941, the UK introduced a postal service specifically designed to meet the needs of a nation at war.[283] The General Post Office supplemented its traditional postal service with the Airgraph Service. The concept of Airgraph was simple. Instead of an actual letter, a tiny photographic negative image of the letter was transported instead by air. Upon arrival, the image was enlarged and printed, put in an envelope and posted locally. Airgraphs reduced the weight of airmail to just 1%, and servicemen and women could exchange letters with their families in a matter of days or weeks rather than months the sea route would take.

Smith's work on using rocket power for transport received the same muted response from the British military authorities that Goddard's work had received from the US government in the 1920s. However, his primary contribution was to demonstrate through experiment the potential of rockets as a practical means of transport.

Smith's Personal Life

Stephen Smith was born on 14 February 1891 in the town of Shillong, then in Assam in the north-east of India. His father Charles William Bath Taylor from Lincoln in Britain had travelled to Assam to work as the Superintendent of the Hoolingurie Tea Estate. His mother, Arabella nee Martin, was the daughter of an English tea planter. Contrary to material available in the public domain today, Smith was not an Anglo-Indian.[284] Although born in Assam, his birth was registered in Britain. He had a sister called Marjorie, who married an Englishman and moved to England. There are not many pictures of Stephen Smith.

One image, probably taken when he was around the age of 30, reveals a clean-shaven, dark-skinned, black-haired handsome man with sharp facial features. His handwritten English is elegant, and he probably spoke English with similar competence. He lived in Calcutta among the Anglo-Indian community in the pre-independent India from the 1920s to his death in 1951. Living among the Anglo-Indians in Calcutta, he was probably seen and treated as one. With tangible connections to both Indian and British roots, Anglo-Indians belonged to neither. They suffered discrimination from the local Indian communities and racism from the British. The question of the identity of the Anglo-Indian community has cast a long shadow over its 500-year long history, and it preoccupies the ever-diminishing community even today.[285] Smith attended St. Patrick's Boys School in Asansol, West Bengal, established originally by Christian Brothers from Southern Ireland for Anglo-Indian boys. During his lifetime, he wrote several books. In 1927, he authored a small book *The World Flyer's Danger Zone* covering the hazards of airmail flights south-east from Calcutta across the Bay of Bengal to Burma and Thailand. He dedicated the book and its proceeds to the widow of Arthur B. Elliott, who was killed on 5 July 1926. He also published *Indian Airways* in three volumes between 1926 and 1930. A final posthumous publication was a revised second edition called *Rocket Mail Catalogue*, which was published under his name in 1955 but authorised by his wife.

After leaving school, Smith worked briefly at the customs department in Calcutta before joining the Calcutta Police Force as a Round Sergeant on 18 March 1913 on a salary of Rs.100 per month.[286] While with the police, he completed his training as a dentist. His time with the Calcutta police was uneventful, and he resigned on 4 December 1914. It was as a dentist that he served in World War I, after which he continued in the profession with a private dental practice based at his home on 25A Elliot Road in Calcutta. This appears to have been his primary source of income after the War. He married Fay Gulner Anne Harcourt on 6 November 1918, whom he had known at least since 1913 when he joined the police force. They had one son, Hector, who did not share his father's interests.

Figure 5-7 Stephen Smith and Fay Harcourt Married on 6 November 1918 in Dhurrumtollah Street Roman Catholic Church Calcutta. Credit Paul Sandford

Smith died in 1951, and his wife survived until 1985. Both are buried at the North Circular Road cemetery in Calcutta near where they lived. There are not many alive today who have a personal recollection of Stephen Smith. Living on the same Elliot Road that Smith lived on, Melvyn Brown remembers meeting Stephen Smith's son, Hector, twice in the late 1970s. [287] Brown has been championing and chronicling the achievements of Anglo-Indians and, in 1991, unsuccessfully attempted to have a part of Elliot Road renamed after Stephen Smith. Smith's son Hector married and had a son and daughter. His son grew up with a learning disability and was sent to an institution. Hector's daughter Gloria emigrated to the UK in 1982 after her second marriage. Hector and his wife joined them in 1987, but Hector did not respond well to the British climate. He caught a cold and died soon after he arrived. Smith's grand and great-granddaughter live in London, but the family no longer has any connection with rockets.

Even with the support of friends in high society, it seems as if Smith was isolated and lacked the necessary support, both practical and moral, to successfully promote rockets as a means of transport. Despite his pioneering work, he did not fully accomplish his ambitions. On 11 April 1935, following a successful firing of his rocket number 54 in the presence of the King of Sikkim, a certificate was awarded to him by Mr C.E. Dudley from the Indian Civil Service. This formal recognition, "certifying the utility of the rocket as a means of transport during floods and landslips", was possibly the only formal acknowledgement of his work in his lifetime with rockets and rocket mail. Smith wanted to realise the potential of rocket power for transporting mail and materials, just as he was witnessing aeroplanes doing so for the first time in history. His limited skills and resources prevented him from making a significant advance in developing rockets as a transport vehicle.

Figure 5-8 Stephen Smith Centenary Commemorative Stamp. Credit Philately World

Smith's son Hector destroyed all his father's rocket-related material soon after his death.[288] Despite his achievements in rocketry and rocket mail, it is in the world of philately that Smith's work is substantially recorded and acknowledged. During the 1920s, he founded the Calcutta Philatelic Club and the Aero Philatelic Club of India (which changed its name to the Indian Airmail Society on 19 January 1930). Smith served as the Indian Airmail Society's secretary for most of the 1930s, during which time he recorded the development of airmail in India in the Society's monthly bulletins.

The bulletins recorded news about scheduled airmail flights within, as well as between, India and the rest of the world. They contained interesting and mundane details, including information on the first balloon flight in India in 1837[289] and the number of airmail flights in the first five months of the Karachi-Croydon route.[290] He noted prices, timetables, quantities of airmail, individual pilots covering hazardous routes and details of air crashes, as well as the routine business of running the Society. Over three decades after his death, in 1989, Smith was inducted into the Hall of Fame by the American Airmail Society.[291] In 1992, a year after the centenary of his birth, the Indian government celebrated his achievements by issuing a stamp and a first-day cover dedicated to Smith and his work.

≺ ◊ ≻

Chapter Six
Vikram Sarabhai: Leadership by Trust

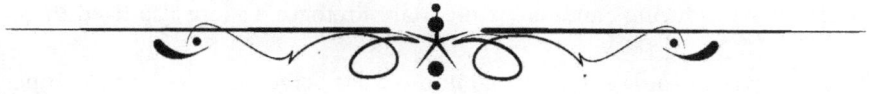

H andsome, charming, wealthy and intellectually gifted, Vikram Sarabhai (1919–1971) had everything going for him. Born with a silver spoon in his mouth, Sarabhai enjoyed a privileged life. He lived in a large house with staff to match, private tutors and laboratories at home to satisfy his childhood curiosity. He had two brothers and five sisters. He was a bright child with a clear scientific interest in building, experimenting and testing. One of his earliest projects was building a toy steam engine. Sarabhai hobnobbed with the incumbent and future prime ministers of India, with scientists of international repute and had personal connections with both of India's Nobel laureates of the time, Rabindranath Tagore and C.V. Raman. Mahatma Gandhi was a family friend; as a child, Sarabhai had joined Gandhi's 200-mile (321.87 km) walk, the Dandi March, to protest against the imposition of tax on the production of salt by the British. In 1961, he hosted Queen Elizabeth and the Duke of Edinburgh when they visited Ahmedabad.

Sarabhai's ancestors were wealthy merchants belonging to the Dasa-Shrimali Jain sect, who had made their fortune in the mid-19th century by lending money to local chieftains and participating in the profitable exploitation of China instigated by the British colonial rulers. The business arrangement involved buying opium in Malan in central India, exporting it to China and then buying Chinese tea and silk to be sold in the UK. China had no interest in Western goods, so demand for opium had to be artificially induced by fostering opium addiction. The cycle of trade was thus concocted exclusively to serve British financial interests. As middlemen, the Sarabhai business benefited, too. In 1880, Sarabhai's grandfather acquired Ahmedabad Calico Printing Company, Ahmedabad's first mechanised cloth printing company. This was not an intentional business acquisition but an inevitable consequence of being the company's largest lender when it failed.

Vikram Sarabhai was born in the family home, The Retreat, built by the Sarabhai family in 1904. It was located in the exclusive area of Shahibaug, north of the city centre in Ahmedabad with the River Sabarmati forming its western perimeter. Shahibaug had once been the extensive garden of a royal palace called Moti Shahi Mahal built by Shah Jahan before he went on to build his more famous creation, the Taj Mahal. Since the time of the British Raj, Shahibaug has attracted well-heeled businessmen, industrialists and professionals. During Sarabhai's time, Rabindranath Tagore also lived there.

Sarabhai was to India's space programme what Sergei Korolev and Wernher von Braun were to the space programmes of the USSR and US, respectively.

Figure 6-1 Vikram Sarabhai with Son Kartikeya and Daughter Mallika. Credit Mallika Sarabhai

Although all national space programmes rely on the contributions of thousands of individuals, history has the tendency to single out one man (and it usually is a man) as the originator above all others.[292] Following Yuri Gagarin's breath-taking accomplishment as the first human in space in April 1961, Sarabhai wrote to the Government of India proposing a space satellite programme for India.[293] Sarabhai's outrageous vision of how a developing nation like India could utilise space technology for social development and also the break-neck speed with which he implemented it earned him the title 'father of India's space programme'. Professor Yash Pal (1926-2017) recalls Sarabhai's confidence about the development space research could bring about in areas of broadcasting and communication and the boldness with which he approached Prime Minister Mrs Indira Gandhi and said, "We want to test whether space could be used for such purposes." She asked, "But where are your rockets and how can you test?" He replied "I have friends; we have scientific friends in America everywhere, and so on. Let's see what we can do."[294]

Education

Today, a foreign education, especially from a recognised university in the US or Europe, is almost a routine expectation for a young Indian from the Indian middle class. This was not so in the pre-independent India. In the India of the 1930s, if you were born into a successful family business, especially as a son, your future was pretty much set. Convention dictated private tuition at home followed by an independent school or college with frequent involvement in the family business during holidays. By the late teens, the son would have had years of on-the-job experience even if he had not sought it. Tradition required that when the time was right, the reins of the business gradually pass from the father to the son in a casual but measured period of handover. But the Sarabhai family were not followers of tradition. As a mill owner, Sarabhai's father, along with his siblings, had travelled extensively and seen for themselves the new industrial processes and technological innovations that were introduced in Europe and the US. Sarabhai's father was fond of Western fashion, music and cars and was the first to own a car in Ahmedabad.[295]

A letter of recommendation from Nobel laureate Rabindranath Tagore, who was also a neighbour at Shahibaug, secured Sarabhai a place at Cambridge to study physics and mathematics. Tagore is to India what Shakespeare is to the UK and Goethe to Germany. In 1937, along with his brother Gautam, Sarabhai set sail for the UK. The route from India to Cambridge to study physics and mathematics was well trodden by the late 1930s. Muhammad Raziuddin Siddiqui (1908–1998), Nobel laureate Subrahmanyan Chandrasekhar, as well as Homi Bhabha, who would later have a direct impact on Sarabhai's future, had all studied at Cambridge. The 1920s and 30s were a heady time for science and the Cavendish Laboratory in Cambridge. It was the period when many of the key discoveries and inventions in physics, the principles of rocket propulsion, continental drift,[296] the internal structure of the atom, were validated for the first time through experiment. Edwin Hubble's detection in 1929 of the redshift phenomenon in distant galaxies demonstrated that the solar system was part of an expanding universe.

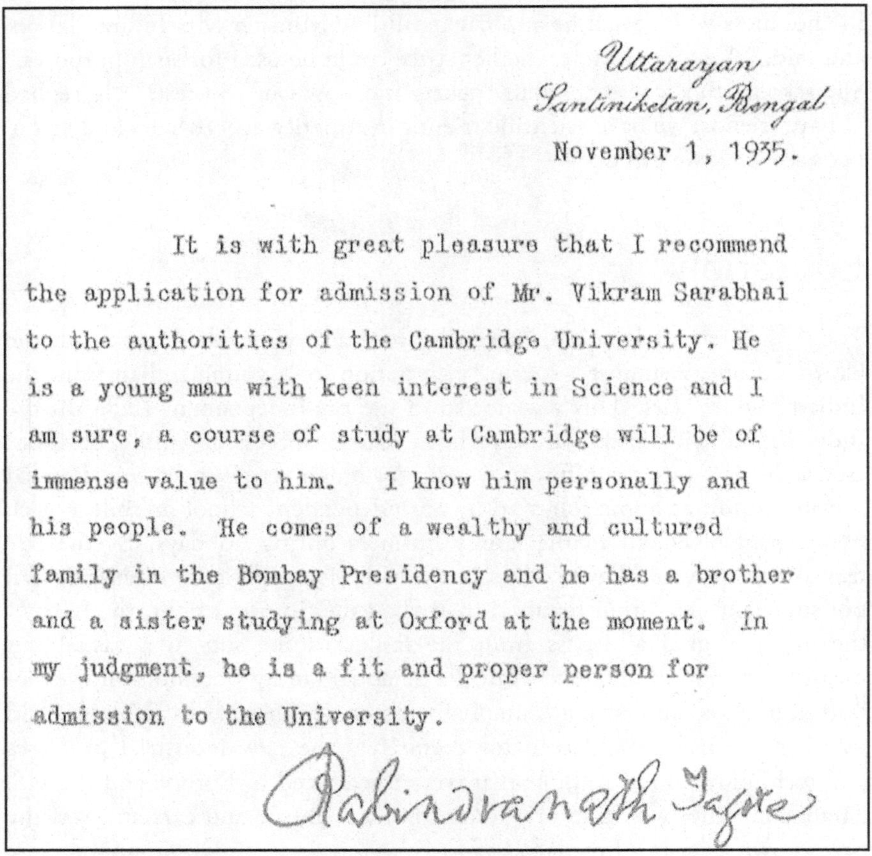

Uttarayan
Santiniketan, Bengal
November 1, 1935.

It is with great pleasure that I recommend the application for admission of Mr. Vikram Sarabhai to the authorities of the Cambridge University. He is a young man with keen interest in Science and I am sure, a course of study at Cambridge will be of immense value to him. I know him personally and his people. He comes of a wealthy and cultured family in the Bombay Presidency and he has a brother and a sister studying at Oxford at the moment. In my judgment, he is a fit and proper person for admission to the University.

Rabindranath Tagore

Figure 6-2 Letter of Recommendation to Cambridge from Rabindranath Tagore. November 1935. Credit Vikram Sarabhai Archives

The Cavendish Laboratory in Cambridge was at the centre of some of the greatest fundamental scientific discoveries in the 20th century. In the 50 years preceding Sarabhai's arrival, all three fundamental atomic particles, the electron, proton and neutron, had been discovered in Cambridge. For the first time in human history, the key components of an atom had been scientifically predicted, detected and quantified by experiment. The atom could finally be understood as a central nucleus containing positively charged protons and neutral neutrons surrounded by a cloud of the lighter, negatively charged electrons. In the decade between 1927 and 1937, scientists at the Cavendish Laboratory won Nobel Prizes almost every other year.

In 1932, John Cockcroft and Ernest Walton designed an experiment to fire high-speed neutrons at atoms of lithium. The resulting impact split the lithium atom and generated helium. This so-called 'splitting' of the atom, or transmuting one element into another, would result in the award of several Nobel Prizes. Others who were honoured with Nobel Prize for their work at Cavendish included Patrick Blackett for developing the Wilson Cloud Chamber, a device that revealed the evidence of positron or the positive electron, another inferred, but until then undetected, particle. Paul Dirac and Niels Bohr, who made profound contributions to the study of nuclear physics, also had strong connections to Cambridge. It was in this environment, rich in scientific accomplishment that Sarabhai came to study, first, mathematics and physics and later, cosmic rays. When he embarked on building scientific institutions in India, he would call upon some of these scientists for professional assistance. Some would also become Sarabhai's friends.

Sarabhai had just completed his undergraduate studies when war broke out in Europe in 1939. His father demanded that he return to India. Sarabhai had concluded that among the two popular fields of research at the time, his interest lay in cosmic rays, not in atomic fission. He wanted to investigate the physics of high-energy cosmic rays entering the Earth's atmosphere from space. However, if he returned to India how would he conduct high-quality research, which was then available only at Cambridge? A solution emerged when Cambridge University provided special permission for him to continue his postgraduate study in India if he was supervised by the Bangalore-based C.V. Raman.[297] In 1930, C.V. Raman had won a Nobel Prize in physics for discovering the inelastic scattering of light. It is probable that C.V. Raman's availability in Bangalore was a key factor in Sarabhai's choice of cosmic rays as the subject for his PhD research.

Cosmic ray research was in its early stages when Sarabhai chose it as a topic of research. The potential impact of cosmic radiation and solar flares on human spaceflight was realised soon after humans ventured into space. Cosmic rays arriving on the Earth from the Sun, stars and other galaxies are deflected by the Earth's magnetosphere or are almost all absorbed by the upper atmosphere. There are no significant consequences for life on Earth but are a hazard to astronauts beyond the protection of the Earth's atmosphere and magnetic field. In 1963, two Soviet spacecraft with a human crew, Vostok 5 and Vostok 6, were scheduled to be launched on 10 and 12 June, respectively. A large solar flare raised radiation risks for the cosmonauts, and the launches were delayed by four days.[298] The Apollo flights to the Moon beyond the safe realm of the Earth's magnetic field provided an opportunity for quantitative research. When the astronauts on the crippled Apollo 13 aborted the mission to land on the Moon and headed back to Earth, they reported seeing bright flashes. The flashes did not come from any particular direction; they did not need to look out of the windows and saw them even when their eyes were shut. These flashes, confirmed through subsequent investigations, were the result of cosmic rays impacting on the astronauts' retinas.[299] It was the study of these cosmic rays that occupied Vikram Sarabhai's future professional life as a scientist.

After relocating from Cambridge, Sarabhai started his PhD in Bangalore. His first scientific paper, entitled Time Distribution in Cosmic Rays, was presented at the Indian Academy of Sciences in 1942. This was the first of the 85 scientific papers he would formally present until his premature death in 1971.[300] Prior to the Space Age, high altitude cosmic ray research used balloons and sounding rockets. In an experiment in 1943, Sarabhai climbed to an altitude of 3.11 miles (5 km) in Kashmir to measure the rate and intensity of high altitude cosmic rays using radiation detectors (Geiger counters) that he had himself modified.[301]

Soon after the war ended, Sarabhai returned to Cambridge along with his wife to finish his PhD. The thesis, entitled Cosmic Ray Investigations in Tropical Latitudes, took just over a year to write and was completed on 24 May 1947. Many senior Cambridge scientists had not survived the war or had been displaced by it. Cambridge could not facilitate his PhD oral examination, so it took place in the drab surroundings of the post-war northern England, in the city of Manchester, under the scrutiny of Professor Patrick Blackett.[302] With his experience of India, Cambridge and cosmic rays, Blackett was ideally placed as Sarabhai's examiner. He had visited India in January 1947 at the personal invitation of Nehru, who was only a few months away from becoming the first prime minister of independent India. Blackett had studied and later done research at the famous Cavendish Laboratory under Ernest Rutherford. In 1937, he moved to Manchester as the professor of physics. Although Blackett did not know it at the time, he, too, would win a Nobel Prize for physics in 1948. Sarabhai successfully defended his thesis during his viva voce in Manchester and was awarded the PhD. His wife, Mrinalini, recalls a letter from Professor Blackett in which he recounted that day "I remember very vividly your splendid red sari contrasting so strongly with the grey gloom so characteristic of the old Manchester labs."[303]

To celebrate this success and the news of his wife's pregnancy, they returned to India, making stops in the Netherlands and France on the way. Once back in India, their pace of life picked up. Sarabhai took on an increasing number of projects, building institutions of international repute. They included Ahmedabad Textile Industry Research Association, the PRL, Indian Institute of Management (IIM) Ahmedabad and Centre for Environmental Planning and Technology.[304] His work on India's space programme was yet to begin.

Sarabhai Family and Gandhi

On 12 March 1930, Mohandas Karamchand Gandhi (1869 – 1948) began a 240-mile (390 km) march on foot from his home in Ahmedabad to the coastal town of Dandi with the intent of making salt without paying the tax that British law demanded. It was long and arduous, but a pivotal act of protest in Indian history. He was accompanied by around 100 supporters, including 11-year-old Vikram Sarabhai.

Upon his return from South Africa, Gandhi had settled in Ahmedabad in 1915. He was welcomed to the city with widespread affection and enthusiasm. He was already well-known in India for his success in introducing some measure of equality for the Indian population in South Africa, though the global status he has today was still a long way away. However, when Gandhi allowed an untouchable to join him in his settlement, the caste-ridden society of Ahmedabad shunned Gandhi. Caste system still dominated the Indian society in the early part of the 20th century and was a dominant force against an egalitarian society. The sense of community Gandhi was attempting to build evaporated along with his funds. Gandhi found himself isolated and deprived of the means of continuing the work of setting up his ashram. An anonymous donor, who was later identified as Ambalal, Vikram Sarabhai's father, provided Gandhi financial relief of Rs. 1,300, thus becoming his benefactor. [305]

In 1917, Gandhi found himself wedged between two influential figures of the Sarabhai family on opposing sides of an industrial disagreement. One party was his benefactor and mill owner Ambalal. The other was Anasuya Sarabhai, Vikram Sarabhai's aunt. Headstrong, intelligent and highly motivated, she had her own unique take on the political issues of the time. Atypical of women in India, she exhibited no fear of authority and did not hesitate to challenge tradition. She had gone to Britain in 1912 to study medicine but soon switched to politics, primarily to avoid animal dissection. She was drawn immediately to two of the most pressing concerns of the time, potential war with Germany and political power for women.

An intense and radical women's suffrage movement had established firm roots in the UK since the late 18th century. During her time in London, she was convinced of the merits of the Fabian belief that social progress comes out of gradual development rather than revolution. She returned to India in 1914, and in her first major public confrontation, she undermined her own family business. In 1917, with support from Gandhi, she led the first strike of textile labourers in Ahmedabad. Her opponents were the mill owners and industrialists of Ahmedabad, principal among whom was her brother and Vikram Sarabhai's father, Ambalal Sarabhai.

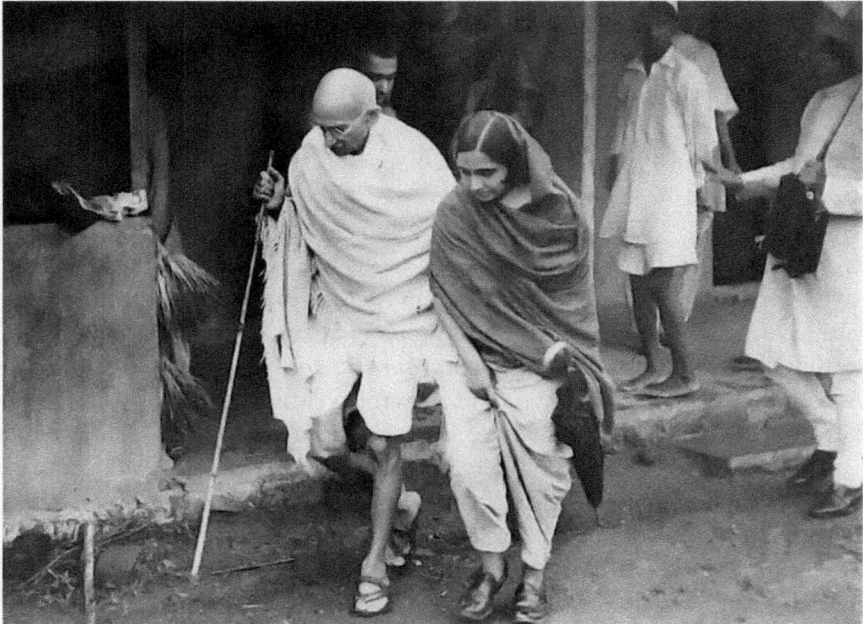

Figure 6-3 Mahatma Gandhi with Vikram's sister, Mridula Sarabhai. 1942. Credit Unknown

As trade and commerce recovered following the end of World War I, trade unionism in Europe flourished, particularly in the mill and mining towns of the industrial heartland of northern England. Deep-rooted matters of social concern resurfaced as the working classes in Europe and India sought to improve their living conditions following World War I. Ahmedabad, with its established connections to textiles, was known as the Manchester of India.[306] As the Industrial Revolution spread around the globe, so did concerns about working conditions, child labour, fair and overtime pay, discrimination, collective bargaining and workers' rights. As technology and processes of industrialisation began to make an impact across the world, they triggered social and political unrest, which soon reached the mill workers in Ahmedabad. Gandhi stuck to the middle ground in the dispute between Ambalal, his benefactor and mill owner on one side, and Ambalal's sister Anasuya, who championed the rights of workers on the other. He had benefited from the generosity of the Sarabhai family and campaigned for a negotiated compromise acceptable to both sides. It was here that Gandhi employed his distinctive non-violent solution of fasting for the first time. He also brought to bear his skills as a trained lawyer to achieve the final settlement. The compromise helped Anasuya in 1920 to establish the Majoor Mahajan Sangh (Ahmadabad Textile Labour Union), which celebrated its 90th anniversary in 2010. Though Gandhi never held a formal political post in the Indian government, his practice of ahimsa or non-violent struggle against colonialism and his assassination in 1948 elevated him as the father of the nation.

The principle of fasting to achieve a specific political goal had been well established in Britain and Ireland and deployed effectively by the suffragette movement.[307] Both Anasuya and Gandhi would have been aware of fasting as the primary tactic used by the suffragette movement in London in their pursuit of women's right to vote. Could it be that Gandhi acquired this overwhelmingly Fabian attrition-like quality of his campaigning style from Anasuya? It is interesting to speculate that in choosing this approach to resolve the conflict between Ambalal and Anasuya, Gandhi was perhaps motivated by the desire to avoid offending his patrons and only later recognised the innate righteousness of non-violent tactic. In August 1942, Gandhi initiated the Quit India campaign and demanded that the British simply withdraw from India in an orderly fashion. He asked for passive resistance, 'do or die', from Indians all over India. Surprisingly, the Quit India campaign did not result in malice against British residents or even British soldiers based in India. Harry Turner (1920–2009), a radar engineer from Manchester, who had helped establish the Manchester Interplanetary Society and had been involved in the testing of rockets prior to joining the army, found himself in Bangalore in late 1945. During a visit to a cinema to watch an Indian film, he noted "The 'Quit India' movement has a strong following here, yet I find the locals decidedly friendly."[308]

Although the Sarabhai family could not escape the dramatic politics of the Indian independence movement during the inter-war years, Vikram Sarabhai expressed no interest in politics. His family, especially his sister and aunt, were deeply involved in politics, but he resisted and maintained his links with the scientific community. World War II had dwindled the British capacity to rule India, and after the War, the independence movement surged. This crucial period coincided with Sarabhai's completing his PhD and embarking on his professional career.

Marriage

Against the backdrop of the tumultuous events leading up to India's independence, Vikram Sarabhai got married to Mrinalini Swaminathan, a Bharatanatyam and Kathakali dancer. Sarabhai had proposed to Mrinalini more than once. The uncertainty that prevailed in the midst of a revolutionary change convinced Mrinalini to finally accept. However, the bride's expectations, "a grand wedding with music and classical dance to entertain the visitors who were coming from Ahmedabad", were not to be.[309] The wedding ceremony took place in the drawing room of the bride's Bangalore home. The journey to Ahmedabad, about a thousand miles (over 1,600 km) north of Bangalore, is a demanding at the best of times but especially problematic at the time of civil unrest. Given that the bride and groom were each from affluent families, the wedding was surprisingly austere: no large-scale celebration, no elegant dresses, no high-profile guests, no exquisite cuisine and no honeymoon.

Sarabhai did not honour the marriage vows administered during the hastily arranged wedding. A couple of years after their marriage, during a social event, his wife introduced Sarabhai to her friend, Kamla Chowdhry, whom she had first met as a dance student in 1941.[310] Sarabhai established and maintained an intimate relationship with Kamla until his death. The Sarabhai family had repeatedly demonstrated that they were not bound by strict traditions and that they took a pragmatic (that is "doing what seemed right") than a sacred view of the bond of marriage.[311] Vikram Sarabhai's brother, his aunt and later his daughter had engaged in unconventional relationships. He even justified the absence of an absolute morality by drawing on the ambiguities inherent in ancient and newer texts, the Upanishads (ancient religious Hindu texts), and even special relativity.

Far from leading a double life, however, the relationship was not secret. At the same time as professing his sincere love for his wife, Sarabhai appeared not to see the inevitable demise of his marriage. Biographer Amrita Shah captures this astonishing shortcoming by observing that his "… inability to anticipate his wife's deep sense of hurt [betrayal], a startling naivety in a man capable of immense complex reasoning."[312] Mrinalini, too, persisted with the marriage and did so without breaking her vows. Despite invitations from admirers, Mrinalini remained loyal.[313]

Peaceful Uses

One of the unique aspects of the space programme in India, unlike the motivations of other nations, was the absence of a military component at the outset. Its primary objective was the advancement of social and economic goals for the people of India. The non-military objective was repeatedly espoused by Sarabhai and others during his lifetime and has frequently been repeated since. However, there have been instances that demonstrated that the objectives of India's space and nuclear programmes are not entirely civilian.

In early 1968, Prime Minister Gandhi announced a programme to develop "a device to help take-offs on short runways by high-performance military aircraft."[314] Following Homi Bhabha's death in 1966, Sarabhai was not only leading the space programme but also heading the DAE. The timing of the meeting and the accompanying secrecy suggest a shift in India's space and nuclear programmes from purely civilian to include a military component. At the time, Indian military aircraft operating under extreme environmental conditions (high altitude, high temperature or geographically difficult terrain) and limited to short runways used Soviet-built rocket motors to assist in take-off. Sarabhai had met with Abdul Kalam and Group Captain V.S. Narayanan at 3:30 am on the same day before the public announcement was made. Sarabhai instructed Kalam and Narayanan to initiate a project to cut the cost of rockets by half and to manufacture rocket motors in India. He gave them 18 months to complete the task. For the first time, Sarabhai had unambiguously crossed the threshold from a civilian space programme and initiated INCOSPAR towards a programme to develop rockets for into the military domain. This contradicted his stated views on keeping the space programme non-military. At a personal level, this would have been a blow to his sense of integrity given his public stance but it was a requirement imposed on him.

While Sarabhai is widely recognised as the father of the Indian space programme, Bhabha is acknowledged as the father of the Indian nuclear programme. Bhabha had his eyes set on India developing a nuclear deterrent and joining the nuclear weapons club along with the US, USSR, UK, France, China and Israel[315]. When Bhabha suddenly died in an air crash in 1966 and Sarabhai took over as Chairman of the DAE and secretary at the AEC, his first goal was to steer India away from Bhabha's vision of an India with a nuclear bomb. Sarabhai's vision centred on an India exploiting civilian uses of nuclear energy.[316] Peaceful Nuclear Explosions, as anachronistic as they may sound, were conducted several times during the 1960s and 1970s. These were tests conducted underground, with limited yield, for national economic development and had no military purpose. India's intention to participate in such Peaceful Nuclear Explosions was thus a politically legitimate goal.[317]

Sarabhai may not have succeeded completely in his goal of redirecting the nuclear programme in India towards a strictly civil one, but his desire for an India free of nuclear weapons was sincere. He had deep roots in the Jain tradition of non-violence and reverence for life. As a follower of Gandhi, he is likely to have been influenced by his values of non-violence, too. Writing in September 1948, Gandhi had insisted "I regard the employment of the atom bomb for the wholesale destruction of men, women and children as the most diabolical use of science."[318]

In public, Sarabhai asserted India's nuclear-weapon-free stance. In private, his position was ambiguous. One of his first decisions in his new role at the DAE and AEC was to formally end the theoretical work on the Subterranean Nuclear Explosion for Peaceful Purposes (SNEPP) programme, for which Bhabha had worked hard to acquire authorisation from Prime Minister Lal Bahadur Shastri.[319] Despite formally ending the SNEPP programme during his period in office, there is evidence that SNEPP activities in India did not stop. AEC engineers continued their design work on a potential nuclear explosive device for India. Amrita Shah details the evidence and concludes "It is possible that Vikram knew, he either had been informed or had discovered it himself, and chose to say nothing."[320]

In 1971, India was caught in the fight between East and West Pakistan. One nation in two parts separated by almost 1,000 miles (over 1,600 km) of Indian territory was an awkward product of independence. On 26 March 1971, the largely Bengali population of East Pakistan declared independence as the new nation of Bangladesh. The Indian government established its allegiance with Bangladesh possibly fearing a massive refugee influx. The Bangladeshi government in exile was offered refuge in Calcutta. In response to a direct military strike by Pakistan on India, India's military retaliated on 3 December 1971. It was inevitable that the relatively small Pakistani force would capitulate to the overwhelmingly large Indian troops supported by native Bengali freedom fighters.

On 16 December 1971, Pakistani forces signed the instrument of surrender to the Indian-Bangladeshi allied forces. The possibility of war had reawakened the desire in some parts of the Indian military for India to develop its nuclear weapons. Sarabhai had established his opposition to

India developing its nuclear weapons capability. If Indira Gandhi was to pursue the nuclear weapons option for India, she had to remove the key obstacle: Vikram Sarabhai. In the last week of November 1971, Prime Minister Gandhi told Sarabhai of her intention to move him out of the DAE and into the new DOS.[321] Perhaps, Sarabhai recognised that his fight for a nuclear-weapon-free India was lost.

Sudden Peaceful Death

Sarabhai was approaching the end of his first decade in a high government office. Moving from the orderly world of science determined by the unchanging logic of mathematics to the scheming, ambiguous world of politics must have been a challenge. However, he stuck to his primary objective of raising the quality of life of the nation's poor through the power of science and technology. Occasionally, this clashed with his duty to meet the requirements of national security and international collaboration. Maintaining this subtle balance was to take its toll. Ever since taking on the mantle of the DAE from Bhabha in 1966, Sarabhai had worked in a stressful environment over long hours. His workload grew with time. One of his earliest recruits, R. Aravamudan recalled "Meetings with him would continue well into midnight. They would start again at dawn the next day and sometimes continue at the airport until he boarded the aircraft. Some meetings took place on the aircraft itself."[322] The intensity of his daily obligations, the pressure of the war in 1971 and the knowledge that he would be moved from his high-level position and possibly forgo ready access to the prime minister could have triggered a burden his body could no longer bear.

On the morning of Thursday, 30 December 1971, Vikram Sarabhai was found dead in bed, having died apparently peacefully in his sleep. A book he was reading lay on his chest. His sudden death at 52 shocked his family and friends and left India in a panic reminiscent of that in 1966 when Bhabha had died in an air crash. Abdul Kalam, an engineer by training, who went on to hold probably the highest political office that any rocket engineer has ever held by becoming the 11th President of India in 2002, describes Sarabhai's pioneering style as a "great example of leadership by trust."[323] Kalam was one of the last persons to whom he had spoken before he died and would have been the next to meet, had he not. Sarabhai had asked Kalam to wait for him at Trivandrum airport. By chance, Kalam was arriving from Delhi following his regular missile panel meeting and Sarabhai was departing to Bombay on the same day. Kalam recalls "I was shocked to the core; it had happened within an hour of our conversation."[324] As Sarabhai had not been ill, his death had not been anticipated. However, his wife, Mrinalini, recalls that "even though he did not confide his work problems to me, I knew instinctively that he was under a great deal of stress."[325] The extent of that stress was common knowledge even beyond the close circle of friends and family. Jacques Blamont (born 1926), who helped in establishing the space programme in France in the early 1960s and assisted Sarabhai in setting up the space programme in India, also considers that Sarabhai's excessive workload contributed to his early death.[326]

Figure 6-4 Vikram Sarabhai R. Aravamudan and an Apollo 11 Moonrock at Thumba in 1969. Credit R. Aravamudan

An anonymous official within the Indian Government indicated that concerns about Sarabhai's health had reached the highest levels; Prime Minister Indira Gandhi was aware of Sarabhai's workload and had remarked that he would die if he went on like this.[327] Sarabhai's legacy is firmly and inextricably woven not only into ISRO's infrastructure but also in the hearts and minds of those who work there. M.G.K. Menon (1928–2016), who took over in the immediate aftermath of Sarabhai's sudden death, turned to Sarabhai's companion and professional partner Kamla Kapur "to analyse Vikram's ideas on the institutional management framework."[328]

In 1972, in a tribute to the man who inspired it, the Thumba Equatorial

Rocket Launch Station (TERLS) was renamed the Vikram Sarabhai Space Centre (VSSC). By then, VSSC had grown into a large, high-technology complex. Additional facilities developed included Space Science and Technology Centre, Propellant Fuel Complex and Rocket Fabrication Facility. These segregated units were merged into a single entity and renamed VSSC in June 1972. Four decades on, ISRO has expanded its footprint throughout India and beyond, but VSSC remains its largest complex.

Figure 6-5 Crater Sarabhai 4.66 mile (7.5 km). Photographed by Apollo 15 Command Module Pilot Al Worden from Lunar Orbit. 30 July 1971. Credit NASA

Most accounts of Sarabhai's life are short, informal and overwhelmingly hagiographical. When the extent of his contribution to the development of Indian space programme is fully assessed, he stands tall and justly deserves the national accolade as the father of Indian space programme.[329] If anything, his enormous accomplishment of promoting the crucial and wider role of science and technology in India's development, not just in the space programme, is yet to be fully appreciated.

Sarabhai's passionate advocacy of "indigenisation and self-reliance" that he imbued INCOSPAR with, from which ISRO emerged in 1969, continues to inform the culture of ISRO and India.[330] Writing in 2001, Professor U.R. Rao insists "Sarabhai's vision on the development of space technology and its extensive application for the betterment of society continues to be the guiding light of our space programme even today."[331] Many of the institutions he established continue to have a transformative impact on India in the 21st century.

⊰ ◈ ⊱

Chapter Seven
First Launch

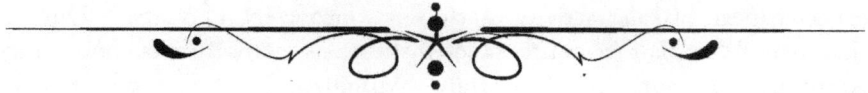

T he true beginnings of the Space Age lie in the International Geophysical Year (IGY) 1957–58 and not the Cold War as is commonly thought.[332] The International Polar Years of 1882–83 and 1932–33 had already demonstrated how international scientific collaboration could be leveraged to learn about the Earth. The IGY was modelled on and shared an objective similar to the International Polar Year programmes, which had aimed to understand the Earth's polar regions through international collaborative science programmes. An enormous investment of time and resources during World War II had contributed to an unprecedented accelerated development in rocket technology in the US and Europe. The IGY provided an opportunity to engage the technology that had been developed a few years earlier for war in the pursuit of peace.

The idea of what became the IGY had crystallised four years earlier during a dinner party on 5 April 1950 held by James Van Allen (1914–2006) at his private residence.[333] A special committee for the IGY had its first plenary session in Brussels between 30 June and 3 July 1953. By then, 30 national academies had responded positively to a programme of research to cover the entire surface of the Earth. India was represented by Vikram Sarabhai.[334] The IGY was formally proposed by the National Academy of Science to the Assistant to the US President Eisenhower in a letter dated 21 April 1954. The letter stated that it would be a "major international cooperative undertaking."[335]

In pursuit of President Eisenhower's desire for international collaboration and enthusiasm for the US as a world leader, a budget of $12.5 million (Rs.5.95 crore) was approved on 2 June 1954 for the IGY.[336] The proposal was taken up by the International Council of Scientific Unions (ICSU),[337] and the scope extended to include not just the north and south poles but the entire planet. The ICSU established the IGY to study 11 areas of Earth Sciences, including cosmic rays, ionospheric physics, longitude, meteorology and solar activity. In 1958, it established the Committee on Space Research (COSPAR) with a focus on international scientific collaboration on space research. Four years later, COSPAR led to the creation of INCOSPAR in India which oversaw all India's space activities until the establishment of ISRO in 1969. Initially, planned for just one year, the IGY was extended to 18 months. It started on 1 July 1957 and concluded on 31 December 1958. IGY was stretched to encompass the expected peak of solar activity that occurs once every 11 years.[338] During this period, Indian scientists and institutions across India, including Amritsar, Jodhpur, Bombay (now Mumbai), Poona (now Pune), Kodaikanal, Dehradun, Gulmarg, Delhi, Madras (now Chennai), Trivandrum (now Thiruvananthapuram) and Nainital, participated in the IGY. They were involved in studying cosmic rays, aurora, meteorology, geomagnetism, ionosphere and solar activity.

Figure 7-1 International Geophysical Year 1957-58. Credit NASA

Eventually, 76 nations took part in the various scientific programmes under the IGY, including the USSR and US, but not China, which was in a state of great upheaval following the communist takeover.[339] It was so successful that when 1 January 1959 arrived, no one wanted the idea of the IGY to end. The spirit of the IGY and the scientific collaborative alliances it helped to nurture was renamed and continued under the International Geophysical Cooperation. The IGY concentrated on understanding the Sun's 11-year cycle. A peak in solar activity on the sun's surface (prominences, flares, sunspots and coronal mass ejections) is followed by a trough, a quiet period 5.5 years later. The ICSU had defined this as the International Quiet Sun Years (IQSY), which was to commence on 1 January 1964.[340] The collaboration continued through the IQSY from 1 January 1964 to 31

December 1965 and examined the Sun during this solar minimum phase of its 11-year cycle.

A decade after the end of World War II, both the USSR and the US had developed nuclear weapons and emerged as competing global superpowers. During the political hostility of the Cold War, the IGY gave rise to a unique environment where science came before politics and facilitated genuine international collaboration. The dark shadow of national self-interest, secret projects and espionage that marked the Cold War was by and large replaced the warm and friendly environment of the IGY. Intentions of putting a satellite into space using rockets were discussed openly during public meetings between international scientists.

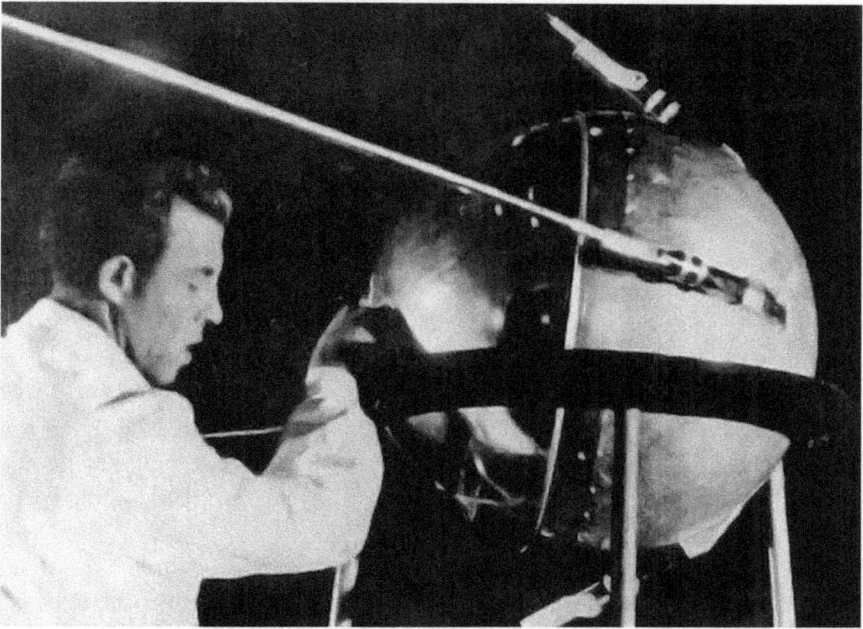

Figure 7-2 Sputnik. First Artificial Satellite Launched by the USSR.
4 October 1957. Credit NASA

In October 1954, during a public meeting of the ICSU in Rome, a resolution was adopted that stated "in view of the advanced state of present rocket techniques, thought should be given to the launching of small satellite vehicles."[341] A White House press statement on 29 July 1955 announced "plans are going forward for the launching of a small, unmanned Earth-circling satellite as part of US's participation in the IGY."[342] Had the US been on schedule, Explorer One would have been launched in November 1957.[343] But the launch was delayed because of numerous problems with the proposed launcher. The problems were not trivial and included the lack of sufficient thrust in the first stage, the need for a redesign in the second stage and the third stage it was simply too heavy.[344] Nevertheless, Explorer One played a key role in hastening the arrival of the Space Age. In hindsight, it looks as if the advance notice by the US that Explorer One would be launched by November 1957 was the motivation for the USSR's dash to launch Sputnik in October 1957.

The USSR had announced in June 1957 their intentions to launch a satellite during IGY, but it did not gain much media attention at the time.[345] Sputnik was launched by the USSR into space on 4 October 1957. On 5 October, the headline in *Pravda*, the leading newspaper in the USSR, was not about the launch of Sputnik but that 'Preparation for winter is an Urgent Task'.[346] The launch of the satellite was a technologically astonishing achievement, and the USSR was completely oblivious to the international response Sputnik gave rise to. The extraordinary political commotion in the US following the launch of Sputnik surprised even the USSR. It was this unexpected, perhaps unwarranted, response to the launch of Sputnik that directly led to the epic rivalry of the Space Race.

As the IGY drew to a close in 1958, the countries that had participated in its numerous collaborative scientific programmes sought out new national structures through which progress could be sustained. Fired by the USSR's success with Sputnik, nations around the globe were eager not to be left out of the new frontier. Before the year was out, on 29 July 1958, the US had established NASA, and on 18 December 1958, the UK had established the British National Committee for Space Research (BNCSR). China, France and Argentina were a few of the other nations that initiated their space programmes.

The possibility of Commonwealth countries collaborating in space research was raised during the COSPAR meeting held in Nice in January 1960. As a result, the Commonwealth Consultative Space Research Committee (CCSRC) was established with the intention of using sounding rockets as an instrument for investigation. A coordinated programme of sounding rockets to be launched from the UK, Canada, Australia, India and Pakistan was initiated. It was intended to provide invaluable insights into the Earth's atmosphere at 80 km. Meaningful participation for India, however, would require a complex set of ingredients. The most significant of which perhaps was the vision of an academically gifted, industrious and a well-connected individual, Homi Bhabha.

Indian National Committee for Space Research

When the Space Age arrived, India was very much a third world nation. Having led independent India for just over a decade, Nehru with bold vision directed Bhabha to ensure India was not left behind in the new era of space. In the absence of any experience of space and completely devoid of infrastructure, the first rocket launch in India would necessarily draw on foreign expertise and resources. Born amidst the Cold War and the Space Race, India's space programme was also driven by national prestige, just like in the US and USSR. Unlike them, however, the space programme in India, from the outset, was wholly non-military and targeted entirely to meet the social needs of its huge population.

In February 1962, Bhabha created INCOSPAR, the Indian counterpart to international Council for Science COSPAR. Nehru was convinced that societies based on science had improved the quality of human life throughout history and that should be the path for the newly independent India. COSPAR proposed an extensive international synoptic (broad global coverage) sounding rocket programme to investigate, through experiment, the atmosphere around the globe in 1962. Sounding rockets were used to gather data on meteorology, ionosphere, solar activity, the Earth's magnetic field and aeronomy. Aeronomy, a term more popular in the past than it is now, is the study of the atmosphere at altitudes above the point where balloons can reach but below where satellites orbit (approximately 50 km–100 km). Sounding rockets fill this gap perfectly.

Vikram Sarabhai was appointed to lead INCOSPAR. Sponsored by the United Nations (UN), INCOSPAR was guided by the shared scientific ideals of the IGY. The primary goal of INCOSPAR was for India to "promote international cooperation in space research and exploration, and in the peaceful uses of outer space."[347] The first step towards this broad and ambitious goal was to establish a team of scientists and engineers who would eventually deliver it. In the absence of domestic space experience, Sarabhai drew on his worldwide contacts to help garner the essential international support to kick-start India's space programme.

To broaden and strengthen the international connections, Sarabhai ensured that India made a strong contribution in the IGY. He organised international events and experiments in India and invited international scientists to participate. Through those connections, he arranged for Indian scientists and engineers to train abroad. Many of his experiments centred on cosmic rays, a subject he was familiar and comfortable with. In one experiment conducted in the Bolivian Andes, Indian scientists collected data on an elementary subatomic particle called Meson, working alongside scientists from Massachusetts Institute of Technology (MIT) and Japan. Mesons are subatomic particles resulting from cosmic ray interactions that occur only in the high altitudes of the Earth's atmosphere. High-altitude vantage points, like the Bolivian Alps are essential to detect them. In addition to sending Indian scientists to scientific institutions in the US and Europe, he encouraged Indian scientists and engineers based outside India to return and collaborate with the experiments hosted in India.

Praful Bhavsar (born 1926) was one of the several Indian scientists whom Sarabhai brought back to India for India's nascent space programme. Bhavsar had completed his PhD under Sarabhai's supervision in 1958 at the PRL and, in the same year, had moved to the University of Minnesota in the US to study cosmic rays. He conducted balloon experiments from Canada, carrying an X-ray spectrometer that he had built to detect high-altitude radiation and concluded that "he had a wonderful time for four years."[348]

Another scientist that Sarabhai successfully persuaded to return to India was U.R. Rao. Rao was a space scientist who had cut his scientific teeth in the 1960s working on solar wind interaction with the Earth's magnetic field before moving to MIT. There, he worked on instruments carried by the early NASA interplanetary probes, Mariner, Explorer and Pioneer. Following multiple requests from Sarabhai, Rao eventually accepted and returned to India in 1966. Initially as a professor at the PRL, later, he held several high-ranking positions in India's space programme, including the Chairman of ISRO. Writing in 2001, Rao asserted that Sarabhai's interest "naturally led him to initiate a dynamic space programme for India, in the process making the PRL the cradle of the Indian space programme."[349] By mid-1962, Sarabhai had the beginnings of a core team of scientists. Half a century later, this team has grown to 16,902 employees in ISRO.[350] For any space programme, a central capability is the capacity to build and launch rockets. INCOSPAR's first task was to identify a suitable location within India to prepare and launch rockets.

Thumba Equatorial Rocket Launch Station

While on the launch pad and immediately after launch, the rocket is a significant hazard to those in the close vicinity of the complex and the ground track of its trajectory. To mitigate this risk, launch sites are usually located in remote coastal locations with the trajectory taking the rocket over unpopulated coastal waters immediately after launch. India has thousands of miles of coastline. By another accident of geography, the Earth's magnetic equator flows over the southern tip of India making it an ideal location to undertake the scientific study of high-altitude cosmic rays and the Equatorial ElectroJet (EEJ) 100 km directly overhead. Such investigation could not be done from other places in the world.

The light from the Sun takes eight minutes to arrive on Earth, but particles emitted during a solar storm can take a couple of days. Upon arrival on Earth, these high-energy particles strip electrons from the molecules in the Earth's tenuous upper atmosphere. These now charged particles make up an eastwardly flowing electric current called EEJ. Because this phenomenon relies on solar illumination, it only occurs on the dayside of the Earth. In some ways, conceptually similar to the jet stream, it is fundamentally different. The EEJ was discovered in the 1920s.[351] It is around 400 km (248.5 miles) wide, 100 km (62 miles) high and consists not wind (air molecules) but moving charged particles thus an electric current. The EEJ flows about 100 km high over the southern tip of India.

Sarabhai had successfully invited P.R. Pisharody (1909–2002), a meteorologist working on cosmic ray research at MIT, to join the nascent INCOSPAR team. Pisharody and E.V. Chitnis, another of Sarabhai's former students, were tasked with locating a site suitable for launching rockets in India. The first potential launch site identified was Vellanathuruthu, which meant 'the sandbar of the white elephant'. Sarabhai's response, Pisharody recalled later, was unequivocal "I'll not have it here at any cost! No white elephant. The Government will not like it, the United Nations will not like it. We won't get it through. I can't. Shift it. Find another place."[352]

With input from NASA personnel, an alternative site 50 km south and 14 km north-west of Trivandrum, the state capital of Kerala, was selected.[353] It ticked all the boxes and moving to that site was a "comparatively smaller" undertaking compared to other options.[354] Called Thumba, it was named after a local plant that grew there in abundance. It was a parcel of land about 2.5 km (1.55 miles) long, squeezed between a railway line and the Indian Ocean and populated by a small fishing community. Thumba is 8.5°N of the equator[355] with up to 140 rainy days every year and temperature ranging between 20–35°C. The EEJ flowed directly overhead and sounding rockets launched from Thumba could easily collect data and provide critical insights into the Sun's 11-year cycle. To help with the assessment, a low-level flight over Thumba was arranged in 1962. Bhabha sitting in the co-pilot's seat of a Dakota and Sarabhai standing next to him got their first aerial view of what would be India's first spaceport.[356]

In 1963, six years after the launch of Sputnik, India prepared to join the small number of countries with had launched rockets into space. A formal announcement on 21 January 1963 in the Parliament of India declared that Thumba, then still a small fishing village, would become India's launch site for sounding rockets. India joined a race that had already started and, with incredible haste, acquired and redeveloped the site along with the accompanying infrastructure before the year was out. Thumba was transformed into TERLS and used to successfully launch a rocket into space from the Indian soil in November of the same year. It was a remarkably expeditious achievement given India's formidable bureaucracy and its starting point of zero experience.

The swift land acquisition was down to the personal intervention of key personnel. Homi Bhabha visited Thumba and used his significant influence to engage the central government officials very early on. K. Madhavan Nair, who at the time was the district collector of Trivandrum, acknowledges the generous support from the central and state governments, as well as the officers in charge of the local departments, which was essential in meeting the unusually tight schedule. He recalls "of all the cases of land acquisition I had to deal with during my tenure as District Collector, that associated with the Rocket Launching Station at Thumba proved to be the most difficult."[357]

Some claimed that its traditions went as far back as St. Francis Xavier's visit to South India in 1554. In the 17th century, the region had been administered by Jesuit priests from Portugal during the Portuguese colonial period. In 1858, Pope Pius IX established the Diocese of Cochin, which incorporated Thumba, following an agreement with the Portuguese government. The largest building at the centre of the land INCOSPAR sought was the church of St. Mary Magdalene. Built at the start of the 20th century, this iconic building was considered by the villagers to be the heart of their community. It was consecrated ground with a Christian cemetery at the rear.

Nair approached the Bishop of Trivandrum, the Right Reverend Dr Peter Pereira, to act as a mediator. As probably the most respected member of the community, the Bishop successfully persuaded the villagers to comply with the relocation request. During a few hectic months in the middle of 1963, the villagers were resettled in new housing in nearby Pallithura. It was not required that all families move. Some of the original families and their descendants still live along the road leading to the church. In 1968, in a speech delivered in the presence of the Prime Minister of India, Sarabhai thanked the Bishop "who spared no pains in supervising the rehabilitation of the displaced people."[358]

The facilities at Thumba for the staff members were basic. In the absence of regular transport services, during the early days at INCOSPAR staff used public train and bus services. Bicycles were used not only for travel but transport, giving rise to one of the iconic images of India's space programme depicting a rocket cone being transported on the back of a bicycle. During this 'Bicycle Era' between 1963 and 1965, there was "a single green-coloured standard van" that served to meet all the requirements of the space programme at Thumba. Bicycles were used for everything else.[359] The church and temporary sheds were used for accommodation, and a canteen at the railway station was the nearest source of catering. One of the trainees that Sarabhai sent to the US recalled that it was "a very far cry indeed from the luxury in terms of equipment and facilities which we had got used to at Washington DC and Wallops Island."[360]

At this time, India's experience with launching rockets was zero. H.G.S. Murthy, the Range Director at Thumba, recalls an apt comment from Sarabhai on the day of the first launch that described the situation in those days.[361] After inspecting the proposed launch site at Thumba, Murthy met with him and asked, "What is it you really want us to do?" Sarabhai was heard to say, "blind leading the blind". Abdul Kalam, who went on to play a significant role in the Indian space programme and later in the national politics, was in charge of Integration and Range Safety Operations for the first rocket into space from India. He recalls the unusual setting for India's first rocket launch facility "The St. Mary Magdalene church housed the first office of the Thumba Space Centre. The prayer room was my first laboratory, the Bishop's room was my design and drawing office."[362] This peculiar and unique setting was also an unlikely unifier of India's diversity; Kalam a Muslim, Sarabhai a Hindu and a Christian church.

During these early days, it was Sarabhai's infinite optimism and his ability to persuade his growing team to share in his bold vision that propelled the project. As Korolev in the USSR and von Braun in the US had recognised, managing politicians and developing public expectations was just as important as the technological challenges. Sarabhai, through the Sarabhai family reputation, had already established contacts within the higher echelons Indian government. Through his travels, initially as a student in the UK and then as a researcher in the US, he had accumulated a substantial list of influential international contacts. He was mainly preoccupied with how the UN could help India start its space programme. The UN's goal was to foster peaceful uses of space research and provide developing nations with practical training and education. In 1963, as the UN looked for an equatorial rocket launch site, India stepped forward as a willing host. With UN engagement, an Indian launch site would satisfy UN criteria for aid and be internationalised from the outset.

Figure 7-3 Former Church now a Museum with model launch vehicles in the foreground. Thumba. Credit Author

Subsequently, Thumba was operated in accordance with the principles of the UN Outer Space Committee, which were broadly defined under four categories: (i) exploration of the upper atmosphere, including neutral particle and ion composition of the ionosphere, (ii) the study of the magnetic and electric fields associated with EEJ and their time variations in relation to solar activity, (iii) study of the meteorology of the stratosphere and mesosphere and (iv) research in selected aspects of astronomy. Thumba especially suitable.[363]

In a formal ceremony at Thumba on 2 February 1968 attended by numerous international representatives, Prime Minister Indira Gandhi dedicated TERLS as an "international range for scientific research open to all UN member states". This was a conclusion to an undertaking made by her father and the Prime Minister at the time, Nehru, in 1962.[364] The UN dedication was in part a response to the support India had received, but it was also an acknowledgement of the unique role that only the UN could have played during the Cold War. The reorganisation following Sarabhai's death, Thumba was subsumed within the VSSC in 1972. In the spirit of IGY and to fulfil the UN obligations for international collaboration, during the 1960s and 70s, Thumba launched sounding rockets not only for India but also for the UK, USSR, France, Germany and the US. By 1968, TERLS received an estimated $3 million (Rs.2.25 crore) for launching sounding Rockets for its international collaborators, the USSR, US, UK. On 9 February 1975, five Petrel rockets were launched in one day to examine ionospheric winds and the EEJ.[365] Several British universities, including Sussex, Birmingham and University College, London, along with UK's meteorology office, used Petrel and Skua rockets with barium and strontium payloads to investigate high-altitude winds.

India's collaboration with the USSR was multifaceted. On 14 May 1970, a Memorandum of Understanding (MoU) was signed in Moscow between India and the USSR "to study the properties and processes which characterise the physical state of the stratosphere and mesosphere". To conduct this study, about 50–70 meteorological rockets were launched annually from Thumba during the initial period of 1971–72.[366] The time for instruments to record data during a sounding rocket launch is very short. To acquire data over a longer period, the India-USSR collaboration included the use high-altitude balloons. Between 1977 and 1981, 67 experiments in extra-atmospheric astronomy were conducted on high-altitude Indian balloons carrying telescopes provided by the USSR.[367]

Although Sriharikota on the east coast is used for all ISRO's commercial and scientific launches, Thumba within the VSSC complex still remains in operation. By mid-2015, 2,345 rockets of all types had been launched from Thumba. Today, it continues with the tradition of launching sounding rockets on the third Wednesday of each month. Currently, ISRO has three operational sounding rockets with different capacities. RH200 from Thumba can deliver a payload of 10 kg to an altitude of 75 km, RH-300 Mk2 from Thumba or Sriharikota can deliver 70 kg to an altitude of 120 km, and RH-560 can deliver 100 kg to 550 km but is only launched from Sriharikota.[368]

Pakistan's Space Agency

A day before India was officially independent, Pakistan became an independent nation. Although the US and USSR were already the undisputed leaders, other nations around the globe were starting to join the race. India took frantic first steps to advance its space technology to avoid being left behind in a relatively new arena. It is also possible that India's haste was motivated by events closer home. On 7 June 1962, a NASA-built two-stage sounding rocket carrying a sodium-vapour payload was successfully launched from Sonmiani Rocket Range in Pakistan to an altitude of about 130 km. It was called Rehbar-I and was followed two days later by Rehbar-II.[369] Located about 150 km (93.2 miles) north-west of Karachi, the Sonmiani Rocket Range and the rocket launch site were set up with the help of NASA engineers from Wallops Flight Facility.[370] Professor U.R. Rao, former Director of ISRO, however, insists that it was the other way around. India had publicly announced the intention to build a rocket launch station in early 1962, and this motivated Pakistan to launch first.[371] With NASA's help, Pakistan initiated its own sounding rocket programme in June 1962.[372]

Both India and Pakistan were part of CCSRC established in 1960. When the CCSRC met for the second time on 29 September 1961, India was represented by S. Chandrasekaran and Pakistan by Abdus Salam (1926–1996) and C.K. Raheem.[373] This was probably the last time that India's and Pakistan's space programmes were on par. Since then, while India has continued to invest in and develop its space programme, the absence of political and financial investment has stunted the growth of Pakistan's programme.

In 2017, Pakistan chose not to participate in India's South Asian Satellite where India offered the free use of a transponder for each of the SAARC member countries. The political and scientific leadership in India were driven by the shared vision of science enabling social and economic growth. This has sustained investment in the space programme since its inception. At that time, Pakistan probably shared the same vision and aspiration. NASA was supporting India and Pakistan by hosting engineers from both courtiers in Wallops for training during from 1962. One Indian engineer recalls an encounter as "we found that they were no different from us in attitude, culture or even language."[374]

Space and Upper Atmosphere Research Commission (SUPARCO), Pakistan's space agency, was founded in 1961 led by Abdus Salam. He was a joint recipient of the Nobel Prize for theoretical physics in 1979 for his work in developing the idea of the electro-weak force, unifying for the first time two of the four fundamental forces of nature, electromagnetism and the weak nuclear force. Unlike Vikram Sarabhai, Salam's roots had no connection to great wealth or social status. Nevertheless, he followed a path similar to Sarabhai's. With outstanding educational accomplishments as a young student, he attracted a scholarship to go to Cambridge and achieved a double-first in theoretical physics and mathematics. After a short spell of working in his home country in the early 1950s, he spent most of his professional life researching and teaching in the UK, US and Italy. Salam was the Chief Scientific Advisor to the Government of Pakistan and a member of the Pakistan Atomic Energy Commission when, under his recommendation, Pakistan's space agency Space and Upper Atmosphere Research Commission (SUPARCO) was established.

Figure 7-4 Space and Upper Atmosphere Research Commission. Credit SUPARCO

Under his leadership, Pakistan established a rocket launch facility and sent several of SUPARCO engineers and scientists for training at NASA installations in the US. Throughout the 1960s, both the US and USSR were investing heavily in their manned Earth observation and interplanetary exploration programmes. To help expand their geopolitical influence, numerous joint programmes were established with countries that had an interest in developing national space programmes of their own. Two of the earliest such joint ventures were between British National Committee for

Space Research (BNCSR), NASA and SUPARCO. Under this agreement, sounding rockets launched from Sonmiani would investigate the upper atmosphere (between 50–150 km) during IQSY of 1965.

The IGY was a period of high solar activity; IQSY was exactly the opposite.[375] The influential chairman of BNCSR, Harrie Massey, had travelled from the UK to attend an INCOSPAR meeting in Ahmedabad in India and was one of the first Westerners to see the initial plans for the Indian launch site at Thumba. During the same trip in September 1962, Massey also visited the Sonmiani launch site in Pakistan.[376] In 1990, with assistance from China and the USSR, SUPARCO launched its first experimental satellite, BADR-1 and BADR-B in 2001. While Pakistan has active programmes of communication and remote-sensing satellites, it has not pursued an indigenous launch vehicle capability as India has.

Today, SUPARCO has its headquarters in Karachi with ground and research stations in Islamabad, Multan, Peshawar and Lahore. Its space projects today are not as far reaching as contemplated at the outset. One notable contributor to SUPARCO was a Polish Air Force pilot, Władysław Turowicz. He was born in 1908 in Siberia, the USSR, and died in 1980 in Karachi, Pakistan. He joined the Royal Air Force in the UK and flew against the Luftwaffe over Europe during World War II. Turowicz moved to Pakistan after the War to help build Pakistan's air force and ended up as SUPARCO's administrator between 1967 and 1970. His military connections and experience in aeronautics allowed him to support Abdus Salam in developing Pakistan's early aerospace and missile technologies. In 1954, Abdus Salam left Pakistan and set up home in Europe although he continued to play a significant role in Pakistan's atomic and space programmes for many years. Salam had the intellectual capacity, vision and the desire to develop Pakistan's space ambitions comparable to that of his counterpart Sarabhai in India. He had the international clout and contacts to engage Pakistan with the UN and the collaborative scientific effort of the IGY, as India had done. But religious prejudice, factional politics and the absence of visionaries in the government prevented him from doing so.

In 1964, Salam established the International Centre for Theoretical Physics in Italy, where he had a base spanning many years. The Centre was Salam's vision to help develop the skills of scientists from developing countries, foster international collaboration and mitigate the brain drain of scientists in developing countries that would otherwise ensue. In 1997, a year after his death, the Centre was renamed the Abdus Salam Centre for Theoretical Physics. Unlike Vikram Sarabhai, who is celebrated as the father of the Indian space programme, Salam has been largely sidelined and forgotten in his own country.[377] Even though his achievements have gone unrecognised in Pakistan, many countries around the world, including India, have awarded Salam honorary doctorates in acknowledgement of his work. TIFR awarded him an honorary fellowship, and he received an honorary doctorate from Amritsar.

Salam had encouraged former ISRO chairman and space scientist Professor U.R. Rao to study cosmic rays, indirectly influencing the space programme in India.[378] Freeman Dyson (born 1923), a theoretical physicist, astronomer and mathematician, considers Salam his hero and goes on to say that he "was great as a scientist, greater as an organiser, greatest as the voice of conscience speaking for the advancement of science among the poorer two-thirds of mankind."[379]

Despite only a day's difference in gaining independence and joining the space race around the same time, India's and Pakistan's space programmes have evolved very differently. Pakistan's smaller economy and its unstable political environment seem to be the primary factors for this difference. India has a larger economy, which is better placed to provide the large investments that space programmes require, and the Indian economy has grown under secular democratic politics that has, largely prevailed since independence. ISRO in India has an annual budget of around $1 billion, whereas SUPARCO in Pakistan operates with approximately $30.6 million.[380] While India has mostly enjoyed political stability, sectarianism and military coups have regularly disrupted democracy in Pakistan resulting in economic, political and social instability. The general election of 2013 in Pakistan was the first instance when one democratically elected government handed over power to another in a peaceful transition. A precarious political and economic environment is not one where national space programme can flourish.

First Launch in India

On Thursday, 21 November 1963, as the Sun set over the southern Indian state of Kerala, an orange glow appeared high in the evening twilight sky. The spectacle was seen by viewers for whom the Sun had just set, but was still shining 100 km above. The glow came from an exploding cylinder of sodium vapour supplied by France, launched aboard a Nike-Apache rocket from the US while a computer from the USSR measured its altitude. This was India's first venture into space.[381]

By the early 1960s, investigation of the high-altitude atmosphere at the edge of space using sodium vapour delivered by rockets was well established. By photographing from different locations, the dissipation over time of the orange sodium vapour cloud, usually a column several tens of kilometres long, scientists were able to calculate the changing magnitude and direction of the winds at the range of altitudes the cloud cuts through. The idea for the first rocket launch with a sodium payload from Indian soil came out of a conversation in the US between Praful Bhavsar and Professor Jacques Blamont, the Technical Director for the National Centre for Space Studies (CNES) established in France by President Charles de Gaulle (1890–1970) in 1961.

As a post-doc fellow in 1962, Bhavsar was assisting Professor Blamont in an IGY project researching high-altitude radiation using balloons at the University of Minnesota in the US. Bhavsar arranged Blamont to meet with Sarabhai in Washington in May 1962, which led eventually to a strong friendship. This was a significant point in Indo-French history. Blamont and Sarabhai's meeting took place against the backdrop of the formal end of French control on parts of India it had ruled for centuries.[382] Half a century on, Blamont recalls fondly that "Sarabhai was playing games between different benefactors" to secure for India the essential resources for its first rocket launch.[383]

Figure 7-5 Battle of Guntur 1780. Credit Charles Hubbell

The launch site Thumba had been secured and brought into operation with assistance from the UN and in the spirit of collaboration of the IGY. Sarabhai acquired a Nike-Apache sounding rocket with launcher from

NASA and a Mi-4 range recovery helicopter, a shaking table for pre-flight tests and a Minsk-II electronic computer from the USSR. CNES from France donated the sodium payload and ground equipment, including COTAL LB Radar,[384] which had a range of 300 km for tracking rocket trajectory.[385] On 14 January 1963, NASA had announced the signing of a MoU with the DAE in India that initiated a programme of collaboration between the US and India. The key scientific objective was to understand, through experiment, the EEJ and upper-atmosphere winds from Thumba, where the geomagnetic equator was overhead. The agreement with NASA included nine Nike-Apache and four Nike-Cajun launch vehicles, along with ground-launching and tracking equipment on an on-loan basis. The US would also provide training to the Indian engineers who would subsequently perform the launches in India. It was up to India to supply four sodium-vapour release payloads, photographic equipment, launch site and facilities, personnel and supporting meteorological data prior to the launch.[386]

As part of the MoU, several Indian engineers received training in NASA centres (including Goddard Spaceflight Centre and Wallops Flight Facility) in the US to help in the preparation and eventual rocket launch from India.[387] R. Aravamudan was a member of the first batch of Indian scientists to visit NASA, and he recalls that "to begin with, we were a group of four: Ramakrishna Rao, Kale, Prakash Rao and me. After about three months, H.G.S. Murthy, Easwaradas and Abdul Kalam joined us."[388]

Figure 7-6 Nike-Apache at Thumba. Credit Professor Praful Bhavsar

In early 1963, Abdul Kalam was surprised to see an intriguing painting hanging in the reception of Wallops Flight Facility. It was a painting depicting the Battle of Guntur, which reminded him of India's long tradition in rocketry. In his biography, Kalam points out that the painting depicted "a fact forgotten in Tipu's own country but commemorated here on the other side of the planet."[389] Except for Stephen Smith's experiments with rocket mail in the 1930s, no rocket development had taken place in India since Tipu almost two centuries earlier. The Nike-Apache was a 7.5 m long unguided two-stage solid-propellant rocket reconstructed from parts originally developed for the American surface to air missile programme. It was, in fact, two distinct rockets one on top of another. A lower stage (0.5 m diameter 3.8-m-long) Apache was fitted to a slimmer upper stage (0.16m diameter and a 3.7-m-long) Nike.[390] Following ignition, the Apache would burn for 3.4 seconds followed by a gap of 16.5 seconds before the second-stage Nike fired for 6.3 seconds. Collectively, that generated sufficient thrust to deliver a payload of around 50 kg to over 200 km when launched at an elevation angle of 80°.[391] The Nike-Apache rocket was one of NASA's most efficient and reliable rockets and was retired eventually in 1980.

Figure 7-7 R. Aravamudan (right) and A.P.J Abdul Kalam (left) integrating payload

On 9 September 1963, NASA launched a sounding rocket taking a 35-kg payload of instruments 106 miles (170.5 km) during a 6-minute-and-40-second flight. This launch of a Nike-Apache from its Wallops Flight Facility off the Virginian coast was in "preparation for the EEJ programme to be conducted from India."[392] Two months later, India launched its first rocket.

Inaugural events, by definition, lack previous history and traditionally attract unexpected problems. NASA delivered the first Nike-Apache rocket to the Indian capital by US military air transport. On its 2,000-km (1,242.7

miles) journey by road south to Kerala, practically the full length of the country, the truck broke down. Jacques Blamont contacted Abdul Kalam in the US, who was receiving training at NASA. He had agreed to provide the payload, a sodium ejector fabricated in his laboratory.[393] Blamont asked Kalam to hand-deliver it from his office in Minnesota to India. Upon arrival in India, the sodium cylinder, an explosive device, was impounded under the Indian Explosives regulations. Eventually, both the rocket and its payload arrived in Thumba, where the two were to be integrated.

During payload integration, it was discovered that the sodium cylinder, the payload, did not fit inside the rocket. The French had been using the metric system, and the Americans used the imperial. The payload diameter was larger by a few millimetres and would require machining. Bhavsar, the project scientist, recalled "The sodium-vapour package was a pyrotechnic. It would explode if the temperature exceeded 800°C." Machining was too risky, so a skilled mechanic Mr Ratilal Panchal was brought to Thumba from Ahmedabad to do the job by hand. Bhavsar recalls that to assure the mechanic of their confidence, "Kalam and I sat with the mechanic while he manually machined the cylinder for comfort."[394] The church was the only large, solid construction within the land acquired by INCOSPAR, and much of the work, including payload integration, was done there. Rocket integration and safety were Abdul Kalam's responsibilities. On the day of the launch, the hydraulic system of the crane on the truck transporting the assembled rocket from the church to the launch platform sprung a leak. With launch time approaching, the rocket was lifted in place physically "using our collective muscle power."[395] Five minutes prior to the launch, a loud alarm was sounded to instruct all personnel to return to the blockhouse. However, P. Kale recalls that two minutes after the alarm, there was a "man still adjusting the launcher", who had to be instructed to evacuate to the safety of the blockhouse.[396]

Two minutes after sunset on Thursday, 21 November 1963, at 6:23 pm, the rocket was launched by an electronic trigger. The first stage ignited and powered the rocket for 3.5 seconds; after a delay of 16.5 seconds, the second-stage Apache ignited and burned for 6.5 seconds. The separation between the two stages occurred naturally by differential forces.[397] As the second stage fired, its acceleration pulled it away from the spent first stage below. Several seconds before reaching the maximum altitude, the sodium-vapour payload was released during the ascent initiating a vertical column of bright sodium vapour. About four minutes after the launch, the rocket achieved its maximum altitude of about 140 km and started its free fall descent into the Arabian Sea. The launch was a dramatic relief and a complete success for the Indian scientists and engineers on their first mission. It was observed by many Keralites, and the Kerala Assembly paused to view the spectacle.[398]

The primary purpose of this flight was to measure the atmospheric winds and turbulence in the Earth's atmosphere between 40 and 200 km. Space is now internationally agreed to begin from a 100-km altitude above the Earth's surface, although technically, Earth's atmosphere continues for several hundred kilometres beyond. The payload was ignited during the ascent leaving a vertical column of sodium vapour. The rate and direction of the subsequent dispersal of this column would help scientists to understand the physics of the upper atmosphere.

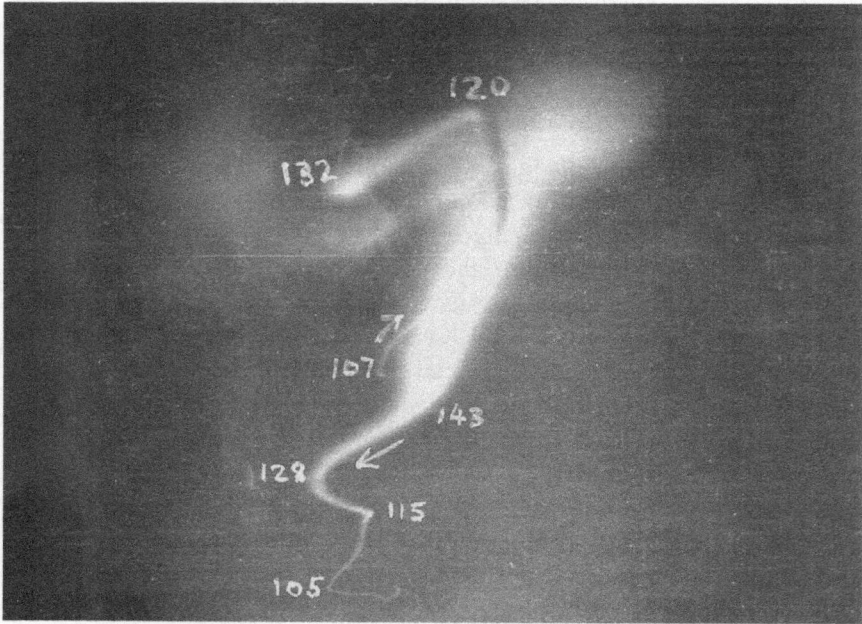

Figure 7-8 Sodium-Vapour Trail. 21 November 1963. Credit *Professor* Praful Bhavsar

As the rocket did not carry a radio transmitter, all the science data were to be captured by radar, photographic camera and visual observations from the ground. Four teams of photographers located in four pre-selected locations around the launch site in Kanyakumari, Palayamkottai, Kodaikanal and Kottayam were to record a series of images that would show the speed, direction, turbulence and the rate of diffusion of the sodium vapour once it was released at its designated altitude. The sky was completely overcast at Kodaikanal and partially so at Kottayam at the time of the launch. Useful pictures were taken from the other two locations.[399] Sarabhai concluded "Putting together information from these stations where the cloud was photographed against the background of stars, it will be possible to gain fresh insight into the complicated problems connected with the Electrojet."[400] The success of the flight was due to critical contributions from the US, the USSR and France. However, except for two French and two American engineers who were available for advice, only Indian scientists and engineers were involved in the launch operations. None had any previous experience of launching rockets into space. International media reports of this achievement were, however, lost amidst another dramatic event that took place on the other side of the world on the following day: the assassination of the US President Kennedy on 22 November 1963.

Figure 7-9 Vikram Sarabhai and Jacques Blamont in Kanyakumari. January 1964.
Credit Jacques Blamont

Indian scientists still in training in the US heard the news of the successful launch at Thumba over the Wallops Flight Facility public address system. News of the President's assassination was announced on the same system. One of the Indian engineers recalled "We were informed of the assassination of President Kennedy at Dallas. What struck us then, as most remarkable, was that the Wallops Range and NASA continued to work as usual and no holiday was declared."[401]

Gradually, the launch of sounding rockets at Thumba became a routine. Not all launches went according to plan. Some exploded on the launch pad. One travelled horizontally and skimmed the sea, temporarily becoming airborne again. Another headed inland, instead of out to the sea, and was recovered from near the Trivandrum Engineering College. India had defined the term 'non-aligned movement' and chose to chart its own course between the US and USSR.[402] By selecting to remain neutral, India had selected to balance its trade, defence and scientific affiliations with both the West and the USSR. In this unique position, Thumba became a neutral place where individuals from different countries could meet. Scientists from nations where relationships were traditionally cold, if not hostile, could pursue friendly, apolitical and productive scientific relationships in Thumba under the UN auspices. Jacques Blamont, who was in attendance during the UN dedication ceremony, recalls that "COSPAR provided the only place in the world where such collaboration, motivated by the noble spirit of IGY could take place."[403] Although COSPAR still exists, since the demise of the Cold War, the role it now serves is no longer as profound as it once was.

Type of Rocket	Total
Dragon	1
Nike-Apache*	36
Centaure I	10
Centaure II-A	2
Judi-Dart*	36
Boosted Arcas I*	6
Skua I	10
Skua II (T)	1
Nike-Tomahawk*	3
Test Rocket (2.7')	29
Rohini 75	15
Rohini 75 Test Rocket	22
Rohini two-stage	1
Menaka I	13
Rohini 100	13
Rohini 125	2
Fibreglass Rocket	5

Table 7-1 The 205 Rockets Launched from Thumba between 21 November 1963 and 31 March 1970. Credit Ashok Maharajah. * US supplied.

Sarabhai's scientific interest in cosmic rays and Thumba's central research theme of investigating the EEJ were far removed from the primary objective of national economic development. E.V. Chitnis, who played a key role in initially selecting Thumba as the launch site and leading the operations there, wrote "on one side you had the technology, but at the same time, whatever you were doing must be relevant to the country's needs. This was the basic idea of Sarabhai when he started the space programme."[404]

Sarabhai's widow recalls "When Vikram started India's space programme, he always spoke of using space for national development."[405]

One Village One Television: SITE

Satellite television is a standard fixture in most middle-class family homes of the 21st century, but it was first made available to some of the most uneducated and poorest people on the planet in 1975. Although he did not live to see it, Vikram Sarabhai established the vision and much of the groundwork for the Satellite Instructional Television Experiment (SITE) programme to beam educational TV programmes to communal television sets in rural Indian villages. The potential for television to entertain had long been established. Could it also be used for education? SITE epitomised Sarabhai's vision of how the space programme and technologies could be exploited for the benefit of the ordinary people of India. By the 1960s, most developed and developing countries had installed a robust national infrastructure of TV studios for making programmes and a national network of transmitters for broadcasting them to the expanding domestic television market. The range of a television transmitter then was only around 40 km and by the late 1960s, India had just one. To cover the vast landmass of India, an overwhelmingly large number of transmitters would be required.

A TV transmitter in the sky was the ideal solution to cover large swathes of the Indian population. Sarabhai also saw an opportunity for developing nations to leap ahead by skipping a network of land-based transmitters and going straight to broadcasting from satellites. Writing in 1971, almost five years before SITE became operational, Arthur C. Clarke, who had come up with the idea of communication satellites in his celebrated paper Extra-Terrestrial Relays in 1945, said "It can be difficult for those nations which have taken a century and a half to slog from the semaphore to satellite to appreciate that a few hundred pounds in orbit can now replace the continent-wide networks of microwave towers, coaxial cables and ground transmitters that have been constructed during the last generation."[406] By chance, Arthur C. Clarke would participate in SITE because his home in Sri Lanka since 1956, fell inside the proposed footprint of the communication satellite.[407]

On 26 January 1967, India conducted one of the earliest trials on the value of delivering education via television to 80 villages around New Delhi. New Delhi was chosen because it was the only city in the country to have a TV transmitter at the time.[408] Known as the Krishi Darshan Programme, it was

155

made in collaboration with All India Radio (AIR), Indian Agricultural Research Institute and the Delhi Administration. Sarabhai emphasised its importance as an investment, not overhead. Post-trial surveys concluded that an average of 400 individuals, mostly from farming communities, benefited from the programmes. Villagers would congregate around a single 30-cm television in open-air tele-clubs. The Sun sets around seven in the evening, and except during the monsoon season, the weather permitted open-air viewing throughout most of the year. The topics covered by the programmes included weed control, fertiliser, high-yield seeds and a few minutes of song and dance to "sweeten the education pill."[409] Exploiting the same concept, in 1969, the British government established the Open University (OU) with the intention of delivering higher education using television broadcast as the primary method of communication. The principle of educating large dispersed audiences through television had been established. In India, millions of lives could be transformed by relatively basic education.

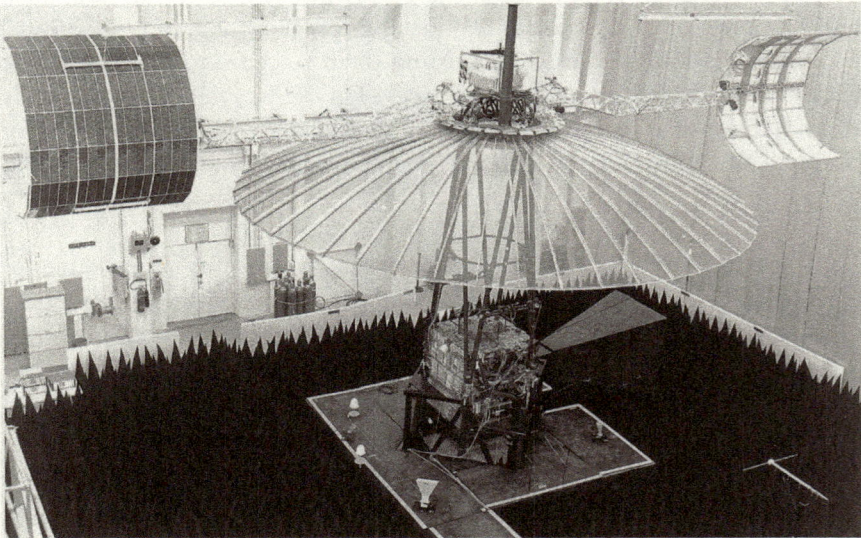

Figure 7-10 NASA's Applications Technology Satellite (ATS-6). Credit NASA

SITE was ambitious. Its goal was to identify a potential solution to help end the cycle of illiteracy, malnutrition and disease caused by poverty. As Clarke put it "Illiteracy, ignorance and superstition are not merely the results of poverty, they are part of its cause."[410] SITE would distribute adult educational training programmes for reading, writing, arithmetic, agricultural methods, hygiene and family planning. It included educational programmes for school children between the age of 5 and 12 during term time and teacher training programmes during holidays. The TV programmes were recorded in SITE studios in New Delhi, Cuttack and Bombay and delivered via magnetic tape to the Earth Stations at Ahmedabad and New Delhi for uplink to the orbiting satellite. Four hours of programmes (1.5 hours in the morning for school children and 2.5 hours in the evening for adults) were transmitted in two slots each day. Participating villages were supplied with modified televisions and a 3-m satellite dish.

SITE delivered educational programmes for one year to around 5,000 villages in six separate states (Andhra Pradesh, Karnataka, Bihar, Madhya Pradesh, Orissa and Rajasthan), in initially four different languages (Hindi, Oriya, Teledu and Kannada) from a satellite parked 36,000 km above the Earth's equator over Kenya in Africa. In addition to NASA and the Government of India, it was supported by various international agencies, such as the United Nations Development Plan (UNDP), United Nations Educational, Scientific and Cultural Organisation (UNESCO), United Nations International Children's Emergency Fund (UNICEF) and International Telecommunications Union (ITU). In the end, the impact of SITE was not restricted to raising literacy in rural India, but it demonstrated the potential of future satellite communication across the planet.

SITE Infrastructure

The satellite used for SITE was not Indian, but American. In the mid-60s, NASA initiated its Applications Technology Satellite (ATS) programme. The plan involved launching seven satellites targeting specific applications, although the last one was cancelled and only six were actually launched. This series of satellites would, in part, be the first implementation of Arthur C. Clarke's idea of using satellites in geostationary orbit (GEO) to transmit television signals to Earth. ATS-6, originally designated ATS-F, was a large satellite with a huge 9-m communication dish with a new innovative three-axis stabilisation offering precision-pointing capability.[411]

During a visit to NASA in December 1964, E.V. Chitnis signed a letter of understanding to participate in the ATS series.[412] Understanding India's requirement for an Earth Station, Chitnis acquired funding from the UN to help build what is today the 14-m diameter Ahmedabad Earth Station (known initially as the Experimental Satellite Communication Earth Station). It was constructed in 87 days and was first tested with ATS-2 in August 1967.[413]

In the first year of its operation following its launch on 30 May 1974, ATS-6 was scheduled to conduct experiments within the US. They included investigating three key areas: (i) conducting Health, Education and Telecommunication (HET) experiments using one-to-one low-cost widely dispersed terminals in the US, (ii) providing television re-transmit capabilities at seven of those one-to-one terminals and testing satellite communications for video seminars on continuing education and (iii) trial telemedicine for outpatient and joint consultation clinics over four remote regions of the US, Appalachia, Rocky Mountain states, Pacific Northwest and Alaska.[414] The medical institutions involved were located in urban cities, like Washington, and remote areas of Alaska, Montana and Idaho.

The ultimate goal was to increase the flow of knowledge between practitioners and the University, broaden educational opportunities and facilitate the study of medicine by physicians and instructors in remote areas. During a three-week period between December 1974 and January 1975, when extreme weather cut-off Alaska from the outside world, ATS-3 and ATS-6 remained the only means of communication.

Fortunately for Sarabhai and India, NASA's objectives for the ATS series included field-testing the concept of broadcasting television programmes to terrestrial receivers via satellite. The orbital characteristics of ATS-6 had already defined a shortlist of three countries that were geographically convenient for field-testing, China, Brazil and India. A satellite transmitter at 36,000 km is a far more convenient solution for small villages dispersed over a large area. Brazil was not suitable because its population was mainly concentrated in a few large cities, which were conveniently served by traditional terrestrial transmitters. China was not suitable from a political perspective, and that left India.

Engaging India also served the American objective to "channel Indian resources down the path of civilian technologies."[415] The experience of Krishi Darshan and the 14-m steerable Earth Station built in Ahmedabad also helped India's case. But there was a problem. The US had sought to set up radio transmitters in India to carry its national station Voice of America, but India had refused. To avoid the potential for another snub, the request had to come from India to the US, something that Sarabhai eagerly accepted.[416]

By the late 1960s, the ATS project and the applications it could host had become much clearer. Three weeks before the Apollo 11 crew, Neil Armstrong (1930-2012), Buzz Aldrin (born 1930) and Michael Collins (born 1930), arrived in Bombay as part of their goodwill world tour, an agreement was formally signed on 18 September 1969 between NASA Administrator Dr Thomas O. Paine (1921–1992) and Vikram Sarabhai, representing the DAE and the Government of India. It was a remarkable deal. India got access for a year to an American satellite costing $180 million (Rs.135 crore) with a then projected reliability lifetime of two years and a mission goal of five years.

ATS-6 was the largest and most powerful communication satellite at the time. NASA budget cuts announced on 27 July 1970 delayed the SITE programme from 1972 to 1974.[417] It was finally launched on 30 May 1974. The orbit insertion went better than planned and saved fuel allowing the original 2-year lifetime to be extended to the full 5-year mission goal. Modern satellite TV dishes are small, but it was not always like that. In 1975, before high-performance digital electronics, satellite TV dishes had to be big. From 36,000 km above the Earth's equator, ATS-6's large 9-m dish generated a footprint that covered the whole of India. A 3-m dish in the footprint could receive a TV signal. At that time, a 3-m dish could easily be constructed from chicken wire by unskilled labour. It brought the cost of reception equipment down from a few thousand dollars to a few hundred.

Figure 7-11 Vikram Sarabhai and NASA Administrator Thomas O. Paine Signing the SITE Agreement. 18 September 1968. Credit NASA

According to the 1969 deal, NASA would loan ATS-6 to India for the SITE project for one year. India, however, needed an Earth Station, a large dish that could be used to send the TV programmes directly to the satellite in orbit for re-transmission back to Earth from its vantage point. A site near Pune was selected for a large steerable antenna. At the time, Indian companies had no experience of building such structures (the 14-m antenna in Ahmedabad was constructed largely with Western technologies and engineers), so the Indian government commissioned Canadian companies to build it. Sarabhai did not agree. According to Amrita Shah, Sarabhai's biographer, his response served to highlight his passionate advocacy for Indian self-reliance "How will the Indians experiment if not in India?"

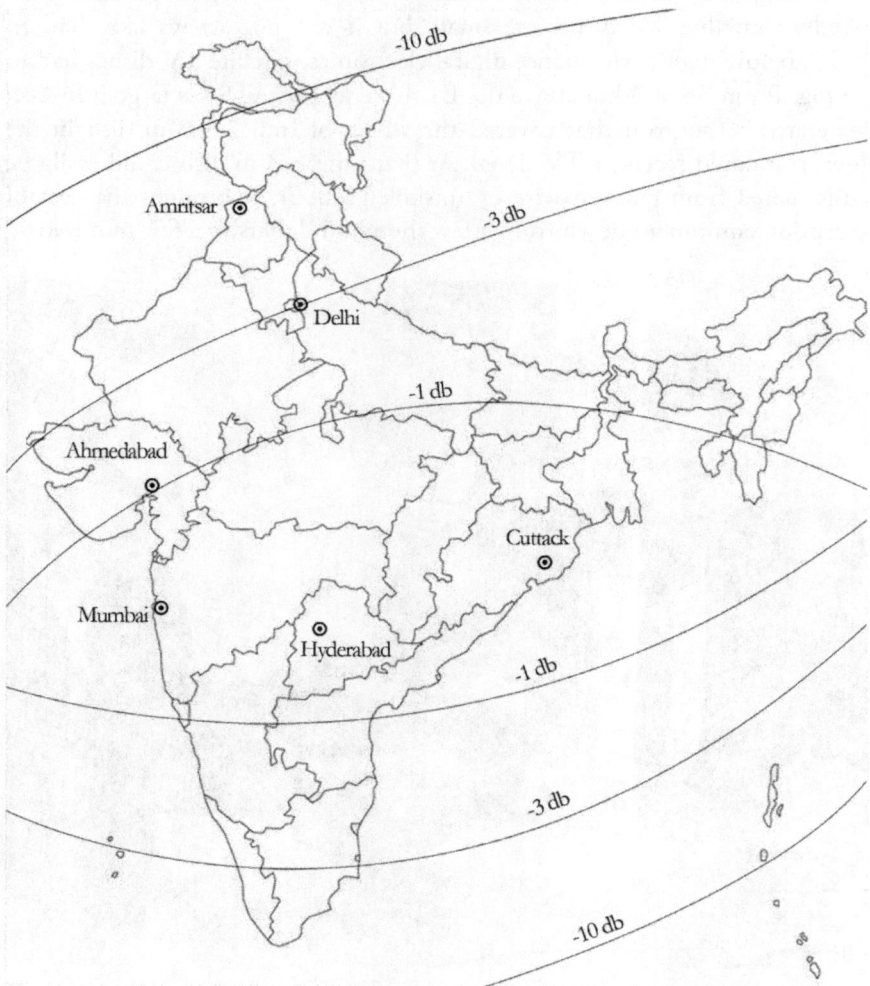

Figure 7-12 ATS-6 Footprint over India. Credit UNESCO

In a rare outburst, he expressed his frustration with the overwhelming bureaucracy "It is all very well for me to hold forth and have a design and vision for India but it has to be translated to this rotten system of the Government of India and its bureaucracy."[418] The government relented, and in October 1969, the Arvi Earth Station (later renamed Vikram Earth Station) opened ahead of schedule. The 29.5-m diameter parabolic reflector antenna was constructed almost entirely within India with the help of a Tata company, TELCO. In the process, India saved $800,000 (Rs. 600 crore) and raised the international profile of both Sarabhai and the Indian space programme.[419]

Prior to SITE, ATS-6 was scheduled to participate in two other experiments. The first, immediately after launch, was designed to test ATS-6 as a communication satellite for remote communities in the US was limited to 121 terminals with a direct connection to ATS-6. A year later, the experiment in India was similar but on an altogether larger scale. ATS-6 was also assigned to participate in the Apollo-Soyuz Test Project between 15 and 24 July 1975. Initially, ATS-6 was in geosynchronous orbit (GSO) at the equator directly overhead the city of Kansas in the US. In preparation for SITE on 20 May 1975, NASA engineers moved it halfway around the world from longitude 94° W to 35° E over Africa from where it had the required coverage over India.[420]

The Apollo-Soyuz Test Project was the first joint manned space programme between the US and the USSR. The mission involved an Apollo spacecraft and a Soviet Spacecraft (Soyuz 19) docking in orbit 250 km above the Earth. The two commanders, Tom Stafford (born 1930) and Alexei Leonov (born 1934), shook hands in the hatch connecting the two spacecraft. It was widely seen as a formal end to the Space Race that had started with the launch of Sputnik in 1957. The ATS-6 satellite covered 130 of the 138 Apollo-Soyuz Earth orbits to relay telemetry, data, voice and TV between the Apollo-Soyuz spacecraft and the USSR and US.[421] This was the first time that an orbiting satellite was used as a relay between orbiting manned spacecraft. A few days after the end of this historic mission, ATS-6 was setup for the SITE programme, which commenced on 1 August 1975 and formally continued until 30 July 1976.

The ground segment of the SITE infrastructure included four Earth Stations, Ahmedabad, Delhi, Amritsar and Nagpur. The primary station in Ahmedabad with a 14-m antenna was used to transmit the television programmes prepared by the SITE studios in Bombay, Cuttack or New Delhi to ATS-6 at 6 GHz for ATS-6 to re-transmit back to Earth at 860 MHz, which could then be picked up directly by 2,000 villages with their

3-m receivers. A further 3,000 villages received the same signal via terrestrial re-transmission using the standard very high-frequency signal via TV masts. Each transmission carried two audio channels facilitating at least two major languages to cater for the diversity of languages among the participating Indian states. With additional microwave relays, the same signal was redistributed to many other villages. The 10-m antenna in New Delhi was responsible for uploading content from the nation's capital, as well as for fulfilling the role of a backup for the primary station in Ahmedabad. The ground station located in Amritsar was not used for uplink (Earth to satellite transmission); it was used for reception (satellite to Earth) only. The Nagpur ground station was fitted with a unique communications beacon that transmitted a signal up to ATS-6 to help it to recover in the eventuality that it lost attitude control.[422]

Figure 7-13 Chicken Wire Mesh Antenna and Television Used for SITE. Credit ISRO

Many Indian homes today have multiple large high-definition flat-screen televisions in a single house. During the SITE programme, a single small communal cathode ray tube television, along with a 3-m dish, was typically available in one village for use by everyone. A custodian was charged with the responsibility of ensuring that the TV and the dish was available at the time of programme transmission. Groups larger than a typical cinema audience would participate. An audience of 1,000 viewing one television was common, but 4,000 was the highest reported. Local conditions (full moon, cultural rituals, festivals, harvest, weather and the time of year) determined the actual attendance.

Howard Galloway, a NASA representative based in Ahmedabad for the duration of the SITE experiment, filed weekly reports on SITE's progress.[423] For one of his reports (16–27 October 1975), he visited 29 villages, including Kapurwala, Hasanpura-Was, Nevta, Charenwala, Iuniywas, Bhavgarh, Bandhya, Dantli, Goner, Achrol and Dhand. He documented problems experienced by the villagers. This included unavailability of electricity or TV sets that did not work but found that in 89% of the visits, the system worked successfully. His reports included the first-hand testimony of those who had watched the programmes, as well as the SITE instructors, teachers and students. The novelty of the teaching approach using the television was particularly well received. He reported "There can be no two opinions about this. The TV is by far the most superior method that we have". The farmers benefited from the agricultural programmes and started using the right kind and quantity of fertilisers and also adopted some improved agricultural practices. One villager, after viewing a programme on pesticides, said that if he had seen that last year, he could have saved his crop. An ISRO SITE assessment recorded similar success "Gain in knowledge and attitude per respondent was the highest in the field of agriculture followed by family planning, animal husbandry and health."[424]

In November 1975, Galloway's report points to the enthusiasm with which the programmes were received by the villagers "Without exception, in all of the villages that we visited, the villagers sat quietly on the ground, on walls, on roofs, in trees, etc., with no talking and with an intense concentration on the screen. This behaviour was not exhibited for my benefit. I know this because until my photoflash went off, the folk were not aware of my presence."[425] SITE introduced change to family and cultural traditions. Women came out of their homes to view the programmes. Cooking and family meal times were adjusted to accommodate the TV schedule. Watching SITE programmes also introduced a social change. Prior to SITE, most social interactions occurred when people went to the local market. TV provided a new source of entertainment and an opportunity to meet friends and family. Programmes on smallpox awareness increased demands for inoculations. Drunkenness and brawls that had been common also decreased. The use of a communal television set introduced an unintentional consequence: it "broke down the age-old caste distinctions."[426] NASA's representative in India also reported a perhaps more profound impact "that some parents did not want their children to become educated in schools because such an educated person was lost to farming. If a family had lots of children, they might waste one on schooling. Hopefully, this seems to be changing."[427]

E.V. Chitnis later took charge of the SITE programme and noted more significant benefits, which potentially had more long-lasting outcomes. In 1983, Chitnis assessed the value of the SITE programme and included the following as some of the more interesting observations:[428]

- The majority of SITE viewers were young men (15–24 years), middle-aged women (25–34 years) and children (below 15 years).

- In two sessions, 48,000 teachers were trained.

- SITE demonstrated that it is possible to teach and instruct illiterate men and women.

- SITE attracted non-media participants. Before SITE, 30% of the males and 63% females in villages had no contact with mass media. SITE reduced these percentages to 10% and 19% for males and females, respectively.

- Small farmers, marginal farmers, landless labourers, illiterates living in thatched mud houses were the majority of people in the SITE audience. The bicycle was their most prized possession.

- The evening SITE transmission consisted of 28% entertainment, 14% agriculture, 14% health and nutrition, 10% national awareness, 10% social problems and 15% news/information and 9% for the rest.

One of the shortcomings that Galloway reported in September 1975 highlights the technological revolution that SITE had introduced. He notes that "most of the villagers have never seen TV or cinema before. They believe that an actor is a person that he portrays. Therefore, when they see two different stories on the same night, with the same actor in different roles, they become confused". Overall, SITE was an outstanding success. A common plea from villagers, Galloway reported, was "please extend SITE by 1 or 2 or more years". Probably, the only group who were not supportive were the school teachers because they felt threatened by the overwhelming success of SITE. The experiment was not extended.

In January 1976, with SITE in full swing, India hosted a 2-week winter school sponsored by the UN in association with UNESCO to share the concepts and technology behind SITE. Seventeen participants from developing countries attended. They included Bolivia, Egypt, Indonesia, Iran, Iraq, Kenya, Kuwait, Malaysia, Nigeria, Pakistan, Philippines, Sudan,

Thailand, Turkey and Tunisia. The Sri Lanka-based Arthur C. Clarke also attended the winter school. Of the 2,400 direct receiving stations, one was made available by ISRO at Clarke's residence.[429] In the 21st century, millions around the globe enjoy the luxury of a satellite-connected television in their living rooms. In 1975, Clarke's house in Sri Lanka was the only private residence with direct satellite reception. Despite being at the southern end of the ATS-6 footprint, he reported good reception.[430]

Despite a large national commitment in effort and resources, the US made ATS-6 available to India for a year. What was in it for the US? Arnold Frutkin (born 1918), Assistant Administrator for International Affairs for NASA, noted in 2002 that SITE was "not only an educational lift to India and demonstrated what such a satellite could do, but it brought money back into the US, through commercial contracts for satellites for a number of years."[99] This was a reference to the satellites of the Indian National Satellite System (INSAT) that India purchased from Ford Aerospace and Communications Corporation in the wake of SITE in the 1980s. Apart from the technical and economic benefits, political advantage was probably the most significant. At a time when the USSR's activities were blanketed in secrecy, SITE reinforced the US's transparency. It also initiated a new market in aerospace products which continue to improve the balance of US trade.[100] Using space technology to get a foot in the door, the US could then achieve its broader, global political objective of promoting democracy.

◄ ◇ ►

Chapter Eight
Inside the Indian Space Research Organisation

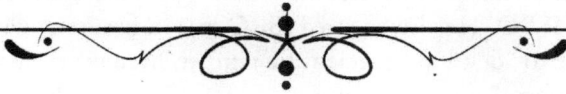

⸎

I n just over 5 decades, India has developed from the ground-up the capability to deliver to its 1.3 billion citizens the modern first-world services that can only be supplied from space. With a beginning epitomised by images of a rocket cone on a bicycle and a satellite on a bullock cart, today India is one of 11 nations that have developed an operational space industry[431], delivering a variety of services to enhance the quality of life for most of its citizens. The growing quality and scale of services include satellite-based search and rescue, tsunami warnings, navigation, television content, communication, scientific exploration and meteorological and remote-sensing services. By the end of the 20th century, ISRO had developed the domestic capacity to build satellites and launch them to orbit and had established the sophisticated ground infrastructure necessary to operate them. The national space programme first imagined by its founders was in place.

The space programme in India had its origins in the 1962 INCOSPAR, inspired by the 1958 COSPAR established by the International Council for Science. INCOSPAR became ISRO and formally came into its current form in 1972 under a reorganisation driven by its second chairman, Satish Dhawan, who took over after Vikram Sarabhai's sudden death.[432] Not only was ISRO's headquarters moved from Ahmedabad to Bangalore (now Bengaluru), but the delinking of India's space and nuclear programmes proposed before Vikram's death was also implemented. ISRO was moved from the DAE the new Department of Space.[433] Under this new arrangement, ISRO's chairman was also required to hold the roles of the secretary to DOS and the chairman of the Space Commission, thereby reporting to the highest political office. The DOS through the Space Commission defined the space policy that ISRO implemented.

Organisationally, this structure gave the ISRO chairman direct access to the Prime Minister, which "was crucial in bringing policy-making and executive functions under one head," reducing the frustrating and time-consuming procedures for which the Indian Civil Service is infamous.[434] The constitution used by the DOS for ISRO was modelled on the concise one-page constitution designed by Homi Bhabha for the AEC.[435] Dhawan oversaw the building of India's first satellite Aryabhata in 1975 and the Satellite Launch Vehicle (SLV-3) in 1980 and firmly implanted Sarabhai's original vision of national development through space technology into the very fabric of ISRO. He also ensured ISRO was in harmony with Jawaharlal Nehru's desire to "develop the scientific temper, humanism and the spirit of inquiry and reform".

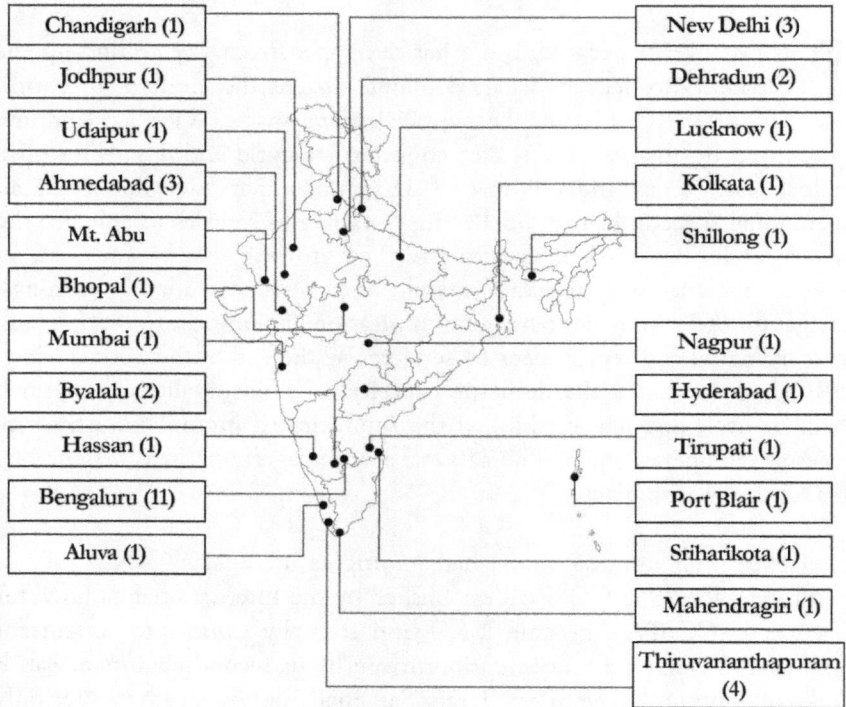

Chandigarh (1)		New Delhi (3)
Jodhpur (1)		Dehradun (2)
Udaipur (1)		Lucknow (1)
Ahmedabad (3)		Kolkata (1)
Mt. Abu		Shillong (1)
Bhopal (1)		
Mumbai (1)		Nagpur (1)
Byalalu (2)		Hyderabad (1)
Hassan (1)		Tirupati (1)
Bengaluru (11)		Port Blair (1)
Aluva (1)		Sriharikota (1)
		Mahendragiri (1)
		Thiruvananthapuram (4)

Figure 8-1 ISRO Centres Providing Scientific, Technical and Administrative Support across India. Credit ISRO

While the DOS primarily serves an administrative role, and is charged with managing the finances, the Space Commission made up of a dozen representatives from other departments of the government, such as finance and defence, and two individuals (scientists, engineers, or academics) from outside the government provide an element of independent oversight. The Space Commission's role is to scrutinise ISRO's proposals for feasibility and compliance with its objectives and government policy. It is a "very powerful body, which has immeasurably helped in evolving and executing a healthy space programme in India with total flexibility, power and full accountability."[436] For example, ISRO's mission to the Moon in 2008, Chandrayaan-1 triggered a lively and lengthy debate because its science objectives were difficult to reconcile with ISRO's vision to harness space technology for national development.[437]

Since its establishment as INCOSPAR in 1962 under India's first Prime Minister, Jawaharlal Nehru, ISRO has evolved in its scale and scope. Today, ISRO employs a 16,000-strong workforce spread across 42 separate centres located across India's 29 states and seven union territories. It has an annual budget of $1.2 billon (Rs. 8,020 crore) and the vision "to harness space technology for national development while pursuing space science research and planetary exploration". ISRO uses this infrastructure to design, build and operate satellites for scientific research, remote sensing, communication, navigation, meteorology, search and rescue and social applications.

Figure 8-2 ISRO Headquarters in Bengaluru. Credit Author

Interestingly, ISRO with a budget of $1.2 billion (Rs. 8,020 crore) and NASA with a budget of $19.5 billion (Rs. 130,000 crore) have about the same number of staff, 16,000 and 18,000, respectively. NASA outsources most of its work to third parties, such as Boeing, Lockheed Martin, Orbital ATK and SpaceX whereas ISRO has struggled to outsource,[438] in part because India does not yet have a sufficiently mature private sector capable of meeting the highly specialised needs for space-qualified components and systems.[439]

Vikram Sarabhai Space Centre

VSSC, located in Thumba 14 km north-west of Thiruvananthapuram in Kerala, is ISRO's cradle. During the 1960s, the site around Thumba grew as INCOSPAR embarked on its long journey to fulfil Sarabhai's vision of an India with an indigenous space programme capable of pushing national development. The number of buildings increased to support additional functions, such as a facility for launch vehicle manufacturing, workshops for developing propellant and laboratories for vehicle testing and integration. VSSC was established in its current form after Sarabhai's death in 1971 by amalgamating the various workshops, activities, teams and laboratories that had accumulated over the first decade of India's space programme. Employing over 5,000 people, VSSC is ISRO's largest centre.[440]

Figure 8-3 Vikram Sarabhai Space Centre. Credit ISRO

After it was founded in 1962, TERLS, which grew into VSSC, was used primarily for launching sounding rockets, a tradition that persists, albeit with reduced intensity. It was established with the aid of the UN and, in February 1968, was formally dedicated as a UN site with a focus on international collaboration. Today, VSSC sprawls over an extended area and encompasses the original Thumba launch site and the church of St. Mary Magdalene used as headquarters in 1962. The church has now been converted into a museum housing many of ISRO's space artefacts. The central office today is located in a modern building, high on a hill overlooking the Arabian Sea. While the Satish Dhawan Space Centre or Sriharikota (also known as SHAR or SDSC-SHAR but most commonly known as Sriharikota) is the centre of attention on a launch day, most of the production and preparation in the weeks leading up to a launch take place at VSSC, 800 km southwest from Sriharikota.

Given that both VSSC and Sriharikota are on India's coastline, transport by sea would appear to be a natural option, but India's strict environment laws forbid it. The rocket stages and the inter-stage sections are designed, built, tested and manufactured at VSSC and transported to Sriharikota by road. Likewise, liquid, semi-cryogenic and cryogenic propellants are produced at the ISRO Propulsion Complex (IPRC), Mahendragiri, transported by road and stored at the Filling Control Centre at Sriharikota. The solid propellant is manufactured locally at Sriharikota. The spacecraft from Bangalore, too, arrives at Sriharikota by road. Once all the elements arrive at Sriharikota, the vehicle integration in the Vehicle Assembly Building or on-site at the First Launch Pad. Much of ISRO's research and development on launch vehicle systems is conducted at VSSC, including overall launch vehicle design, structures, aerodynamics & flight mechanics, polymers and composite materials, avionics, guidance systems, thermal coatings and all testing/qualification and system integration for each flight. A Hypersonic Wind Tunnel facility was opened in 2017 at VSSC to support supersonic flight.[441]

VSSC also houses a Supercomputer for Aerospace with GPU Architecture (SAGA). It was originally set up in 2011 and is regularly upgraded to support the R&D computational requirements. Each launch vehicle programme has its project teams which bring together all the systems to realise the launch vehicle. Initial work on future ISRO programmes for human spaceflight, reusable launch vehicle-technology demonstrator (RLV-TD) and air-breathing engines are also conducted at VSSC. VSSC fabricated the aircraft-like structure for the RLV, completely in-house. The technology of Lithium-Ion batteries used to power spacecraft in orbit was developed in VSSC and is now being outsourced for mass production for commercial use by Indian industry. Up-scaled version of these batteries is being used to power electric buses for public technology – a spin-off technology. Space Physics Laboratory, initially conceived by Sarabhai, is now an autonomous laboratory and a part of VSSC. SPL was instrumental in developing scientific payloads for Chandrayaan, Astrosat and Mangalyaan missions.

Space Applications Centre

While attending the 1968 Committee on the Peaceful Uses of Outer Space (COPUOS) in Vienna, Sarabhai had concluded "one of the most striking things to emerge has been appreciation of the great potentiality of remote-sensing devices, capable of offering large-scale practical benefits."[442] By 1972, Sriharikota had been set up, Sarabhai's plans for the SLV-3 launch vehicle was in progress, and U.R. Rao had started building India's first satellite, Aryabhata. The prospect of Indian-built satellites launched from Indian soil by Indian launchers was becoming imminent.

Figure 8-4 Experimental Satellite Communication Earth Station Built in 1966.
Credit ISRO

In September 1972, ISRO's chairman Satish Dhawan created the Space Applications Centre (SAC) by consolidating the various ISRO specialist units in Ahmedabad into a single centre. Professor Yash Pal, a senior professor, specialising in cosmic rays at the TIFR, was appointed SAC's first director. SAC was tasked to build remote-sensing sensors, subsystems and applications for Indian satellites. A year after Sarabhai's death, the elements necessary to realise his vision of an indigenous space programme were beginning to come together.

Just as Professor Pal was starting his role at SAC, the US launched the Earth Resources Technology Satellite (ERTS) designed to look at the US's natural resources, including forests, rivers, minerals and geology and help monitor floods and environmental pollution. Among its many achievements, ERTS (later Landsat 1) discovered an island of almost 3,000 km², located off the east coast of Canada.[443] By 1972, India had been independent for just over two decades and did not have a clear picture of its natural resources. ERTS demonstrated the potential of space-borne systems. Space technology was the most cost-efficient mechanism to help identify and quantify the natural resources across the huge Indian landmass. SAC's primary purpose was the development of space-borne and airborne instruments and associated applications that can be exploited for national development and societal benefits.

Figure 8-5 George Joseph Explaining to Prof. Dhawan (extreme left) and Prof. Yash Pal the Operation of the Multispectral Scanner Inside a Dakota Aircraft. 1976. Credit Dr. George Joseph

Pal recognised the importance of engaging all stakeholders from the outset to ensure that the applications met the needs of all its users. The multidisciplinary approach that Sarabhai had built into ISRO at the outset, Pal implanted within SAC. He engaged specialists from different fields; they included scientists, engineers, application developers and social scientists. Just as Sarabhai had no experienced rocket scientists in India when he embarked on a programme to build rockets, Pal did not have a pool of experienced specialists for building space grade sensors, remote-sensing systems and applications. He turned to TIFR and invited three scientists, Dr George Joseph (born 1938), a cosmic ray physicist; Dr Baldev Sahai, a nuclear physicist; and Mr D. S. Kamath, a computer specialist. None of them had experience with remote sensing. Again, like VSSC in 1963, SAC made use of makeshift facilities. In the absence of optics laboratories, initially, the remote-sensing team at SAC converted a kitchen within their residential apartment into a dark room and set up an optical bench.[444]

The journey towards SAC had begun in 1966 with the Experimental Satellite Communication Earth Station, a 14-m dish antenna, established in

Ahmedabad by Sarabhai. Located at the centre of SAC, the antenna remains in operation today. The Earth Station was established with the assistance of the International Telecommunications Union of the UN, and at inception, its primary function was to train engineers from India and beyond. It continues that tradition today by hosting a 9-month post-graduate diploma course in satellite meteorology and communication run by the Centre for Space Science and Technology Education in Asia and the Pacific for students from the Asia-Pacific region. SAC has been at the heart of most space-borne applications ISRO has built since its inception. Prior to building space-borne instruments, SAC conducted remote sensing from high-altitude balloons. In 1975, for example, a platform housing multiple Hasselblad cameras with a variety of films was carried by a huge balloon (around 6840 m^3) to 27-km altitude.[445] The following year, ISRO's first multispectral scanner was tested within a modified aircraft prior to deployment in space.

Much of the groundwork for the SITE programme in 1975 was done at SAC. After SITE, SAC initiated the Satellite Telecommunications Experiments Project (STEP). STEP used a transponder on the Franco-German satellite called Symphonie that was moved to 49° over the Indian Ocean for two years between 1977 and 1979. ISRO's first experimental communication satellite in GEO called the Ariane Passenger Payload Experiment (APPLE) was fabricated at SAC.

While Ford Aerospace and Communication Corporation in the US built the first four satellites (INSAT 1A to INSAT 1D) of INSAT, engineers at SAC designed and the five satellites in the second series (INSAT 2A to INSAT 2E). SAC has been in the vanguard of building sensors used by all ISRO's Earth observation (EO) satellites. The first experimental EO satellite, Bhaskara-1, carried a 1-km resolution TV camera and a three-channel Satellite Microwave Radiometer called SAMIR. SAC introduced multiple charge-coupled devices (CCD) to deliver high-resolution images of 5 m for Indian Remote Sensing IRS-1C. It was the first time that such a high resolution was made available in a satellite for civilian use. The cameras that ISRO used in recent missions to image the Moon and Mars, like SAMIR and the high-resolution CCD, have evolved from this early technology developed at SAC.[446] In addition to the hardware, SAC has been driving the development of applications that can best exploit the data generated by ISRO's now extensive space-based infrastructure. Some of these applications are summarised below.

- Communication: Communication transponders are the primary communication devices that facilitate the telecommunication connection between Earth and space. This may be telephone calls, direct-to-home transmission (satellite TV), search and rescue, or Global Positioning System (GPS)-Aided GEO-Augmented Navigation or satellite navigation (or satnav) service. SAC designs and develops the subsystems that comprise communication transponders. It also undertakes R&D of software applications and models for monitoring coastal erosion, desertification, coastal sediment transport and shoreline vulnerability to storm surges and tsunamis.

- Agriculture: For over two decades, SAC has been making Crop Acreage and Production Estimation forecasts for important crops using satellite remote-sensing data for the Ministry of Agriculture. In 2007, it moved to a more inclusive model forecasting agricultural output using space, Agro-meteorology and Land-based Observations. Through these data, India recognised the impact of climate change in the air quality of its major cities, as well as rural communities. SAC has been involved in a series of projects that attempt to use remote-sensing data to help develop long-term ecological records of Indian Himalayas, establish a geospatial database and identify and characterise mangrove ecosystems.

- Environment and Meteorology: Satellite data continues to be used to develop models for understanding the connection between the polar environment and Indian monsoon. Specifically, ISRO's Satellite with ARgos and ALtiKa (SARAL), a joint Indo-French mission, collects high-resolution altimetric measurements of the sea-surface elevation. These data help generate surge forecasts and inform monsoon prediction models. INSAT-3D data provides regular and systematic information on the state of the marine ecosystem of Indian exclusive economic zone at synoptic scales to policymakers, as well as those who rely on this region for their livelihood. Sea-surface wind measurements, along with Oceansat-2's Ocean Colour Monitor, have helped SAC refine models used to generate the Potential Fishing Zone advisories generated daily for fishermen along the substantial Indian coastline. This technique has now been transferred from SAC to the Indian National Centre for Ocean Information Services.

- Space technology for national development: A meeting in New Delhi identified 170 projects that include a wide range projects to exploit space technology that include real-time forest fire alert system, support the development of 100 Smart Cities across India, coral reef mapping to monitor coastal pollution, support safety and planning for Indian railways & roads and river water quality and river bank erosion management.447

- Climate Change: Using data from multiple remote-sensing satellites, SAC has developed state-of-art techniques for the analysis and forecasting of weather and climate on a regional and global level. In the backdrop of worsening greenhouse gas emissions and increasing global temperature, India has shown its willingness to cooperate at an international level. During a meeting of 11 of the world's largest space agencies in New Delhi in April 2016, India proposed a "virtual remote-sensing satellite constellation for the BRICS nations" to help them jointly meet the Paris Agreement on Climate Change.448

ISRO Satellite Centre

ISRO Satellite Centre (ISAC) is a modern hi-tech facility for building, integrating and testing communication, remote-sensing and science satellites. It emerged from the site used by Professor U.R. Rao's team to build India's first satellite, Aryabhata, in 1975. All satellites share a basic design for subsystems for electrical power, communication, temperature control, attitude control and navigation sensors. Each satellite is a 3-D jigsaw puzzle constructed by integrating multiple systems by several teams belonging to disparate establishments based potentially in different cities working to a single overall design.

It is at ISAC that these components and subsystems come together and are tested as a single integrated spacecraft for the first time. ISAC has grown, and today, it incorporates the ISRO Satellite Integration and Test Establishment and the Laboratory for Electro-Optics Systems. Also located within ISAC is a Comprehensive Assembly, Test and Thermo-Vacuum Chamber used to test a fully assembled satellite in space-like conditions. The large chamber can physically accommodate a satellite in its final integrated state and then subject it to a space-like environment. The air in the chamber can be removed, and temperature increased or decreased to simulate the ambient conditions of space, where the satellite will operate. Typically, the temperature in LEO fluctuates between +150°C on the day-side and -150°C 45 minutes later on the night-side of the orbit. Most spacecraft rotate in orbit to minimise the variation in temperature. A spacecraft is tested rigorously in this environment to qualify components and systems that will operate in the extremely fluctuating conditions that prevail in space.

A spacecraft's power supply is the key factor that determines its capacity to fulfil its mission objectives. ISAC, in cooperation with Bengaluru-based

Bharat Heavy Electricals Ltd, has streamlined the production of solar panels used on ISRO spacecraft. Almost all modern spacecraft operate on a battery recharged by modern lightweight solar panels that are folded during the launch but unfurled once in space.[449] It is this battery that provides the power for maintaining the satellite's subsystems and the payload, which may be transponders for communication satellites (Geosynchronous Satellite (GSAT) series), atomic clock and radio transmitters (Indian Regional Navigational Satellite System (IRNSS), or camera and sensors (Mars Orbiter Mission). Each satellite's power system is designed to meet its unique requirements. Its solar panel efficiency and battery capacity must be sufficient to meet the power requirements for its mission objectives. The solar panels must be large enough to recharge the battery while the spacecraft is in the sun to sustain it through the period when it is not.

Typically, most of the mass of a rocket at launch is the propellant necessary to get the payload into orbit.[450] The final orbit does not remain stable but is subject to minor gravitational fluctuations, potential east-west variations as a result of solar radiation or a reduction in altitude as a result of atmospheric drag. For example, the ISS is located in an orbit of 400 km, but over time the tenuous atmosphere causes drag and lowers the orbit (by around 50 m per day) to such an extent that its orbit has to be boosted a few times every year. Navigation sensors and control algorithms developed at ISAC allow ground operators at the Master Control Facility (MCF) to determine where a spacecraft is and calculate with precision any adjustments that may be required. These critical manoeuvres require short engine burns using the onboard attitude control thrusters to reposition the satellite.[451] Efficient algorithms minimise the propellant required for station-keeping (maintaining the desired orbit) and thus maximise the operational lifetime of a spacecraft.

Liquid Propulsion Systems Centre

The Liquid Propulsion Systems Centre (LPSC) is responsible for the R&D for control and management of liquid and cryogenic propellants, control valves, transducers, liquid engines and complete launch vehicle stage. It is also responsible for developing the launch vehicle stages for launch from Earth to space and the propulsion systems used by a spacecraft's onboard engines once in space. LPSC originally consisted of three sites: LPSC Valiamala, LPSC Mahendragiri and LPSC Bangalore. On 1 February 2014, LPSC Mahendragiri was renamed as a separate entity, the ISRO Propulsion Complex (IPRC). The other two sites remain within the LPSC.

Part of VSSC's integration team is based within the LPSC campus at

Valiamala to support subsystem level integration of launch vehicle stages at VSSC. Integration of stages with liquid engines is coordinated both by VSSC and LPSC. LPSC Bangalore develops the thrusters that are fitted to the spacecraft (IRS, GSAT and INSAT) as part of its onboard propulsion system. Once the launch vehicle has delivered a satellite to space, its onboard propulsion system takes over. This system is responsible for manoeuvring the satellite to its final operational orbit, an operation that is typically overseen by ISRO's MCF in Hassan. In support of the Mars Orbiter Mission (MOM), a duplicate Liquid Apogee Motor (LAM) in Valiamala was used to test commands on Earth before they were sent to for Mars. This was particularly important for the critical Mars Orbit Insertion manoeuvre. This testing provided confidence that the MOM LAM would fire after the long pause between leaving Earth and arriving at Mars.

Figure 8-6 ISRO Propulsion Complex, Mahendragiri. Credit ISRO

LPSC is tasked with designing, building and delivering liquid engine propulsion systems for complete rocket stages. It is also responsible for developing engines, propellant tanks and the associated control units that regulate the combustion of propellant during the launch. Combustion requires fuel, oxygen and ignition. On Earth, the atmosphere provides the oxygen, but rockets that operate in space must take the oxygen supply with them. A variety of rocket fuels and oxidisers are used, and each has associated attributes of performance, handling, cost, manufacture, transport, storage and efficiency.

As a launch date approaches, propellants matching the launcher requirements are made available at Sriharikota. For instance, the four stages of the Polar Satellite Launch Vehicle (PSLV) alternate between solid (stages

1 and 3) and liquid (stages 2 and 4) propellants. The same solid propellant is used in the first and third stages, while the second and fourth stages use different liquid propellants.[452] The spacecraft within the launch vehicle also requires a propellant to reach and maintain orbit. The solid propellant is produced onsite at Sriharikota. All other propellants are manufactured at three geographically separated sites.

ISRO Propulsion Complex

IPRC is located in Mahendragiri, 350 km south-west from Sriharikota, and is responsible for leading-edge research, development and testing facilities for liquid, cryogenic and semi-cryogenic propulsion technologies.

Development work on semi-cryogenic technology is conducted at IPRC. Although a late starter, ISRO is expecting to test semi-cryogenic engines by the end of 2017. IPRC undertakes a series of activities, which include the production and supply of liquid and cryogenic propellants for launch vehicles and spacecraft. The testing activities at IPRC include:

- Structural testing in a pressure chamber that can accommodate hardware up to 5 m in diameter to simulate space-like environment

- Testing of launch vehicle subsystems, such as turbopumps, injectors, gas generators, gas bottles and umbilicals

- Development of semi-cryogenic engines and production of cryogenic propellants

- High-altitude testing of cryogenic engines

- Modelling thermal structures and computational fluid dynamics analysis of components and subsystems.

ISRO successfully designed, built and used a cryogenic stage on the Geosynchronous Satellite Launch Vehicle (GSLV) Mk2 that placed India's communication satellite GSAT-14 into Geosynchronous Transfer Orbit (GTO) on 5 January 2014. All cryogenic engines used up until then on the GSLV Mk1 were acquired from the former USSR. Cryogenic propellants are so far, the most efficient type of propellant available for launch vehicles. Engineers measure propellant efficiency in terms of Specific Impulse, abbreviated to I_{sp}. The I_{sp} increases from solid, liquid, semi-cryogenic to cryogenic propellants.[453] Cryogenic technology introduces extremely low temperatures, and the engineering challenges they bring make rocket science, rocket science. The most efficient propellant is a hydrogen (fuel) and oxygen (oxidiser) combination, but both being gases demand large storage tanks that define the size and shape of the launch vehicle.

However, once cooled, they turn to liquid. In their liquid form, they occupy much less space and are easier to transport and store. Oxygen liquefies at -183°C and hydrogen at -253°C. Nothing can be colder than -273°C (absolute zero). Rocket engines using cryogenic technology present engineering challenges of enormous proportions. Storage tanks need to operate at low temperatures, intricate plumbing is required to transport the fuel from the tanks to the combustion chamber, pumps must operate at 4,200 revolutions per minute, valves must switch on or off with microsecond precision, and the temperature of the combustion chamber must be maintained at 3,000°C without catastrophic destruction. It is at IPRC that these engines are tested and qualified.

ISRO Telemetry, Tracking and Command Network

The significant number of national and international communication links required to support ISRO's space operations come together like the lines of a spider's web in Bangalore at the headquarters of the ISRO Telemetry, Tracking and Command Network (ISTRAC). Initially, established as the ground segment for the IRS satellite in 1982, ISTRAC now incorporates all ISRO's communication resources and continues to grow. It is the hub for communication with spacecraft in orbit or deep space. Over the years, ISRO's communication infrastructure had developed on an ad-hoc basis but is now optimised as a single integrated entity. ISTRAC is responsible for supporting space vehicle launches, guiding satellites into their designated orbits and managing them during their lifetime once there.

Telecommunication transponders are built to high standards to operate at fixed frequencies. Poor design or quality of components may allow frequency drift, which could result in designated recipients being unable to receive data. Worse, radio interference could result in the loss of a service from a nearby satellite. ISTRAC's telemetry, tracking and command (TT&C) facilities consist of ground stations located throughout India and beyond, including Bengaluru, Lucknow, Sriharikota, Port Blair, Thiruvananthapuram, Mauritius, Brunei, Antarctica Ground Station for Earth Observation Satellites (AGEOS) and Biak (Indonesia).[454] A network of ground-based radar systems is used to track launch vehicles from the moment of launch through to orbit insertion.

Figure 8-7 ISRO's 32-m Antenna at Byalalu. Credit Author

In 2008, ISRO launched Chandrayaan-1 its mission to the Moon. ISRO designed and built the spacecraft, which also carried a host of instruments from several international partners, and launched it from Sriharikota. To detect radio signals arriving on Earth from the Moon 380,000 km (236,121.05 miles) away, a large antenna was required. The site was selected from three shortlisted against key criteria.[455] A few days before the launch of Chandrayaan-1, on 17 October 2008, ISRO inaugurated its flagship deep space antenna, a 32-m fully steerable dish at Byalalu close to Bangalore, which is now the nucleus of the Indian Deep Space Network (IDSN), an integral element of ISTRAC. The Byalalu facility is the centrepiece of

IDSN and the ISRO Navigation Centre (INC), a communications hub for the IRNSS, which came into full operation in 2016. In addition to the 32-m antenna, Byalalu also operates a fully steerable 18-m antenna. The two antennae were built and are operated to international standards and can interoperate during periods of collaboration with space agencies of other nations. Byalalu has been instrumental in operating ISRO's MOM, which has been much more challenging than operating the Moon mission given the increased distance.

To support missions with special requirements, such as Space Capsule Recovery Experiment (SRE) and RLV-TD, ISTRAC has also developed an airborne TT&C provision as a rapid deployment asset. A 4.6-m transportable TT&C station can also be used from a ship or a temporary ad hoc location. It was deployed during the SRE-1in 2007, launch of MOM in 2013 and the 2016 RLV-TD mission.

Today, ISTRAC is the backbone of ISRO's national, international and extra-terrestrial communication systems. It is an extensive infrastructure operating at international standards. ISTRAC is an observer agency within the Consultative Committee for Space Data Systems (CCSDS), which publishes standards to promote interoperability and cooperation between space agencies. This allows ISTRAC to ensure that its infrastructure conforms to CCSDS standards allowing it to work with other nations that also comply with CCSDS standards. ISTRAC has supported space missions from Malaysia, Taiwan, Germany, France, the US, Thailand, Canada, Italy, Norway, Japan and China.[456] To support Indian remote-sensing satellites, ISTRAC has established a Telemetry, Tracking and Command (TT&C) stations in Norway, Sweden, the US, Brazil and Brunei and is currently in negotiations with Malaysia and Vietnam.

Master Control Facility

To manage the INSAT series of satellites, India's first series of satellites positioned in GSO, the Master Control Facility was established in 1982. Today, it handles all ISRO satellites in Earth orbit and beyond. Once a satellite is delivered to space by the launch vehicle, engineers at the MCF activate the deployment of solar panels and through a series of engine burns shepherd the spacecraft to its final orbit. MCF is located in a large 17.2-hectare (0.172 km/0.107 miles) picturesque site 8 km north of Hassan close to the historic monument of Halebidu from the 12th century Hoysala Empire. ISRO preferred it to be located in Bangalore. However, to minimise interference that may arise from high-power C-band transmission, a minimum distance of 500 km from the nearest similar transmitter had been mandated by the International Telecommunications Union. Intelsat happened to have one in Jaffna, Sri Lanka, and Bangalore was within that range, so an alternative location had to be found.[457] The current location was also ideal because its low population and minimal urban development contributed to low radio noise, which was helpful when listening to faint radio signals from spacecraft thousands or millions of kilometres away.

Figure 8-8 Main Building of Master Control Facility, Hassan. Credit ISRO

The site was selected in the early 1980s and became operational with the launch of INSAT-1A on 10 April 1982. The MCF complex consists of three full-motion and about a dozen limited-motion antennae. The full-motion antennae can be rotated 360° and can be tilted from 0° to 90° in elevation so that they can look at and track satellites in any part of the sky. They send commands to the spacecraft and receive data from them. The full-motion antennae are mostly used during launch, but once the satellite is in its prescribed orbit, the limited-motion antennae are used. In mid-2016, the MCF was responsible for managing 19 satellites (INSAT 3A, 3C, 3D, 4A, 4B and 4CR; Kalpana-1; GSAT 6, 7, 8, 10, 12, 14, 15 and 16 and IRNSS 1A, 1B, 1C and 1D).

From its location, MCF can track any launch with its apogee over the Indian Ocean. The MCF consists of a Mission Control Centre, where operational decisions are taken, a Spacecraft Control Centre, where commands to the satellites are issued and several Earth Stations, which consist of antennae that send and receive signals between the spacecraft and the Earth. When established, MCF had two Earth Stations and a single Spacecraft Control Centre. While most of the equipment was produced in India, the Spacecraft Control Centre was initially fitted out by Ford Aerospace Communications Corporation because it had built the first four INSAT satellites.

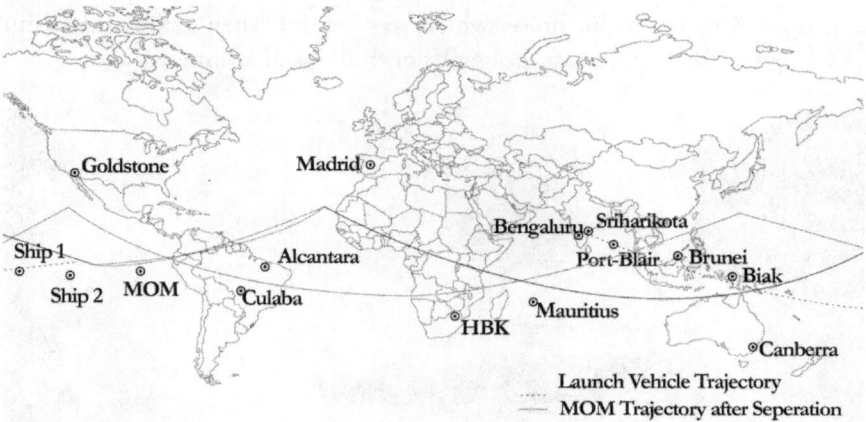

Figure 8-9 Ground Stations Used for Tracking Mars Orbiter Mission Including International and ISTRAC Centres. Credit ISRO

In 2005, MCF operations were boosted by the addition of another smaller facility in Bhopal to support the MCF. MCF Bhopal consists of a Spacecraft Control Centre, an Earth Station and a power complex. With one 11-m and three 7.2-m antennae, MCF Bhopal is responsible for managing three of the seven IRNSS satellites. It also offers some level of redundancy. However, MCF Bhopal's smaller size allows it to operate only a smaller number of satellites compared to MCF Hassan. S.V. Shivakumar, a former MCF Director, regards the MCF as the "hottest seat in ISRO", given its responsibility of managing so many high-value assets.[458] The telemetry data received by MCF from each spacecraft consists of around 800 parameters (for example, the power generated by the solar panels, state of charge of the battery, quantity of remaining propellant, the temperature of the units housing the electronics, location and attitude in orbit) transmitted once every second. Computers at MCF display these data in a usable graphical form in real time.

Over time, a spacecraft can drift from its designated position. The cause of the drift can be a combination of the accumulative gravitational influence from the Sun and the Moon, solar-radiation pressure, the impact of high-speed particles emanating from the Sun and the non-uniform gravitation field of the Earth caused by the Earth not being a perfect sphere. These forces can shift a satellite in a GEO left or right (in longitude) or up and down (in latitude). It is the MCF's responsibility to manage this displacement. In the case of INSAT series of satellites with a mass of about 1,000 kg, an engine burn using 4 kg of fuel is required approximately every 80 days to maintain latitude. Corrections in longitude require more frequent (perhaps once every two weeks) but shorter burns using much smaller (0.5 kg) quantity of fuel. These corrections can take up to four hours, and although the satellite is moving, services are not affected.[459]

In addition to maintaining orbit, MCF must maintain the satellites' orientation. The antennae of satellites must point to the Earth with high precision. From a height of 36,000 km, an error of just 0.1° will shift the signal by 600 km along the Earth's surface, potentially resulting in a loss of signal for many users.

MCF also has to manage periods of solar eclipse. The duration of an eclipse varies through the year with the longest being 72 minutes around the equinoxes. The solar panels of a satellite cannot generate electricity for the duration it is in the Earth's shadow. It is critical that the battery sustains the satellite during this period of darkness. The capacity of the battery is largely mandated by this requirement. Further, the temperature of the satellites drops when in shadow, so heaters are necessary, further increasing the demand for battery power. To compensate, some services or subsystems may have to be switched off temporarily.

Recovering Costs - Antrix

By the early 1990s, ISRO's operations were sufficiently mature to engage with the blossoming commercial market in space. In September 1992, the Government of India established Antrix Corporation Limited. Antrix is an anglicised version of the Sanskrit word *antariksh*, meaning space or sky. It is headquartered at Antariksh Bhavan within the site of ISRO headquarters in Bengaluru. An autonomous body, Antrix's primary objective is to negotiate commercial space services with national and international clients, which are ultimately delivered by ISRO's centres. As ISRO's marketing agency, operating under the DOS, Antrix markets and delivers customised services, which include data from satellites, transponders and launch services for national and international clients.

Data from remote-sensing satellites are has a commercial value. Using a variety of spectral bands (wavelengths of light that are in the optical, infrared, ultraviolet or radio and spatial resolution (level of detail), these satellites capture information of the Earth's atmosphere, land and oceans. These images can be exploited for numerous purposes, including meteorology, agriculture, forestry, irrigation, mining, fishing, tracking sea currents and l surveillance.

Figure 8-10 Indian Remote-Sensing Satellite RISAT-1. Credit ISRO

Remote-sensing satellites make use of a Sun-synchronous polar orbit that typically has an inclination of 98° at an altitude of 800 km. Each pole-to-pole orbit takes about 100 minutes, moving slightly eastward each orbit. The whole globe is covered in a period of 16 days, and the cycle gets repeated. As the number of India's remote-sensing satellites grew, so did the quantity of data collected from around the globe. ISRO could potentially generate income from selling these data to countries that did not have an EO programme of their own, just like the US's Landsat programme had been doing since the mid-1980s.

The American Landsat Programme had been designed with precisely this objective in mind. Landsat had its origins in NASA during the early 1970s but was transferred to National Oceanic and Atmospheric Administration (NOAA) under presidential directive 54 issued by President Jimmy Carter (born 1924).[460] He also instructed NOAA to transfer Landsat operations to the private sector. In 1985, with five Landsat satellites launched, but only two, Landsat 4 and Landsat 5, were operational at the time, NOAA selected the private sector entity Earth Observation Satellite Company (EOSAT) to exclusively operate and market Landsat 4 and 5 data on a commercial basis and build Landsat 6 and 7 for future launch.

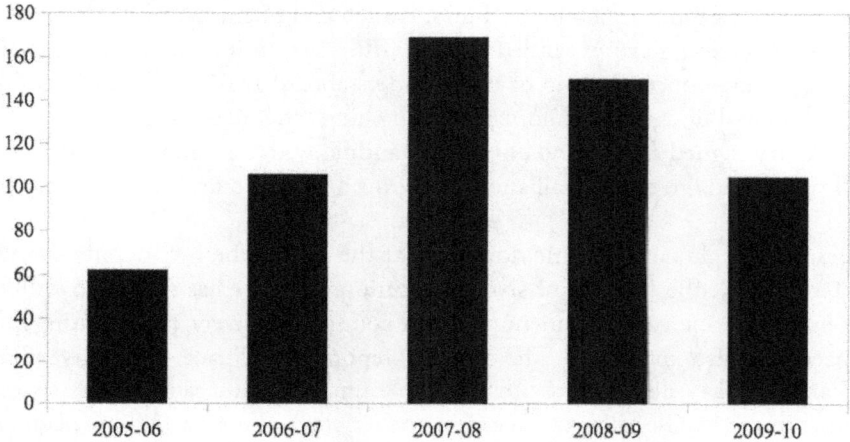

Figure 8-11 Profit After Tax Rs (Lakhs) Antrix Corporation Limited. Credit Antrix

ISRO's satellites, like all EOSAT satellites, collect data of all the countries under their orbit. In October 1993, EOSAT's Landsat 6 failed to reach orbit and was lost. By chance, at that time, ISRO had two EO satellites IRS-1A and IRS-1B in orbit and significant experience in operating them. EOSAT approached India to help fill the gap. As ISRO's chairman, Professor U.R. Rao signed an agreement for EOSAT to have exclusive worldwide rights to IRS imagery. ISRO helped to establish a ground station in Oklahoma in the US to receive IRS data, the first one outside India.[461] EOSAT had an existing global network of 16 ground stations to receive Landsat data. The agreement that Rao signed required EOSAT to pay ISRO $0.6 million (Rs.1.89 crore) per year for every ground station that received IRS imagery. The ground station in Oklahoma was inaugurated in May 1994. In its publicity material, EOSAT declared that "the country that gave the Taj Mahal to the world has now delivered a highly sophisticated satellite capable of producing high-resolution imagery over our planet."[462]

From this beginning in 1994, ISRO has expanded its offering with additional IRS satellites and, in 2014, was generating approximately $20 million (Rs.123 crore) per year. By the end of 2014, 11 EO satellites were operational in orbit, Resourcesat-1 and 2, Cartosat-1, 2, 2A, 2B, RISAT-1 and 2, OCEANSAT-2, Megha-Tropiques and SARAL. Antrix was established to generate income from ISRO operations to help recoup some of the costs and provide a source for further investment. Speaking in 2011, former ISRO Chairman U.R. Rao stated "When we set up Antrix, our vision was clearly to delink the commercial and business side from day-today governmental functions. Moreover, we wanted to professionalise marketing activities and expand the reach of ISRO's products and services across the world."[463] Today, India is one of an increasing number of nations with an indigenous capability to build, launch and operate satellites in orbit. Antrix's services have expanded beyond IRS data to include technological consultancy services, leasing of transponders aboard the GSAT and INSAT satellites and designing, building and launching satellite services. Over time, ISRO has signed memoranda of understanding with 28 countries that have allowed Antrix to develop alliances and grow a portfolio on the global stage.

Despite the global economic downturn at the end of the first decade of the 21st century, the funding of space institutions globally has remained either constant or increased. The number of countries actively participating in space has also increased. The OECD report The Space Economy at a Glance 2014, lists more than 40 countries with an active space programme.[464] Globally, the space sector employs almost a million people, and growth is expected to continue in the short and medium terms. Although the space sector is healthy with strong growth potential, Antrix does not necessarily increase its profits year on year. Infrastructure investments, loss of a satellite or a launch failure can have a significant impact on its bottom line. For instance, the 16.52% decrease in profits in 2009–2010 was attributed "mainly to the capacity constraints."[465]

In a formal answer to a written question in the Parliament of India on 17 July 2014, the Minister of State for Science and Technology and Earth Science quantified Antrix's income since its establishment in 1992.[466] A total income of Rs. 4,408 crore ($709 million) came from (i) direct reception of data from IRS satellites to national and international clientele and (ii) leasing of satellite transponders on-board. INSAT/GSAT satellites used for communication or TV broadcasts. ISRO has also launched 40 satellites for customers from 19 different countries, generating a further $74 million (Rs.445 crore). The minister also outlined Antrix's future objectives: (i) expanding the data and direct reception services of the Indian remote-sensing satellites to international clientele, (ii) enhancing leasing of satellite transponders to Indian customers, (iii) increasing launch services for foreign satellites on-board Indian launch vehicles and (iv) enhancing marketing of satellites and subsystems.

ISRO has an established track record of building and launching satellites successfully. In February 2017 ISRO launched a record 104 satellites of which Cartosat-2D was the primary payload and 100 smaller foreign satellites less than 10 kg. The income helped to recover about half of the launch costs and brought the tally of commercial satellite launches by ISRO to 209. Even though accurate and unambiguous statistics on income generated from international commercial launches are not available, the global revenue from commercially procured satellite launches in 2013 was $5.4 billion (Rs. 32,111 crore). According to one source, the US's share accounted for almost half (45%) of this revenue.[467] The US dominated the market in producing the spacecraft launched in 2013. Of the 107 satellites launched, the US claimed the lion's share of 70%, Europe 17%, China 5%, Russia 3%, Japan 3% and all the other nations combined accounted for the remaining 2%.

Although these figures do not necessarily illustrate this, the space programme in Russia could be considered the most commercial of all. Since the retirement of US's Space Shuttle, the Russian Soyuz is the only route for astronauts to reach the ISS. The Soyuz is operated by the state-owned Roscosmos State Corporation for Space Activities. In the context of such a vast global commercial space market, ISRO's market share is smaller than it should be. One analyst suggests that India is still recovering from the strict sanctions regime introduced in 1998 following its nuclear tests.[468] Intriguingly, the same sanctions were the motivation for India to develop its indigenous capability for building and launching satellites. ISRO is gradually building its international reputation as a reliable integrator and launcher of medium-sized satellites. The significant potential for growth in the international market for satellite services has been established. Three key factors determine the extent and speed at which Antrix can grow: i) government procedures: in their current form, they do not encourage small and medium-sized Indian companies to engage with ISRO, ii) infrastructure: the existing infrastructure requires several months of lead time between rocket launches. Cutting project lead time by building additional launch facilities would not only increase capacity but make it more competitive in the international market. The building of a second vehicle assembly unit at Sriharikota is designed in part to meet this need iii) timeline for operationalising the GSLV Mk3.

Establishing routine access to space using its own heavy-lift vehicle is probably ISRO's most pressing goal. With it, not only will India enter the market for launching heavier geosynchronous communication satellites but it will finally become self-sufficient. With the success of first developmental flight of the GSLV-Mk3 in June 2017, ISRO is expected to end its reliance on Ariane 5 to launch its communication satellite at the cost of around Rs 400 crore ($73 million) sometime in 2018. Despite being a government agency operating under multiple layers of traditional government bureaucracy, Antrix has been successful in its objectives for the same reason that ISRO has been. ISRO is refreshingly praised as an example of how public organisations in India can succeed in the private sector. One essential ingredient for this success was articulated in 2008 by Antrix Managing Director K.R. Sridhara Murthi "neither ISRO nor Antrix experienced any political interference in their functioning and [they] were left to follow their own professional objectives and styles in the running of the organisations."[469]

Year	Rs (Lakhs)	USD (Million)	EURO (Million)
2007–2008	12,679.57	20.56	0.894
2008–2009	14,031.38	24.428	2.412
2009–2010	13,228.97	12.159	5.911
2010–2011	8,077.64	17.998	4.086
2011–2012	3,135.54	2.199	0.629
11th FYP Total	51,153.1	77.344	13.932
2012–2013	11,157.01	3.99	13.097
2013–2014	17,470.66	5.578	17.598
12th FYP Total	28,627.67	9.568	30.695
Grand Total	79,780.77	86.912	44.627

Table 8-1 The Foreign Exchange Earned by Antrix. Credit Press Information Bureau of India 26 November 2014[470]

As a product of a democratic nation, ISRO is charged with ensuring diversity in its workforce. Compared to other areas of Indian society, business and industry, ISRO has promoted equality of opportunities between its diverse national communities and especially for women. Women make up only around 10% of its workforce of about 1600 majority of whom are based at the VSSC. Throughout its half-century of existence, Inia's space programme has been innovating and growing. That continues as it ventures into new areas, such as interplanetary exploration, observing the universe from space, satellite navigation services, developing a lunar lander & rover and building human spaceflight capability. Given political and economic support, the facilities and capabilities of ISRO's forty-two centres around India are likely to continue that tradition of innovation and growth.

◁ ◈ ▷

Chapter Nine
Sriharikota. India's Spaceport

O n Friday, 18 July 1980, an Indian-built satellite was delivered to Earth orbit by an Indian-built rocket launched from India. The outrageous ambition of men with a vision from two decades earlier had been realised in spectacular style. The launch took place from Sriharikota located in Andhra Pradesh. Sriharikota has been the starting point for all of India's spacecraft to the Earth orbit, the Moon and Mars. On 5 September 2002, Sriharikota was renamed as Satish Dhawan Space Centre (SDSC or sometimes as SDSC-SHAR but Sriharikota is still in common use) in memory of Satish Dhawan, who had succeeded Vikram Sarabhai as the chairman of ISRO. Today, Sriharikota, as it is more commonly known, is India's primary and only spaceport.

Figure 9-1 Sriharikota on India's East Coast. Credit Google

Located 100 km (62 miles) north of Chennai on India's east coast, Sriharikota is a barrier island as large as the city of Chennai with a maximum width of 8 km (4.9 miles), comparable to Merritt Island at NASA's Cape Canaveral. Typically, 10 m above sea level, Sriharikota located at 13.74°N and 80.2°E, sandwiched between Pulicat Lake and

Buckingham Canal on the West and the Bay of Bengal on the East. Like Merritt Island, much of the environment of Sriharikota is a nature preservation area. Pulicat Lake abounds with rare and beautiful birds, including pelicans, kingfishers and pink flamingos. A tangible risk of a tsunami as the one in Asia in 2004 and Japan in 2011, however, makes Sriharikota vulnerable given that the two launch pads are only a few hundred meters from the sea and only a few meters above sea level. Sriharikota was first formally announced as a rocket launch site in the 1968-69 annual report by the DAE.[471]

Sarabhai had entrusted E.V. Chitnis with the responsibility to locate a suitable launch site. In March 1968, E.V. Chitnis and U.R. Rao were conducting X-ray balloon experiments in Hyderabad. They were notified that winds were too high and no balloons could fly for up to four days, so they decided to use this time to identify a site on India's east coast suitable for launching large satellites of substantial mass into Earth orbit. To make use of the energy from the Earth's rotation, rockets are launched in the easterly direction. TERLS, located on the west coast, was not ideal because the launch vehicle would fly over inhabited areas. A launch site on the east coast ensured that the launch vehicle flew mostly over water. Based on the selection criteria used by Germany, the US and Japan, Chitnis and Rao quickly identified Sriharikota, and the first visit to this site took place a month later.[472] In a convoy of six jeeps, Sarabhai visited Sriharikota in May 1969. At that time, there was only a dirt road covered by casuarina branches and Palmyra leaves.[473]

On 9 October 1971, Sriharikota became operational with the launch of RH-125. The RH-125 (short for Rohini from Hindu mythology) was one of India's first indigenous solid fuel-propelled rockets with a diameter of 125 mm. Almost a decade later, India's first satellite launched successfully by its own launcher SLV-3 on 18 July 1980 was also called Rohini. From the 125-mm diameter of RH-125, today, Sriharikota builds rockets with a diameter more than 3 m and launches satellites to LEO, polar orbit, geostationary transfer orbits and to interplanetary space.

Compared to Thumba, Sriharikota has the capacity to launch larger satellites to higher orbits, is sufficiently large to cater for all the ground-based facilities required for a spaceport and has scope to cater for future growth. As India's premier launch site, it is now probably ISRO's best-known centre, drawing global audiences during live TV broadcasts of satellite launches for science, EO or deep space missions.[474] Establishing this site to full operations was one of the final actions that Sarabhai completed prior to his death in December 1971. It was formally founded in 1969 and commissioned three years later, complete with facilities for preparing rockets for launch and with ground support for the telemetry and tracking required for sounding rockets. Based on imagery from their spy satellite, the US concluded in a then Top-Secret memo, now declassified, that Sriharikota was a "possible joint India/Soviet project supporting a Soviet synchronous satellite."[475] Sriharikota is an integral element in Vikram Sarabhai's strategy paper published in 1970 Atomic Energy and Space Research: A Profile for the Next Decade 1970–1980.

By the late 1960s, success in space by the US and USSR but also of nations, such as China, Japan and France, had made a political and economic case for India's space programme. Scientifically competent individuals, like Bhabha and Sarabhai, with support from political leaders, like Jawaharlal Nehru and subsequent prime ministers, initiated a space programme that hastens India's journey towards a developed economy. While Thumba was established at a time when India had zero experience and nothing more than an aspiration for a national space programme, Sriharikota and its infrastructure were built on the success of Thumba and designed to accelerate India's capabilities in space.

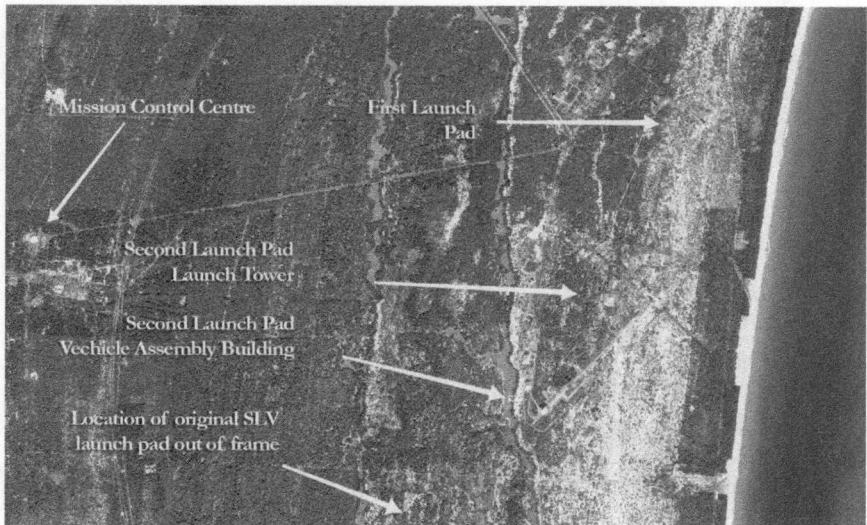

Figure 9-2 Sriharikota. Credit Google Earth

Sriharikota's infrastructure started with a satellite-tracking and communication system required to communicate with a satellite in orbit. One of the key ingredients was a computer that could calculate in real time the position of a moving satellite. The Electronics Corporation of India Ltd (ECIL), which had been formed in 1967, supplied Sriharikota a Trombay Digital Computer (TDC-12) that had been built at the Atomic Energy

Establishment, Trombay.[476] The TDC-12 supported Aryabhata when it was launched in 1975. Later, ECIL imported an IRS 55 computer from France, which was used for the development of the SLV-3 launch vehicle, and built TDC-316-I for remote telemetry ground stations, such as Car Nicobar Island, and TDC-36316-II for the two Bhaskara satellites that followed Aryabhata. As the number of sites and computers grew, ISRO called on the German Space Research Organisation to assist with connecting them into a single network.[477] With an annual budget of Rs.550 crore ($100 million) and 2,000 permanent and an additional 2,500 contract staff,[478] Sriharikota is responsible for functions including:

- Managing and operating the Mission Control Centre (MCC)

- Integrating launch vehicle and payload, testing and launching up to the designated transfer orbit[479]

- Managing range operations using data networks between MCC, launch pads and the remote ISRO centres; operating six radars for communication and launch vehicle tracking; optically tracking launch vehicle during launch and establishing the single timebase (timing signal) at high precision for use by all sites for synchronisation

- Manufacturing approximately 700 tonnes of solid propellant annually in two separate onsite manufacturing plants

- Conducting tests and establishing quantitative mechanical and ballistic characteristics of the solid propellants produced.

- Using a thermo-vacuum chamber to assess solid propellant properties for use in space.

At the outset, the ground infrastructure was still basic. A. Aravamudan recalls a unique story of how Sarabhai exploited any opportunity to hasten the development of the space programme. In the late 1960s, the European Launcher Development Organisation (ELDO) abandoned its plan to develop a European launcher.

The UK was a key partner in ELDO, and Australia, a former British colony,[480] had huge swathes of mostly uninhabited land that UK used as a rocket-test range for ELDO. So, when the ELDO programme ended, Australia had large quantities of equipment associated with launching rockets that suddenly became redundant. It was destined for destruction. Sarabhai sent his engineers to Adelaide in Australia to acquire a complete satellite telemetry ground station for a fraction of its original cost. It was shipped to Madras (now Chennai) and eventually "installed at Sriharikota as the first satellite telemetry ground station."[481] Initially, a rocket sledge facility for testing the effects of vibration, acceleration and deceleration on rockets, spacecraft and their subsystems was planned at Sriharikota but was never built. The sledge was designed to be powered by a rocket motor to horizontally push a payload on a high-speed dual rail track to simulate launch conditions. During tests, sensors, transducers and high-speed cameras would have recorded data to inform design decisions and evaluate prototypes. In 2014, a 4-km-long rocket sledge was opened in Chandigarh, Punjab, by the Defence Research and Development Organisation (DRDO). ISRO may make use of this facility in the future as it gradually advances towards its human spaceflight programme.[482]

The only transport link that Sriharikota has is a single road about 6 m wide that connect the spaceport to the mainland across Pulicat Lake. It does not have air or sea transport links, except for helipads used mainly for VIP transport. The rocket stages, inter-stages and liquid and cryogenic propellants, along with the payload spacecraft, are delivered to Sriharikota via this road. Onsite facilities include a cricket field, temple, school, auditorium and an open-air theatre. To accommodate ISRO's international commercial customers, Sriharikota has a large accommodation complex complete with restaurants offering multi-national cuisines, recreation facilities and a sophisticated medical facility, including a 60-bed hospital. In addition, routine first aid and ambulance provision, enhanced medical services, including specialist doctors, are on standby during scheduled launches.

Mission Control Centre

The heart of the Sriharikota complex is the MCC, appropriately a space-age design building resembling a flying saucer. It was inaugurated by the President of India in 2012.[483] Prior to launch, the launch vehicle is can be considered as a very large fuel tank. A safety procedure used by most space agencies is to evacuate the immediate area around the launch vehicle. The MCC is located 6 km away from the two launch pads and connected by underground cables (copper and fibre) for data, control and audio and video communication. It houses two Launch Control Centres (LCC1 and LCC2) adjacent to each other. Each desk is supplied with two monitors, headsets, and power and communications links connected to LCC1 and LCC2.

During the days leading up to the launch Information from all the disparate sources associated with a launch is brought together in the MCC. This includes payload data (for example, in-situ spacecraft battery status, fuel tank temperature), communication (for example, live video of a launch vehicle and safety updates from the range officer), control (for example, remote liquid refuelling), launchpad status and range safety.

Figure 9-3 Mission Control Centre, Sriharikota. Credit ISRO

The MCC can support multiple missions at the same time. Data from various sources arrive at the MCC via a 1.8 Gigabit Ethernet based network. Some of the sources include

- Range Instrumentation facilities that include Tracking, Telemetry Telecommand systems, CCTV systems, Optical Tracking systems, Technical Photography.

- Dedicated meteorology facilities that include weather radar and microwave radiometer to provide real-time data on Cloud.

- Telemetry data coming into the MCC is used by the Range Safety officer to make critical decisions on launch or no launch as well as self-destruct decisions when the launch vehicle loses control.

- Real-time images from cameras on the Launchpad, launch vehicle, spacecraft, and the surrounding lightning towers are consolidated in the MCC.

Redundancy is built into MCC's infrastructure. All underground cables connecting MCC to the two launch pads are duplicated. Both LCCs have direct connections to both of the launch pads. Each LCC has built-in redundancy, incorporating an uninterruptible power supply and backup diesel generators.

Figure 9-4 Second Launch Control Centre. Credit ISRO

The wider picture from ISTRAC, which provides the necessary clearance from downrange monitoring stations, international partners and air traffic control also comes into the MCC. During the countdown, these data are displayed on monitors in real-time and used for go/no-go decisions by the various teams. One of the monitors is reserved for the ISRO chairman, who is typically present on the day of the launch.

First Launch Pad

The launch pad known today as the First Launch Pad (FLP) came into operation in 1993. The one built in 1979 was known as the SLV launch pad and was used to launch SLV-3 and the Augmented Satellite Launch Vehicle (ASLV), but it was decommissioned in 1994.

The FLP has a 76-m high 3,000-tonne mobile service tower (MST) that embraces the launch tower with movable platforms providing access to each one of the four stages of a PSLV. MST is used to integrate the launch vehicle prior to the launch, and it is designed to accommodate a launch vehicle up to 5-m taller than the PSLV.[484] It is also designed to withstand the Indian monsoon rains and wind (up to 230 km/h) during September to November each year. All the elements of the launch vehicle, stages, inter-stages, strap-on motors and the payload, are brought directly to the FLP by road. A 60-tonne crane atop the MST moves the launch vehicle elements from the road transporter onto the launch pad for integration. The PSLV consists of four stages and the external strap-on motors. The complete launch vehicle is assembled vertically, one stage at a time, from the bottom up. The top of the FLP incorporates a clean room used to manage the integration of the payload in the final stage.

Figure 9-5 Site of the First Launch Pad. Credit Google

Once the launch vehicle is complete, umbilical connections for pneumatic, electrical and propellants are connected and tested. The MST is then moved 150 m away at 7.5 meters per minute on twin rails.[485] All personnel retreat from the launch pad at this time leaving the launch vehicle attached to a steel umbilical tower. Subsequent propellant loadings, control and system checks and eventual launch are all done remotely from the MCC 6 km away.[486]

Like rocket launch pads around the world, FLP is protected by three anti-lightening towers. Lights, 72 cameras in flame-resistant cases and other equipment are attached to these towers that surround the launch pad. These cameras provide live imagery to the MCC before, during and immediately after the launch. The FLP was constructed during a period of sanctions forcing ISRO to develop in-house components that it otherwise would have obtained commercially in the open market.[487] The four large bogies on 16 wheels designed to carry the enormous weight of the MST required "special efforts" to build and test in time for the first PSLV launch

Figure 9-6 First Launch Pad with PSLV-C18. 12 October 2011. Credit ISRO

Many of the private Indian industries that were engaged at that time are still involved in supporting ISRO.[488] The FLP has been used for ASLV, PSLV and GSLV Mk-2 launches.

Second Launch Pad

The Second Launch Pad (SLP) is the newer of the two launch pads at Sriharikota. A final report for establishing the SLP was submitted in 1996. The government sanctioned the payment of Rs.289 crore ($81 million) of Rs.316.11 crore ($88 million) in June 1997. The SLP was completed and became operational on 5 May 2005 with the launch of PSLV-C6 and has been used for several launches since then, including Chandrayaan-1 in 2008 and MOM in 2013. In addition to providing redundancy, the SLP was specifically designed to launch the larger GSLV-Mk3/LVM3 (ISRO initiated the name change from GSLV-Mk3 to LVM3 but the name has not taken and ISRO now appears to be reverting to GSLV-Mk3 but LVM3 name persists in numerous documents and online).

Unlike the FLP, the launch vehicle is not integrated at the launch pad but inside a building a kilometre away. The SLP complex incorporates a Solid Stage Assembly Building (SSAB, a fixed Vehicle Assembly Building (VAB), a mobile launch pedestal (MLP), a kilometre-long twin rail track connecting the VAB to the launch pad, an umbilical tower, a water-based acoustic dampener and a jet deflector immediately below the launch pad.

SSAB was established to allow integration of the two S200 solid strap-ons of GSLV-Mk3/LVM3. A short distance from the SSAB (58 m high, 55 m long and 40 m wide) is the larger VAB (83 m × 40 m × 32 m). A rail track connects the SSAB, VAB and the launch pad. Another track has been added connecting the SVAB to the SLP.

Figure 9-7 Site of the Second Launch Pad. Credit Google

An additional MLP has been developed specifically for the GSLV-Mk3/LVM3. This heavy-lift launch vehicle consists of a first stage with two solid strap-on motors, a large liquid second stage and a cryogenic third stage. The complete vertically integrated launch vehicle exits the VAB through 20 horizontal doors and is transported on a 16-wheel MLP (19.5 m × 19.5 m, 700 tonnes). The journey from the VAB to the launch pad along a 14-m wide, 1-km-long twin rail track can take several hours. The launch pad has a 70-m tall umbilical tower used for control, communication and fuelling.

Three vertically repositionable and swivelling access arms provide access to the vehicle while at the umbilical tower. The SLP is surrounded by four 120-m high towers to protect the launch vehicle against lightning. Once again multiple cameras are mounted on the towers to remotely monitor events from the MCC during the preparation and launch. SRO's largest launch vehicle GSLV-Mk3 introduced a concern with noise and vibration. The first stage consisting of two solid stage boosters with 210 tonnes of solid propellant in each one, is one of the largest solid stages in use around the world today. The aluminium particles emanating from the exhaust of the two large solid motors generate strong vibrations, increasing the risk of structural damage particularly to the sensitive components in the payload.

To mitigate the hazards of vibration and reduce the noise at launch, a water-based acoustic dampener was installed in 2013 in the SLP. The 80-m high water storage tower, which has a capacity of 50,000 tonnes of water, is emptied in 20 seconds at launch. The energy of the exhaust from the solid boosters is used to turn the water into steam, adding to the dramatic spectacle of the launch. This acoustic dampening system has produced the desired reduction in vibration with a tangible reduction of the noise at the moment of launch to one-quarter of the original (as engineers measure it, an attenuation of 6 dB).

Figure 9-8 Second Launch Pad Noise and Vibration Suppression System. Credit ISRO

Sriharikota is located in a region where storms and cyclones are common. Unlike at FLP, the launch vehicle for the SLP is integrated on MLP within the VAB and then moved to the launch pad. This fixed VAB provides launch vehicle assembly and payload integration environment for all weather conditions. If required, perhaps under extreme weather conditions, the assembled launch vehicle can be wheeled back from the launch pad to the VAB for safety, maintenance or reconfiguration. This approach also reduces the time the pad is occupied. The FLP and SLP are supported by onsite storage, transfer and servicing facilities for all the propellants ISRO uses for its launchers, solid, liquid and cryogenic. Solid propellants are integrated into the solid stages at the time of manufacture, months in advance. Liquid and cryogenic propellant filling operations are conducted only at the launch pad during the hours leading up to the launch.

Figure 9-9 GSLV-D5 on the MLP moving towards the SLP. Credit ISRO

Fuel loading is managed remotely by the MCC only after the integrated launch vehicle stands alone and all personnel have been evacuated from the launch pad. Propellant loading is conducted by the Filling Control Centre (FCC). Propellant loading is a critical step in preparing for the launch.

Propellants consititutes most of the mass of a launch vehicle which is comsumed by the time the spacecraft arrives in its orbit.

The mass of the payload is typically less than 1% of the total mass of the fully loaded launch vehicle. The liquid fuel used by PSLV and GSLV is loaded in the final hours just before the launch. These fuels are delivered to the launch vehicle in colour-coded pipes, red for propellant (UH25) and yellow for the oxidant (N_2O_4). Currently, cryogenic propellants (LOX and LH2) are manufactured at IPRC in Mahendragiri and transported using specialised tankers by road to holding tanks at Sriharikota. ISRO's cryogenic engine CE-20 has a capacity of 27 tonnes of combined LOX and hydrogen. While offsite manufacture and transport is practical at the present, Sriharikota will need onsite facilities to produce both as quantities rise in the future.

Prior to launch, the FCC loads the GSLV's third stage propellant from these holding tanks. To prevent high-pressure build-up in the hydrogen tank, some hydrogen is allowed to escape. However, to keep the propellant tank full, the hydrogen tank is continually topped up during the final seconds prior to the launch. During the launch, when the cryogenic stage kicks in, turbo-pumps running at around 40,000 rpm deliver the propellants from the onboard tanks to the combustion chamber.

Solid Propellant Booster Plant

The Solid Propellant Booster Plant (SPROB) was commissioned in March 1977 and is responsible for the manufacture of the solid propellant on an industrial scale. Vasant Gowariker (1933–2015) was influential in establishing SPROB at Sriharikota following his work in setting up the Propellant Fuel Complex at Thumba in 1972.[489] Gowariker was one of the scientists Sarabhai had brought back to India from the UK in the early days of Thumba. Some of the propellant processes and formulations used by ISRO today were first developed at Thumba. ISRO experimented and developed materials used in propellants, such as bonding agents, resins, plasticisers, inhibitors and insulators. In pursuit of self-sufficiency, ISRO acquired from foreign nations some of the plant equipment needed for high production volumes, typically 500 tonnes per year.[490] SPROB is one of the largest facilities within the Sriharikota complex comprising two independent plants for the production of the solid propellant to meet ISRO's needs. It is charged with the production of Hydroxyl-Terminated Polybutadiene (HTPB), a propellant that ISRO has a long history of producing for its launchers, including the Rohini series of sounding rockets,

SLV-3, ASLV, PSLV and the GSLV. ISRO has been gradually increasing the quality and types of propellants and the scale on which it manufactures them. During the 1960s, the solid propellant grain (a cylindrical unit of solid propellant produced by the plant) increased from the 75-mm diameter, weighing 4.5 kg for the RH-75 to the 2.8-m diameter, weighing 125 tonnes used by the PSLV.[491] HTPB is an established composite propellant used by space agencies of other nations, including the US, Russia and Japan, as well as an increasing number of commercial organisations.

Once manufactured, HTPB is safer to handle, store and transport compared to liquid and cryogenic propellants. The process of HTPB manufacture is still hazardous. The HTPB propellant is a mixture containing oxidiser (ammonium perchlorate, 68%) and fuel (aluminium powder, 20%). Combining the oxidiser within the propellant permits combustion outside the atmosphere in the vacuum of space. HTPB is a paste when manufactured and is cured (hardened) at temperatures of up to 150°C. Although not an explosive, it is a highly combustive material that, once ignited, cannot be stopped.

Worldwide, there have been many instances of fires and explosions in the production of propellant, resulting in damage and loss of life. Abdul Kalam recalls from his days at Thumba a particularly dramatic escape "trapped in this inferno, Sudhakar, however, did not lose his presence of mind. He broke the glass window with his bare hands and literally threw me out to safety before jumping out himself."[492] The incident Kalam refers to did not involve HTPB, but an incident on 23 February 2004 at SPROB did and cost the lives of six staff, including three engineers. The incident occurred during the transport of a rocket motor from SPROB building 117 to another. Building 117 was completely destroyed.[493] For safety reasons, the SPROB facility is located away from other Sriharikota installations, and individual SPROB buildings are dispersed over a large 25 km^2 SPROB complex isolated from each other for the same reason.

Also, located on-site at Sriharikota is the Static Test and Evaluation Complex (STEX). Through a variety of tests at STEX, ISRO measures combustion characteristics. For every batch of propellant produced, a sample is test-fired to quantify its physical and propulsive characteristics. When used in external strap-on boosters, it is critically important that the propellant in each booster combusts at a known and consistent rate to meet the mission launch profile. Inconsistent combustion between strap-ons can result in differential forces leading to a loss of directional control. To analyse the structure of the solid booster casing, SPROB uses a high-power X-ray source. This meticulous level of inspection provides assurance on the physical integrity of the booster casing before it is used in flight. Further, while most rocket motors used for spacecraft manoeuvres in space use liquid propellant, some use a solid propellant. To establish baseline characteristics for space-borne firings, a sample of the solid propellant is tested within a thermo-vacuum chamber situated at Sriharikota.

Local Propellant Facilities

A fully loaded rocket is a precarious and dangerous entity. To minimise risk, liquid propellants are loaded only in the final few hours prior to the launch. Each of the launch pads has a local storage facility for liquid propellant UH25 (75% unsymmetrical Dimethylhydrazine or UDMH and 25% hydrazine hydrate) and N_2O_4 (dinitrogen tetroxide) from where the loading takes place. Sriharikota has a bulk storage facility for up to 400 tonnes located within the complex but at a safe distance from the launch pads. On 19 August 2013, the launch of the GSLV-D5 mission using a GSLV Mk-2 developed a propellant (UH25) leak in its second stage, 14 minutes prior to the scheduled launch at 16:50. The leak was triggered during the second-stage pressurisation phase that occurs near the end of the launch sequence. Once the leak was detected, the launch was immediately aborted, but by then 750 kg of the propellant had leaked, contaminating other parts of the launch vehicle and the immediate vicinity of the launch pad.[494]

UH25 and $N_2O_4$are known as hypergolic fuels. In addition to being efficient propellants, they exhibit a very desirable attribute in that they require no ignition mechanism. They spontaneously ignite upon coming into contact with each other. Had there also been a leak of the oxidant N_2O_4 at the same time, it could have resulted in a devastating explosion on the launch pad? The launch vehicle, payload and the launch pad itself could have been destroyed.[495] This potentially disastrous incident with far-reaching consequences was averted by the prompt implementation of the procedure established to address such a scenario. In addition to safely draining the nearly 350 tonnes of propellant from the problematic second stage, the liquid fuel from the four strap-ons and the cryogenic fuel from the third stage was also removed. This had to be done while ensuring that the small pyrotechnic devices used for stage separation were not triggered. The situation was managed professionally, and SLP was made safe. It was "more than defusing a bomb", recalled the ISRO chairman.[496] The SLP was back in normal use after six days. Eventually, all four of the strap-ons and the problematic second stage were replaced, and the GSLV-D5 Mission was launched successfully on 5 January 2014.

Electric Propulsion

A typical two tonne communication satellite requires typically around 25 kg of chemical propellant for station keeping (maintain its precise place in orbit) for each year of its operational life which is typically around 12 to 15 years. On arrival at its orbital slot, a communication satellite will have about 300 kg of propellant. It is the depletion of this propellant that ultimately marks the end of the satellite's mission. Electric propulsion has the potential to reduce or replace this propellant.

Traditional engines use combustion to create thrust (accelerate and expel particles at high speed). Electric Propulsion System (EPS) creates thrust by using electricity from solar cells to create thrust (accelerate and expel particles at high speed using electrostatic or electromagnetic force). EPS produce very low thrust compared to chemical propellants but is very much more efficient with an Isp of around 1500 to 3000 compared with an Isp of 450 of Cryogenic engines. The low thrust requires EPS to be active continuously for days whereas chemical engines operate for seconds or minutes. For interplanetary spacecraft, EPS has to operate continuously for weeks, months or years. One of the earliest spacecraft fitted with an operational EPS using the inert gas xenon as fuel was NASA's Dawn mission launched in 2007 to explore the asteroids Ceres, Vesta and others in the solar system. Dawn's EPS accelerates the spacecraft from 0-60 mph in 4 days but uses only about one kg of fuel.[497] Although not suitable for use at launch, EPS is a remarkably attractive propulsion mechanism once in space.

Instead of EPS, ISRO uses the terminology, Stationary Plasma Thrusters (SPT). ISRO first attempted to test SPT on GSAT-4 in 2010 but the launch was not successful. The 2,230 kg GSAT-9 launched on 6 May 2017 is configured with the chemical propellants for station keeping. In addition, it has 80 kg supply of xenon and 4 SPTs as a technology demonstrator.[498] ISRO engineers will test the SPT technology on GSAT-9 and acquire operational experience with view to increasing the use of SPT in the future. Whereas GSAT-9 SPT is a technology demonstrator, chemical thrusters provide the primary source for station keeping, ISRO is designing GSAT-20 where only SPTs are used for station keeping. GSAT-20 is scheduled for a launch in 2018. Electric propulsion has been used for many years by other space agencies. Hundreds of the current operational spacecraft use electric propulsion. It is expected that all new spacecraft from 2020 onwards will be equipped with electric propulsion systems.

Launch Dynamics

Any launch site selected is a compromise of some or all the above factors. India's first launch site, Thumba, was not entirely unpopulated at the outset but it was on the coast. Further, Thumba was designed for sounding rockets and launched only to sub-orbital (go up and then down again) flight over the Arabian Sea.

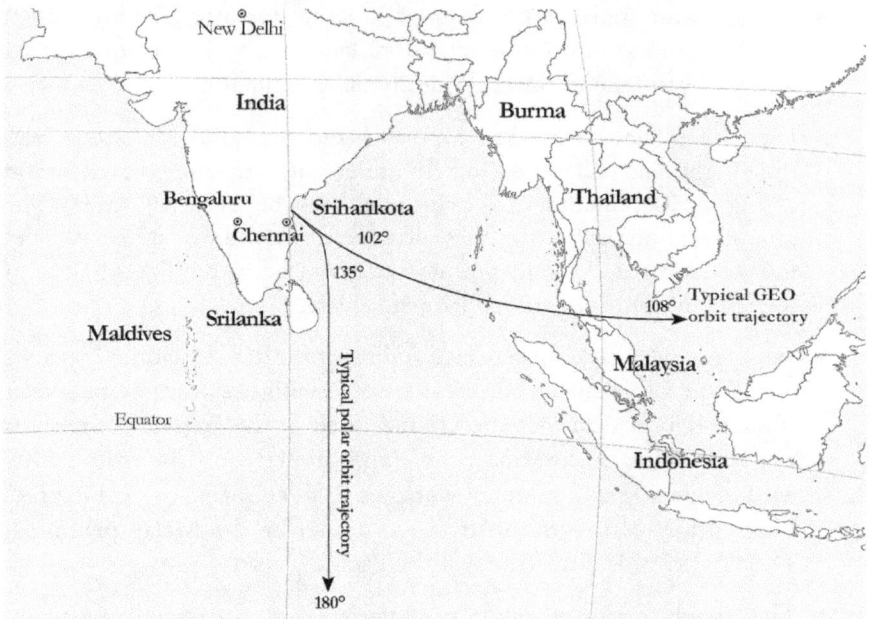

Figure 9-10 Launch Trajectories from Sriharikota. Credit Bhushan Hadkar

The ground track of the sub-orbital flights did not cover the Indian land mass. A special road was built when Thumba was commissioned, and it has rail links, too.499 The trajectories of all launches from Thumba took the rockets away from land and over to the Arabian sea. Thumba's location, near the south-western tip of India, launching to a polar orbit is not an issue but it is not suited for equatorial (easterly) launches, as the launch vehicle would have to fly over populated areas. Satellites orbit the Earth in two distinct orientations, pole to pole (north-south) known as a polar orbit and west to east in line with the Earth's rotation known as an equatorial orbit. Each category, polar and equatorial, has multiple variations and there are other orbit types too. A polar orbit requires launch in a north or south direction, whereas a due east launch is required for an equatorial orbit. An ideal due east launch from the Earth's equator can take advantage of the Earth's maximum rotation and pick up a 'free' 465-m/s velocity (or a lower

207

value if the launch site is further away from the equator[500]). Establishing an actual launch site is subject to a number of safety, logistical and technical considerations.

- The launch site and the area immediately around it must be located away from populated zones and yet offer convenient transport links for the supply of launch vehicle parts, components, propellants, as well as staff who work there.

- An optimum polar orbit requires a clear due north/south launch corridor, and an optimum equatorial launch requires an unhindered due east/west launch corridor, ideally on the equator.

- Launch vehicles are tracked (optically and by radio) for safety, real-time flight data collection (to help understand unexpected events) and taking measurements that help calculate the eventual orbit. The ground stations collecting these data from ascending launch vehicles must be located along its ground track and be in the line of sight of the transmitters onboard the launch vehicle.

- Staging (where each rocket stage is discarded after the launch, once its propellant has been consumed) is a well-established feature of a rocket launch. It is critical that the ground track of the launch trajectory is unpopulated to ensure that people and property are safe from falling used rocket stages. Equatorial launches (PSLV/GSLV) discard the first stages prior to the ground track encountering the Malay peninsula. The final stage is discarded over the Pacific Ocean.

- The launch trajectory should not interfere with the sovereign airspace of neighbouring countries. India is a signatory to the International Liability Convention and thus legally liable for any loss of life or damage in neighbouring countries resulting from its space programme.

The barrier island topography of Sriharikota provides a natural separation from the larger populated towns nearby, such as Chennai. It has, over the years, developed its infrastructure to meet its annual operational goals, although the key transport link is a single road leading from the mainland to Pulicat Lake. However, Sriharikota is located at 13°N on India's eastern coastline, and local geography enforces launch constraints. From Sriharikota, Sri Lanka in the south gets in the way for a polar orbit and Malaysia and Indonesia (including the busy shipping lanes and offshore oilfield) in the east for an equatorial/geostationary orbit. Thumba was conceived and built between 1962 and 1963 to explore the EEJ at a time when India had no rocket launch experience, and the capability to launch satellites into orbit was a distant aspiration. Although not suited for equatorial orbits, Thumba fulfilled its objectives at the time. To place payloads into equatorial orbit, India needed a launch site on the East coast.

Figure 9-11 International Launch Sites around India. Credit Adapted from Federal Aviation Administration Compendium 2016

To overcome these constraints, a typical polar orbit trajectory for a PSLV launch is not due south at 180° (the most efficient) but 135° with a course correction after launch vehicle to manoeuvres around Sri Lanka. Similarly, a typical (each mission has its unique requirements) GSLV launch trajectory for an equatorial orbit is not due east at 90° but 108° with a course correction once the launch vehicle is beyond Malaysia. These enforced initial launch trajectories and course corrections decrease the efficiency of any launch from Sriharikota. To compensate, additional propellant is required which results in a reduced payload mass or decreases the satellites orbital altitude. At present, between the two launch pads, ISRO can schedule up to eight launches every year. While the VAB is occupied for several weeks for each launch, the launch pad is engaged for only around ten days. To cater for the expected increase in launch operations additional VABs and launch pads are required. A formal ministerial announcement in the Rajya Sabha on 12 March 2015 indicated that a Third Launch Pad (TLP) would be built at Sriharikota, but no firm timescales have been established. In the meantime, construction of the second VAB (SVAB), finally got underway in early 2015 when funds dedicated to the project of Rs.130 crore ($20 million) was announced.[501] It will have its rail track to the SLP. SVAB will be used to prepare existing and future launch vehicles.

It is expected to become operational around the end of 2017.[502] Since its establishment, Sriharikota has conducted just over 50 launches of substantial payloads to Earth, Lunar and Martian orbits. When eventually built, the TLP at Sriharikota will be subject to the same restrictions of launch trajectory as the exiting launch pads. ISRO contemplated the idea of establishing another launch site south from Sriharikota from where PSLV could be launched due south, directly into polar orbit, the proposal was initially dropped but the lobbying continues.[503] One proposal for a TLP would have placed it near the southern tip of India (Kulasekarapattinam in Tamil Nadu), which could have increased the launch capacity of a PSLV by about 300 kg.

Figure 9-12 Proposed Second VAB. Credit Adapted from ISRO

This increase in capacity would come from the more southerly (closer to the equator) location and a reduction in the post-launch manoeuvre required to avoid countries in the ground track of its trajectory.[504] In addition, the site is conveniently located close to IPRC. Scientists, engineers and politicians have been championing the Tamil Nadu site for the TLP but appear to have lost out for the present.[505] As India's space operations continue to evolve, perhaps the Kulasekarpattinam site will be commissioned in the future.

≺ ◇ ≻

Chapter Ten
ISRO's Rockets

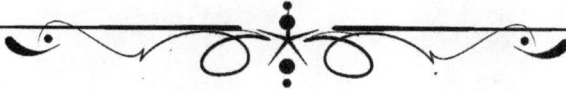

S pace is only a 100 km away, less than the distance to Chennai from Sriharikota. Officially designated as the Karman line[506], an altitude of 62 miles (100 km) is internationally recognised as the boundary between Earth and space. A spacecraft at an altitude of 100 km or above is in space. As with the Earth's equator, there is no visible indicator marking the Karman line. At an altitude of 100 km, the Earth's atmosphere is so negligible that wings cannot generate lift, and the absence of oxygen prevents combustion engines from working, so flight by aircraft is technically impossible. Getting to space is about velocity not altitude. Rockets, or launch vehicles, are the only means by which a spacecraft can be delivered to space with the required velocity to stay there. Unless a spacecraft has a velocity of at least 5 miles/s (8 km/s), it will fail to maintain orbit and fall back to Earth. It is the job of the launch vehicle, the rocket, to give the spacecraft this initial velocity essential for it to reach and stay in space.

The heavier the payload or higher the orbit, the larger the rocket has to be to carry the additional fuel. India's primary launch vehicle is just over 300 tonnes, while European Space Agency's (ESA) Ariane 5 is about 780 tonnes. The American Saturn 5 rocket that took three men a quarter of a million miles to the Moon weighed 3,000 tonnes. Saturn 5 was and remains the largest and most powerful rocket ever flown (USSR's N1 rocket was equally powerful but not successfully flown).

The first goal that ISRO (then INCOSPAR) set itself following the successful launch of the Nike-Apache rockets in 1962 was to develop its own rockets. In the years that followed, young Indian rocket engineers gradually developed the technology and techniques to produce rocket fuel and build rockets. They also built the infrastructure necessary to launch, track and communicate with a spacecraft beyond the Karman line. While the USSR and US engaged in the space race and developed the technology for human spaceflight and the Moon landing, most other nations engaged in the more prosaic use of rockets (initially sounding rockets) for the scientific investigation of the Earth's upper atmosphere. During the early 1960s, many nations were preoccupied with their commitments to the IGY (1957-58). Experiments using sounding rockets spread around the world to places, including India, Argentina, Canada, Japan, Pakistan, Denmark, Norway, France, Germany, Spain, Sweden, the UK, US and the USSR.[507]

Figure 10-1 ISRO Family of Launch Vehicles. (Left to Right) SLV-3, ASLV, PSLV, GSLV Mk2, GSLV-Mk3. Credit Wikimedia Commons

Sounding rockets were used to investigate (to "sound" out) the upper regions of the atmosphere made accessible with the advent of rockets. They could study the regions of the Earth's atmosphere above where balloons could go and below where satellites orbit. India's first sounding rocket was the RH-75 weighing just 7 kg. It was first tested in November 1967. RH was the abbreviation for the name Rohini and 75 its diameter in millimetres. RH-75 used a cordite, a mixture of nitro-glycerine and nitrocellulose, manufactured in cylinders with a hollow core in a factory in Tamil Nadu and transported by road to Thumba. Fuelling the RH-75 required sliding the cordite cores into an aluminium casing and inserting a black-powder based igniter at the base.[508] Between 1967 and 1995, eight versions of the Rohini were developed, each successively more powerful, with a capacity to deliver a heavier payload to a higher altitude.[509] Over 1,200 RH-200 were launched from Thumba and Sriharikota for science programmes, such as Monsoon Experiment, Equatorial Wave Studies and Dynamic Middle Atmosphere.[510] In November 1997, an Indian-built sounding rocket was launched from outside India for the first time. A RH-300 Mk11 was launched from Svalbard Rocket Launching Range in Norway. At a latitude of 78°N, Svalbard is one of the northern-most launch sites on Earth. [511]

Propellant Type	Typical Isp	Characteristics
Solid	250	Cheap, relatively easy to manufacture, store and transport. Rocket-motor technology is not complex. Its main disadvantage is that once ignition starts, it cannot be stopped or regulated. Typically, HTPB, a common solid fuel used in the first stage of the PSLV, has an Isp of 247.
Liquid	335	An engine using liquid propellant can be started and stopped and restarted multiple times. This ability to regulate flow and thus control thrust is a necessary requirement to place spacecraft in a precise orbit and later help maintain that orbit.
Semi-Cryogenic	370	Semi-cryogenic rocket propellant is one of the oldest, and still very common, a combination of fuel and oxidiser. The fuel is typically kerosene, and liquid oxygen at -183°C is the oxidiser.
Cryogenic	455	The most efficient rocket engines in common use today. A cryogenic engine uses LH2 at -253°C and LOX at -183 °C, which is a formidable engineering challenge.

Table 10-1 Relative Differences between Propellant Types

These sounding rockets did not represent profound technological achievements. They provided the initial experience necessary on the road to a national space programme. During the late 1960s, Indian scientists and engineers went to study and train in France, and French engineers came to Thumba to help set up plants for the local manufacture of the Centaure rocket stages and its propellant. Sarabhai envisaged this as a necessary phase for a transition from permanent reliance on foreign nations and insisted that "space technology was acquired under licence from abroad only as a means to buy time."[512] Between 1965 and 1988, 81 Centaurs were launched from Thumba; all but 10 had been manufactured in India under licence.

Engines that use solid, liquid, semi-cryogenic and cryogenic propellants have a respectively higher Isp. For example, to generate 1 tonne of thrust, 3.8 kg of solid, 3.5 kg of liquid or just over 2 kg of cryogenic propellant is required per second.[513] Each propellant type has unique characteristics, ease of manufacture, storage, transport, cost and hazards associated with its handling. Electric propulsion is the most efficient with an Isp of over 1000 but has low thrust. Whilst it can manoeuvre spacecraft in orbit slowly over time with much greater efficiency then chemical propellant, it lacks the required thrust for use at launch.

Starting in the 1960s with rocket motors that used solid propellants, India has now developed rocket engines that use liquid, semi-cryogenic and cryogenic technologies, too. [514] Engineers measure the efficiency of a rocket motor or engine using the concept of Specific Impulse (Isp). One Isp is the thrust generated by 1 kg of propellant in 1 second of combustion measured in units of seconds.[515] The higher the Isp of a rocket engine, the more efficient it is. The more efficient the engine, the less the propellant required to place a given payload in orbit.

Launch Vehicle	SLV	ASLV	PSLV	GSLV
Period of Development	1972–1983	1982–1994	1982–2017	1991–2017
Mass at Launch (tonne)	17	40	320	640
Length × Width (m)	22 × 1	24 × 1	44 × 2.8	51 × 4
Max Payload (kg)	40	150	1,400	2,500
Number of Stages	4	5	4	3
Propellant Type	Solid	Solid	Solid and Liquid	Solid, Liquid and Cryogenic
Guidance	Open Loop Inertial	Closed Loop Inertial	Closed Loop Inertial	Closed Loop Inertial
Orbit Type and Altitude (km)	LEO 400	LEO 400	LEO and SSPO 800	GTO 36,000
Total launches (mid 2017)	4	4	40	17

Table 10-2 Key Characteristics of ISRO's Launch Vehicles up to mid- 2017[516]

In January 2014, ISRO overcame the final challenge and successfully flight-tested its own cryogenic engine with the launch of GSLV-Mk2 that carried

GSAT-14. From the tiny RH-75, ISRO has gone on in every successive decade to develop a launch vehicle with enhanced capabilities, SLV-3 in the 1970s, ASLV in the 1980s, PSLV in the 1990s and GSLV since the 2000s. It has not been a smooth, clean serial development. There have been overlaps, with many teams working on multiple technologies simultaneously and that approach continues today.

ISRO has two operational launch systems, PSLV and GSLV, and is actively developing new systems for the future. Larger rockets are required to take heavier payloads to higher orbits. There are broadly three types of Earth orbits used by ISRO's satellites, LEO at 200–2,000 km parallel with the Earth's equator, a pole to pole Sun-synchronous polar orbit (SSPO) at 600–800 km and GEO at 36,000 km also parallel to the equator. GSLV has placed a few payloads into orbit, but with almost 40 successful launches, the PSLV is ISRO's current launcher of choice. The LVM3, first launched in December 2014 to a sub-orbital flight, matured into the GSLV-Mk3 and in June 2017 successfully placed the GSAT-19 in orbit. The Unified Launch Vehicle (ULV) and Heavy Lift Launch Vehicle (HLV) are ISRO's next evolutionary step in launch vehicle design. Still, in the early design stage, the ULV and HLV will use semi-cryogenic and cryogenic technologies. The initial configuration is designed to deliver six tonnes and more to GTO and eventually replace the PSLV and GSLV in the coming decades.

Inertial Guidance System

Rocket engines provide the power to place a spacecraft in orbit. One of the many onboard subsystems is the Inertial Guidance System (IGS), which is the directional control system ensuring that the satellite is delivered to the designated orbit with high precision. IGS technology is highly classified since it can be used by rockets or missiles. Developing IGS technology was one of the many "formidable problems" that ISRO had to overcome.[517] IGS uses the basic principle that if the starting place and time are known, and all changes in speed, direction and duration are measured during the journey, then position and distance to the desired destination can be calculated at any stage in the journey.

In the past, IGS relied only on mechanical components engineered with high precision, analogue signals and mechanical actuators. Today, they are likely to be digital (commonly found in smartphones), but the principle remains the same. Three accelerometers detect motion in each of the three directions, east-west, north-south and up-down. A set of three specially positioned gyroscopes can detect rotation in each of the three planes. Like a compass floating in a bath of water on a moving ship, these sensors are fixed to a platform within the launch vehicle to maintain the original orientation irrespective of the direction of the launch vehicle. As the launch vehicle moves from take-off to its destination in orbit, electronic signals generated by the sensors are sent to the onboard computer. The computer detects where the launch vehicle is, knows where it should be, dynamically calculates the required directional change and activates the directional control mechanism to achieve it.

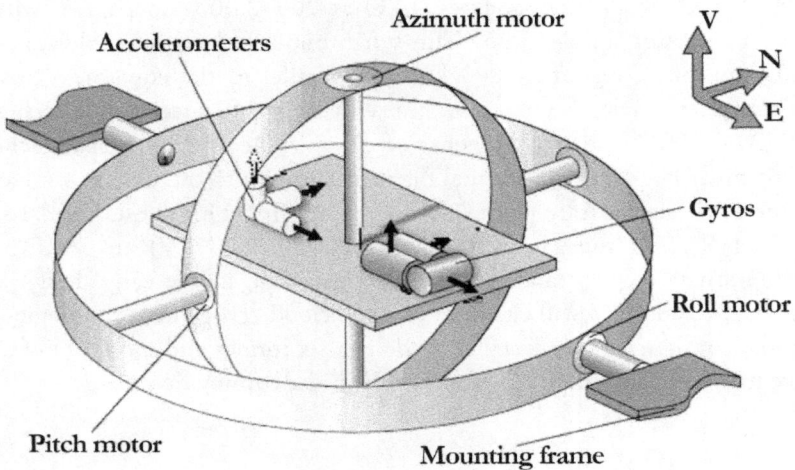

Figure 10-2 Inertial Guidance System. Credit GEC Marconi

There are four primary control techniques used as steering mechanisms for launch vehicles, secondary injection thrust vector control (SITVC), Flex Nozzle Control System, Engine Gimbal and reaction control system (RCS). SITVC is used on a solid stage at the nozzle where the high-temperature exhaust is escaping. Small quantities of strontium perchlorate are injected at particular points around the nozzle to create a differential thrust to modify direction.[518] An alternative to SITVC is to implement a flex nozzle where the nozzle is not fixed and can move to change the direction of flight.[519] Engine gimbaling is similar to flex nozzle but is a more sophisticated technique, where the whole nozzle can be moved, like a ship's rudder, to modify the direction of flight.

The RCS is a mini rocket engine (also known as a vernier engine) with either a rotatable nozzle or multiple fixed nozzles pointing in different directions. The direction of the launch vehicle is controlled by firing this small engine in a specific direction. Since directional control is required from the moment of launch to the moment of orbit insertion, each rocket stage has its own individual control mechanism that manipulates the direction of the flight in three axes. The single IGS system to which each of the stages is electronically connected is located in the equipment bay of the final stage, along with the computer that receives input and issues instructions to control the direction of flight to each successive stage.

Figure 10-3 Attitude Control. Left: SITVC used by PSLV-XL booster for roll control. Centre: PSLV Second Stage engine gimbaling. Right: GSLV-CUS with Two Vernier Engines (circled). Credit Adapted from ISRO

To arrive at the prescribed orbit, the launch vehicle must be physically steered during the entire flight by the computer-controlled IGS. As ISRO evolved its launch systems, it also developed its competence in the IGS and the four control systems it operated, SITVC, flex nozzle, engine gimbaling and RCS. The SLV-3 and ASLV used solid propellants and relied on SITVC and small RCS for directional control. The second and fourth stages of the PSLV use liquid propellant employing engine gimbaling. The GSLV's cryogenic upper stage (CUS) uses two vernier RCS.

Satellite Launch Vehicle (SLV-3)

As the first men to walk on the Moon returned to Earth, Vikram Sarabhai was setting down his vision for India's space technology in the new decade. In his landmark paper published in 1970, he articulated in detail his plans for India to develop the indigenous capability to build and launch satellites during the decade 1970–1980. This capability was essential if India was to benefit from applications of space in the fields of communication, meteorology and remote sensing.

At the outset, ISRO had relied on the USSR, US or ESA to launch Indian satellites. By the late 1960s, ISRO had already attained a significant level of competence and confidence in producing and using solid fuel. The development of liquid propellant within ISRO was then at an early stage.[520] The SLV-3 was India's first launch vehicle with a capability to deliver a significant payload to orbit. ISRO had already developed the capacity to design and build satellites. With SLV-3, India was able to place them in orbit and for the first time attain self-sufficiency. For the next step ISRO would also need to develop unique materials and methods of construction requiring advanced aerospace engineering.[521]

By the time Sarabhai published this paper, a feasibility study and much of the preliminary work for SLV-3 had already been completed, so he was able to share some of the specifications. Through the new launcher's name, ISRO reaffirmed its aspiration. It wanted to build its own capability for launching satellites.[522] Following Sarabhai's sudden and unexpected death on 30 December 1971 and the consequent reorganisation, Satish Dhawan eventually took over the role of Chairman of ISRO. He appointed Abdul Kalam as Project Director for the SLV-3 programme, which formally kicked off on 3 November 1973.

SLV-3	Stage 1	Stage 2	Stage 3	Stage 4
Burn time (second)	49	40	45	33
Thrust (tonne)	56	30	10	2.9
Length × width (m)	10 × 1	6.4 × 0.8	2.4 × 0.8	1.4 × 0.7[523]
Weight of propellant (tonne)	10.8	4.9	1.5	0.4

Table 10-3 Key Design Features of SLV-3

The step from a sounding rocket to a launch vehicle capable of placing a significant payload in Earth orbit was one of ISRO's earliest challenges. The US had used the same approach with the Viking/Vanguard and France with Diamant/L3S; for India, it would be SLV-3. SLV-3 comprised 44 major subsystems, 7,000 electrical components and 25 km of internal wiring.[524] As Sarabhai had outlined, the final specification of SLV-3 was a four-stage, solid fuel, 22-m high rocket with a diameter of 1 m. Weighing 17 tonnes, it was designed to launch a 40-kg spacecraft into LEO.

SLV-3 was broadly based on the American Scout rocket. Its name derived from the acronym, Solid Controlled Orbital Utility Test rocket, with which countries outside the US were very familiar. The US had invited experiments from other nations that they launched aboard a Scout. The public offer was announced in The Hague at the March 1959 meeting of the COSPAR.[525] Thus technical details of the Scout were more readily available than others launch vehicles. While SLV-3 was not an exact copy of the Scout, a Scout-like design was a natural choice given its success rate of 96% and the experience that Indian scientists had gained during their brief time in the US.

Around 50 industries and institutions from outside ISRO were required to develop and build the major subsystems of SLV-3. The new set of technologies included designing, building and deploying modern instruments and subsystems typically found in all launch vehicles. They include

- Inter-stages (small sections between stages)

- Heat shield (usually considered for re-entry, but the speed during ascent is sufficiently high to require one too.) Also known as the Payload Fairing.

- Self-destruct mechanism (a safety precaution should the launch vehicle veer off course)

- Stage-separation mechanism using flexible linear explosive cords

- Gyroscopic stabilised platform for IGS

- Kevlar epoxy motor casings[526]

- Silica and carbon composite nozzles

- Multi-source tracking for range safety

Each stage of the SLV-3 had an independent system for directional control. The first stage used the SITVC for the first 17 seconds after launch and then an electro-hydraulically operated fin control after that. The second and third stages used a red fuming nitric acid (RFNA) and hydrazine-powered RCS, and the fourth stage was spin-stabilised. SLV-3 used two distinct methods for stage separation.[527] The separation between the first two stages was carried out by initiating the flexible linear charge located between the stages, and a ball-lock separation system was employed for the third and fourth stages.

The initial SLV-3 plan incorporated a sub-orbital launch using only two of the four stages (S2 and S4) prior to the first orbital flight. However, following a successful RH-560 launch, SLV-3 Project Director Kalam was convinced by his colleague Abdul Majeed's suggestion that it would be beneficial for the project to scrap the S2+S4 sub-orbital flight and go instead straight for the orbital flight.[528] Following a single board meeting, the sub-orbital flight was dropped, and the SLV-3 programme was fast-tracked to proceed with the first orbital experimental flight. The first launch attempt of the SLV-3 in August 1979 failed, and the launch vehicle was lost 317 seconds after launch. Once the Failure Analysis Committee (FAC) completed its investigation, Kalam described the painful failure "it was found that even while the rocket was on the launch pad, a full 8 minutes before the take-off, all the RFNA had leaked out. At T minus 8 minutes, a warm-up test was done on the second-stage RCS, as scheduled. A solenoid valve that was designed to close after the warm-up test did not close.

The resulting leak led to the complete depletion of the RFNA.[529] After launch, the first stage worked as planned, but without a working RCS on the second stage, the launch vehicle lost directional control. The flight was terminated, and the whole vehicle was lost prior to the ignition of the third and fourth stages. In July 1980, the second SLV-3 successfully placed India's first satellite, the Rohini Satellite (RS-1), in Earth orbit. Almost two hours after launch, RS-1's orbit brought it over the horizon in India, and a tracking station in Trivandrum (now Thiruvananthapuram) detected its signals.

Date	Payload	Result	Notes
10 Aug 1979	Rohini Technology Payload (RTP) 35 kg	Experimental flight. Failure.	A faulty valve caused a fuel leak in the second-stage RCS eight minutes prior to launch, resulting in loss of control and the flight was terminated. The vehicle crashed into the Bay of Bengal 317 seconds after launch. The payload was designed to monitor and provide telemetry on the performance of the launch vehicle.
18 Jul 1980	Rohini RS-1 35 kg	Experimental flight. Success.	RS-1 delivered to the orbit of 305 × 919 km with an inclination of 44.7° and remained in orbit for 20 months.
31 May 1981	Rohini RS-D1 38 kg	Development flight. Partial Success.	One of the SLV-3 stages underperformed, and although RS-D1 arrived in orbit, it was much lower than the intended altitude and did not last long. The planned orbit was 296 × 834 km, but the actual orbit achieved was 186 × 418 km with an inclination of 46°. The orbit decayed after 9 days.
17 April 1983	Rohini RS-D2 41.5 kg	Development flight. Success.	RS-D2 was placed into an orbit of 371 × 861 km with an inclination of 46°. It was in operation for 17 months and remained in orbit until April 1990. Its primary payload, a smart sensor camera, took over 2,500 pictures.

Table 10-4 SLV-3 Launch History

With this success, India became the sixth nation with the capability to design, build and launch its own satellites. Each of the four SLV-3 flights saw gradual improvement. Eighteen major improvements were implemented in the fourth flight, the most significant of which was the improved fibre-reinforced material used in the manufacture of the casing of the fourth stage, which reduced its weight from 26 kg to 16 kg. This single modification allowed an increase in the fuel it carried from 270 kg to 320 kg, which increased the orbit to 919 km from 438 km. During its short lifetime, SLV-3 made four flights, the first two were classed as experimental and the other two as developmental. The first flight failed, the third was mostly successful and the other two were recorded as entirely successful.

Sarabhai had established the principle that buying ready-built satellites and launch services from foreign countries was not sufficient; India needed to build the capability within India. Speaking at the press conference soon after the third SLV-3 launch, Dhawan stated "the cost of the total programme, in which 10,000 bright Indians have worked, we could hardly have bought two satellites" and insisted that "if you don't build your nation yourself, nobody else is going to come and build it for you."[530] The success of the fourth SLV-3 launch attracted widespread international attention and was a morale boost for the nation, as well as the engineers involved.

Prime Minister Indira Gandhi was present at the launch and commented "For me, it was a special thrill. Although I am well above 60, I have not lost the sense of wonder and marvel at what man can achieve."[531] Satish Dhawan continued to carry the torch of nation-building, crystallising what Sarabhai's imagination had woven into the space programme. A year later, Dhawan and Kalam received two of India's highest awards, Padma Vibhushan and Padma Bhushan, respectively.[532]

The fourth SLV-3 was so successful that the two subsequent SLV-3 launches planned were abandoned so that work on the next launch vehicle, the ASLV, could begin earlier. Through the development of SLV-3, ISRO proved its capability in successfully developing a launcher and placing an operational satellite in orbit. ISRO's next launcher, the ASLV, needed an enhanced launch capacity to put larger satellites into higher orbits.

Augmented Satellite Launch Vehicle (ASLV)

The Government of India approved the ASLV in 1981 (a year after the first successful launch of the SLV-3) to validate new technologies needed for future launch vehicles, such as strap-on boosters, canted nozzles, a bulbous heat shield to carry a large payload, vertical integration of the launch vehicle and inertial guidance system for navigation. The need for vertical integration also motivated the development of the FLP with a mobile service tower at Sriharikota. The FLP is still in active service.[533]

One of the key design changes in ASLV compared to SLV-3 was the use of strap-on boosters. To test strap-on (SO) technology before the first ASLV test, ISRO launched a two-stage hybrid launch vehicle, SO-300-200, constructed from a RH-300 core stage with two RH-200 strap-ons. This launch on 16 October 1985 allowed ISRO engineers to build and test canted nozzles, explosive bolts for strap-on separation and asymmetric ignition, strap-on (RH-200) ignition at launch and ignition of the core stage (RH-300) during flight.[534]

ASLV was conceptually a simple modification of SLV-3. It was SLV-3 with two strap-ons. The first stage of SLV-3 was augmented with another SLV-3 first stage on each side as strap-on boosters.[535] Although ASLV used solid fuel for propulsion in all stages, it used a small quantity of liquid fuel for directional control.[536] While SLV-3 had four stages, ASLV had five. However, to maintain consistency with the existing naming convention, the two strap-on boosters were referred to as stage zero or simply the booster stage. The rest of ASLV was pretty much SLV-3. At launch, ASLV weighed 40 tonnes, stood 24 m high and had a total flight duration of 640 seconds to reach the orbit. Compared to the SLV-3, the ASLV payload capability increased to 150 kg to LEO of 400 km. At launch, only the two strap-on boosters of ASLV ignited. The core stage ignited at T+49 seconds, just after the boosters were exhausted. The booster separation occurred at T+55, while the first stage was active. At T+1 minute and 35 seconds, the first-stage propellant was exhausted and the stage discarded. The second stage ignited and burned for 36 seconds. During this phase, with ASLV is in space at an altitude of 120 km, the heat shield was discarded.

There was a pause after the third stage as ASLV coasted for 294 seconds. Then, the final stage ignited. This fourth stage burned for 32 seconds and generated the spin required for payload stability. After a total flight of 10 minutes and 40 seconds, the spinning satellite separated from the launcher and entered orbit.[537] The propellant used in all stages of ASLV was solid but of two distinct types, polybutadiene acrylonitrile (PBAN) and High Energy Fuel (HEF-20) developed by ISRO.[538] HEF-20 is a more efficient propellant than PBAN. PBAN was used in the first two stages so that its weight was discarded early in the flight, and HEF-20 was used in the final two stages, just as in SLV-3.

In total, four ASLV launches attempted to place a satellite of the Stretched Rohini Satellite Series (SROSS) carrying a science payload into LEO. All ASLV flights were designated development flights and therefore were identified with the prefix D. The SROSS series (SROSS A, B, C and C2) satellites were small solar-powered, spin-stabilised satellites with a mass of around 106 kg with two science instruments to collect data on the plasma within the Earth's ionosphere and cosmic gamma-ray radiation. The primary objective of SROSS-A was to test and evaluate the performance of the ASLV launcher itself, so it did not have a science payload. Instead, it had an optical reflector so its position could be located using ground-based lasers to assist with trajectory determination.

SLV-3		Stage 1	Stage 2	Stage 3	Stage 4
Burn Time (second)		49	40	45	3
Propellant (tonne)		8.6	3	1	0.26
Length × Width (m)		10 × 1	6.4 × 0.8	2.5 × 0.8	1.5 × 0.7
Lift-off Weight (tonne)		10.8	4.9	1.5	0.4
ASLV	Stage 0	Stage 1	Stage 2	Stage 3	Stage 4
Burn Time (second)	46	46	36	43	32
Propellant (tonne)	10 × 2	10	3.8	1.2	0.4
Length × Diameter (m)	10 × 1	10 × 1	6.4 × 0.8	2.5 × 0.8	1.4 × 0.65
Lift-off Weight (tonne)	2 X 8.6	8.8	3.1	1.1	0.3

Table 10-5 A Comparison of SLV-3 and ASLV[539]

ASLV-D1, the first ASLV, was launched at 12:09 IST on 24 March 1987. However, 46 seconds after launch, the mission was lost. The two strap-ons (stage 0) were exhausted, but the core stage did not ignite as scheduled. The FAC identified the fault with the electrical signal for the ignition of the first stage. The electrical ignition signal for the first stage did not reach it. Why that happened remained unclear. Perhaps, the vibration at launch or the strong high-speed winds at 10-km altitude dislodged connections resulting in an open circuit, or perhaps, there was an inadvertent short circuit in both the primary and backup ignition circuits, or maybe, there was a random malfunction of the safety circuit that arms the ignition system. In its report, the FAC recommended an additional redundant ignition system to prevent a repetition.

Just over a year later, the ASLV-D2 flight also failed at about the same time in the flight (around T+50 seconds). After a successful booster phase, the first stage ignited, but then the launch vehicle lost attitudinal control and broke up.

Launch Date	Payload	Result	Notes
24 March 1987 ASLV-D1	SROSS-A. 150 kg. Carried two retro-reflectors for tracking.	Development flight. Failure.	The first stage failed to ignite. The launch vehicle achieved a maximum altitude of only 10 km.
13 July 1988 ASLV-D2	SROSS-B. 150 kg. Carried a West German optical scanner and ISRO's gamma-ray burst experiment.	Development flight. Failure.	The booster phase ended about one second prematurely. The launch vehicle lost directional control and broke up.
20 May 1992 ASLV-D3	SROSS-C. 106 kg. Carried a gamma-ray burst experiment and a retarded potential analyser experiment.	Development flight. Partial Failure.	Partial failure of the fourth stage resulted in a lower than planned orbit of 391 km × 267 km with an inclination of 46°. The satellite operated from 25 May 1992 until re-entry on 15 July 1992.
4 May 1994 ASLV D4	SROSS-C2. 115 kg. Carried enhanced versions of a gamma-ray burst experiment and a retarded potential analyser experiment (a plasma detector).	Development flight. Success.	SROSS-C2 was placed in an orbit of 600 km × 430 km with an inclination of 45° as planned and remained in orbit until 12 July 2001. It discovered 12 gamma-ray burst sources.

Table 10-6 ASLV Launch History

Following two failures in succession, two committees, one within ISRO headed by S.C. Gupta and another with a broader national remit headed by Roddam Narasimha were established to investigate the cause and make recommendations. The FAC concluded that the second failure was not a repeat of the first. This time, the second stage had ignited, but after a brief, fatal delay. The investigation revealed that the strap-ons had burned out 1.5 seconds earlier than expected.[540] The transition from the first stage to the second occurred at 10 km, where the wind gusts were strong. While each of the stages had a SITVC control mechanism, during those 1.5 (some sources indicate 0.5) seconds, neither was actively controlling the launch vehicle. Like a car without a driver, the launch vehicle deviated and no longer pointed in the direction of travel. At high-speed the formidable aerodynamic forces resulted in catastrophic failure.[541]

Additional redundancy for booster separation and a Real-Time Decision System (RTDS) to prevent a 'no control zone' were recommended. The RTDS would continually monitor the thrust of the strap-ons and ignite the first stage once the thrust of the strap-on drops to a pre-specified threshold. Fins for vertical stabilisation used in SLV-3 but removed initially for ASLV were restored. The inclusion of the RTDS and other recommendations resulted in a reduction of payload capacity from 150 kg to 106 kg.

ASLV D-3 was launched in May 1992 and placed SROSS-C in orbit but lower than the intended one. A problem in the final stage prevented SROSS-C from attaining the expected orbit of 938 km by 437 km. It only managed a lower orbit of 391 km by 267 km. Although tenuous, the atmosphere caused sufficient drag for SROSS-C to re-enter two months after the launch. The fourth and final launch, ASLV-D4, in May 1994 placed SROSS-C2 in the intended orbit. It carried enhanced science payloads and exceeded in all its mission objectives. SROSS-C2 had a planned mission lifetime of only six months but operated for four years. During its operational life, SROSS-C2 discovered 12 gamma-ray sources and was included in NASA's Third Interplanetary Network.[542] The ASLV programme provided engineers with the opportunity to understand and experience deploying and operating a Closed Loop Guidance System (CLGS), which constantly monitored the launch vehicle's position, velocity, direction and attitude.

It maintained the desired trajectory by commanding course changes using the SITVC and RCS systems. The CLGS used accelerometers and gyroscopes on the launch vehicle to provide real-time information for navigational manoeuvres during ascent. SLV-3 had used an Open Loop Guidance System (OLGS), in which the launch vehicle was pre-programmed with a set of manoeuvres prior to launch.

The technology for OLGS as used in launch vehicles is almost identical to that used by early missiles. Initially, India had acquired assistance with OLGS from France; the US had refused.[543] Unlike CLGS, OLGS operated in the absence of real-time feedback. Once launched, the course changes were conducted as programmed without catering for trajectory variations that may result from inconsistent thrust or dynamic atmospheric turbulence. The CLGS could deliver the satellite to orbit with high precision. The SLV-3 D2 flight using OLGS had delivered the RS-D2 satellite within 62 km of the designated apogee (furthest point of the orbit from Earth), whereas the fourth flight of ASLV, using CLGS, delivered the SROSS-C2 satellite to within 18 km of the planned apogee.[544]

Figure 10-4 ISRO's First Hybrid Launch Vehicle with Strap-On, SO-300-200, to Test Strap-On Technology. 16 October 1985. Credit ISRO

ASLV was also ISRO's first launch vehicle to be integrated vertically. Today, the PSLV and the GSLV are integrated vertically. By the end of the ASLV programme, ISRO had demonstrated all the key capabilities to build, launch vehicles, and place satellites in orbit with precision. In addition to operating boosters and getting two SROSS satellites into orbit to return scientific data, ISRO engineers had acquired critical experience and confidence by investigating problems and developing solutions for each of the first three ASLV flights that had not gone as planned. Had the ASLV programme been 100% successful it would not have generated the valuable learning experience that it did. In the process, the engineers developed a deeper conviction in their own abilities. A thorough investigation of the first ASLV flight failure probably could not have prevented the second, as the causes were unique. Lessons learnt from ASLV were implemented in PSLV, including an improved autopilot that ensured that the second stage kicked in once a deceleration was detected indicating that the first stage burn was coming to an end. ISRO, however, needed to design a launch vehicle with greater thrust as satellites for use in EO or communication were heavier and needed to get to higher altitudes than ASLV could muster. ASLV was still relying only on solid propellant technology. The next significant step for ISRO was to grapple with the technology of liquid engines.

Polar Satellite Launch Vehicle (PSLV)

The target in Sarabhai's vision was to develop a launch vehicle by the end of the 1970s capable of putting a 1,200-kg satellite into GEO at 36,000 km.[545] He had given the task of developing more powerful rockets to the Space Science and Technology Centre, which was established close to Thumba and continues to flourish today. Even though the development and launch of four SLV-3 between 1979 and 1983 and four ASLV between 1987 and 1994 provided ISRO with the experience of placing progressively heavier launch payloads using larger launch vehicles, it could only manage small payloads (SLV-3 40 kg and ASLV 135 kg) to LEO.

Eventually, ISRO fulfilled Sarabhai's target of 1,200 kg with the third iteration of its satellite launch vehicle resulting in the PSLV, which has since been the mainstay of ISRO launch vehicles for over two decades.[546] One of the earliest references to the PSLV outside India was made in a US secret service report where it was referred to as the Polar Space Launch Vehicle. The evidence came from US spy satellite imagery of what looked like a static display of PSLV mock-up components on the ground of Sriharikota.[547] The PSLV stands 44.4 m high with a diameter of 2.8 m and depending on its configuration weighs either 229, 296 or 320 tonnes at launch. It was designed to take an Indian Remote Sensing (IRS) satellite, typically 1 tonne, to a 900-km SSPO from Sriharikota. The most powerful variant of the PSLV, the PSLV-XL, can deliver 1.4 tonnes to GTO.

The key innovation in the PSLV, a four-stage launcher, was the introduction of liquid propellants. PSLV uses solid and liquid propellants in alternating stages. Not only are liquid propellants inherently more efficient than solid propellants, but unlike engines that use solid propellants, liquid propellant engines can be regulated, stopped and restarted multiple times. It was essential for ISRO to master this capability. The additional control and thrust available only through the technology of liquid engines were essential to achievingt its ambition for larger payloads to GTO and from there to Geosynchronous Equatorial Orbit (GEO) required for communication satellites.

During the 1960s and early 1970s, France assisted ISRO engineers to gain key skills and experiences in solid rocket motor technology through the construction under licence of the Centaure rocket in India. However, ISRO's total experience in using liquid fuel was negligible. Small amounts of liquid propellants were used for the operation of the SITVC and RCS systems that were deployed on SLV-3 and ASLV. What ISRO needed was to design, build and deploy complete stages using only liquid fuel, something that in the 1970s was completely outside ISRO's capability. By chance, France made an offer ISRO could not refuse.

Sarabhai had established the tradition of sending Indian engineers and scientists to reputable institutions in the US (Wallops Island, MIT, Goddard Space Flight Centre) to gain hands-on experience in launching rockets when Thumba was first established. The tradition continued throughout the period of the Cold War and even into the early 1990s. Indian engineers went to French Guyana Space Centre in South America for training in handling launch vehicles, range safety and radar tracking; to Deutsche Forschungs-und Versuchsanstalt für Luft-und Raumfart in West Germany for training in high-altitude test facilities, pulse code modulation telemetry and wind tunnel testing; to France for developing the Centaure rockets programme with Sud-Aviation[548] and to Russia for training in cryogenic engine technology.[549]

Sarabhai had first engaged France through Jacques Blamont to supply the sodium payload for India's first sounding rocket in 1963. Dhawan continued that relationship and invited CNES President J. F. Denisse to visit ISRO in January 1973. A seven-member delegation visited Ahmedabad and Trivandrum, and a joint commission set up two working groups, one on TV satellites and the other on launchers.[550] In the following year, Dhawan visited France and concluded a unique agreement that would allow ISRO to acquire liquid Viking engine technology in exchange for hardware and labour rather than cash. This agreement was a product of the working groups established during the CNES President's visit. In July 1973, France had led the replacement of the European launcher programme called

Europa with Ariane. As its primary backer, France had to develop Ariane on a tight budget and timeline and needed the support of a large number of technically skilled personnel.[551] The agreement involved India helping France in developing Ariane in exchange for liquid engine technology transfer.[552]

The Viking engine for the Ariane launcher was being developed by the French company Société Européenne de Propulsion (SEP). ISRO would eventually customise and build Viking in India and rename it as the Vikas engine for use on the PSLV. The unusual agreement required ISRO to provide 100 person-years to work at SEP. This unique arrangement was the product of India's inability to make payments in hard currency and the supportive relationship that had been cultivated with France over many years. SEP would decide how and where 75 of the 100 person-years would be deployed, and ISRO would direct the remaining 25. Thirty-Five Indian engineers were sent to the SEP facility in Vernon to work on system development, quality control and programme management. In parallel, 25 engineers based at VSSC worked on design fabrication, integration and testing of the Viking engine technology.

The license fee for the Viking technology was included in the deal. No money changed hands. The difference was met by ISRO providing CNES 10,000 space-qualified pressure transducers.[553] The agreement ended in March 1978. By then, the Viking technology designs and documentation had been transferred to ISRO, and the engineers sent to France had returned having gained considerable experience.[554] The Vikas engine used by PSLV's second stage evolved from this experience. The second stage uses a single Vikas engine, 40 tonnes of UDMH as fuel and N_2O_4 as an oxidiser. The fourth stage deploys two L-25 engines using liquid fuel, Mono-Methyl Hydrazine (MMH)) as fuel and mixed oxides of nitrogen (MON) as the oxidiser. This use of liquid engine technology in the second and fourth stages is a key innovation in PSLV. The larger masses of the PSLV stages also required a more active stage separation system. To ensure collision-free stage separation, the first stage employed eight and the second stage four retro rockets.[555] In addition to the larger physical size and the mix of solid- and liquid-fuel stages, the PSLV incorporated a complex network of electronic systems that required 40,000 connections with 20 km of cable. Between the four stages, PSLV had 450 avionics packages and subsystems, which included navigation, guidance, pyro system and telemetry.

Originally, the PSLV was designed as a three-stage launch vehicle, stage 1 and stage 3 using solid fuel and stage 2 based on liquid fuel. The final stage of a launch vehicle is responsible for precisely inserting the payload into its designated orbit. To achieve that precision, the final-stage engine has to have the capacity to regulate its power, stop completely and restart. This level of control is only possible with liquid engines. ISRO recognised that the PSLV needed an additional fourth stage and it had to be powered by a liquid engine. The fourth stage was added after the project had been approved.

PSLV's liquid fourth stage used two engines, but unlike stage 2, they were not the new Vikas engines. Instead, they were a modified reuse of the existing RCS used for roll control of the first stage. The redesign involved increasing the thrust from 6.5 kN to 7.5 kN and extending the operational time from 110 seconds to 420 seconds. Since the fourth stage would kick in once the launch vehicle was in space, it would also need to ignite and operate in a vacuum. ISRO's comprehensive testing over five years made PSLV's fourth stage the most tested stage at that time.[556] Further innovations since the 1990s when the PSLV first became operational have increased the maximum payload to near 1,850 kg from the initial 1,200 kg. PSLV is now capable of placing payloads in GEO and not just the SSPO.[557] The PSLV was responsible for launching all seven of the IRNSS satellites, and with the aid of gravity assist, it was used to get spacecraft to the Moon and Mars.

PSLV-XL	Stage 1 Solid	Stage 2 Liquid	Stage 3 Solid	Stage 4 Liquid
Burn Time (second)	103	148	112	525 (max)
Propellant Type and Mass (tonne)	S138 HTPB	L40.7 UH25+ N₂O₄	S6.7 HTPB	L2 Liquid MMH+ MON[558]
Length × Width (m)	20.3 × 2.8	12.8 × 2.8	3.6 × 2	2.7 × 2
Lift-off Weight (tonne)	138	42	7.2	2.5
Directional Control (Pitch and Yaw)[559]	SITVC	Engine Gimbal	Flex Nozzle	Engine Gimbal

Table 10-7 Typical PSLV-XL Specification

ISRO has developed a shorthand to describe the propellant type and the quantity in each stage for all its launch vehicles, not just the PSLV. A letter denotes the propellant type, solid (S), liquid (L), semi-cryogenic (SC) and cryogenic (C), and a number the quantity in tonnes. For example, an S40 stage has 40 tonnes of solid propellant, an L110 stage has 110 tonnes of liquid propellant and so on. The first PSLV launch on 19 September 1993 failed. It failed immediately after the ignition of the third stage.

Following its investigation, the FAC identified three causes:[560] (i) during the second-stage separation, two of the retro rockets did not function as commanded indicating a break in the electrical circuit, so the separation did not go as planned; (ii) during the second stage, yaw (side to side) error grew and was not corrected by the autopilot; and (iii) the pitch (pointing up and down) error had increased because of an error in the autopilot's software that failed to detect an overflow condition. Despite the failure, most of the telemetry indicated that many of the subsystems operated as designed. All PSLV launches since then have been successful.

Through the 1990s, a further four launches placed IRS satellites in orbit, which is what the PSLV had been designed for. During its more than two decades of active use, the PSLV has evolved into three, increasingly powerful, configurations: (i) PSLV Regular, also known as PSLV Generic (PSLV G); (ii) PSLV Core Alone (PSLV CA), without strap-ons and (iii) PSLV XL. Three key developments have increased PSLV's capacity to deliver 1,600 kg to SSPO of 600 km or 1,850 kg to GTO.

Figure 10-5 Three Configurations of PSLV. Regular, Core Alone and XL (What looks like strap-ons in the Core Alone configuration are SITVC fuel tanks, which are present in all three configurations). Credit ISRO

The first-stage solid motor capacity has increased from 125 tonnes to 139 tonnes, the second-stage has increased from 40 tonnes to 42.5 tonnes, and the longer strap-ons used for the XL version has increased the strap-on capacity from 10 tonnes to 12.8 tonnes. By mid-2016, the PSLV XL's higher payload capacity had made it the most used variant, PSLV XL 13 flights, PSLV CA 11 flights and PSLV G 12 flights. The first stage of PSLV is made up of five segments (each 2.8 m in diameter and 3.4 m long) and six strap-on boosters that use HTPB propellant. Each strap-on is 1 m in diameter and either 10 m or, in the XL variant, 13.5 m in length. At launch, the first stage and two of the strap-ons (or four depending on the payload) ignite within the first 0.6 second. The first stage is steered using SITVC for pitch and yaw and RCS for roll. The SITVC system consists of four sets of six valves (24 in total) located symmetrically around the nozzle of the first stage injecting strontium perchlorate into the flow from the nozzle. This provides the differential pressure to manipulate pitch and yaw sufficient for the PSLV to steer away from Sriharikota to its destination. The launch trajectory is designed to minimise the hazard to life and property along the ground track. For polar orbit the initial launch angle of around 135°S is gradually increased to 180°S to avoid Sri Lanka. For equatorial/GTO orbit launch angle is around 105°S and the rocket stages are discarded in the Indian ocean and the final stage in the Pacific Ocean.

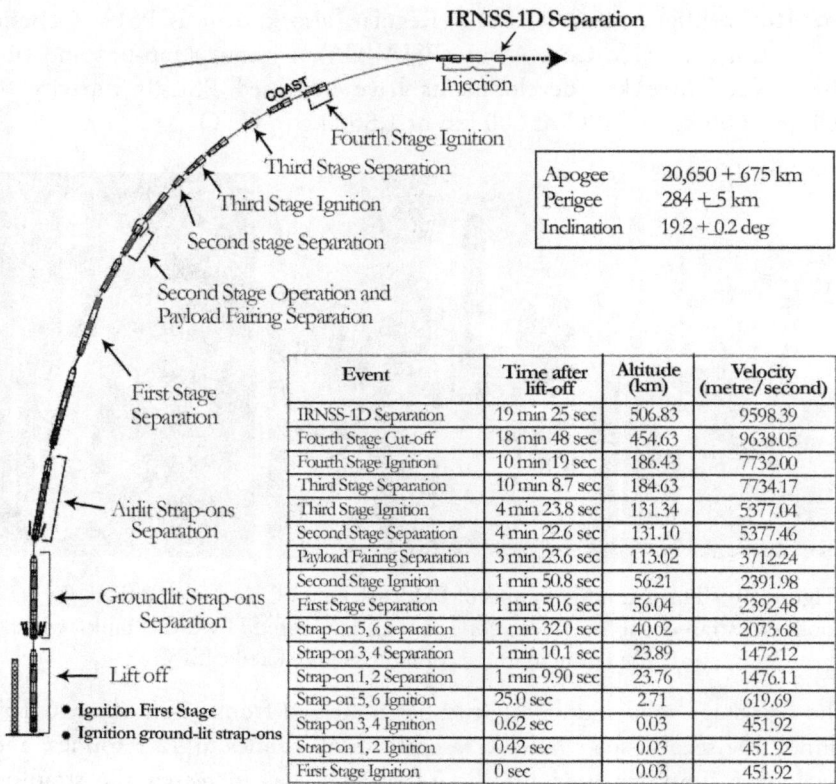

Apogee	20,650 + 675 km
Perigee	284 + 5 km
Inclination	19.2 + 0.2 deg

Event	Time after lift-off	Altitude (km)	Velocity (metre/second)
IRNSS-1D Separation	19 min 25 sec	506.83	9598.39
Fourth Stage Cut-off	18 min 48 sec	454.63	9638.05
Fourth Stage Ignition	10 min 19 sec	186.43	7732.00
Third Stage Separation	10 min 8.7 sec	184.63	7734.17
Third Stage Ignition	4 min 23.8 sec	131.34	5377.04
Second Stage Separation	4 min 22.6 sec	131.10	5377.46
Payload Fairing Separation	3 min 23.6 sec	113.02	3712.24
Second Stage Ignition	1 min 50.8 sec	56.21	2391.98
First Stage Separation	1 min 50.6 sec	56.04	2392.48
Strap-on 5, 6 Separation	1 min 32.0 sec	40.02	2073.68
Strap-on 3, 4 Separation	1 min 10.1 sec	23.89	1472.12
Strap-on 1, 2 Separation	1 min 9.90 sec	23.76	1476.11
Strap-on 5, 6 Ignition	25.0 sec	2.71	619.69
Strap-on 3, 4 Ignition	0.62 sec	0.03	451.92
Strap-on 1, 2 Ignition	0.42 sec	0.03	451.92
First Stage Ignition	0 sec	0.03	451.92

Figure 10-6 Launch Profile of the PSLV-C27 IRNSS-1D. Credit ISRO

The precise launch trajectory is dependent on the payload and its final destination. As an example, PSLV-C27 took nearly twenty minutes to place IRNSS-1D in orbit. The journey started at T=0, with first stage ignition and lift-off occurred 0.6 seconds later when four strap-ons ignited. At T+25 seconds the two-remaining strap-ons ignite. Within 90 seconds after ignition, all strap-ons were exhausted, and drop away. This approach of staggered ignition of strap-ons ensured that the PSLV's speed was not high and the dynamic pressure on PSLV and its payload was within safe limits while it traversed the densest part of the Earth's atmosphere.

At T+1 minute and 51 seconds at an altitude of 56 km, the second stage is ignited. This stage was powered by the ISRO-built Vikas engine using the hypergolic liquid propellant UDMH and N_2O_4. Vikas, the Indian variant of SEP's) Viking engine, was rated to burn for a longer period for the stretched second stage. During the second-stage burn, the payload fairing, which included the heat shield, completed its job of protecting the payload as it ascended through the atmosphere and separates at an altitude of 113 km. Stage two is steered using its engine gimbaling control (EGC) systems until the propellant is exhausted, and discarded at T+4 minutes and 22 seconds, and one second later the third stage is ignited.

The third stage took about six minutes to burn its almost seven tonnes of solid HTPB propellant before it also separated and burned up during re-entry through the Earth's atmosphere. The fourth stage had the critical role of placing the payload precisely in the required orbit. It used a more efficient liquid propellant, MMH with MON-3, and had two identical engines instead of one. Immediately after the third-stage separation, the twin engines in the fourth stage were ignited. Following a burn of about 10 minutes, the payload was ejected. The PSLV successfully delivered its payload IRNSS-1D to the planned transfer orbit and the PSLV has completed its mission. IRNSS-1D uses its onboard engine to get to the final orbit.

Through its collaboration with SEP in France, ISRO had obtained the know-how to build the critical systems that comprise the liquid-fuel engine, the gas generator, turbine, turbo pump and the combustion chamber. Many of the materials, components and tools required to make them, along with the manufacturing technique, were part of the technology transfer agreement. The PSLV programme took over a decade from the initial design before it demonstrated its first successful flight. In the ISRO infrastructure, the facility at Valiamala (about 30 km from Trivandrum), originally a firing range for the police service, became a key centre for designing, building, testing and developing the PSLV. SLV-3 and ASLV enabled ISRO for the first time to launch satellites to Earth orbit. However, mastering the liquid engine technology was the single technical advance that brought India to a point where it could claim to be self-sufficient in space.

With the PSLV, India had all the elements launch vehicles, satellites, infrastructure and professional expertise to deliver the space-based services as Sarabhai had imagined decades earlier. The liquid engine capability of the PSLV and its larger capacity was essential for ISRO to reliably place payloads for national development in designated orbits with high precision. The capability was then quickly expanded to include payloads with commercial potential and wider needs, such as space re-entry vehicles, science and interplanetary exploration. Satisfied by its competence in solid and liquid engines, ISRO then turned to the epitome of rocket science, the cryogenic engine technology.

Geosynchronous Satellite Launch Vehicle

At the outset, PSLV had sufficient thrust to place a 1 tonne satellites into polar orbit. Since then, through a series of enhancements, ISRO has extended this capability to 1.8 tonnes. Even before the PSLV became operational, the Indian government had approved the second series of INSAT in 1987. INSAT-2 weighing around 2.5 tonnes was larger and heavier than the IRS satellites. Getting the INSAT class communication satellites to the higher geosynchronous orbit was beyond the capability of the enhanced PSLV. To get INSAT-2 to orbit, ISRO had to commercially engage foreign launch service providers, predominantly ESA's Ariane 5. In the process, undermining one of its key objectives. Self-sufficiency. Almost a decade and a half before the INSAT-2 was approved, the Geosynchronous Satellite Launch Vehicle (GSLV) had been proposed in November 1972 with the modest goal of placing an 800-kg spacecraft in GTO.[561] The design incorporated solid, liquid and cryogenic stages, and the intention was to have the first GSLV launch by 1981.

Cryogenic engines rely on LOX and LH2. In their liquid form, oxygen and hydrogen are very cold and a formidable challenge for engineers. Though the GSLV at 50 m is not much larger than the PSLV at 44.5 m, it incorporates the next evolutionary step, cryogenic technology, which provides the most efficient thrust a launch vehicle can generate. For ISRO, "it was clear that the last stage of the GSLV had to be a cryogenic stage."[562] This third stage was to provide half of the total 9.7 km/s velocity necessary to attain GTO. Most countries that developed the cryogenic engine technology (LOX + LH2) had also developed semi-cryogenic engines (LOX + kerosene). Although ISRO had proposed developing a 7.5-tonne semi-cryogenic engine in the early 1970s during the PSLV design phase that never happened. The story of the development of cryogenic engine has been one of ISRO's most complex challenges involving economic sanctions, political intrigue, cost overruns and a series of delays. It started with the acquisition of the liquid-engine technology from France, which "created a culture in the liquid propulsion group which favoured import of technology". This fateful decision, according to investigative journalist Gopal Raj, demanded "a heavy price" that ISRO paid when the time came for designing the GSLV.[563] Although it shares some characteristics with the PSLV and is not much bigger, the three-stage GSLV is a more powerful launch vehicle than the four-stage PSLV. The first stage is composed of a solid core with four liquid engine strap-ons. The second is uses a liquid propellant. The innovation lies in the third stage with a cryogenic engine using LOX as oxidant and LH2 as fuel.

	GSLV Mk1	GSLV Mk2	LVM-3/GSLV-Mk3
Operational During	2001– 2010	2010- mid-2017	2014- mid-2017
Number of Flights	6	5	1 sub-orbital flight in 2014. 1 to GEO in May 2017
Payload to GTO (tonne)	1.82	1.95 – 2.35	4 – 4.5
Lift-off Mass (tonne)	414	414	600+
Max Height × Width (m)	49.13 × 2.8	49.13 × 2.8	43.5 × 4

Table 10-8 Overview of GSLV Configurations and Launch History

To accommodate the GSLV, an additional launch pad, the SLP, using a new approach in integrating the launch vehicle was proposed in 1988. Rather than transporting individual components and assembling the launch vehicle at the launch pad using the Mobile Service Tower, the launch vehicle would be assembled in a VAB, and transported to the launch pad complete requiring only the propellant. Since the initial design, the GSLV has evolved into a variety of configurations and sub-configurations. GSLV Mk1, which is no longer in service, came in three variants, Mk1, Mk1+ and Mk1 (F06). All Mk1 variants, including the last Mk1 to be flown in 2010, used cryogenic engines purchased from the USSR/Russia. By 2010, ISRO had developed its own cryogenic engine, which is the distinguishing feature of Mk2. By 2010 11 GSLV had been launched - 6 Mk1 (2 failed, 2 succeeded and 2 partially failed), 5 Mk2 (4 successful and 1 failure). Of the two heavy-lift attempts, LVM-Mk3 suborbital flight in 2014 and 1 GSLV-Mk3 in 2017 were successful. GSLV variants come in a number of specifications. The launch profile between the Mk1 and Mk2 is similar. The Mk1 used the 3rd stage cryogenic engine from the USSR and the Mk2 used one developed by ISRO in India.

The following sequence describes the GSLV Mk2 GSAT-14 mission launched on 22 February 2014. At launch, the strap-ons of GSLV Mk2 ignited first at T-4.8 seconds. Collectively, the four strap-ons, each using a Vikas 2 engine, lack sufficient thrust for the vehicle to take-off. If for any reason all four do not ignite and provide the designed consistent thrust, the onboard computer controlling the launch sequence has time to switch them off and abort the launch.

Strap-on	Stage 1	Stage 2	Stage 3
Mk. 1. Three variants a, b and c. Total flights=6. Now Retired			
4 x L-40H	S125	L39.3	CE-7.5
UH25 and N_2O_4	HTPB	UH25 and N_2O_4	LOX + H2
Mk2 Total. Flights=3			
4 x L-40H	S139	L39.3	CE-12.5
UH25 and N_2O_4	HTPB	UH25 and N_2O_4	LOX + H2

Table 10-9 Typical GSLV Mk1 & Mk2 Specification.

As the strap-ons performance was nominal, the core stage was ignited at T=0, and the launch was committed since the solid core stage once ignited cannot be stopped. GSLV's first-stage directional control is provided by EGC on the four-liquid strap-ons. The four nozzles are controlled dynamically by the computer to maintain a predefined course stored in its memory. The strap-ones EGC is so effective that directional control for the core stage is not required. High-speed winds in the Earth's atmosphere, especially between 10 and 30 km, and the potential inconsistent thrust from the solid core stage are potentially the most unpredictable forces that IGS has to manage. The core stage exhausted its 138 tonnes of solid propellant at T+1 minute and 47 seconds, but the strap-ons continued for another 40 seconds. Since the strap-ons were strapped to the first stage, they carried the dead weight of the now spent core stage. At T+2 minutes and 29 seconds, at an altitude of 70 km, the strap-ons also used up their propellant and shut down. The second-stage engine was ignited about 2 seconds prior to the separation of the first. The flexible linear shaped charge was used for stage separation, and exhaust from the second-stage engine helped push away the spent first stage with the attached strap-ons.

The second-stage Vikas 4 engine was optimised for the vacuum environment of space. Stage two also used EGC for directional control. By the time the second stage kicked in, the launch vehicle was in space and no longer vulnerable to the strong, unpredictable external forces present in the upper atmosphere. At T+3 minutes and 46 seconds, while travelling at just over 3 km/s at an altitude of 115 km, the payload fairing that protected GSAT-14 against the Earth's atmosphere became superfluous in the vacuum of space and was ejected, preparing the satellite for deployment. At 4.9 km/s, which is around half the velocity necessary for orbit, the second-stage Vikas 4 engine shut down at T+4 minutes and 49 seconds and separated from the third stage three seconds later with a pyro-actuated cullet release mechanism. The launch vehicle, then just 10 m long, was a quarter of that at launch.

The final stage using LH2 and LOX fired at T+4 minutes 53 seconds and burned for over 12 minutes, longer than both of the previous stages combined. In the absence of the Earth's atmosphere and building on the momentum provided by the previous stages, the third stage achieved the required orbital velocity of 9.7 km/s before exhausting its 12.5 tonnes of propellant and shutting down. The third cryogenic-stage directional control was provided by two small swivelable vernier engines powered by LH2 and LOX. Seconds after the third-stage shutdown, the payload GSAT-14 was pushed out by spring thrusters mounted at the separation interface into a GTO of 80-km perigee with an apogee of 35,975 km. This highly elliptical GTO orbit was circularised over next several days to the 35,975-km orbit by ISRO's MCF in Hassan. Once it arrived in its operational slot, GSAT-14 started providing its service as a communication satellite.

Launch Vehicle Mark 3 (LVM3)

On 18 December 2014, ISRO conducted the Crew Module Atmospheric Re-entry Experiment (LVM3-X/CARE), an experimental sub-orbital flight powered only by the first and second stages. The LVM3-X/CARE payload consisted of a 3.7-tonne mock-up of a crew module as the payload. It was launched using only the boosters and the core stage into a sub-orbital flight to an altitude of 126 km.

LVM3-X	Strap-on X2	Booster Stage	Stage 3
Burn Time (s)	130	200	Not used
Propellant Type	HTPB	Liquid UDMH + N_2O_4	Simulated
L×W (m)	25 × 3.2	17 × 4	13.5 × 4
Propellant Mass (tonnes)	207	110	Simulated

Table 10-10 Suborbital flight on 18/12/2014. LVM3-X/Care

The third stage cryogenic engine was not present for the flight with LOX and LH2 replaced by liquid and gaseous nitrogen of equivalent weight. The crew module splashed down near the Andaman and Nicobar Islands in the Bay of Bengal 20 minutes and 43 seconds after launch. The launch of

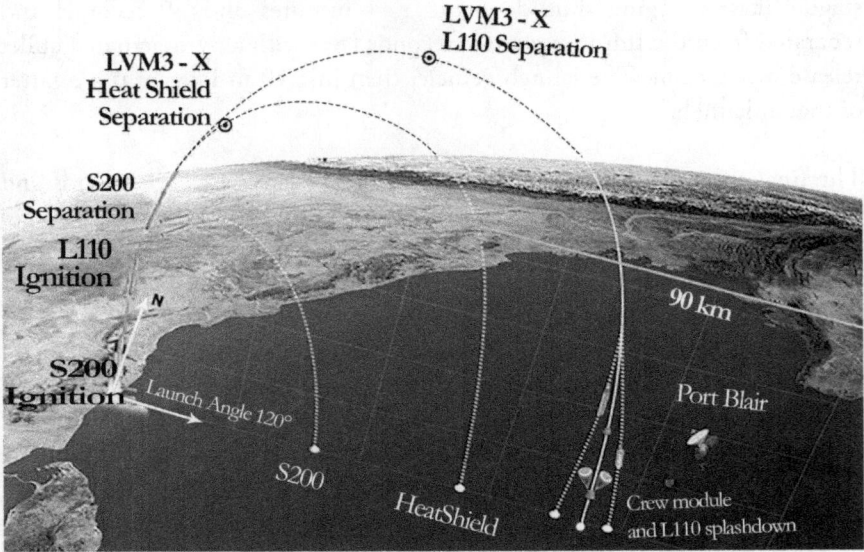

Figure 10-7 LVM3-X/Crew Module Atmospheric Re-entry Experiment Flight Profile. Credit ISRO

LVM3-X/CARE concluded with a twin-parachute controlled splashdown in the Bay of Bengal. The module was recovered immediately after splashdown by the Indian coast guard. The primary objective of the LVM3-X/CARE was to test the large solid strap-ons used on ISRO's latest launch vehicle.[564] The scope of this test flight included assessing the flight dynamics, the core module subsystems, three-axis-control manoeuvres for re-entry, heat shield, parachute systems, splashdown and recovery. The other key objective was to test the strap-on boosters and the re-entry characteristics for a module which could potentially form a part of a future human spaceflight programme.

GSLV-Mk3

ISRO's next generation launch vehicle initially called the Launch Vehicle Mark 3 or LVM-3, ISRO appears to have returned to the original naming convention and is now called the GSLV-Mk3 even though it has no evolutionary connection with GSLV Mk2. GSLV-Mk3 made its maiden flight on 5th June 2017 successfully placing GSAT-19 in GEO. GSLV-Mk3 is India's heavy-lift launch vehicle albeit with a capacity to launch about 4.5 tonnes to GTO. Other heavy-lift vehicles (i.e. Ariane 5, Delta 4 Heavy and Long March 5) have about twice that capacity. GSLV-3 uses 2 solid strap-ons as the first stage, liquid propellants for the second stages and has a cryogenic third stage.

The extent to which ISRO can enhance GSLV/LVM3 is limited. The GSLV-Mk3 was designed over two decades ago. To catch-up with larger capacities of heavy-lift vehicles used by other nations, further technological developments are required. Incorporating semi-cryogenic fuel for the liquid stages and introducing some level of recovery and reuse of the boosters or stages are some of the potential options that ISRO is contemplating.[565] To master heavy launch capability, ISRO needs to develop larger cryogenic engines.

	Stage 1	Stage 2	Stage
Burn Time (s)	140	263	643
Propellant(Tonnes)	S205 X 2	116	28
Propellant type	HTPB (solid)	UH25 and N_2O_4 (Liquid)	LOX + H2 (Cryogenic)
Length × Width (m)	26.2 × 3.2	21.39 X 4	13.5 × 4

Table 10-11 GSLV-Mk3-D1/GSAT19 Specifications

Operationalising this heavy launch system will not only end ISRO's reliance on foreign launch services for its communication satellites but allow ISRO to consider new missions including the exploration of the solar system, human spaceflight, space station and provide greater access to the growing commercial space launch market.

Future Launch Vehicles

Most launch vehicles around the world still use the expendable single-use technology that initiated the Space Age more than half a century ago. The rocket technology that took Yuri Gagarin into space was not very different from that used to take astronauts to the ISS in the 21st century.[566] Engines for launch vehicles which operate at a higher efficiency are cryogenic and semi cryogenic engines. ISRO plans to replace the current launcher family, PSLV, LVM and the GSLV with a single modular design. At one stage referred to as the Unified Launch Vehicle (ULV) but may be renamed in the future. Using a combination of fuel types (solid, liquid, semi cryogenic and cryogenic) each ULV flight will be customised to meet the mission objectives.

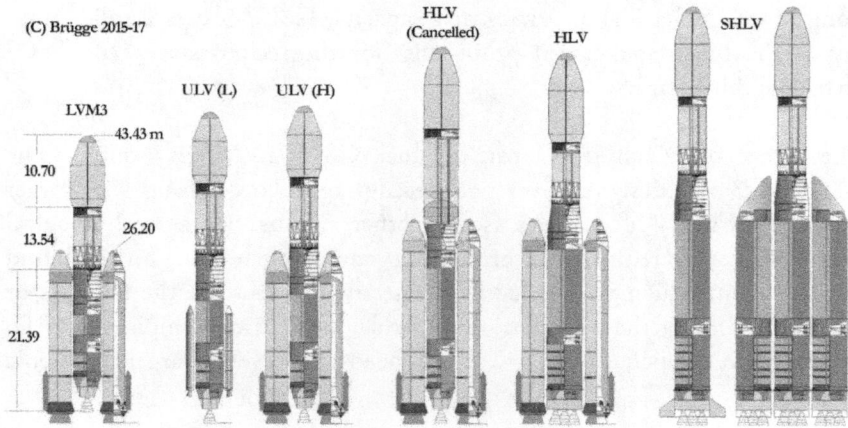

Figure 10-8 Future Launch Vehicles. Based on fact and informed speculation.
Credit Norbert Brugge[567]

Historically, when a new engine technology is introduced, ISRO initially engages in an international agreement before becoming self-sufficient. France was involved with the solid and liquid engine development and the USSR, at least briefly, with the cryogenic engine technology. To engage with the semi-cryogenic technology, in 2006 ISRO signed a framework agreement with the Ukrainian manufacturer Yuzhnoye on a joint project called Jasmine.[568] In this agreement, ISRO would develop the Yuzhnoye RD-810 engine as the SCE-200 (semi-cryogenic engine with a thrust of 200 tonnes) for future launch vehicles. GSLV-Mk3 that placed GSAT-19 in orbit on 5 June 2017 used ISRO's CE-25 cryogenic engine as the third stage. The CE-25 uses an 'Open Cycle' design that uses a small amount of fuel to in a separate combustion chamber sometimes known as a pre-burner chamber.

The exhaust from this chamber is used drive turbo pumps that deliver the LH2/LOX into the main combustion chamber. Developing this engine was a huge achievement following over two decades of development. ISRO is now beginning to develop a Semi-Cryogenic Engine with a large capacity of (SCE-200).[569] Semi-cryogenic engines use LOX and kerosene rather than LOX and LH2 as propellants. ISRO plans to use its own formulation of kerosene that is referred by some as Isrosene.

A future ULV would have solid strap-ons, a semi-cryogenic second stage and with CE-25 cryogenic third stage with an increased payload to GTO of

10 tonnes. In 2015, India signed a further MoU with the Russian Federation. The remit of this MoU is not clear but likely to include progress with the SCE engine technology.[570] The current design of ISRO's future launch vehicles will continue to use solid boosters (up to S250), a semi-cryogenic first stage (using the SCE-200) and a larger cryogenic second stage (up to CE-60). ISRO plans introduce semi-cryogenic technology for the first time with the ULV and HLV but probably not until the next decade.

Version	Booster	Stage 1		Stage 2	
LVM3[571]	2 × S-200	L-110 (UDMH/ N₂O₄	Vikas-X 2	L-25 (LH2/LOX)	CE-20
ULV Light	6 × S-13	L-160 (Kerosene/LOX)	SCE-200	L-30 (LH2/ LOX)	CE-20
ULV Heavy	2 × S-200	L-160 (Kerosene/LOX)	SCE-200	L-100 (LH2/LOX)	CE-20
HLV	S-250	5 X L-200 (Kerosene/LOX)	SCE-200	L-100 (LH2/LOX)	CE-60
SHLV	5 X SCE-200	5 X L-200 (Kerosene/LOX)	SCE-200	2 x L-100 (LH2/LOX)	CE-20

Table 10-12 Potential Future Launch Vehicles. Credit Norbert Brugge

ISRO's designs for launch vehicles in the future do not involve a radical new technology but envisages an increasingly larger first stage (solid), enhanced second stage (semi-cryogenic) and third stage (cryogenic). Given that semi-cryogenic propellants were first used in the German V2 missiles during World War II and in the launch of Sputnik in 1957, ISRO is joining late. Just as with the PSLV and GSLV, multiple configurations are envisaged for ULV and HLV. Not all initial designs make it beyond the design phase.

The typical cost of placing 1 kg payload in Earth orbit is about $20,000 (Rs.0.13 crore). Although the fuel used by a launch vehicle is around $300,000 (Rs.2 crore), the cost of the single-use launch vehicle is about $60 million (Rs.400 crore).[572] Reusable launch vehicles are a key area for development but ISRO is currently at a very early stage. The potential future launch vehicles listed above is a product of informed speculation based on public presentations by senior ISRO personnel rather than formal policy announcements. [573]

Reusable Launch Vehicle (RLV)

The USSR had briefly demonstrated its capability in reusable space vehicle with Buran, but it was the US, through the space shuttle, that proved launch vehicle reuse was technically possible although the economic benefits were not as substantial as first envisaged. While still in its early stages, India too is working on its Reusable Launch Vehicle (RLV) and the 'two-stage-to-orbit' (TSTO) concept. A TSTO concept relies on a vehicle with two stages to get to orbit, one of which may be expendable.

Figure 10-9 Reusable Launch Vehicle Technology Demonstrator. Credit ISRO

In January 2007, ISRO conducted Space Capsule Recovery Experiment usually abbreviated to SRE-1. For the first time ISRO recovered the vehicle it had sent to space. This demonstrated ISRO's ability to perform experiments in space, de-orbit the spacecraft for a safe re-entry and recover it intact from a predefined location after splash down. The SRE-1 was a small capsule of 550 kg and conducted two onboard experiments during its 12 days in Earth orbit. Through SRE-1, ISRO proved its subsystems for navigation, guidance and control at hypersonic re-entry, thermal protection system, parachute deployment and recovery following re-entry. While ISRO demonstrated its capability to recover a spacecraft after launch, it was only the payload, not the launch vehicle that was recovered.

ISRO's vision of a RLV is a TSTO space shuttle, like a delta-winged spaceplane. The winged design was the product of almost 4,500 wind tunnel tests. Launched vertically by a single rocket motor (first stage), the RLV is designed to use its onboard engine (second stage) to arrive in orbit. Once the in-orbit operations are complete, the RLV is designed to return to Earth using the same engine to de-orbit and land on a runway as a traditional aircraft. In ISRO's phased development programme, the RLV did not have engines on-board for the initial series of tests. In May 2016, ISRO conducted its first RLV test. It was launched using a single solid booster from the FLP and returned simulating an unpowered glider approach to a splashdown at sea with no plans for recovery.

A research paper identified some of the RLV technology objectives "The key aerodynamic and aerothermodynamics design aspects are optimum heat flux, heat load, load factor, less than 4 g deceleration, sufficient payload bay, down-range and cross-range capability, good longitudinal and lateral directional aerodynamic stability and adequate control surface effectiveness."[574] The reference to a deceleration of less than 4 g is typically associated with human spaceflight. The maximum acceleration of GSLV-Mk3 is also limited to 4 g. While the flight characteristics of RLV design may cater for human passengers, currently all ISRO's RLV experiments fly completely autonomously without the capacity to support human crew.

Figure 10-10 RLV-TD at Launch. 23 May 2016. Credit ISRO

The first time ISRO publicly displayed the design of its RLV was at the bi-annual Aero India International Aerospace Exhibitions in Bangalore (now Bengaluru) in February 2009. Initially, ISRO intended the first flight to be in 2011, but a series of delays followed. In its annual report in 2014, ISRO outlined its intentions stating "solid booster motor (HS9) has been positioned at Satish Dhawan Space Centre, Sriharikota. Assembly of the vehicle is in progress towards RLV-TD HEX-01(Reusable Launch Vehicle Technology Demonstrator Hypersonic Flight Experiment) mission targeted originally for launch during the first half of 2015."[575] It took place a year later.

At 7 am local time on 23 May 2016, ISRO launched a 1.75-tonne 6.5-m long (one-sixth scale) model of its version of a reusable space shuttle from the FLP in Sriharikota. The RLV-TD was launched vertically on a sub-orbital trajectory using a single 9,200 kg solid fuel booster based on the 1 m by 10 m strap-on used on the PSLV. Typically, PSLV boosters achieve a velocity of about 0.8-1 km/s, which is about Mach 3. The one used to launch RLV was a customised version reaching a higher speed before burning out and separating after 91.1 seconds at an altitude of 56 km. The space shuttle had also used boosters, but they were in parallel with the shuttle, whereas the RLV was on top of the booster increasing the overall length and thus the length-to-diameter (L/D) ratio. This was an important parameter that can increase the vehicle's dynamic instability, but this first and successful flight demonstrated the success of ISRO's calculations, simulations and modelling.

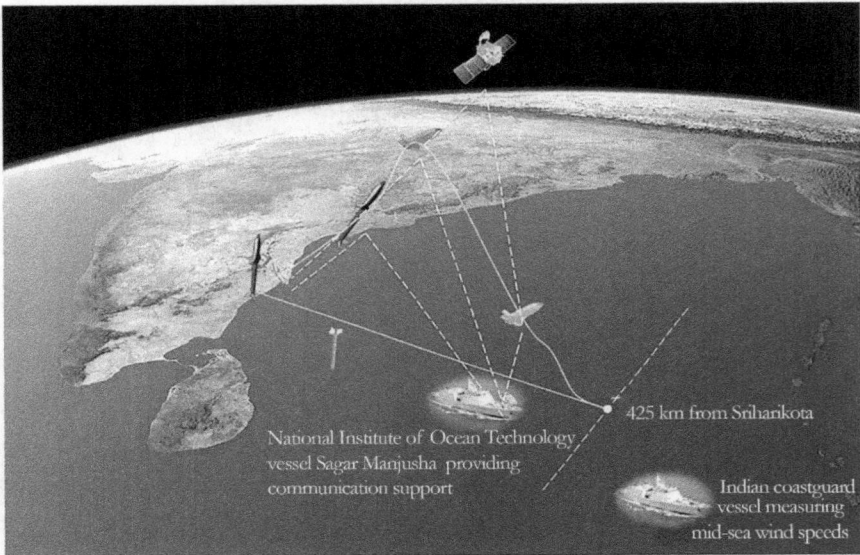

Figure 10-11 RLV-TD as Tracked by Ship, Satellite and Ground Station at Sriharikota. Credit ISRO

Directional control during ascent was maintained by the booster's four fins, as well as the SITVC control system used by ISRO for its solid fuel booster. The SITVC fuel (strontium perchlorate) is usually contained in what looks like a smaller version of the strap-on strapped vertically on the booster, but in this instance, it was stored in a toroidal tank located at the base of the booster. Following the launch, the RLV ascended to 65 km and started its supersonic re-entry at Mach 4.78 (5855 km/h). It did not cross the 100-km boundary considered the threshold for space, so this was not a space mission, which is perhaps why ISRO kept a low profile throughout the mission, publishing its first press release immediately after the mission was complete.

The RLV-TD mission verified the protection provided by the 1,200 thermal tiles against the 1,000°C temperature during re-entry. Onboard sensors collected a variety of data, including load, acceleration, temperature, air speed and pressure. There probably were cameras onboard, too, but ISRO has not released any in-flight images from RLV-TD.[576] During the flight, RLV-TD was tracked using a C-band beacon installed in the vehicle. Data was transmitted in real time using the onboard S-band transmitter to the ground stations at the Sriharikota. ISRO was assisted by the Indian coast guard, and the National Institute of Ocean Technology's research vessel Sagar Manjusha provided ship-borne data relay along the flight path.[577]

RLV-TD was tracked to 800 m above sea level by ship-borne telemetry at 2 MHz and also at 1 KHz data from INSAT right up to the point of impact.[578] All mission objectives were successfully completed. As RLV-TD was not expected to survive the impact unlike SRE-1 in 2007 and the LVM3-X/CARE mission in 2014, no recovery was planned. The vehicle, however, did survive landing on the water, and ISRO reported that it was intact and floating.[579] Images of the vehicle floating on the sea were captured by helicopter 20 minutes after touchdown but were not published. Had provisions for recovery been in place, RLV-TD could have been recovered (at least some of the larger parts. It is unlikely the vehicle survived the impact intact) and returned to VSSC for further analysis.[580] In this first flight, the RLV-TD had no onboard engine, so during the return journey, it operated as a glider, just as the US space shuttle design. The onboard navigation and guidance systems controlled the descent autonomously from 40 km altitude down to a gentle splashdown. RLV-TD had two vertical tail planes offset from the centre and canted (like the US's secret X37B unmanned spaceplane).

The space shuttle had just one fin and a split-rudder assembly doubling as a speed brake. The descent profile was controlled by operating the rudders on the canted tail planes, the elevons on the trailing edge of the wings and an RCS. Although there was no runway in this instance, the glide path simulated an airstrip approach that will be used in future missions when a runway is available. Twelve minutes and 50 seconds after launch, the vehicle splashed down in the Bay of Bengal, 412 km from the launch site. In the post-flight press release, ISRO confirmed mission success stating that "autonomous navigation, guidance & control, reusable thermal protection system and re-entry mission management have been successfully validated."[581]

ISRO planned to develop its RLV systematically through four distinct phases using its scaled-down experimental delta-winged RLV. The hypersonic flight experiment (HEX) of the first phase has now been completed. This will be followed by a landing experiment (LEX), return flight experiment (REX) and the scramjet propulsion experiment (SPEX).

The second phase, the landing experiment, involves landing on a runway and recovering the vehicle for the first time. This would not be a sub-orbital launch but a release from a mother aircraft from high altitude. The third phase, return flight experiment, combines the first two, a sub-orbital launch followed by a return to Earth and landing on a runway. The fourth phase, the scramjet propulsion experiment (SPEX), involves an active scramjet propulsion to get to Earth orbit before returning to a runway landing. Through these tests, ISRO intends to master hypersonic aero-thermodynamic characterisation of winged re-entry body, along with autonomous mission management to land at a specified location.

The RLV-TD was built by about 500 ISRO engineers at the VSSC over five years at an estimated cost of about \$15 million (Rs.99.5 crore). Despite the success of the first test in May 2016, the TSTO concept based on a reusable spacecraft is still more than a decade away. The next key stages of RLV require two critical elements that ISRO is yet to fully develop, scramjet engine technology and a 5-km runway at Sriharikota. The next three tests of landing on a runway, refurbishment and reuse, this can only take place once the runway for the RLV is available. ISRO has no runway although there is ample space to build one. A detailed report for the construction of a 5-km runway at Sriharikota was completed in 2011, but the work is yet to start.[582] The other critical dependencies are special types of combustion engines that operate only at supersonic speed, ramjet and scramjet engines.

Scramjet

Most of the mass of a rocket at launch pad is in the propellant required to acquire the high speed to achieve orbit. The propellant in all its forms, solid, liquid or cryogenic, consists of an oxidiser and fuel, which are brought together for combustion. Typically, rocket fuel mixture ratio has more oxidiser than fuel, and the oxidiser is the heavier of the two. For example, the huge propellant tank used by the space shuttle contained 100 tonnes of LH2 and 629 tonnes of LOX. If an engine were to use oxygen from the air, it could reduce launch weight significantly. Even though there is no oxygen in space, the first 50 km of Earth's atmosphere has sufficient oxygen, and it is this oxidiser supply that is used by ramjet and scramjet engines. Since atmospheric oxygen is available during ascent and descent, ramjet and scramjet technology lends itself for reusable space launch vehicles. Because oxygen is picked up from the atmosphere during flight, the total mass at launch is considerably reduced. Instead of using, for example, four stages of PSLV to get to orbit, a design using a ramjet/scramjet combination can do it with just two stages.

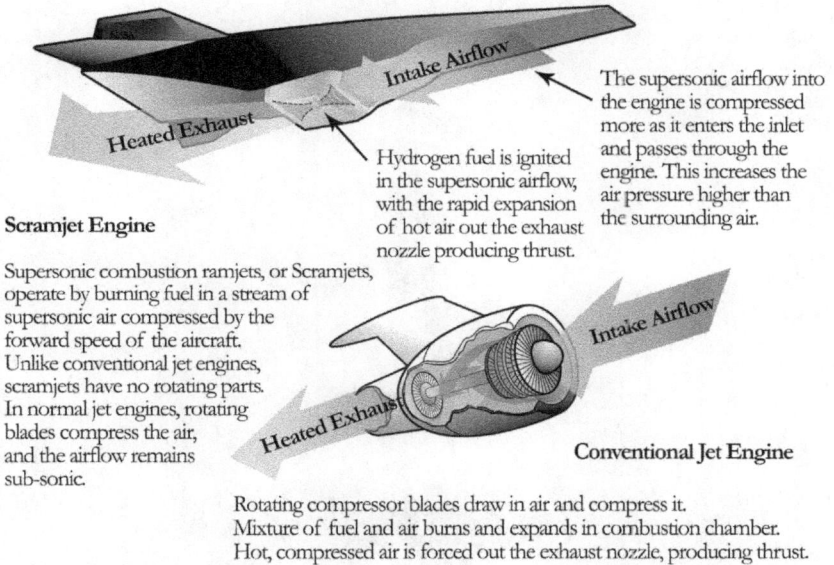

The supersonic airflow into the engine is compressed more as it enters the inlet and passes through the engine. This increases the air pressure higher than the surrounding air.

Hydrogen fuel is ignited in the supersonic airflow, with the rapid expansion of hot air out the exhaust nozzle producing thrust.

Scramjet Engine

Supersonic combustion ramjets, or Scramjets, operate by burning fuel in a stream of supersonic air compressed by the forward speed of the aircraft. Unlike conventional jet engines, scramjets have no rotating parts. In normal jet engines, rotating blades compress the air, and the airflow remains sub-sonic.

Conventional Jet Engine

Rotating compressor blades draw in air and compress it. Mixture of fuel and air burns and expands in combustion chamber. Hot, compressed air is forced out the exhaust nozzle, producing thrust.

Figure 10-12 NASA X-43 Scramjet versus Traditional Jet Engine. Credit NASA

Ramjet and scramjet engines are combustion engines with no moving parts and operate at supersonic (at least 324 m/s or Mach 1) and hypersonic (Mach 5 and beyond) speeds. The difference between a ramjet and scramjet is the speed at which it operates. The air in a ramjet combustion chamber is slowed to sub-supersonic speed before combustion, but in a scramjet, combustion occurs with air at supersonic speed. A scramjet (Supersonic Combusting Ramjet)) is a ramjet that operates in a higher speed regime than a ramjet. Ramjet technology has been in use for decades employed mostly in military missiles worldwide. India already deploys ramjet technology in its Brahmos and Akash missiles. However, operationalising ramjet and scramjet technologies for space vehicles is a major engineering challenge.[583] These engines rely on the enormous pressures created by their supersonic speed to compress the air entering the combustion chamber. ISRO is developing a ramjet-scramjet combination in a Dual Mode Ramjet (DMRJ) for its RLV.

Once built, it will operate between Mach 3 to Mach 9. ISRO has been developing its ramjet and scramjet technologies in the Air Breathing Propulsion Project for several years using sounding rockets.

Figure 10-13 Advanced Technology Vehicle Development Flight. One of the Two Passive Scramjets Can Be Seen in the Middle of the Launch Vehicle. 3 March 2010. Credit ISRO

On 3 March 2010, ISRO conducted its first advanced technology vehicle (ATV-D01) experiment. A pair of passive scramjets were attached to the second stage of a two-stage sounding rocket. Launched from Sriharikota, the ATV-D01 solid fuel booster with a launch mass of 3 tonnes reached the intended speed of just over Mach 6 at an altitude of 46 km. ATV-D01 sustained the critical hypersonic speed of Mach 6 for seven seconds during the four-minute experimental flight. A second development flight, ATV-D02, was conducted on 28 August 2016. This time, two active scramjets were attached to the side of the second stage of the RH-560, which operated during the hypersonic flight phase of ATV-D02. During the second stage burn of the RH-560, both scramjet engines were generating thrust by using oxygen from the atmosphere for combustion. Albeit for just five seconds, achieving combustion at hypersonic speed is a significant technological feat.[584]

An RLV can have a profound impact on reducing the cost of placing satellites in orbit. Broadly, 85% of the cost of a space launch is the launch vehicle, 5% is fuel, and 10% is the payload. Reusing the launch vehicle offers the best option in cost-saving. If successful with RLV technology, ISRO will be able to reduce the cost of space launches by at least a magnitude, from $20,000 (Rs.0.13 crore) per kg to Earth orbit to $2,000 (Rs.1.3 lakh). This is the ultimate ambitious objective for ISRO's RLV design. RLV is the first stage in ISRO's ultimate goal of developing a TSTO spaceplane. While the RLV is a civilian ISRO project, a separate but similar spaceplane called Avatar is being developed by the DRDO. Ultimately, Avatar is designed to be an autonomous single-stage-to-orbit-vehicle with the capability of launching from a runway, placing its payload into orbit and returning to Earth, on the same runway. Organisationally, ISRO and DRDO are separate organisations. However, transfer of technology between the two would not be unusual as similar arrangements are practised in other countries. Other space agencies have attempted to or are planning to make use of spaceplane technology.[585] NASA's space shuttle was retired after 135 missions, primarily because of the high-risk nature of its flight regime, but also because it did not bring the cost-savings initially envisaged. ESA, Russia, Japan and China have also experimented with spaceplanes, but none have shown significant commitment to the routine deployment of spaceplanes in the near future. Scramjet technology is still highly experimental. Although further scramjet tests are planned, ISRO, too, is not 100% committed to the RLV or scramjet technology.[586] Private industry. innovations from, for example, Skylon in the UK and Dream Chaser in the US may change this.

⊰ ◈ ⊱

Chapter Eleven
Struggle with Cryogenic Technology

N ational space programmes are designed to provide specific space-based services but, in practice, serve a more subtle and complex role. They are an expression of national prestige, advance strategic capability and offer a potential for geopolitical influence. In addition, they play a role in the national economy and also offer an opportunity to compete on the fast growing international commercial space market, which was worth close to $7 billion (Rs.20,300 crore) in 1993[587] and has grown to $323 billion (Rs.2,041,038 crore) in 2015.[588] Half a century after its inception, India's space programme still operates a single launch site, relies primarily on a single operational launch vehicle with low launch capacity and still turns to ESA to launch all its large communication satellites. Why has India's space programme not advanced at the pace envisioned by its founders?

The answer may lie in ISRO's troubled history with cryogenic-engine development. The absence of this key technology and the resulting lack of heavy-lift launch capability has held back the pace of its space programme. In its absence, ISRO struggles to meet its domestic demand for transponders, is limited in its ability to take on challenging science missions and can offer only a restricted range of commercial launch services. Notwithstanding ISRO's sustained progress over the last decade, it would have made greater strides more quickly with a heavy-lift launch capacity. For example, ISRO launched the first of a series of seven IRNSS satellites in 2013 and the last in 2016. The first satellite in the constellation was already a quarter of the way through its shelf life before the constellation was complete.

With a cryogenic-engine powered heavy launch vehicle, ISRO could have launched two or more IRNSS satellites on a single launch vehicle. An operational heavy launch vehicle is also a prerequisite for India to progress with its lunar lander and rover missions and its human spaceflight programme. Further, having only a single launch site (Sriharikota), two launch pads, one VAB (SVAB is due to come online in 2018) and predominantly a single operational launch vehicle, the PSLV, determines an upper limit to ISRO's operations. The construction of an additional VAB is already underway, but developing a heavy launch vehicle has been one of ISRO's messy and enduring challenges.

In October 1990, the Space Commission approved the heavy-lift launcher GSLV complete with a Cryogenic Upper Stage (CUS) to launch the heavy INSAT satellites that ISRO was building. Since then, it has made slow progress and is finally expected to have an operational heavy-lift vehicle by about 2020. However, even when India's heavy-lift launch vehicle is finally complete, it will still only deliver around 4 tonnes to GTO, a launch capability that is at least a decade old. Currently, ESA's two decades old Ariane 5 can place 6.5 tonnes in GTO, and the proposed Ariane 6 (scheduled for 2020) has a capacity twice that.[589]

Cryogenic Engine Technology. Buy or Build?

Engines using cryogenic propellants are much more efficient than rocket engines powered by solid or liquid propellants. They require the least amount of propellant to deliver a spacecraft of a given mass to a predefined orbit[590] and provide the highest thrust through sustained combustion of LH2 and LOX. Cryogenic engines have almost 200% efficiency over liquid-fuel engines typically using UDMH and N_2O_4.

However, the very low-temperature engineering required (for example, storage tanks, pipes, pumps, valves) to handle LOX at -183°C and LH2 at -253°C is a major technical challenge. Developing such technology from scratch is not only costly but can take decades to develop and operationalise. Bhabha and Sarabhai had recognised the value of international collaboration with technologically advanced nations when developing new technologies to avoid having to start from scratch. Such collaborations had led to close connections between India and the USSR, US, Japan and France, and India benefited from these connections in the development of sounding rockets, multi-staged launch vehicles, satellites and solid and liquid propulsion systems. A similar arrangement for cryogenic-engine technology transfer with a country that already had it was a natural solution for India. The US, Japan, China, France and Russia have developed and successfully deployed cryogenic engines for their launch vehicles. Approximately 18 cryogenic engine types have been developed across the world, but only seven have been fully tested in flight.[591] The US had developed large cryogenic engines (C-100) capable of delivering a thrust of 100 tonnes (approximately 1,000 kN) for the Apollo missions in the 1960s.

Later, CE-200 engines were developed for the space shuttle main engines. When the Viking liquid engine agreement concluded in 1978, France offered the cryogenic technology as used in HM-7 to ISRO with a price tag of Rs.1 crore ($150,000).[592] ISRO declined the offer, a decision that even ISRO's far-sighted and visionary chairman at the time, Satish Dhawan regretted. Later Japan, the US and after that initial offer France, too, refused to engage in a commercially viable deal that included cryogenic-engine technology transfer to India.

Over a decade later, when the Indian government approved the heavier INSAT series of satellites in 1987, which mandated a heavy-lift launch vehicle, the quest for cryogenic engine started again. However, this quest took place against the backdrop of a tumultuous period in recent history. The late 1980s and early 1990s saw the end of the Cold War, the reunification of Germany, a redrawn map of Europe and the two superpowers reduced to just one.

India's Cryogenic Engine

India's first foray into cryogenic technology started in 1970, led by V.R. Gowariker, who had been instrumental in establishing the solid propellant provision in Thumba. He established a small team to develop the Cryogenic Technology Project for rocket engines based on LOX as oxidiser and LH2 as fuel. That was the end goal. They started with a semi-cryogenic fuel pair (LOX and Kerosene), and the plan was to move on to LOX and gaseous hydrogen and eventually to LOX and LH2.

Despite the limited access to LOX and LH2 in India during the early 1970s, this small team of ISRO engineers successfully met several of the challenges of cryogenic-engine technology.[593] By the end of 1971, a semi-cryogenic engine was tested for a few seconds.[594] However, a decision was taken in 1974 to disband the Cryogenic Technology Project. The premature death of Sarabhai and the focus on the development of liquid-engine technology with support from France had shifted the priorities within ISRO. Development of liquid propellant engines that eventually became the Vikas engines used in PSLV shifted the culture within ISRO away from semi-cryogenic and cryogenic technologies. In retrospect, the seminal 1974 decision resulted in ISRO's long struggle with cryogenic-engine technology, from which it is only now emerging. In December 1982, six months after the PSLV design was finalised and approved, Dhawan established the Cryogenic Study Team. Its remit was to come up with a cryogenic stage replacement for PSLV's liquid fourth stage.[595] A year later, the team produced a 15-volume report recommending the development of a cryogenic engine for use by a future launch vehicle with the capacity to launch to GEO. It was from this work that the GSLV emerged. Dhawan's successor, U.R. Rao, secured Rs.16.30 crore, most of which was used to establish a Cryogenic Engineering Laboratory in 1986. A decade after it had been abandoned, India returned to developing cryogenic-engine engine. Within three years, this laboratory had built and tested small-scale, water-cooled cryogenic engines using LOX and gaseous hydrogen. Following this tentative success, ISRO added a 25-tonne (CE-25) cryogenic engine to its development programme, in addition to its 12-tonne (C-12) engine.[596]

Figure 11-1 Cryogenic Engine CE25 used on GSLV Mk3. Credit ISRO

The progress in the 1980s was on sub-scale 1-tonne engineering models. This work allowed ISRO engineers to experiment with different designs and subsystems. One of the key requirement was an engine cooling subsystem. The primary combustion chamber operates at a very high temperature for efficiency. However, to prevent extremely high temperatures that may result in catastrophic destruction, the engine must be cooled when operating. During the 1980s, ISRO engineers used a variety of techniques, including a heat sink, water cooling and using one of the propellants as a cooling agent instead of water. This work was done by 1989, which helped ISRO engineers finalise some of the design parameters, fabrication processes and configuration techniques for the eventual full-scale engines. But ISRO's work on cryogenic-engine technology stalled once again. This time because of an imminent deal with the USSR that included cryogenic-engine technology transfer.

Russian Roulette

The USSR had developed a cryogenic engine, the KDV-1, for its Moon mission during the 1960s, which had been cancelled. The engines had been built but never flown and, therefore, had not generated a return despite the investment. Also, the space programme in USSR had peaked several years earlier, and the demand for these engines no longer existed in USSR.[597] The political unrest in the late 1980s led to the break-up of the USSR. India sought something that Russia possessed and had no plans to use. A potential deal with India would not only be good for India, but it also made commercial sense for Russia.

The Russian offer included two cryogenic C-12 KDV-1 engines, cryogenic technology transfer including the building of a third C-12 engine in India, training of Indian engineers in Russia, transfer of documents, support with specialist materials and components and assistance during the testing and development phase. India signed this agreement in January 1991 for Rs.230 crore ($128 million) with Glavkosmos, the Russian Space Agency founded in 1985 under President Mikhail Gorbachev.[598] A year after the agreement was signed, the role of Glavkosmos was transferred to a private company the Russian Federal Space Agency.[599] By going with Russia and its cryogenic-engine technology, ISRO had negotiated a deal that cost an order of magnitude less than other offers.[600] There were two reasons why ISRO was able to conclude a favourable deal. The primary one was that KDV-1 had not been tested in spaceflight and, at the time, Russia had no plans to use them. Secondly, the engines Glavkosmos would deliver were not complete. They contained only functional elements, such as pumps, combustion chambers, valves and actuators. It was up to ISRO to provide the interfaces and electronic controls and make the engines operational.[601]

India signed the deal with the USSR in January 1991, but on 26 December of the same year, the USSR formally ceased to exist, and the Cold War came to an end. In May 1992, using the sanctions under the Missile Technology Control Regime (MTCR), the US imposed a 2-year ban on all US-licensed exports, imports and contracts between India and Russia.[602] Despite representations from the Indian and Russian leaders, the US would not rescind. In October 1993, Russia evoked the force majeure clause and told India it could no longer service the contract to transfer cryogenic technology as had been agreed.

The agreement with Russia was in place for almost three years.[603] For the most part over that time, the contract played out as planned. India made the payments and Russia delivered on its obligations. Many ISRO engineers received training in Moscow, and a significant part of the technical drawings for the cryogenic engine were transferred to India. ISRO scientists and engineers were based in Russia developing the engines using Russian test facilities. India, at the time, had very little cryogenic infrastructure.

Since Russia was no longer able to deliver the technology transfer element of the agreement, the agreement was renegotiated over several days in December 1993 in Bangalore (now Bengaluru). Originally, Glavkosmos had agreed to deliver two working C-12 engines, two mock-up engines, as well as the technology transfer. In the amended agreement, two more working engines (now a total of 4) were added in the place of technology transfer. To buy time to develop its GSLV, ISRO purchased an additional three engines (now a total of 7) for a $9 million (Rs.28.3 crore). A schedule was agreed for the delivery of all seven working engines plus two mock-ups that would start in 1997 at the rate of one engine every six months. The forced renegotiation of the agreement included a requirement that India make payments in USD and not Rupees. According to ISRO, this change in currency alone increased the cost of the agreement by 30%. However, an independent assessment concluded that the price had doubled.[604]

During the mid-1990s, ISRO overstated, or at least inaccurately assessed, how much it had learned from the Russian technology transfer that had been completed before the agreement ended. Further, ISRO over-optimistically projected that an indigenous cryogenic engine could be built by 1998. Independent assessment by an investigative journalist who had sought out views from ISRO engineers involved in building the cryogenic engine concluded that a flight worthy engine was a decade away.[605] The consequences for ISRO were not only the increased cost and a long delay in mastering the cryogenic-engine technology but it commitment ISRO to an increasingly outdated design.[606]

When the CE-25 engine was first proposed in the 1990s, the average mass of a communication satellite was increasing from about 1 tonne to over 2 tonnes. Today, payload capacities of heavy-launch vehicles have increased to around 10 tonnes, and some can achieve over 14 tonnes. The cryogenic-engine technology agreement between ISRO and the USSR was signed in January 1991, first suspended and then ended in October 1993. The events surrounding the deal resulted in one of the most controversial phases in ISRO's history. It generated an atmosphere of political distrust, suspicion of industrial espionage and allegations of secret deals in the pursuit of commercial advantage and national defence. The instrument that caused the break-up of the agreement, the Missile Technology Control Regime

(MTCR) had been in place only for about 4 years.

Missile Technology Control Regime

The MTCR was established in 1987 and was made mandatory under US law in 1990. It was designed to limit the spread of missile technology that could use unmanned vehicles to deliver large payloads (of half a tonne) of traditional explosives over a distance of 300 km. Countries with this capability clubbed together to ensure that others could not acquire it. The nature and type of equipment that fell in the scope of MTCR were vague and broad. It was not just missile technology, but anything that could be used in support of missile technology. Once a component or a system was deemed to be in the scope of MTCR, it would be subject to export controls. The US was the only and final arbiter. Components or systems that were defined to be "dual use" were particularly problematic. It prevented India from developing its space programme and other industries, including aerospace, avionics, manufacturing, computing and transport.

ISRO's solid and liquid stages in development from the 1970s possessed the capability of carrying half a tonne of explosives over a distance of 300 km. India was already in breach of MTCR when it was established in 1987 but had gone unchallenged. Further, by early 1990s, India already had a well-established missile programme. The US had first become aware of India's capability to launch missiles as early as 1974.[607] A national space programme, in every nation where it has emerged, has had a strong connection with national security. India, despite its frequently stated primary objective for societal development was no different. In January 1975, M.G.K. Menon who had temporarily stepped in to lead ISRO while waiting for Satish Dhawan to return from the US and take up the post, said "I tried to tailor the policy to have a commonality of technology so that without too much more effort you could produce an entire family of missiles with the same basic pieces like a Meccano or Lego set from which you could build all sorts of things."[608] Following his work leading the SLV-3 programme, Abdul Kalam moved from ISRO to DRDO to develop India's intermediate range ballistic missile and concluded that "SLVs and missiles can be called first cousins: they are different in concept and purpose, but come from the same bloodline of rocketry."[609]

The USA was most likely motivated to invoke MTCR following India's nuclear tests in mid-May 1998 followed by those in Pakistan's two weeks later. The tension of a new nuclear arms race in Asia raised anxiety around the world. The US's focus returned to India's space programme and the role it could play in a potential regional arms race. During testimony before the Committee on Science in the House of Representatives, Professor Gary Milhollin stated that India's SLV-3 was an exact copy of the US Scout rocket and that the first stage of the SLV-3 was also the first stage of Agni, an intercontinental range ballistic missile developed by India. He went on to assert that "India's biggest nuclear missile is an international product" with the second stage from the USSR, the liquid-fuel technology from France and the guidance control system from Germany.[610] India was clearly in breach of the MTCR and had been for many years. ISRO's infrastructure and experience had matured to a capability sufficient for military use. However, the US had made no effort to stop the Glavkosmos agreement that had been concluded openly just 15 months earlier.[611]

Above all, the case for a cryogenic engine powering a missile for military use was not a logical one. Even though the cryogenic engine capability at the heart of the India-USSR deal took India beyond the capacity limit set by MTCR, cryogenic technology is practical only for vehicles readied and launched immediately and not for weapons that must be on standby for an extended period of time. Missiles of the Cold War in the US or USSR had not used cryogenic rocket engines. Commercial considerations, however, may have also played a part in the US's insistence that India comply with MTCR.

By the end of 1991, the USSR had ceased to exist. The US saw itself as the Cold War victor and, perhaps, felt strong enough to assert its new-found authority. With a long history of achievements in space and founded on the principles of liberal economic principles, the US had become the dominant player in the commercial space market. ISRO had already demonstrated its expertise and confidence in designing, building, launching and operating spacecraft for scientific research, remote sensing and communication. Its world-class satellites for EO, Indian Remote Sensing Satellites (IRS-1A and IRS-1B), were already in orbit and their data was subject to commercial agreement with EOSAT on the international stage. In addition, ISRO was making good progress with its launcher programme. The Satellite Launch Vehicle, SLV-3, programme was already complete, the ASLV was almost complete, and the PSLV was in advance design. ISRO's capability to routinely build and launch satellites was imminent. Consequently, ISRO was "posing a serious commercial threat to established space powers."[612] The US may have seen ISRO as a new competitor in the emerging commercial space market.

ISRO was surprised that the US used the MTCR to impose sanctions. However, given the tumultuous events of the early 1990s, ISRO should have "read the writing on the wall with respect to the MTCR and its consequences for importing cryogenic technology". Someone did. S. Chandrashekar, based in the Launch Vehicle Programme Office at the ISRO headquarters in Bangalore, saw the peril of ISRO's decision to engage with the USSR for cryogenic-engine technology. Chandrashekar insisted "The cryogenic engine deal clearly violated the provisions of the MTCR, and so sanctions demanded by the US law were inevitable. I was convinced that the USSR, in the political and economic situation it found itself, was not going to be able to withstand the US pressure. But no one in ISRO was willing to listen to me."[613]

The US asserted that the agreement with Russia for cryogenic engines and technology transfer was designed to enhance India's military programme and established a set of international sanctions against India. The sanctions demanded that many high-tech space-qualified components, such as thermal blankets and inertial sensors could only be used in Indian Satellites if those satellites were launched from Western countries. India was not new to sanctions. It had been subject to international sanctions following its first nuclear test conducted in 1974, in the deserts of Rajasthan.[614] The termination of the India-Russian deal would prevent, or at least severely delay, the rate at which India would progress with cryogenic technology and thus develop a heavy-lift capability.

The US went further and extended the embargo to agreements for technology that had already been signed but not delivered. A potential for military application was considered sufficient to evoke MTCR sanctions. It was a blunt instrument. The effects of the embargo even extended to transactions completed long before it was imposed. For example, a tape recorder for use aboard IRS-1C had been returned by ISRO to Lockheed Martin in the US for repair, and it became subject to the embargo. A satellite cannot always transmit data immediately after it is captured for many reasons. It may not be over a ground station, or it may not be possible to orientate the antenna to point to the ground station or even available electrical power may be limited at the time, so data is stored temporarily and transmitted later. The tape recorder (modern satellites use solid state storage) is critical. By recording data, IRS-1C would be able to transmit it to Earth when a ground station came in view. With insufficient time before launch to develop an alternative, ISRO planned to launch IRS-1C without a tape recorder, knowing that some data would be lost and mission efficiency would be undermined.

The Chairman of ISRO, Professor U.R. Rao, went to the US to remonstrate over the numerous sanctions that impacted several ISRO projects. In an hour-long meeting with the Vice President Al Gore (Albert Arnold Gore Jr., born 1948) on 11 June 1993, Rao reiterated that the agreement with Russia had already been in place for over two years, insisted on the civilian nature of India's space programme and underscored the inappropriateness of the cryogenic engine for missiles. Gore highlighted the close connection between ISRO and India's DRDO. Several of ISRO's senior scientists and engineers, including Dr Abdul Kalam, had gone from ISRO to the DRDO. Rao countered that while up to six scientists had gone to DRDO, 10 had gone from ISRO to NASA, and as a democracy, India was in no position to dictate where individuals should go.[615] However, Gore insisted that India terminate its cryogenic-engine agreement with Russia, but he did offer to reconsider the impact of the retrospective embargo on a case by case basis.

With unexpected support from Frank George Wisner (born 1938), the Under Secretary of State, the embargo on the tape recorder for IRS-1C was rescinded, and it was delivered to ISRO in time for launch from Russia in December 1995.[616] The US did not exclusively pick on India to impede its space programme through sanctions. Following its decision to temporarily withdrew from active participation in NATO in 1966, France was subject to US export controls.[617] In part, it was this history of US-France division that motivated France more than any other country in the West to continue to support India during the sanctions period.[618] In the 1960s, US technology transfer to Japan was restricted. Inadvertently, this resulted in Japan benefiting.[619] Since 2011, NASA has been prohibited by the US Congress from engaging in bilateral agreements or cooperating in space with China.[620]

Although the relationship between India and the US improved following a state visit by President George Bush to India in 2006, the sanctions remained in place until President Obama's visit in 2010. During the two intervening decades, there were several attempts by India to breach the embargo not only in pursuit of its space programme but also its missile agenda. In May 2007, the FBI indicted two Indian scientists behind a Singapore-based third party called Cirrus for attempting to "illegally procure sensitive technology for India's ballistic missile programme" for VSSC and Bharat Dynamics. This technology could be used in missile guidance and firing systems.[621] In 2009, Germany was processing an export request for extruded aluminium plates to India. Knowing that the US had denied India something similar in the past, Germany asked the US through a secret cable if that denial had been under MTCR and if Germany should also refuse. The US confirmed that its denial of the extruded aluminium plates had not been under the scope of the MTCR and that Germany could fulfil the export order, US recommended that Germany should not. The US's response goes on to unambiguously articulate its reasons for evoking MTCR in the first place. It indicates that India's military programme was not the sole target of the sanction, stating "the U.S. does not support India's missile or space launch vehicle programme."[622]

In one remarkable instance, a US companies Mayers and Thiokol violated US's own restrictions and provided India with chemical mixer units that the Ukrainian government had stopped a Ukrainian company, Yuzhnoye, from supplying to India under the MTCR restrictions.[623] Despite this, the Ukrainian government sought clarification from the US regarding another request from India for blueprints for a semi-cryogenic (LOX and kerosene) rocket engine. Yuzhnoye had sought and received assurances that India would not use it for purposes other than peaceful uses for space. The Ukrainian government informed the US that if this contract were denied, India would most likely acquire an equivalent design from Russia anyway. The US did not prevent this deal from progressing, and ISRO is currently developing the Yuzhnoye engine for its Unified Launch Vehicle series of launchers under the project name SCE-200 (semi-cryogenic engine 200 tonnes).

The MTCR restrictions, along with others imposed by the US following India's nuclear tests in 1998, were formally lifted in November 2010. Restrictions on technology associated with nuclear weapons remained, but nine specific entities that included ISRO and DRDO were allowed to import a range of technologies that had been prohibited.[624] The President's visit in 2010 was followed by that of a delegation of two dozen US companies in the quest for Indian orders in February 2011, ushering in a fresh wave of commercial opportunities for US companies.

Commercial Space Services

By 2015, the global space industry was estimated to be $323 billion (Rs. 2,041,038 crore).[625] This valuation includes the cost of spacecraft, launch to orbit, ground segments and the services provided from space. Worldwide, there were 86 space launches to Earth orbit in 2015, 22 of which were classified as commercial valued at $2.1 billion (Rs. 14,000 crore). For three years, between 2013 and 2015, ISRO launched 28 satellites for nine international customers, which generated about $88 million (Rs.585.4 crore).[626] ISRO markets itself as a low-cost supplier of access to space. In the absence of an operational heavy-lift launch vehicle, ISRO's market share will remain low.

Space programmes are expensive, and where possible, national space agencies recoup their investment through commercial operations. It was not just India that lost out with the imposition of MTCR restrictions and the termination of the agreement with Russia. As ISRO's progress slowed, many American firms, too, lost out on lucrative contracts for high-tech components and systems. The international commercial interdependency surfaced in 2006 during a meeting between NASA Administrator Michael Griffin (born 1949) and ISRO Chairman Madhavan Nair at the ISRO Satellite Centre in Bangalore. Griffin offered to fly an Indian astronaut within the NASA astronaut programme. No Indian citizen had been in space for over two decades. Despite this, Nair declined on the basis of the cost but insisted that "Co-operation with NASA and procurement of components from US vendors is a top ISRO priority."[627]

Figure 11-2 GSLV Mk3 with Second VAB in construction in background. June 2017. Credit ISRO

Today, ISRO has a mature and efficient satellite building, launching and operating infrastructure. It has also demonstrated its competence in building satellites for the Moon, Mars and science missions. ISRO's first large communication satellites, the INSAT series, incorporated functionality for meteorology, broadcast, as well as communication, and were built by Ford Aerospace Communication Corporation. It was during the building and testing of the first INSAT (INSAT-1A) in California that ISRO engineers experienced American sanctions first hand. Although ISRO scientists were based in California during the build phase, their access to the technology going into INSAT, for which they had paid, was limited. These restrictions were a standard part of a typical commercial contract to protect the owners' intellectual property and not necessarily related to sanctions, but the Indian scientists involved did not see it in that light. These restrictions placed by the State Department were an 'eye opener' and added to ISRO's resolve for self-sufficiency.[628]

Region	Commercial Launches	Percentage	Revenue $ (million)
Russia	5	23	289
The US	8	36	617
China	0	0	0
Europe	6	27	1066
India	2	9	66
Japan	1	5	113
Total	22	100	2151

Table 11-1 Commercial Launches undertaken in 2015. Credit FAA Annual Compendium 2016

By the time U.R. Rao stood down as chairman of ISRO in March 1994, he had initiated a drive to engage Indian industries to produce key space components to meet ISRO's needs. In addition to reducing the dependency on foreign countries, this initiative would benefit ISRO with a shorter lead time and lower costs and also help boost the local economy. Industrial partners, including Godrej and Machine Tool Aids & Reconditioning, were contracted with the development of the C-12 engine. ISRO embarked on partnerships for the building of H2/LOX plant and test rig in Mahendragiri, modifying the SLP to accommodate the GSLV and building the facilities at Sriharikota for handling and storing cryogenic propellant. Having used up 6 of the 7 engines acquired from Russia, in January 2014, ISRO used its own design of cryogenic engine CE-7.5 and successfully placed a communication satellite, GSAT-14, in orbit.

It had taken ISRO almost a quarter of a century to achieve this milestone. The US first launched the RL10 with cryogenic engine in 1963, Russia the KVD-1 in 1967(but not flown in space), China the YF73 in 1975, France the HM-7 in 1979 and Japan the LE5 in 1997.[629] ISRO's long journey to master the cryogenic engine technology was the consequence of the ISRO's strategic decisions, geopolitics including the termination of the Russian cryogenic engine and technology transfer agreement and especially the US-initiated sanctions under the MTCR. The international restrictions were, however, instrumental in ISRO's becoming self-sufficient sooner than it would have otherwise. However, today only 40% of the CE-25 engine is made in India. The quantities of the other 60% required are so small that building production facilities in India are not cost-effective.[630] ISRO is planning to announce a new, better-funded initiative to engage not just academic organisations or private sector businesses, but any organisation without restriction that is willing to participate in the risky area of R&D for new materials and systems for ISRO's future launch vehicles.[631]

In the past, it was clear which nation was responsible for designing, building, launching and operating a spacecraft. The modern picture is complex where spacecraft, launch services and support are procured on a commercial basis from multiple nations. For example, the three satellites in the Disaster Management Constellation (DMC-3) built and operated by UK's Surrey Satellites Ltd were launched by India and procured by a Chinese company, Twenty-First Century Aerospace Technology Company Ltd.[632] Which nation is responsible for a launch and whether it is commercial in nature are further complicated by the capability to launch from mobile platforms; rocket engines built in one nation are used to power launch vehicles of another. In modern complex commercial arrangements, it is not always clear how commercial are the commercial launches procured by government agencies.

The sensitivity of data around commercial space activities forces a level of ambiguity in the nature of the conclusions that can be drawn. While ISRO's ability to compete in the commercial launch market has not seen the growth it once expected, the MTCR sanctions accelerated the domestic production of many of the materials and systems that it would otherwise have imported. Over the last quarter of a century, India has had significant successes in space, missions to the Moon and Mars and the regional navigational system (IRNSS). It also made significant progress in terms of economic and technological achievements. As recognition of its greater influence on the world stage, India sought and was accepted as a member nation of MTCR in June 2016.[633]

≺ ◇ ≻

Chapter Twelve
Satellites and Saris

W e had come all this way to study the Moon and what we discovered was the Earth, Bill Anders (William Alison "Bill" Anders, born 1933), one of the three men on Apollo 8 and the first to go to the Moon, told space historian Andrew Chaikin.[634] The true merit of going to space lies in the unique view of the entire Earth that cannot be seen from anywhere else. The numerous EO, meteorology and communication satellites continue this journey of discovery by observing the Earth every day. Countries, including Japan, Canada, Australia, India and Indonesia, have space programmes that maintain a strong societal connection. From the outset, India has exploited the product of space-borne assets for societal development with the intention to "harness space technology for national development."[635]

Referring to the planned Satellite Instructional Television Programme (SITE) programme, Arthur C. Clarke said "In 1975 there will be a new Star of India. Though it will not be visible to the naked eye, its influence will be greater than any zodiacal sign."[636] Since the mid-70s, India has had many more 'stars' (artificial satellites) that look down upon the vast Indian territory and provide communication, weather and EO information essential for 21st-century societies.

Between 1979 and 1994, in parallel with developing its launch vehicles, first the SLV-3 and then the ASLV, ISRO built and launched eight Rohini Satellites to help develop its indigenous capability in satellites and launch vehicles. Four satellites with a mass of around 35 kg were launched by SLV-3, and the other four, each with a mass of just over 100 kg, by the ASLV. They were all experimental satellites designed to evaluate the launch vehicle, capture scientific data and help develop the ground infrastructure.

In the process, teams of ISRO scientists and engineers developed both the satellite operational infrastructure and the experience required to operate it. Over time, ISRO has designed and built ever larger and more complex satellites and expanded the infrastructure to support them. The INSAT series of satellites were initially built by Ford Aerospace and launched by NASA or ESA. They were, however, operated entirely from India. Ford also built INSAT-1C and INSAT-1D, and they were launched by NASA or ESA.

Satellite	Launch Date	Launcher/ Country	Description
Aryabhata	19 April 1975	C-1 Interkosm os/USSR	A science payload of 360 kg at an orbit of 563 × 619 km, it operated for the first four days and then partially until March 1981. Two of its three instruments returned data. Re-entered Earth atmosphere on 10 February 1992 after 3,500 orbits.
Bhaskara-I	7 June 1979	C1- Interkosm os/USSR	Science payload of 442 kg was looking at the Earth's atmosphere and oceans, it operated successfully for over a year and re-entered Earth's atmosphere in 1989.
Rohini Technology Payload	10 August 1979	SLV- 3/India	The payload of 35 kg was dedicated to measuring the in-flight performance of the launch vehicle SLV-3, the first Indian launch vehicle capable of placing a satellite in orbit. The second stage failed, and the launch vehicle was lost into the Bay of Bengal.
Rohini RS-1	18 July 1980	SLV- 3/India	First success for SLV-3. The satellite was in orbit (305 × 919 km) and operated for nearly two years.
Rohini RS-D1	31 May 1981	SLV- 3/India	The remote-sensing payload of 35 kg was successfully placed in orbit (186 × 418 km) but in a lower orbit than intended, and the satellite re-entered nine days later.
Ariane Passenger Payload Experiment (APPLE)	19 June 1981	Ariane- 1/ESA	India's first experimental communication satellite launched free by ESA's then experimental launcher, Ariane. Its required orbit of 36,000 km located at 102°E longitude and its weight of 670 kg was well beyond the capability of the SLV-3. Initial orbit was not as planned but it operated for over two years.
Bhaskara-II	20 Novemb er 1981	Interkosm os/USSR	Experimental remote-sensing satellite was similar to Bhaskara-1 in weight and orbit. It was declared operational after receipt of 300 television images of the

			Indian subcontinent. Some telemetry was still being received until it re-entered on 30 November 1991.
INSAT-1A	10 April 1982	Delta Launch Vehicle/US	India's first multipurpose communication, broadcast and meteorology satellite. Built by Ford Aerospace in the US and launched by an American Delta launcher, it operated for a few months and failed prior to formal handover to India.
Rohini RS-D2	17 April 1983	SLV-3/India	The final flight of SLV-3 delivering 41.5 kg to orbit (371 × 861 km). The mission was a complete success. RS-D2 re-entered Earth on 19 April 1990, after 17 months of operation. The SLV-3 programme came to a premature end to allow an earlier start on the development of the next launch vehicle, the ASLV.
INSAT-1B	30 August 1983	Space Shuttle	Identical to INSAT-1A and ordered from Ford at the same time, it operated successfully for longer than its planned seven years from 72°E for most of its operational life and from 93°E in 1992 and 1993 until decommissioned in August 1993. As part of the decommissioning activity, it was raised to a graveyard orbit to clear the orbit slot for a future satellite.

Table 12-1 ISRO's First 10 Satellites

INSAT-2A was built by ISRO and launched in 1992. Since then, India has built all its satellites, as well as satellites for other nations. By mid-2016, ISRO had built, operated or launched more than 80 satellites. Four decades since the first satellite, India has one of the largest national programmes of operational satellites providing remote sensing, navigation, meteorological, navigation and communication services from Earth orbit. This journey began with Aryabhata.

India's First Satellite: Aryabhata

As Vikram Sarabhai was putting the final touches to his plan to build and launch India's first satellite in his strategic document, Atomic Energy and Space Research: A Profile for the Decade 1970-80, he received an unexpected offer.[637] In a letter received in April 1971 by Prime Minister Indira Gandhi, the Russian Academy of Sciences offered to assist India in the exploration of space.[638] During a meeting in New Delhi in the following month, Vikram Sarabhai asked U.R. Rao to present ISRO's plans for an Indian satellite to the Soviet Ambassador Nikolai Pegov. After Rao's presentation, the Ambassador offered to launch India's first satellite provided it was heavier than China's first satellite.[639]

In a series of meetings that followed in New Delhi and Moscow, the USSR appeared to backtrack and suggest that India build a scientific instrument to be carried onboard one of their satellites. Vikram Sarabhai had tasked Rao to pursue the offer from the Soviet Union with only these words "do what you think is best for our space programme."[640] Having already flown several instruments on US satellites (Pioneer, Mariner and Explorer series), Rao wanted more than just another instrument in space. Convinced by Sarabhai's vision of using space technology for social development, he saw this as an opportunity for India to bootstrap its satellite programme.

A satellite launched by the USSR would entail zero launch cost, and the mass of a satellite that a USSR launcher could deliver to orbit would be much more substantial than the 30-kg planned for India's first launch vehicle, the SLV-3, which was then still under development. By August 1971, Rao secured a deal 'in principle': India would build a satellite, and the USSR would launch it. Suddenly, in December 1971, Sarabhai died. Rao had had a long relationship with Sarabhai having completed his PhD under his supervision in the late 1950s. This loss seemed to ignite a deeper commitment within Rao to fulfil the task.[641] ISRO appeared to re-energise following Sarabhai's demise, just as President Kennedy's death intensified NASA's resolve to land a man on the Moon. Perhaps, Rao's professional duty to fulfil Sarabhai's vision became his personal sacred mission.[642]

Further meetings in New Delhi and Moscow eventually led to an agreement that was signed on 10 May 1972 by academician Mstislav Keldysh (1911–1978), President of Russian Academy of Sciences, and Prof. M.G.K. Menon the interim ISRO chairman following Sarabhai's death.[643] India would build a satellite that the USSR would launch for free by the end of 1974. With a price tag of Rs.3 crore ($3.95 million), which included one crore in foreign exchange ($1.3 million), the project got the go-ahead from the Prime Minister in mid-1972. The project had a tight deadline to design, build, test and have India's first satellite ready for launch by the end of 1974.

This was a huge undertaking. India had no experience in building a satellite.

It had no qualified personnel. It had no national technology industries with experience in building space-qualified components or subsystems, and it had no infrastructure to operate a satellite. Professor Menon was not convinced that India should proceed despite the offer of a free launch. Rao, however, was optimistic. He told Menon "If we want to go into space, this is the first real opportunity we have. As we are getting a free launch, we can start with a minimum amount of money and then build upon it. You just can't go straight to operational communication satellites or remote-sensing satellites till you have successfully built at least a couple of experimental satellites and established your capability to build complex satellites."[644]

Rao was probably ISRO's most experienced and perhaps the only scientist in India capable of building space-qualified instruments. He had returned to India in 1966 from the US, having worked at MIT as a post-doctoral student and as an Assistant Professor at the University of Dallas in Texas. As a cosmic-ray scientist and joint principal investigator, Rao had been intimately involved in designing and building experiments and instruments for experiments that were launched on several NASA spacecraft, including Pioneer 6, 7, 8 and 9 and Explorer 34 and 41, during the 1960s. Collectively, they explored the solar system and 'space weather' by detecting and measuring interplanetary magnetic fields, energetic particles and plasma.[645]

The Prime Minister had provided the go-ahead to design and build India's first satellite, and a launch date had been set by the USSR. The clock was ticking. Despite the reluctance from Thumba, a decision was taken to locate the satellite project in Bangalore (now Bengaluru) rather than Thumba. This was primarily to facilitate the short timescale to which ISRO was now committed. Bharat Electronics Limited (BEL) and other technology companies that would help deliver the electronics and the technical systems for the satellite project were located in Bangalore.[646] By September 1972, a site consisting of four sheds of 5,000 sq. ft. was acquired in the Peenya Industrial Estate in Bangalore. It was fitted out with a clean room, thermal vacuum chamber, laboratories and a workshop. It was initially staffed with about 40 mostly young engineers, none of whom had any experience in building satellites. This beginning in 1972 has evolved into today's ISAC with over 2,000 engineers and state-of-the-art facilities.

The frantic activity with which the project began continued with a remarkable degree of enthusiasm and round-the-clock activities through to the last few months prior to the launch. Some engineers even delayed their weddings (the average age of the engineers was 25), and none of them had seen a satellite before they embarked on building one. To minimise delays from red tape, Rao initiated a new procurement mechanism to purchase components that would bypass the traditional bureaucratic route.[647] With authority to place orders on the spot, Rao and his team visited the UK, US and France for specialised equipment and components that were not available in India. In addition, Rao used his contacts in the US to loan, on a replacement basis, components otherwise difficult to acquire to help build the initial engineering models.

Figure 12-1 Aryabhata Prior to Assembly. Credit ISRO

The satellite design was to be relatively simple. It would be a spin-stabilised, 26-sided quasi-sphere with solar panels on 24 sides generating 46 watts of power to run the three onboard scientific experiments to detect and measure neutrons from the Sun, cosmic X-rays and gamma-rays. An essential requirement for a spin-stabilised satellite is that it is balanced around the spin axis. A specialised dynamic balance is used to detect uneven mass distribution around the axis of spin. A dynamic balance had been ordered from Germany, but it was delayed. With the launch date approaching, Rao's team designed and built a dynamic balance in just three months.

Sriharikota was to be the primary ground station for all communication with Aryabhata when in orbit, but there was no back-up communication channel. Two months before the launch date, two lavatories behind one of the four sheds at the newly acquired site in Bangalore were converted into an operational ground station complete with a fully steerable yagi array antenna.[648] Despite the challenging start, India's first satellite was completed on time. Initially called the Indian Scientific Satellite Programme, three names for the satellite, Mitra, Jawahar and Aryabhata, were shortlisted in the final board meeting before launch. Aryabhata was a fourth-century mathematician-astronomer, Mitra meant friendship reflecting the India-USSR collaboration, and Jawahar meant the spirit of independence. A month before the launch, the name Aryabhata was selected by Prime Minister Indira Gandhi.[649]

Figure 12-2 Professor U.R. Rao. August 2013. Credit Author

With its science package of three instruments, Aryabhata had a mass of 360 kg. It was placed in an orbit of 563 × 619 km by the Soviet Launcher Cosmos 3M from Kapustin Yar Cosmodrome on 19 April 1975.[650] Immediately after the launch, Aryabhata's telemetry was received at the USSR-based Bears Lake communication centre. Thirty minutes later, the first signals were detected at Sriharikota in India. The electricity supply connection to the third, ionosphere experiment, failed. Two of Aryabhata's three instruments, the X-ray astronomy and the solar neutron and gamma-ray experiments, returned data. Aryabhata captured X-ray data from Cygnus-X1, a well-known galactic X-ray source, it demonstrated satellite-tracking technique using tone-ranging and conducted an experiment to carry voice communication and electrocardiogram (ECG) data between Sriharikota and Bangalore. It also collected weather data (temperature, wind speed, pressure, etc.) from weather stations from remote weather stations around India.[651] This experiment was conducted in collaboration with the meteorological department of India. After orbit 41, a fault occurred in the power supply.[652] Some experiments had to be switched off, but the rest of the spacecraft continued to function normally. This action led some journalists to incorrectly report that the mission had failed and ended.[653] Aryabhata, however, operated successfully beyond the targeted six months.[654] It demonstrated that Indian scientists and engineers were capable of designing, building and operating a satellite in space. Through the success of Aryabhata, ISRO acquired a level of confidence that can only be attained from the achievement of a goal that many believed to have been impossible. The NASA administrator visiting India in 1972 expressed his astonishment with the progress made in one year, later saying "I never thought you could do this."[655] Even the USSR that had initiated the project "doubted that India could build a satellite in 2 1/2 years."[656]

ISRO entered the Space Age at a time when, in India, there were "more bullock carts than cars" on the roads.[657] From empty sheds on an industrial estate, Rao and his team of novice engineers had done it in 31 months. Aryabhata's lasting legacy was the new-found confidence that Indians in India could design, build and operate a functioning satellite in Earth orbit.

Earth Observation: Bhaskara and IRS

Earth Observation (EO) refers to any activity of acquiring information about the Earth's surface, sea and atmosphere, including temperature, density, chemical composition, humidity, wind speed and direction. The technology used to acquire this information is collectively referred to as remote sensing.[658] These two terms, Earth Observation and remote sensing, are frequently used interchangeably. Although EO can be done using aircraft or balloons, modern techniques predominantly involve satellites in approximately 800 km polar orbit.[659] Satellites in GEO orbit can also undertake EO but are less common. India's EO programme began very early in its space programme with the launch of its second satellite, Bhaskara-1, on 7 June 1979.

Arthur C. Clarke's reference to the "new Star of India" referring to the communication satellite used in the SITE programme probably better represents ISRO's EO, rather than communication, series of satellites since they have had a greater impact, albeit not as obvious, on the lives of ordinary Indians.[660] India has an immense and challenging geography with diverse and vast natural resources that include glaciers in the Himalayas, deserts in Rajasthan and tropical forests in Kerala, which can only be monitored efficiently by modern space-based technology. Space assets are critical to meet the challenges of climate change, increasing food production for a growing population and building infrastructure (such as electricity supply and road and rail transport links) for national economic growth.

India's domestic constellation of EO satellites guides Indian farmers on when the use fertilisers, where to dig wells, helps fishermen with information on when and where to go fishing, assists mining companies in locating minerals and metal and provides early warning to citizens who live in flood-prone areas. It helps the local and national governments to prepare for typhoons and the monsoon season and allows authorities to quantify the impact of the devastation caused by natural disasters to aid rescue and target resources in their aftermath.

A terrestrial remote sensing experiment in early 1970 proved that India could benefit from developing satellite-based EO function that several nations were already exploiting, paving the way for ISRO's EO satellite programme.[661] Over a 5-day period in 1970, photographs were taken of coconut plantations in Kerala using two 70-mm Hasselblad cameras from a helicopter flying at 300 m. Upon processing the images, it was clear that this technique could be used to detect early signs of coconut wilt disease and help reduce the $2 million loss suffered by the Travancore-Cochin region annually.[662] More than half a century on, India has one of the world's most sophisticated programmes for EO to help monitor and manage the varied resources in its huge sprawling land mass of 3.3 million sq. km and its extensive coastal zone.[663]

This first remote-sensing experiment in India drew on international support just as the first rocket launch to space from Thumba had done in 1963. The Hasselblad cameras used were made in Sweden, the helicopter was provided by the USSR, the Ektachrome film came from the UK, and the key scientific advisor was American. The project was conducted under the auspices of a UN resolution that encouraged those nations that had expertise in EO to share with those that did not.[664] Having lost the race to the Moon, the USSR focused building a station in space and on asserting its political dominance on Earth. By the time of this UN resolution (16 December 1969), the US had successfully completed President Kennedy's goal of manned flights to the surface of the Moon twice.[665] Just as the US and USSR had vied to entice non-aligned nations during the early 1960s, they again offered support with EO as a vehicle to grow their respective geopolitical influence.[666]

Bhaskara 1 and 2

Two days after the launch of Aryabhata, ISRO signed an agreement with the USSR to launch the next.[667] Bhaskara-1 was built rapidly partly facilitated by using parts developed for Aryabhata. The Bhaskara-1 design mandated that the Aryabhata's stand-by structural model be used as the basic mechanical platform.[668] It was built once more by Professor U.R. Rao's team in Bangalore but with input from all the major ISRO Centres. It was also spin-stabilised and placed into a near-circular orbit at an altitude of 534 km. Initially called the Satellite for Earth Observation, it was formally named Bhaskara-1 after launch on 7 June 1979.[669]

Figure 12-3 Bhaskara-1 Undergoing Testing in 1979. Credit ISRO

Two weeks after the launch of Bhaskara-1, SAMIR was switched on and collected data, but a fault with a high-voltage electrical connection prevented both cameras from working for over a year. When switched on, air trapped within the high-voltage supply caused arcing (corona), resulting in large-scale electromagnetic interference. The cameras were switched off to prevent other electrical subsystems being affected. As ISRO engineers had predicted, over time the trapped air escaped. Eleven months after launch, the cameras were activated, and the first images of India by an Indian satellite from orbit were captured and transmitted; the mission was considered a success.[670]

The primary objective for Bhaskara-1 was EO of the Indian landmass and the surrounding ocean, but it also served to help ISRO develop its processes, internal organisation and supporting infrastructure for gathering, analysing and managing the nation's natural resources. Bhaskara-1 was formally switched off in March 1981 and disintegrated during re-entry in 1989.[671]

Bhaskara-2 was similar to Bhaskara-1 in design, size, mass, instrumentation, launch and orbital characteristics. Learning from Bhaskara-1, the high-voltage supply to the camera onboard Bhaskara-2 was redesigned. The TV camera operated from the outset as expected as did the SAMIR instrument, which had three instead of two frequency channels.[672] Bhaskara-2 was launched on 20 November 1981 and collected data successfully until 1984 when it ceased to operate. It burnt up during re-entry on 30 November 1991. Between them, Bhaskara-1 and Bhaskara-2 captured more than 2,000 images covering the complete Indian subcontinent.[673] With the Bhaskara programme, ISRO demonstrated its capability in designing and operating EO satellites. It was now ready to embark on a full-fledged national EO programme.

IRS-1A

Indian Remote Sensing Satellite IRS-1A launched in 1988 was ISRO's first fully functional remote-sensing satellite with two high-resolution Linear Imaging Self Scanner (LISS) cameras, LISS-1 with a 2-m resolution and LISS-2 with a 36-m resolution. LISS cameras used digital CCDs (Charged Couple Devices) as image sensors. CCDs were first developed as image sensors in 1969. NASA developed CCDs for use in space by 1975. ISRO's first CCD was tested in a laboratory in 1980 and in space in 1988 on-board IRS-1A. When ISRO developed IRS-1A, the PSLV was not yet operational, so ISRO sought launch assistance from the USSR. As the first remote-sensing satellite, it was subject to numerous simulations, structural, engineering and flight modelling and extensive cycles of reviews and tests.

Once IRS-1A was built and the final tests completed, two ISRO engineers flew with the flight model to the Cosmodrome in the USSR aboard Aeroflot IL-76 cargo aircraft on 24 January 1988. On 17 March, it was launched into a 900-km SSPO with 99° inclination by the Vostok launcher of the USSR. Once in orbit, first the solar panels, then LISS-1 and LISS-2 were activated.

Name	Date of Launch	Description
Cartosat–1	5 May 2005	Also known as IRS-P5, it was the first Indian satellite to provide in-orbit stereoscopic images using two panchromatic cameras sensitive in the visible spectrum. Images from it were sold commercially. It exceeded its planned lifetime of 5 years.
Cartosat-2	10 Jan 2007	Has a high-resolution panchromatic camera offering 1 m resolution. It was launched at the same time as the SRE-1, Lap-lan Tubsat for Indonesia and Peuensat-1 for Argentina. It exceeded its planned lifetime of 5 years.
Cartosat-2A	28 April 2008	Was launched on PSLV-C9, along with an Indian mini-satellite and eight nano-satellites from Japan, Canada, Germany, Denmark and the Netherlands.Cartosat-2A is equipped with an agile camera platform and able to image selected areas under its orbit more frequently.
RISAT-2	20 Apr 2009	India's first dedicated reconnaissance satellite with day and night all-weather monitoring capability. In the press, it has been referred to as a spy satellite, in part, because its launch was not televised live like usual. It was called RISAT-2 because it incorporates Radar Imaging technology. RISAT-2 was acquired at short notice from Israel. RISAT-1 was in development at the time but not ready for launch hence RISAT-2 preceded RISAT-1 to orbit.
Oceansat-2	23 Sep 2009	Was designed to provide continuity of service for users of the Ocean Colour Monitor (OCM) instrument on Oceansat-1 launched in 1999. Oceansat-2 has two other instruments in addition to the Ocean Colour Monitor (OCM-2).
Cartosat–2B	12 Jul 2010	Identical to Cartosat-2A with a panchromatic camera capable of imaging a strip of 9.6 km with a resolution of about 1 m.
Resources at-2	20 Apr 2011	A remote-sensing satellite intended to continue with the remote-sensing data services provided by Resourcesat-1 to global users.
RISAT-1	26 Apr 2012	A remote-sensing satellite designed for applications in agriculture and management of natural disasters, like flood and cyclone. The sub 1m imaging resolution supported strategic applications.
SARAL	25 Feb 2013	Satellite with ARgos and ALtiKa (SARAL) is a joint mission between the Indian and French space agencies to monitor sea surface elevation and sea circulation.

Table 12-2 ISRO's Earth Observation Satellites Operational in mid-2016

A slight tweak was required to the orbit to ensure Sun-synchronisation, and IRS-1A was declared operational three weeks after launch. It would cross the equator at 10:30 am local time during ascending node, which brought it to the same point on Earth every 22 days. It operated until July 1996. IRS-1A was entirely an Indian project except for a few components that had to be imported.[674] The National Remote Sensing Agency (now known as the National Remote Sensing Centre) in Hyderabad was equipped with ground-receiving systems for payload data reception and storage and processing facilities to meet the needs of IRS-1A. IRS-1B was launched in 1991 and was identical to IRS-1A. It operated successfully until 20 December 2003. By mid-2016, ISRO was operating nine EO satellites in SSO (Resourcesat-2; Cartosat-1, 2, 2A and 2B; Risat-1 and 2; Oceansat 2 and SARAL), three EO satellites in GEO (INSAT-3D, Kalpana-1 and INSAT-3A) and one in equatorial orbit (Megha-Tropiques).

Remote Sensing Instrument

In the almost four decades since Bhaskara, not only has the number of satellites in orbit increased but also the complexity and sensitivity of the onboard instruments. These instruments are designed to detect and measure four physical attributes seen from space, spectral (variation in colour), spatial (degree of resolution), temporal (changes over time) and polarisation (a property of reflected light that carries information about the physical and chemical attributes of the reflecting source). Collectively, these instruments allow scientists to identify and quantify characteristics, such as the temperature of the air or sea surface, salinity, soil moisture, sea ice, the amount of water in the atmosphere, wind speed and direction. Remote-sensing technology can be passive or active.

The passive technology uses cameras or radio receivers to analyse the electromagnetic radiation reflected by the Earth. The active technology involves the satellite illuminating the Earth using an onboard radio or light source (Light Detection and Ranging (LIDAR) or Radio Detection and Ranging (RADAR) systems) and analysing the reflections. LIDAR and RADAR are conceptually identical; LIDAR operates in the visible range of the Electromagnetic Spectrum and RADAR in the radio (microwave) range. Instruments onboard ISRO's EO satellites include:

Passive instruments

- SAMIR: Satellite Microwave Radiometer designed to measure liquid water content in the atmosphere, water vapour and ocean surface characteristics. It was first used in Bhaskara-1 in 1979.

- MSMR: Multi-frequency Scanning Microwave Radiometer designed to measure wind speed and direction. It was used in Oceansat-1, also known as IRS-P4, in 1999.

- MADRAS: Microwave Analysis and Detection of Rain and Atmospheric Structures designed to measure ice particles in cloud tops. It was used in Megha-Tropiques in 2011.

- SAPHIR: Sondeur Atmosphérique du Profil d'Humidité Intertropicale par Radiométrie designed to determine humidity by measuring water vapour distribution associated with convection and vertically between cloud layers. A joint venture between ISRO and CNES, it was used in Megha-Tropiques in 2011.

- OCM: Ocean Colour Monitor designed to look at the ocean using visible and near-infrared light in the range of 400–800 nm. Radiation from sea water in this range is characterised by a small-scale content of the reflecting source. This includes suspended particulates, minerals, chemical compounds and phytoplankton. It is used in Oceansat-2 launched in 2009.

Active instruments

- SAR: Synthetic Aperture Radar designed to measure soil moisture using radar backscatter. With a resolution of up to 1 m, it could also measure Earth's surface albedo (the measure of how well a surface reflects the light shining on it; for example, a snow-covered surface reflects most of the light so has a high albedo), sea ice sheet topography, land surface imagery, ocean-dynamic topography, sea ice cover, soil moisture at the surface and vegetation type. It was used in Risat-2 in 2009.

- ALTIKA: Altimeter using Ka band transmission. It is used to measure humidity, wind speed, ocean surface wind speed, sea level and ocean wave height. Used in SARAL, the joint ISRO and CNES mission in 2013.

- Scatterometer: A radar technique used to measure ocean surface wind speed and direction. It is used in Oceansat-2 launched in 2009.

Over the last three decades, key innovations in the development of both passive and active sensors have enhanced the number and range of attributes and the accuracy with which they can be measured. The spatial resolution three decades ago was much lower but sufficient to detect man-made structures, such as roads, railways, buildings and bridges. Modern technology offers higher resolution that can detect individuals, vehicles and group activities from space. It can provide images or real-time video.

Global coverage and the orbital configuration that brings the satellite over the same point repeatedly every few days or weeks have made remote-sensing data a unique resource for research and applications, such as cartography, agriculture, geology, urban planning, forestry and surveillance. A third of India's population lives in about 8,000 towns and cities. Satellite information plays a key role in the urbanisation that is expected to double in the next 20 years. Data from satellite technology is a powerful resource that contributes to the economic growth of a nation. Since satellites cover all the globe and not just the countries that launch them, data collected can also be sold to countries that do not have satellites of their own.

Data from Earth Observation Satellites

The US recognised the commercial value of the data it had been capturing since the 1960s. The EO programme was pioneered by the US's Landsat-1 in 1972. Presidential directive 54 in 1979 formally directed NASA to pass on the operational management of its three Landsat satellites to the private sector. In 1985, the Earth Observation Satellite Company (EOSAT) took on the responsibility for Landsat operations. It assumed responsibility to grow the Landsat constellation on a commercial basis. Eight Landsat satellites were launched between 1972 and 2013, of which one failed to reach orbit. In mid-2016, only Landsat-7 and Landsat-8 were operational. In October 1993, ISRO signed an agreement with EOSAT that gave the latter exclusive worldwide rights for a fee of Rs.1.8 crore ($0.6 million) for each ground station it built to receive IRS data.

Within India, the government established two organisations in parallel during the 1970s and 1980s to manage the EO data, the National Remote Sensing Centre (NRSC) and the National Natural Resources Management System (NNRMS). NRSC has gone through a number of reorganisations since being established, and today, it has a large campus in Hyderabad with an Earth station about 60 km away at Shadnagar. It is tasked with acquiring, processing and disseminating satellite data from Indian satellites, as well as foreign satellites with which India has agreements. It also provides 24 × 7 support for the disaster monitoring service. Its commercial function includes providing user-customised geospatial solutions and sale of low, medium and high-resolution images captured by the various instruments on Indian EO satellites.

NNRMS makes available EO satellite data of India's natural resources to those government agencies and ministries that can make direct use of it. NNRMS was first proposed as a centre to manage this data in 1983 and operated under the auspices of the DOS. It is now an integrated national management system responsible for managing a central repository of EO data used to sustainably exploit India's natural resources for national interest. Since its initial founding, NNRMS has expanded to support weather forecasting, disaster management, environmental monitoring, infrastructure development and urban planning. Over the years, NNRMS has developed three specific resources from the data it has accumulated.

An inventory of mapping data for forests, wastelands, surface-water bodies, wetlands, coastal land use, groundwater targets and urban land use. Geographical Information Systems (GIS) incorporating satellite images for developing applications, such as crop-production estimation, land and water resources optimisation in watersheds, coastal zone regulation, environmental and landslides hazard impact analysis. Large GIS databases of state-specific software tools to support unique planning and governance needs in a consistent, effective and standard approach across India. Loss of lives on a large scale, whether through flood, earthquake, drought or famine, has been a regular feature of the Indian subcontinent throughout history.

The 2004 Indian Ocean and the 2011 Japanese tsunamis, captured in harrowing high-definition images, help illustrate the immense destructive power of nature. In part, this loss has vindicated the application-centric thinking of the founders of India's space programme. A wide variety of sensors on meteorological satellites, the advent of space-age technology for early warning systems and space-based search and rescue have for the first time enabled a mechanism to mitigate the consequence of natural disasters. During the Bhola Cyclone in November 1970, Bangladesh lost half a million people. In the1999 Odisha cyclone, the loss of life was reduced to around 10,000 because of improved space-based monitoring. With the incremental enhancement of the space infrastructure, only a relatively tiny number of fatalities were reported during cyclone Phailin in October 2013 because around 10 million people had been evacuated by the time it arrived.[675]

Benefits of India's EO programme are substantial but intangible and often unquantifiable. For example, it is impossible to quantify the value of lives saved through the Indian Tsunami Early Warning System, preventing crop failure and famine through irrigation made possible by finding groundwater using space-based sensors, mitigating cyclone damage by timely forecasts modelled on years of EO data and the commercial potential of a new regional navigation system to the national economy. Some specific examples of where ISRO's EO capability has or can mitigate the impact of natural disasters in India include:

- Uttarakhand flash floods: The sudden deluge between 16 and 17 June 2013 and the resulting landslides swept away roads, houses and infrastructure. Following the inevitable death and destruction, space-based assets supported the relief activities, which included making available 12 satellite phones, five transportable satellite terminals to facilitate video conferences, and voice and data to facilitate recovery activities in the immediate aftermath. Subsequently, 2,400 landslides were mapped from space to aid long-term repair.[676]

- Tropical Cyclone Phailin: A significant and severe weather event was forecasted several days in advance. Phailin was monitored by INSAT-3D as it approached the Bay of Bengal making a landfall on the coast of Odisha on 12 October 2013. The close monitoring and early warning prompted large-scale evacuation limiting the loss of lives to around 44. Fourteen years earlier, the casualties were around 15,000 with an estimated damage of $2 billion.[677] The huge reduction in the impact of Phailin was the result of space-based assets, associated ground elements, as well as the prompt action to engage relief services.

- Groundwater: Most of India relies on groundwater. up to 90% for rural domestic use and 60% for irrigation. Locating groundwater can be costly and time-consuming. A nationwide map identifying groundwater has been produced using space-based remote sensing and hydro-geological and geophysical investigations on the ground. Over 300,000 bore holes were drilled across the country with 90-95% success in locating groundwater.[678]

- Land reclamation: The Ministry of Rural Development initiated a project to detect and map wastelands, that is, land that could potentially be used for agricultural or other purposes but is currently unused. Using remote-sensing data, a wasteland map for the whole of the country was generated, which characterised 16% of India (46.72 mha in 2009) as wasteland.[679] Using this data, the Ministry of Rural Development is implementing wasteland development activities across the country. Forecasting natural disasters: ISRO scientists continue to develop models and establish early warning systems for parts of India with challenging geography. For example, alerts for heavy rainfall/cloudburst in western Himalayas, rainfall-triggered landslide alerts for the pilgrimage route in the Uttarakhand region, early warning for floods in critical locations in flood-prone regions of

Assam and short-term models for snow-melt run-off into rivers, including Alaknanda, Bhagirathi, Yamuna, Sutlej, Beas and Chenab basins.[680]

As the more developed economies have demonstrated, space-based services enhance both the quality of life and the national economy. As ISRO's infrastructure and experience grow, so does the potential for introducing new imaginative space-based solutions. In 2015, ISRO actively considered 170 projects that could have a direct impact on improving the lives of ordinary Indians. Three of ISRO's recent projects include:

- Heritage site conservation: Working with the Ministry of Culture, ISRO plans to document heritage sites and monuments of national importance by creating a database of heritage sites from data collected by Cartosat-1, Cartosat-2 and Resourcesat.

- Mining Surveillance System: A space-based system to detect illegal mining. India has 3,843 mining leases of which 1,710 are working, and 2,133 are non-working mines. The software will automatically scan satellite images for around a 500-m range and trigger an alert on detecting unauthorised activity.[681]

- Unmanned railway crossings: Enhance safety by equipping trains with GPS Aided GEO Augmented Navigation (GAGAN) receivers as used by aircraft. Barriers can be lowered as a train approaches and retracted once it has passed. This space-age technology would boost the safety of India's 150-year-old railway network.[682]

EO has played a key role in helping India on its journey towards economic development. Industries, such as mining, oil, raw materials, farming, fishing and agriculture, have directly benefited. This technology is instrumental for India's strategy of sustainable growth. For example, Potential Fishery Zone forecast is issued every three days. On cloudless days, it is issued daily and during the breeding and spawning periods of 45–60 days, advisories are not issued at all.[683] The impact of services, such as search and rescue, tsunami warning system, urban planning, groundwater detection and post-disaster crises management, are probably beyond quantitative measure, but all play a profound role in fulfilling ISRO's stated goal to harness space technology for national development.

ISRO initially secured the key experience of building and operating satellites in LEO through Rohini, Aryabhata and Bhaskara series of satellites. It still had no experience of communication satellites that could provide communication services with an equal potential to transform society. The larger mass and more distant GEO (36,000 km) orbit was beyond ISRO's reach until the PSLV was operationalised near the end of the 20th century. In 1975, ISRO leveraged another free offer, this time from the ESA, to launch a communication satellite to GEO.

Ariane Passenger Payload Experiment

In 1975, the ESA published an open offer to the international community for a free launch to GEO as part of the development programme for its new launcher, Ariane.[684] Coming a few months after the successful launch of Aryabhata, the timing was perfect for ISRO. Of all the types of satellites, communication satellites provide perhaps the most tangible of services, such as direct-to-home (DTH) TV and radio broadcast and international telephone coverage. A key requirement for a communication satellite is its orbit, GEO. ISRO's SLV-3 launcher was too small to deliver a satellite to GEO, and the launch sites of USSR that ISRO had used (Kapustin Yar at latitude 48.5°N and Baikonur at 45.6°N) were located too far north for efficient launch to GEO.[685] Although technically possible (and such launches do take place), launch sites in Russia are not best placed for launch to GEO. To efficiently reach GEO, launch sites on or near the equator are used.[686] ESA's launch site located at the Guiana Space Centre 5°N of the equator is ideal for GEO.[687]

A free launch offer from ESA was an invaluable opportunity for ISRO to experiment with a communication satellite. ESA's offer came with the understanding that it would not guarantee success and the supplier had to accept the loss if the launch failed and the spacecraft was destroyed. Insurance is seldom available for an experimental launcher. ESA had designated Ariane's first four flights as experimental. By 1976, ISRO and 72 other organisations from around the world had submitted proposals for ESA to launch their home-built communication satellites. ISRO's was among those selected.[688] A formal agreement was signed between ESA and the Government of India in May 1977. On 19 June 1981, India's first GEO satellite was placed in orbit by ESA's launcher Ariane, and it operated successfully for over two years. This very first experimental communication satellite was called the Ariane Passenger Payload Experiment (APPLE), and it was used for television and radio broadcast.

The Soviet offer to launch Aryabhata had provided ISRO with a steep learning curve for building and operating remote-sensing satellites. APPLE would do something similar for the more complex technologies associated with communication satellites. ISRO signed an agreement with ESA on 24 October 1977 with a planned launch date about three years later. Delays at ESA increased this interval by about a year. This 4-year period became another intense learning cycle in which ISRO consolidated it's operational, organisational and technical competences.

During the late 1970s, Indian industry did not have the capacity to build all the space-qualified components and systems ISRO required. To generate electricity, Aryabhata had used body-mounted solar cells. Since the surface area was small and only a part of it could be facing the Sun at any one time, the power it could generate was limited. Additional services of a communication satellite required increased power. GEO satellites are further from Earth, and the nature of communication services makes greater demands on power. Body-mounted solar cells would not suffice, and solar panels that could continually face the Sun would be needed for APPLE.

What	When
The initial offer from ESA for free launch in late 1975.	ISRO's proposal is accepted by ESA and agreement signed on 24 October 1977
Initial design	Designated GEO location at 102°E Operational life of 2 years Ni-Cd battery of 240 watts and two solar arrays.
Total mass	672 kg (380 kg propellant)
Dimensions	1.2 m diameter and height
Launch date	19 June 1981. GTO 201 km by 36,206 km
Transmitter	2 C-band transponders
Designated operational	13 Aug 1981. Final GEO orbit of 35,500 km to 35,900 km
Mission terminated	19 September 1983

Table 12-3 Overview of Ariane Passenger Pay Load Sequence of Events

The lack of some facilities generated some inventive and anachronistic innovations. In the absence of electromagnetic testing facilities, a bullock cart with rubber tyres provided a "non-magnetic environment to conduct the antenna test in an open field to remedy the TT&C link problem caused by the impedance-matching problem. The solution was found quickly at the cost of Rs. 150 ($19) for hiring the cart."[689] India did not have the capability to fabricate space-qualified solar panels or rechargeable batteries. Consequently, the solar panels for APPLE were acquired from Spectrolab, a US company, and the 240-watt Ni-Cd battery was supplied by Saft of France. APPLE used momentum wheels and RCS for attitude control. One of the momentum wheels was manufactured at VSSC, and the other came from Teldex in what was then West Germany.[690]

The Hamilton Standard RCS used for attitude control came from the US. ISRO had developed a unique passive thermal control system using paints, multi-layered insulation blankets, optical solar reflectors and specially treated surfaces for Aryabhata and used that design approach once more in APPLE.[691] In its final form, the 672-kg APPLE was a three-axis stabilised cylinder with a diameter and height of 1.2 m, topped by a 0.9 m parabolic antenna for the two C-band transponders and two solar panels for recharging a 240-watt battery. It was designed for a lifetime of two years.

Figure 12-4 APPLE Electromagnetic testing on Bullock Cart 1981. Credit ISRO

ISRO developed APPLE using the three (Structural, Engineering and Flight) model philosophy. ESA had insisted that a structural model of APPLE be kept in France. Should ISRO not be able to produce the flight model of APPLE, ESA would then proceed with the structural model instead to allow ESA to test its satellite-deploying mechanism. Had the structural model been flown, ISRO would have suffered "considerable humiliation."[692] APPLE was completed in time and tested in India by the joint effort of several ISRO teams.

APPLE was flown out of Bangalore 45 days prior to launch accompanied by a team of ISRO engineers.[693] While the solid-fuel apogee motor was transported to France by aircraft, the liquid-hydrazine fuel made in Thumba for the reaction control engines was transported to Guiana by ship in specially constructed containers. Ariane's launch capacity allowed ESA to launch more than one satellite at the same time. APPLE was launched with two other spacecraft both from ESA, an engineering test vehicle called Capsule Ariane Technologique (CAT-3) and a weather satellite METEOSAT-2[694]. In the launch configuration, APPLE was positioned in the middle with the primary payload METEOSAT-2 above and CAT-3 below in the Ariane payload bay. The first four flights had been designated experimental by ESA. The first flight of Ariane 1 in December 1979 had been a success, but the second flight in May 1980 had failed. APPLE was launched on Ariane's third experimental flight from ESA's Guiana Space Centre in South America on 19 June 1981. Sixteen minutes after launch, the Ariane launcher first released METEOSAT-2 followed by APPLE.

Ariane put APPLE in a GTO of 201 km by 36,206 km with an orbit of 10 hours and 40 minutes. From this highly elliptical orbit, it was up to APPLE's own rocket (known as an apogee motor) under ISRO engineers' command to move the satellite from GTO to GEO to its designated longitude of 102°E and a circular orbit at 35,900 km above the Earth's equator with a period of 24 hours. The onboard engine used by APPLE for this manoeuvre was a modified fourth stage of ISRO's SLV-3 launcher. The apogee motor with 314 kg of solid propellant fired for 33 seconds, but it was insufficient, taking APPLE only to an orbit of 31,000 by 35,800 km. The 16 small thrusters designed for attitude control and station-keeping were used to make up the deficit.[695] This was the first time an Indian-built satellite was successfully moved by ISRO from GTO to GEO.

Once in GEO, APPLE experienced another problem. When the command was issued to deploy the two solar panels, only one deployed. Following unsuccessful attempts (spinning, de-spinning and selective heating) to release the stuck solar panel, it was decided to proceed with the mission using the reduced power of 140 instead of 240 watts. In addition to the loss of power, APPLE's ability to dissipate internally generated heat was limited by the solar panel in its stowed position. Consequently, some of the internal subsystems were operating at a high temperature of 50°C, close to their operational limit.

This was of particular concern during the periods of equinoxes in March and September when APPLE would receive maximum solar radiation. To prevent damage from overheating, ISRO engineers would change APPLE's orientation for four hours daily during these critical days to help reduce the internal temperature. This was done at night to minimise operational impact.[696] Despite the additional demands on fuel for these unplanned manoeuvres, APPLE exceeded its target operating lifetime of two years by three months. On 13 August 1981, the Indian Prime Minister symbolically handed over APPLE to the Department of Telecommunications from an ISRO ground station in New Delhi. During the event transmitted live using APPLE, she thanked the ISRO staff watching at the SAC in Ahmedabad. During its 27 months of active operations, APPLE was used to conduct experiments, provide live TV coverage of national events (such as the Indian Air Force (IAF) Fire Power Demonstration), transmit educational courses, as well as broadcast 600 hours of TV programmes.[697]

In the haste to get APPLE operational, the APPLE Mission Control Centre was established as part of the ISTRAC at Sriharikota. Post APPLE, this facility was moved to Bangalore. The total number of staff at ISAC increased to beyond 600. To accommodate them and prepare for the additional growth that INSAT-2 would demand, funding was made available to move ISAC from leased units in an industrial estate to a new purpose-built centre near the old Bangalore airport where it remains today.[698] The overall cost of APPLE, including the launch and communication support service, was estimated to be about Rs.12 crore ($15 million). In practical terms, the value of APPLE to ISRO was much more. APPLE validated ISRO and its engineers' capability to operate communication satellites in GEO from the associated ground-based infrastructure.

APPLE provided ISRO with the opportunity to hone skills and technologies unique to communication satellites, including raising the orbit from GTO to GEO, using solar panels and 3-axis stabilisation, station-keeping and operating the ground infrastructure necessary to manage the operational services for a communication satellite.[699] As India's first true communication satellite, APPLE was entirely different to Aryabhata in size, function and complexity. APPLE became the test bed that allowed ISRO to learn to design, build, test and operate communication satellites. APPLE's success was the combined effort of a large team of 2,500 engineers, scientists and administrative personnel, with support from industry and institutions. The average age of the team was 27, and just as Rao had discovered with Aryabhata, the absence of previous experience was not necessarily a disadvantage. If they were not capable of achieving this goal, they did not know it. APPLE was an ideal stepping stone for ISRO's next major undertaking, the INSAT programme.

Communication Satellites

In his 1970 paper, A Profile for the Decade 1970-80, Vikram Sarabhai asserted "A most important practical application of space research is the use of satellites for telecommunications."[700] ISRO has since then developed and deployed a series of INSAT satellites that has become the touch point through which most Indians receive the benefits of space technology, even if they are unaware that the services they use rely on space assets. The idea of INSAT was first contemplated by Vikram Sarabhai as early as 1968, as recalled by E.V. Chitnis "We were convinced that India should have a satellite INSAT with television communication, telephonic television and meteorology."[701] The first public announcement of INSAT was made by Vikram Sarabhai in his presentation, INSAT, A Strategy for Development, at the Bombay National Electronics Conference in 1970.[702] The vision at this early stage was for INSAT to deliver for India what satellite technology was delivering for developed countries, long-distance telecommunication, meteorological services and television broadcasts.

Although commercial satellite TV services were available from the 1960s, the cost for private users was prohibitive even for many in the developed nations. A satellite was an elegant alternative to the complex, expensive and time-consuming traditional network of TV antennas on the ground. With its unique location in orbit, a satellite TV broadcast could cover the whole of India from space. For India, with its huge and varied landmass, skipping this traditional network of antennas and going straight to satellite TV transmission made practical sense and was consistent with Bhabha and Sarabhai's vision. By leapfrogging some of the incremental steps in the technological evolution, India could catch up with the pace of development in the West.

India got first-hand experience in satellite-based TV broadcast through the SITE, a pilot project supported by NASA to deliver educational satellite TV programmes to rural Indian villages. This ran for a year between 1 August 1975 and 31 July 1976. An unintended (at least from India's perspective) consequence of SITE was to raise India's expectation that India's INSAT would be modelled on ATS-6, that is, have the capability to transmit TV broadcasts and provide national telecommunication, as well as meteorological services. At the time, ATS-6 was a state-of-the-art spacecraft built by the most advanced space agency in the world, and ISRO was the newest kid on the block.

INSAT 1 Series

A technical specification for INSAT came out of an ISRO scientists interdepartmental meeting in Ahmedabad in 1972.[703] The prevailing approach was to deliver each function, telecommunication, TV broadcast and meteorology, through a dedicated satellite. The INSAT design combined all three into one, cutting duplicate costs, including that of launch. In part, the complex design reflected the interests of three government departments, Department of Telecommunication, responsible for telephony; AIR, responsible for TV and radio broadcasts; and the Department of Civil Aviation, responsible for meteorology. These government departments were funding it, and each one wanted INSAT to satisfy its respective operational needs. Described as a "crowded Indian bus", INSAT was forced to serve three key roles simultaneously.[704] They included: (i) telecommunication services for mainland India and offshore islands (ii) DTH television and radio and (iii) meteorological full Earth images every half hour with up to 2.5 km resolution using visible and infrared sensors.

Communication and TV broadcast functions were readily accepted. However, the benefits of a meteorological payload were not immediately clear. Given that India's agriculture contributed 50% of its gross national product and 10% of that was lost annually "due to bad weather, unseasonal rain and wrong forecasting",[705] the significance of monitoring the weather from space was soon recognised, and a meteorological payload was also included. Consequently, INSAT was the most complex single non-military satellite designed at the time.

Building and operating something as large and sophisticated as INSAT was a new venture for ISRO. Although ISRO had built and operated satellites, like Aryabhata and Bhaskara 1 and 2, INSAT was too large and sophisticated to build or launch. In 1983, ISRO's first launch vehicle, SLV-3, with a very simple solid-fuel design had demonstrated that it could place a 40-kg satellite in LEO. However, INSAT-1A and 1B with a mass of 1,152 kg each were well outside ISRO's capability to launch to GEO. PSLV was still over a decade away. A third party would have to be engaged on a commercial basis to get INSAT started.

Initially, ISRO procured consultancy from ESA in drafting the Request for Proposal, which was issued at the end of 1977 to global vendors for tender. The key payloads included:

- Two high-power transponders for television broadcasting to community TV sets, distribution of TV programmes for rebroadcasting and radio networking.

- Twelve transponders for telecommunications.

- Meteorological EO instrument (VHRR)for imaging the Earth in the visible (2.5-km resolution) and thermal infrared (10.5-km resolution) spectra and measuring wind speeds as low as 3 miles/s.

- Data relay transponder for meteorological data collection from unattended data-collection platforms.[706]

Two vendors responded to India's tender, Hughes Aircraft Corporation and Ford Aerospace and Communication Corporation (FACC). Ford was selected, and an agreement was signed at the Indian Embassy in Washington DC in July 1978 for INSAT-1A to be delivered 28 months later and INSAT-1B a year after that. The cost of the two satellites excluding launch was $60.7 million (Rs.50.5 crore).[707] Throughout its short history, the Indian space programme had always practised a buy, learn and then build strategy. ISRO built its first solid-stage launcher, SLV-3, with assistance from NASA, its liquid-fuel engine technology came from France and a limited cryogenic-engine technology from the USSR. The agreement with FACC had no element of technology transfer, but it did incorporate training to allow ISRO engineers to manage and operate the satellites in orbit, once they were launched. Gaining this knowledge on how to operate satellites in space was a key element in ISRO's plans for self-sufficiency.

The satellites would be launched from the US either on a Delta rocket or on the Space Shuttle, which in the late 1970s was concluding its testing phase and approaching the first launch. The final specification to which FACC built INSAT-1A included:

- Mass at launch: 1,152 kg

- Expected operational lifetime: 7 years

- Instruments in payload: for communication, meteorology and TV and radio broadcasting services

- Stabilisation: 3-axis stabilised

- Power: deployable solar arrays

- Orbital location: 74°Eat GEO

- Launch: from the US in April 1982 using Delta 3920 launch vehicle

During the handover phase of INSAT, 25 Ford Aerospace engineers temporarily relocated to ISRO's MCF in Hassan 200 km west of Bangalore.[708] Following launch on 10 April 1982 from Cape Canaveral, INSAT-1A was successfully piloted by the FACC and ISRO engineers to its designated position of 74°E in GEO with a plan to gradually test and commission all the onboard systems. INSAT-1A was formally commissioned in June, and for a short time, it operated successfully. However, it soon developed issues. INSAT-1A had solar panels on one side, so it required a long boom on the opposite side (called a solar sail) for balance. Despite successfully deploying the solar sail and resolving a problem with an antenna, INSAT-1A lost its orientation. To regain control, the onboard thrusters of the RCS were fired to re-orientate the spacecraft. A faulty propellant valve failed to close after the burn and unused fuel leaked out. Eventually, the entire supply of fuel was exhausted, something from which the spacecraft could not recover. INSAT-1A was deemed to have failed and was officially deactivated on 6 September 1982. The cost of INSAT-1A and its launch was insured, and ISRO recovered the associated costs.[709]

In contrast, INSAT-1B was an outstanding success. With it, India started to receive the space-based services that ISRO's founders had envisaged two decades earlier. It was launched at night on 30 August 1983 by the Space Shuttle STS-8. The Space Shuttle's first night launch was driven by INSAT-1B's requirements.[710] In the attempt to draw customers away from ESA's Ariane launcher, the US offered a heavily discounted launch price on the Space Shuttle, and India took full advantage. Despite initial problems with the deployment of solar panels and the solar sail, INSAT-1B was declared fully operational in October 1983. During the first two years of INSAT-1B's operation, ISRO consolidated its ground infrastructure and built up the technical skills of its engineers in operating spacecraft in GEO. INSAT-1B introduced to India the nationwide space-based services common today, including:

- Increased capacity for long-distance telephone calls: 4,000 two-way voice circuits for 70 long-distance telecommunication routes.

- High-speed communication for commercial entities, such as The Hindu newspaper, Oil and Gas Commission and the National Thermal Power Corporation.

- An experimental satellite-based telegraphy network for rural locations in the north-eastern India

- Nationwide communication services: 400 Earth stations provided communication across India for isolated communities, including offshore islands.

- National weather service: a collection of meteorological data centres with a primary centre in New Delhi and 22 secondary centres across India distributed meteorological information, including wind and cloud cover over land and the Indian Ocean, for farming, transport, fishing communities and the general public daily. Subsequently, this data supported monsoon modelling and prediction.

- Automatic weather data collection: INSAT-1B collected data from remote locations, including rivers and oceans. ISRO provided 1,000 Automated Weather Stations (AWS), one of which was located in the Indian Base Station in Antarctic: The Indian Meteorological Department provided another 679 AWS and 969 Automatic Rain Gauge stations.

- Early Warning System: 100 receivers for the disaster warning system were located in the cyclone-prone areas of Andrea Pradesh and Tamil Nadu.

- Relay of national TV across India: Television programmes from the national broadcaster Doordarshan, initially limited to the capital and large metropolitan areas, were retransmitted via INSAT-1B providing a wider national coverage not possible without a satellite.

- Wider educational reach: Programmes retransmitted for universities, schools and 2,000 community-based direct broadcast (SITE like provision) televisions.

- Increased television content: the number of TV stations grew from 11 to 250 in the first three years of INSAT-1B operations.[711]

INSAT-1B performed as designed, exceeding it's planned seven years of life and was eventually terminated in August 1993 when all its onboard propellant was consumed. To ensure that there was no break in the space-based services inaugurated by INSAT-1B,

ISRO ordered INSAT-1C in 1983 to replace the lost INSAT-1A with a launch scheduled for 1987. In the absence of INSAT-1A, INSAT-1B was potentially a Single Point of Failure. If INSAT-1B failed, no single satellite could replace it. Its functions would have to be met by leasing of services from multiple satellites already in orbit. The meteorological function could be replaced by images from Japan's or European satellite. Communication services could be delivered by Canadian or US satellites but TV was a challenge. TV broadcast would require technical modifications to receivers in ground stations in India to accommodate the different frequencies in use.[712] In addition to avoid this single point of failure and to cater for growth in demand, an order for INSAT-1D was signed with FACC in 1985 for launch in 1989. With this provision, ISRO bought more time to build its infrastructure for launching INSAT-class satellites.

Had the schedule gone to plan, INSAT-1C would have been launched by an Indian astronaut from the Space Shuttle. In the end, INSAT-1C built by FACC was launched by ESA's Ariane launcher in 1988. It was not successful. A fault with the power supply a month after launch and other failures terminated the mission a year after launch. INSAT-1D launch was delayed by two incidents. It was damaged during launch preparations when a crane failed, and it was damaged again during the San Francisco earthquake of 1989. It was repaired each time and successfully launched in June 1990. It operated successfully for over a decade until May 2002 when it lost attitude control. INSAT-1D was the last of the INSAT-1 series.

Figure 12-5 INSAT-1B at the Kennedy Space Centre. 2 June 1983. Credit NASA

The INSAT-1 series comprised of four satellites, of which two failed. An analyst working in the US on behalf of the INSAT-1 series concluded that the INSAT-1 design was fundamentally flawed. FACC had "sold India a failed proposal originally rejected by the US Air Force on the basis that it was too risky and potentially prone to failure."[713] The complex operational demands for navigation, communication and meteorology services were undermined by inherent design weaknesses. INSAT-1B and INSAT-1D remained operational for the planned lifetime but only because of the additional effort from Indian satellite operators.

INSAT 2 Series

The INSAT-1 series was built by FACC in the US. In 1984, U.R. Rao took over as chairman of ISRO, and he was best placed with the vision and hands-on experience to shift ISRO's strategic goals to build large complex satellites in India. Starting with INSAT-2 launched in 1992, ISRO has built its own spacecraft ever since. The five satellites in the INSAT-2 series (INSAT-2A to INSAT-2E) were similar to the INSAT-1 series but provided enhanced services, including higher-resolution sensors for meteorology, larger number of transponders for communication and extended TV coverage to include Southeast Asia and the Middle East. In addition, the INSAT-2 series included new services for search and rescue,[714] mobile telephone support and dedicated channels for business customers. INSAT-2C and INSAT-2D[715] did not carry any meteorological payload, but INSAT-2E was fitted with an enhanced resolution camera for meteorology and was compatible with US satellite-service provider Intelsat. From 36,000 km, the sensors for the meteorological payloads in the INSAT satellites provided images of around a kilometre resolution of the constantly changing weather patterns over India.

However, INSAT was not ideal to fully explore India's natural resources, unlike satellites optimised for EO that operate from a lower orbit and can provide higher-resolution images and information sufficiently detailed to drive government. At the outset, ISRO engineers lacked self-confidence in the face of limited facilities and zero experience in satellite building. The group that built IRS-1A wanted to call it "pre-operational" and the INSAT-2A team for the first INSAT to be built in India wanted to use the suffix TS for Test Satellite, that is, INSAT-2TS. In each case, their faith in their ability grew, and the modified naming convention was dropped.[716] All satellites in the INSAT-2 series and beyond have been built and operated entirely by ISRO in India. Although the contract with FACC for INSAT-1 excluded any technology transfer, building on the experience of Aryabhata, Bhaskara, APPLE and SROSS, ISRO developed sufficient in-house the expertise for designing and building satellites and honed its satellite-production process, as well as quality control and operational procedures.

The ISAC in Bangalore is the hub where spacecraft are realised. Starting with a basic structure, subsystems are gradually added for control, power, telemetry and communications, and eventually, a complete satellite emerges. The system for attitude control (reaction wheels, gyroscopes and momentum wheels) and antenna reflectors are produced at VSSC in Thiruvananthapuram. Transponders and cameras used for meteorological imaging (visible and infrared) are produced at the SAC in Ahmedabad. The propulsion system for the large apogee motor that takes satellites from the initial GTO to GEO and the liquid-fuel-based RCS used for attitude control and station-keeping are produced at the LPSC. These systems and components are integrated as a single spacecraft at ISAC.

After the first decade of building its own satellites, ISRO's techniques, technologies and procedures had achieved a substantial level of maturity. For example:

- 30% of Aryabhata's mass was solid metal. Replacing solid metal with honeycomb structures brought it down to just 7.5% for INSAT-2.

- Aryabhata, Bhaskara and SROSS used body-mounted solar cells. Since APPLE onwards, ISRO has used solar panels. Solar panels are more efficient in generating electricity, even though designing, building, deploying and operating them rely on complex subsystems.

- The increased amount of time in eclipse (in shadow during each orbit), satellites in LEO require a 50% higher battery and solar array capacity compared to satellites in GEO or GSO, which are in shadow only around 90 times a year.[717] Initially, ISRO procured solar panels but later developed in-house capability to manufacture space-qualified solar panels with a capacity to handle high-degree of temperature variations (from -100°C to +100°C) 240,000 times. It has also built the capability for nickel-cadmium batteries that could withstand extreme temperature variation and reliably manage multiple recharge cycles.

- Application Specific Integrated Circuits helped to reduce the quantity and complexity of wiring. The resulting miniaturised electronic units and mass reduction extended a satellite's operating life.

- Traditionally, solar cells have been silicon-based. ISRO moved to gallium-arsenic solar cells that have greater efficiency.

- Nickel hydrogen (Ni-H) batteries, which have a much longer life than the usual nickel-cadmium batteries.

Enhanced communication capacity was a common feature on everything that followed INSAT-1. INSAT-2A had twice the communication capacity of INSAT-1A. INSAT-1A payload included two C- and three S-band transponders and VHRR. INSAT-2A contained 18 C-band and two S-band transponders, VHRR and a search and rescue payload. INSAT2-A was designed with an upper limit on its mass and physical size. Although US's

Delta or ESA's Ariane 4 could launch heavier spacecraft, INSAT-2's mass specification of 2,000 kg was determined by ISRO with an eye on launching its own GSLV-Mk1 (designed for a payload of 2,000 kg) when it came online. During the 1990s, GSLV was still in development. Over time, the mass of communication satellites has increased, along with capacity, to well beyond the 2,000 kg of ISRO's 1990s design.

Since the start of the INSAT programme, the number of satellite-based societal, commercial and strategic applications have increased. New services included Mobile Satellite Service (MSS), Satellite-aided Search and Rescue, early warning systems for flood, typhoon and tsunami and fleet-monitoring applications. The demand for satellite-based services, especially in the consumer sector, has seen particularly strong growth. The number of registered TV channels has grown to 821 as has the number of DTH operators; between them, they serve a customer base of 50 million users.

The demand for MSS has also grown. MSS is a mobile cellular network, but the repeaters are in space, which allows a small portable handheld device to communicate from anywhere on Earth using satellite services. MSS payload was flown on INSAT-3C. Low speed satellite-based internet access from Very Small Aperture Terminals (VSATs) is used in the financial, transport, and predominantly in the shipping sectors. In India, the VSAT user base has grown to about 1,70,000 installations.[718] Since it was established in 1969, ISRO has built, launched or operated over 80 unique spacecraft, almost half of which are currently in operation. India's EO programme, including its weather satellites, is contributing to modernise its national infrastructure. Communication satellites have driven growth in media, industry and the business sector. Yet, with only 260 transponders in C, Ext C and Ku bands, the demand still significantly exceeds supply. Through its science, student and education satellites, ISRO hopes to foster interest and develop the skilled personnel it will need in the future.

Education and Defence

Satellite-based technology is ideally suited to bringing education to large numbers of individuals distributed across a vast geographical area. Disparate communities can be involved in a shared learning experience (for example, lectures by specialist speakers and from academic establishments) via satellite. This technology is cost-effective, and its novel delivery technique can itself enhance learning compared to traditional methods. The Educational Satellite (EDUSAT) launched in 2004 has stimulated the demand for tele-education. ISRO's ambition is to cover primary, secondary and higher education using this technology. Currently, it has established about 80 networks across the country connecting 56,164 classrooms with 4,943 satellite interactive terminals, where students can interact with the speaker, and 51,221 receive-only terminals covering the full range from the primary to professional education.[719]

ISRO learnt from some missions more than others. The Technology Experiment Satellite (TES), launched in 2001, incorporated new technologies, including the attitude and orbit control system, using a single propellant tank, high-torque reaction wheels, solid-state data recorder, improved satellite position system and miniaturised power and communication subsystems. It was built in just two years motivated by the 1999 incursion by Pakistan's military into the Indian-administered section of the Kargil sector of the state of Jammu and Kashmir.[720]

Access to slots in the GEO is determined by the International Telecommunications Union. India has the following GEO slots, all east of the Greenwich Meridian (32°, 48°, 55°, 72°, 74°, 82°, 83°, 93°, 102° and 111°). At present, India has more GEO satellites than slots. To accommodate them all, several are co-located, sharing the same GEO positions. To ensure safe separation, each satellite is maintained at the centre of an exclusive 150-km box of space. Although this is convenient as it eliminates the need to acquire additional GEO slots, MCF operations must work to a higher degree of precision to avoid collisions.[721]

Satellite Assisted Search and Rescue

ISRO is a member of the international COSPAS-SARSAT programme that provides distress alert and location service through Search and Rescue satellite systems. COSPAS-SARSAT was established in 1979 when the USSR's COSPAS (Cosmicheskaya Sistyema Poiska Avariynich Sudov, meaning Space System for Search of Vessels in Distress) and Search and Rescue Satellite Assisted Tracking (SARSAT) developed jointly by the US, France and Canada were merged. Since then, COSPAS-SARSAT has grown. It now covers 60% of the Earth's land surface and is comprised of 11 satellites and 76 ground stations in 50 countries.

Distress signals transmitted from anywhere on Earth by emergency locator transmitters on aircraft, ships and those owned by private individuals configured to use the internationally recognised 406 MHz frequency are detected by the COSPAS-SARSAT satellite network in orbit. A space-based search and rescue system consists of four elements: (i) the transmitter or beacon, which when activated initiates an alert signal[722] (ii) an orbiting satellite capable of detecting that signal and relay to the ground (iii) and a ground station capable of receiving that signal and (iv) a ground-based rescue coordination authority to perform a rescue on the ground, air or sea. Globally, there are an estimated 1,770,000 beacons, of which 1,380,000 are uniquely registered against aircraft, ships or stand-alone units for personal use.

From September 1982 to December 2014, 39,565 persons were assisted through 11,070 search and rescue incidents. In India, the statistics for up to 2013 record that the COSPAS-SARSAT supported the rescue of 1,917 lives in 75 incidents.[723] A total of 42 countries offer services such as user, ground or space segments in the COSPAS-SARSAT system. Each international region is responsible for its component of COSPAS-SARSAT operating to a single global standard. In September 2015, COSPAS-SARSAT reported the operational infrastructure as:

- Five satellites in low polar orbit, also known as Low Earth Orbit Search and Rescue (LEOSAR).

- Seven satellites in GEO (for example, INSAT-3D).

- 54 Local User Terminals (LUT), Earth station antennas, receiving signals transmitted by LEOSAR satellites.

- 23 LUTs, including one in Bangalore and another in Lucknow, receiving signals transmitted by GEOSAR satellites.

- 31 Mission Control Centres, including one in Bangalore, distributing distress alerts to SAR services.

- 1.7 million 406-MHz beacons estimated to be in service worldwide.[724]

India's first contribution was INSAT-1D that carried a COSPAS-SARSAT payload in 1990.[725] All satellites of the INSAT-2 series, except INSAT-2C, also carried a COSPAS-SARSAT payload. None of the INSAT-1 or INSAT-2 satellites are operational today. By the end of 2016, INSAT-3D and INSAT-3DR had a COSPAS-SARSAT payloads. In India, 13,300 beacons had been registered by 700 user agencies by March 2013. Users of COSPAS-SARSAT in India on ships, aircraft or personal beacons (any that operate at 406 MHz) can trigger alerts once activated.[726] Two LUTs based in Bangalore and Lucknow listen in to the satellites for alerts. The LUTs are supported by the Indian Mission Control Centre (INMCC), also in Bangalore, that relays the coordinates of the beacon generating the alert to the Coast Guard, Navy and Air Force or ground-based civilian rescue organisations responsible for carrying out the rescue.[727] Rescue operations in India are coordinated with four rescue centres located in Mumbai, Chennai, Delhi and Kolkata and operated by the Airports Authority of India via the INMCC in Bangalore. The LUTs located in India provide coverage for the Indian Ocean region and seven other countries, Bangladesh, Bhutan, Maldives, Nepal, Sri Lanka, Seychelles and Tanzania. The Indian INMCC/LUTs had started operations in 1986 and, by 2002, were involved in rescuing about 1,300 individuals.[728]

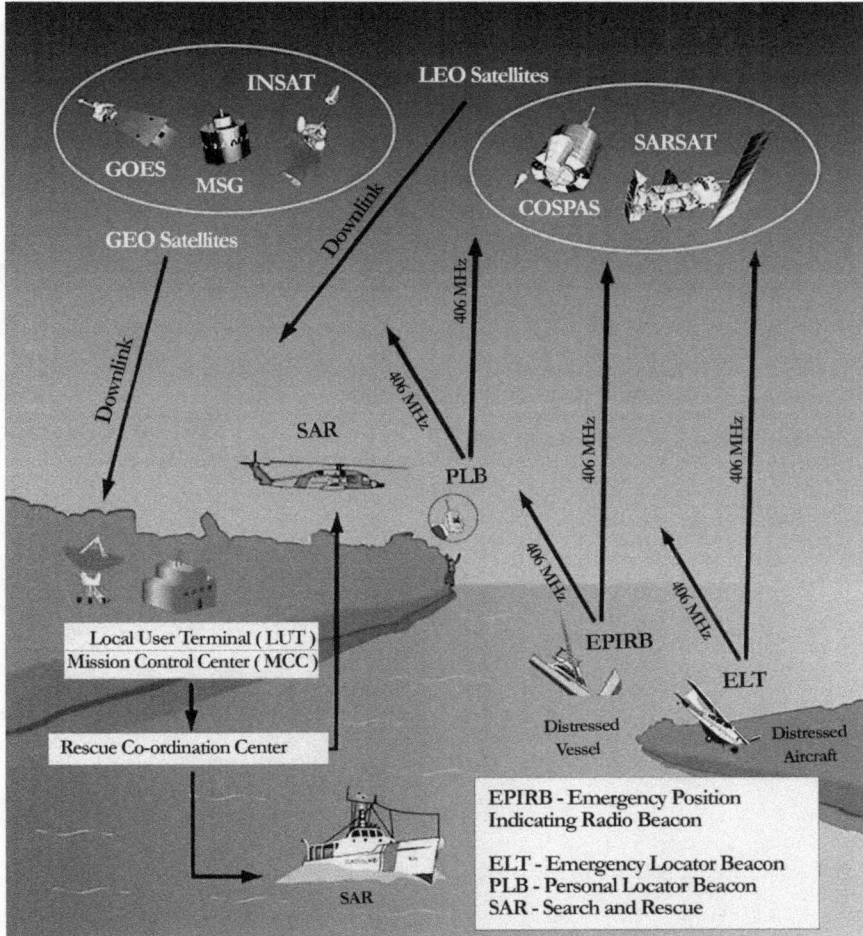

Figure 12-6 COSPAS-SARSAT System. Credit COSPAS-SARSAT

ISRO now has the complete infrastructure, on the ground and in space, to provide end-to-end services required for operating space-based assets. Although international collaboration has been and will continue to be an element of its activities, ISRO is self-sufficient. The services delivered by

ISRO's satellites, built in and operated from India, touch the lives of most of its 1.3 billion population daily. More than any other element of India's space programme, its satellite programme is a realisation of the vision of its founding fathers, Jawaharlal Nehru, Homi Bhabha and Vikram Sarabhai.

⊰ ◈ ⊱

Chapter Thirteen
Indian Regional Navigation Satellite System

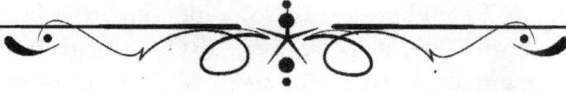

The concept of a satellite-based navigation service arose from observing the world's first satellite, Sputnik. Researchers at the Applied Physical Laboratory in Baltimore recognised that if measurements could be made of Sputnik's orbit as it passed overhead, its entire orbit could be calculated. Dr Frank T. McClure (1916–1973) went further and concluded that the converse was also true.[729] If the position of the satellite in orbit was known, then the location of the observer on Earth could be calculated. In 1960, the first navigation satellite was launched. It was called Transit or NAVSAT and used by the US Navy to provide location information for its nuclear-armed submarines.[730] Today, there are over a hundred satellites in Earth orbit designed to provide precision location anywhere on Earth at any time.

A navigation satellite in space incorporates a high-precision onboard clock that continuously transmits signals to Earth from which its position can be calculated. Every satellite navigation (satnav) receiver, whether in a mobile phone, a tablet or a car, also has an accurate built-in clock. The satnav receiver measures the time difference between when the signal was sent by the satellite and when it was received. By combining the signal from at least four satellites in space and its in-built clock, the receiver can calculate its position anywhere on the surface of the Earth.

In 2006, the Government of India approved the Indian Regional Navigation Satellite System (IRNSS) with the objective of providing independent regional GPS-like services for national applications across mainland India and 1,500 km beyond its coastline.[731] Unlike the well-established American GPS, IRNSS is not a global but a regional system, even though it is frequently lauded in India as India's GPS. On 28 April 2016, India launched the seventh and final satellite, IRNSS-1G, of the IRNSS constellation. The first satellite, IRNSS-1A, was launched on 1 July 2013. Following the successful launch of IRNSS-1G, the Prime Minister of India, who had watched the launch from his office in New Delhi, renamed the collection of navigational satellites as Navigation Indian Constellation

(NavIC).[732] It was a culmination of a project initiated almost a decade earlier.

In the Union Budget presented to the Parliament on 29 February 2008, the Indian Finance Minister P. Chidambaram had made a budgetary allocation of Rs.270 crore ($67.5 million) for IRNSS. Speaking in early 2014, Kopillil Radhakrishnan (born 1949), Chairman of ISRO, estimated the total cost of the IRNSS programme, consisting of seven satellites in orbit and two on standby on Earth and the associated ground infrastructure, at Rs. 3,260 crore ($527.9 million). He stated that each IRNSS satellite would cost Rs.150 crore ($24.3 million); nine would cost Rs. 1,350 crore ($218.6 million).

The ground-support infrastructure would cost an additional Rs. 1,000 crore ($161.9 million), and the cost of each launch would be Rs.130 crore ($21 million), totalling Rs.910 crore ($147.4 million) for seven launches.[733] The launch vehicle used for IRNSS was the PSLV-XL. The PSLV is ISRO's primary launcher of satellites to LEO, SSPO/SSO and GTO and comes in three configurations with the XL being the most powerful. It was used for the first time to launch India's Moon mission Chandrayaan-1 and since then several times, including the MOM. With its six strap-ons, the PSLV-XL could take the 1,432-kg IRNSS satellite to its eventual slot in GSO.

Unlike the navigation satellite constellations of the US, China, Russia and ESA, which have a global coverage from medium Earth orbits at altitudes between 19,000 and 23,000 km, all seven IRNSS satellites are located in a combination of GEO and GSO. The PSLV-XL delivered each of the 1.4-tonne satellite initially to a sub-GTO orbit (283 km × 20,630 km) 20 minutes after launch.[734] Once the solar panels were deployed, the engineers at ISRO's MCF in Hassan used the satellite's onboard liquid apogee motor (LAM) to deliver it to its designated GSO or GEO at 35,786 km.[735]

The IRNSS constellation facilitates a variety of applications within the public and private sectors in India. It is configured to provide two levels of service. A Standard Positioning Service (SPS) with a 20-m resolution over the Indian Ocean and less than 10-m resolution over mainland India and a Restricted Service (RS) with an encrypted signal offering an undisclosed level of higher resolution available only to authorised users. [736]

In addition to the traditional uses of satellite-based navigation, such as navigation aid for drivers in urban areas, hikers and travellers in remote areas, IRNSS is also called upon to assist India's dairy farmers, commercial aviation industry, its expansive railway network and the Archaeological

Survey of India. It will be used for commercial services, such as fleet management, location-based advertising, land surveys, real-time tracking, as well as broader regional and national services, including:

- terrestrial, aerial and marine navigation

- disaster management

- integration with mobile phones

- precision timing

- mapping and geodetic data capture

NavIC will enable a multitude of business opportunities in India. In time, these businesses and their supply chain will provide a significant boost to the national economy. A recent UK government report determined that the Global Navigation Satellite System makes a daily contribution of $1.263 billon (Rs.7798 crore) to the UK economy. Given the larger economy, NavIC will inevitably have a larger impact in India once the service is imbedded within the national economy.[737] IRNSS implements two services typically associated with developed nations, the RS for use by the Indian security services and through GPS Augmented Navigation (GAGAN) navigation support for the civilian aviation sector. These and other disparate uses of space technology continue to drive India's modernisation plans, pulling it into the 21st century, and reflect how ISRO is fulfilling the ambitious aim of its founders to use space technology for national development.[738]

The NavIC/IRNSS system has three distinct system segments, space, ground and user. The space segment consists of the seven satellites in orbit. The ground segment consists of ground facilities used to manage the navigation satellites in orbit. The user segment consists of the physical receiver unit that calculates the position on Earth using the timing information embedded in the navigation signal from the orbiting satellites and the applications that use them.

Space Segment

The seven satellites in orbit and the two on standby on Earth ready for launch when required make up the space segment. All seven satellites hover over or near the equator with a direct view of mainland India at all times. Three of the satellites are in GEO directly over the equator at 32.5ºE, 83ºE and 129.5°E and remain stationary in the sky. The other four satellites are in GSO arranged in pairs at longitudes 55°E and 111.75°E inclined at 29° to the equatorial plane and 180° out of phase with each other. One satellite is almost directly over New Delhi at 35,786 km with three each on either side. The number of satellites in orbit and on standby is likely to increase with time.[739] The architecture of all nine satellites is identical with an expected lifetime of around 12 years.

Unlike the satellites in GEO, those in GSO are not stationary but scribe the figure 8 (a little like the analemma that the sun scribes over a year) in the sky every 24 hours. Between them, the seven satellites have a footprint over a rectangular area between latitudes 30°N and 30°S of the equator and longitudes 32.5°E and 129.5°E. This is the region of India's regional satellite system.

IRNSS Satellite	Longitude (E)	Orbit	Orbital Inclination	Launch Date
1A	55.0°	GSO	29° (±2)	1 July 2013
1B	55.0°	GSO	29° (±2)	4 April 2014
1C*	83.0°	GEO	± 5°	15 October 2014
1D	111.75°	GSO	29° (±2)	28 March 2015
1E	111.75°	GSO	29° (±2)	20 January 2016
1F	32.5°	GEO	± 5°	10 March 2016
1G	129.5°	GEO	± 5°	28 April 2016

Table 13-1 IRNSS Orbits. Credit ISRO (*New Delhi is at longitude 77° E from where IRNSS-1C is nearly overhead)

Each satellite has three rubidium clocks, a total of 21 amongst the 7 satellites in orbit. By late 2016, ISRO had started to detect problems with some of the rubidium clocks. By mid-2017, it was reporting that five of the 21 clocks had failed. ISRO has not been the only victim of the failure of rubidium clocks. The European GPS constellation, Galileo has also suffered from a similar clock failure.[740] Although the satellites continue to function, the timing signals of required precision are not available and thus NavIC as a system cannot operate as designed. All three clocks had failed in IRNSS-1A by June 2016 and attempt to find a fix or a workaround had not been successful. ISRO attempted to replace IRNSS-1A with IRNSS-1H launch on 31 August 2017 but the mission was not successful. The payload faring did not detach stranding IRNSS-1H inside the 4th stage of the PSLV. Following a failure analysis report, another launch will be scheduled. Eventually, an additional three will also join the NavIC constellation bring the total number of satellites in orbit to 11.

Ground Segment

The seven satellites in orbit are supported by a ground segment consisting of facilities for ranging (determining the satellites' position in space), network timing (synchronisation with a master clock), spacecraft control (monitoring and maintaining satellite health and position) and data communications.

IRNSS ground segment

■ ISRO Navigation Centre in

▲ A IRNSS Range and Integrity Monitoring Stations (IRIMS) perform continuous one way ranging of the IRNSS satellites and are also used for integrity de-

● IRNSS CDMA Ranging Stations (IRCDR) carry precise two way ranging

● IRNSS Network Timing Centre (IRNWT) at Byalalu generates, maintains and distributes IRNSS.

★ Spacecraft Control Facility (SCF) controls the space segment through Telemetry Tracking & Command networks. In addition to the regular TT&C operations, IRSCF also uplinks the navigation parame-

IRNSS Data Communication Network (IRDCN) provides the required digital communication backbone to IRNSS network.

International Laser Ranging Stations (Il-RS) is used to calibrate the IRNSS orbit determined by the other techniques.

Figure 13-1 IRNSS Ground Segment. Credit ISRO

The data communication network consists of 15 sites across India operating 24/7 in real time. ISRO's 32-m fully steerable antenna at Byalalu, is the primary element in IDSN. On 28 May 2013, Byalalu formally became ISRO Navigation Centre (INC) and the centre of IRNSS's ground segment. In addition to the INC, the ground segment includes;

- IRNSS Satellite Control Facility: It controls the space segment from the ground through TT&C network. It also uplinks the navigation parameters generated by the INC. Each satellite transmits its navigational signal to the Earth in the L5 band and the timing information in S-band.

- IRNSS Range and Integrity Monitoring Stations: Perform continuous one-way ranging of the IRNSS satellites to determine and help maintain the positional integrity of the IRNSS constellation.

The ground segment is responsible for maintaining and operating the IRNSS constellation during its designated lifetime. At least 15 sites across India have direct communication links to the orbiting satellites. The ground segment has four key responsibilities: (i) calculating with high precision the location of each satellite in its orbit using two independent mechanisms, radio and optical lasers, each satellite is equipped with corner-cube reflectors; (ii) ensuring that the satellite's onboard rubidium atomic clocks are synchronised with the master caesium clock located at INC; (iii) maintaining the quality of the radio signals transmitted by each satellite for use by the ground-based receivers and (iv) correcting, when required, the orbit of each satellite on command from the IRNSS Satellite Control Facility.

New elements of the ground segment have been built, or existing capability expanded, to facilitate IRNSS. An IRNSS Control Centre, along with one 11-m antenna and four 7.2-m antennae, has been added to the Satellite Control Facility at Hassan with the responsibility of controlling IRNSS constellation. An additional 11-m and three 7.2-m antennae are being constructed at the backup Satellite Control Facility in Bhopal.[741]

User Segment

The IRNSS user segment is concerned primarily with the design and construction of receivers that will receive the navigation signals from the satellites. The receivers are then embedded within stand-alone devices, including mobile phones, tablet computers, dedicated handheld navigation devices and satnav devices for use in cars, ships and aircraft.

An IRNSS receiver usually operates on a single frequency (L5 at 1176.45 MHz or S-band at 2492.028 MHz). It can also operate in a dual mode on both frequencies at the same time. Single- and dual-band receivers are capable of receiving both the SPS (accuracy of 20 m) and the RS, but only authorised users have access to decryption keys necessary to access the RS.

An IRNSS satellite receiver calculates its position with an accuracy of 20 m on the Earth's surface from the signals it receives from IRNSS satellites in their orbits 35,786 km away. For this to work, the position of the satellite in orbit and the timing signal it transmits must be known with high precision. The rubidium clock at the heart of each satellite is the central component of a navigational satellite. Though not as precise or as expensive as a caesium atomic clock, rubidium atomic clocks are used by high-end data centres, central television transmitter stations, cell phone base stations, as well as navigation satellites.

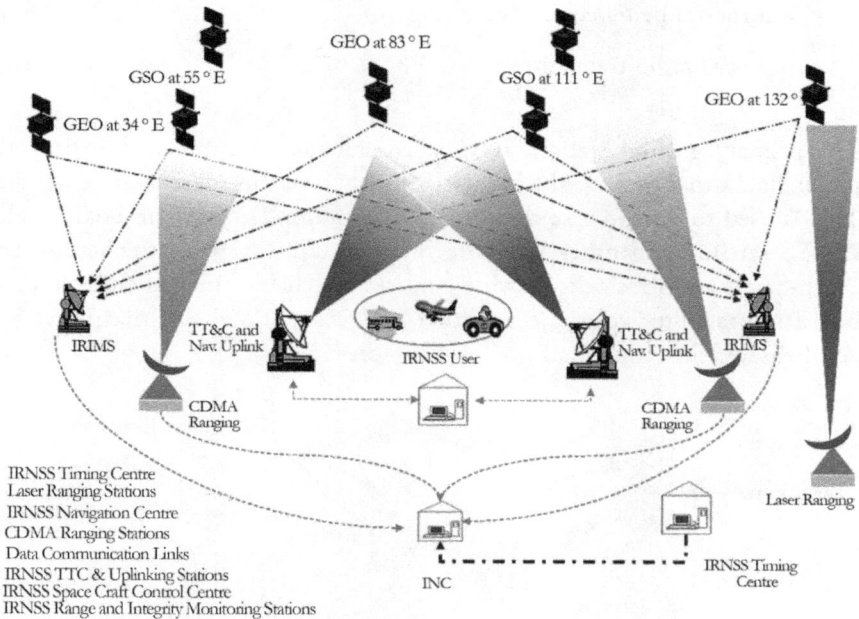

Figure 13-2 IRNSS Architecture. Credit ISRO

ISRO has designed the IRNSS architecture for interoperability with existing international standards.[742] The specific characteristics of the navigation and timing signals that ISRO has selected ensure that the IRNSS system is interoperable with the US's GPS and Europe's Galileo systems developed by the ESA. China and Russia have formally agreed to cooperate with their global navigation systems, BeiDou and GLONASS, respectively.[743]

Navigation Satellite

Ultimately, the primary purpose of IRNSS is the broadcasting of navigation and timing signals by each satellite. To maintain operational status, each IRNSS satellite's position must be known to a high precision, and the onboard atomic clock must be continuously synchronised with the one at INC. For redundancy, the navigation and timing signals are duplicated in the L5 and S bands. These signals received from multiple satellites by a receiver on Earth (for example, a mobile phone) are then used to calculate with precision the receiver's position, altitude, direction and speed.

Each satellite has three onboard components to serve this requirement

- three high-precision rubidium atomic clocks for timing

- a corner-cube reflector for ranging

- onboard radio transmitter.

The primary ground station, INC at Byalalu, has caesium and hydrogen maser clocks that provide the highest accuracy stratum-1-level timing signal. This is used to synchronise the rubidium (stratum-2) clocks on-board each IRNSS satellite and other IRNSS ground stations.[744] While the stationary Earth-based atomic clocks provide consistent, reliable time, the clocks on-board the satellites are subject to two key relativistic effects that must be compensated for to maintain the required precision.

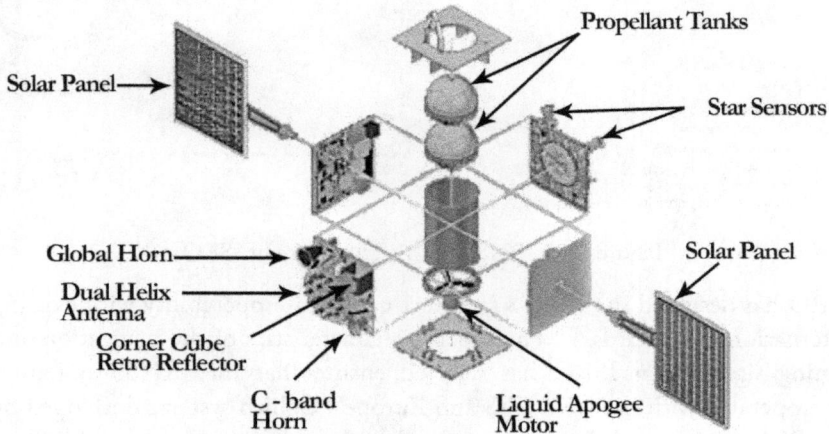

Figure 13-3 IRNSS Subsystems Showing the Central Location of the Corner Cubes. Credit ISRO

The speed of the satellite at 3 km/s in orbit slows the onboard clock due to Special Relativity, and the satellite's distance from the centre of the Earth and the resulting weaker gravitational field speeds it up as described by General Theory; consequently, onboard clocks require daily adjustment. These are tiny variations but are critical to achieving an accuracy of 20 m on the surface of the Earth. The onboard clocks require a regular correction typically in the order of 5 to 10 nanoseconds (nanosecond = thousandth of a millionth of a second). Even though this compensation is largely built in prior to launch, the clocks still require daily monitoring and adjustment.[745] Before a navigational satellite can provide a precise location to a receiver on Earth, the precise location of the satellite in space must be known.

One of the techniques used to measure the position of a satellite in orbit is to direct pulses of laser light from Earth and detect the reflections from it.

The corner cubes attached to the satellite facilitate this independent mechanism (using light, not radio) for determining its position in space. Very short laser pulses are beamed from the Earth to the satellite, and the corner cubes reflect the pulses back to the source. Corner cubes are designed to reflect light back in the direction of the source, irrespective of the direction it comes from. The time between sending the pulse and its arrival back on Earth is a measure of the satellite's distance. Corner-cube retro-reflectors are used not only on navigation satellites but also on other spacecraft for range determination; they have also been used to determine the Moon-Earth distance.[746]

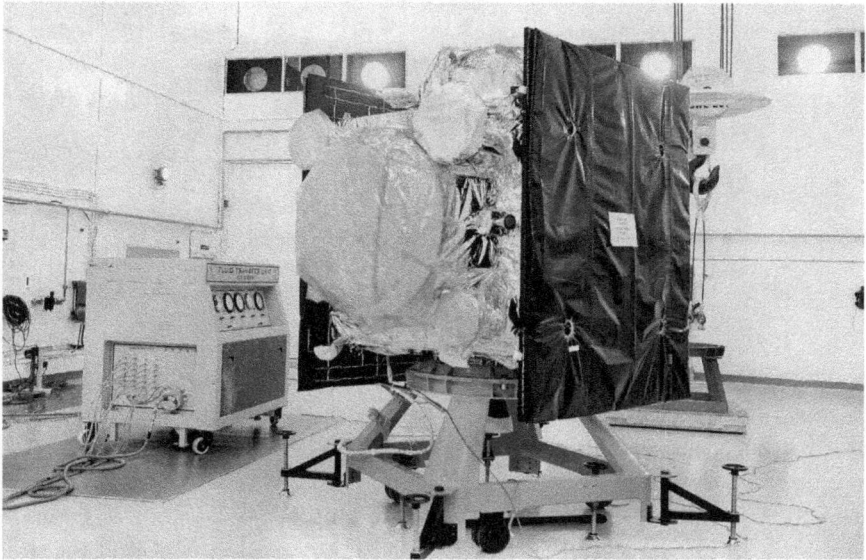

Figure 13-4 IRNSS-1A. Credit ISRO

Each IRNSS satellite has an array of 40 corner cubes manufactured by ISRO's Bengaluru-based Laboratory for Electro-Optics Systems. Each 38-mm cube is designed to provide the position of the satellite with a precision of 5 mm over a distance of 35,786 km.[747] A Satellite Laser Ranging (SLR) station, like an astronomical observatory, is usually located on high ground away from populated regions. An SLR station has a high-energy pulsed laser, a high-precision timer and a small telescope connected to a sensitive photon detector. The photon detector detects the reflected laser pulse from the satellite after its journey to and from the satellite. Modern digital systems allow the time-of-flight (interval between the sending and receiving) of the laser light to be measured with high precision in the order of 10 picoseconds (1 picosecond = 1 million million or 10^{-12}). Although not as efficient as at night, the SLR is used to successfully, locate satellites in orbit during either day or night. The concentration of charged particles in the Earth's ionosphere, through which the satellite's signals must travel, is the largest error source for the radio signals.[748] Ranging by use of laser pulses is not affected by the charged particles in the ionosphere.

Figure 13-5 Mount Abu Satellite Laser Ranging Station. Credit ISRO

All SLR lasers currently in use around the world are considered unsafe for the human eye for the entire path of the laser pulse from the ground station to the subsequent reflection. SLR stations are thus established well away from airports and common flight paths. Even so, SLR-operating procedures require that, prior to their use, checks are conducted to verify that no aircraft is operating in the area. Most stations use a traditional ground-control wide-field tracking radar that follows the laser beam around the sky. If an approaching aircraft is detected, the laser is switched off automatically. SLR stations are highly computerised, and some are automated to operate entirely without human intervention. Automated SLR stations operate to a predefined schedule, self-calibrate for each session, search for satellites using a telescope and shut down if the weather (cloud/rain/snow/wind) exceeds predefined thresholds. ISRO has established two laser stations. One is located at Mount Abu, between Jodhpur and Ahmedabad in northern India, and the other is 2,000 km away in Ponmudi in Kerala in southern India.[749]

Just as IDSN cooperates with international partners for Earth to space radio communication, India's SLR stations cooperate internationally through the US-based NASA-operated International Laser Ranging Service (ILRS). A worldwide network of SLR stations is coordinated through the ILRS. The timing signals received from the atomic clocks of navigation satellites are used as a source of high-precision time to synchronise clocks by terrestrial service providers, such as financial services systems (ATMs, stock trading and banks), traffic management (road, rail, marine and aviation) and utility providers (electricity and water). The services of navigation satellites are now considered a utility in modern societies. Modern satnav receivers are remarkably sensitive and can pick up the signal from a satellite orbiting 36,000 km away. However, that signal is weak and vulnerable to deliberate or unintentional interference, jamming or spoofing.[750] One of the critical services that IRNSS will eventually offer is a high precision location service for the civil aviation sector.[751]

GAGAN: GPS Aided GEO Augmented Navigation

The GAGAN system provides satellite-based location service similar to IRNSS but is currently independent of it. Named after the Hindi word for sky, GAGAN is a joint project between the Airports Authority of India and ISRO to bring modern communication, navigation, surveillance and air traffic management to Indian airspace. GAGAN is an example of a Satellite-Based Augmentation System (SBAS). The primary objective of SBAS systems is to provide information (position, velocity and time) with a high level of accuracy, availability, integrity and reliability to the civil aviation sector.[752] Satnav precision of about 10 m is sufficient for road vehicles, where a driver can clearly see the road and road signs and make a correction on the move. For commercial aviation, it is not sufficient.

GPS or IRNSS data alone is not sufficiently precise for the aviation industry's internationally recognised Safety-of-Life Service criteria as defined by the International Civil Aviation Organisation (ICAO). Commercial airliners coming in to land at 200 km/h in poor visibility or at night require high accuracy positional information around 10 times a second. SBAS provides this enhanced positional accuracy by overlaying the existing GPS/IRNSS signals with dynamic and reliable corrections from the ground station whose position is known with high precision.[753]

GAGAN provides positional information of known accuracy to civilian aircraft while at the terminal, en-route and approach. On 30 December 2013, GAGAN was certified only for en-route operations by the Indian government's regulator, Director General of Civil Aviation. On 21 April 2015, it was reclassified for precision approach. GAGAN provides Non-Precision Approach (positional information of horizontal accuracy of 18 m) over all India and a Precision Approach (with positional information of 16 m horizontal and 20 m vertical) over most of the Indian landmass on nominal days, determined primarily by the variability of the ionosphere through which the IRNSS signal must travel.

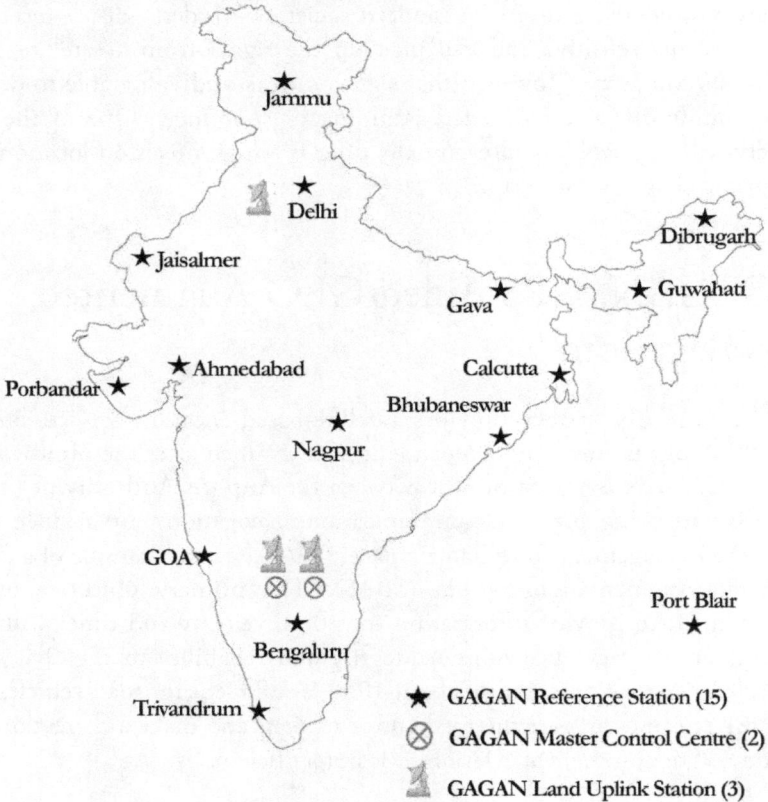

Figure 13-6 GAGAN Ground Stations. Credit Airports Authority of India[754]

It is built with a level of integrity such that only one in 10 million approaches may receive misleading positional information.[755] It has a built-in self-assessment mechanism that alerts the pilot if the quality of the signal is poor and its calculations are not reliable. An SBAS system relies on three components, the US's GPS network, a national network of ground stations and typically three satellites in GEO directly overhead the landmass in which it operates. GAGAN's space segment consists of three Indian communication satellites, GSAT-8, GSAT-10 and GSAT-15, in GEO over India. GSAT-8 and GSAT-10 are active, while GSAT-15 is in stand-by mode. The extensive GAGAN infrastructure consists of three key elements.

- Indian Reference Stations: that continuously determine the precise positions of the US's GPS (26,000 km and moving at almost 4 km/s) and Indian NavIC (36,000 km and moving at 3 km/s) satellites. Since the exact positions of the 15 reference ground stations are known to an extremely high degree of precision. These measurements are used to determine the precise positions of the satellites in orbit dynamically.

- Two Indian Master Control Centres (INMCC): in Bengaluru to which the reference stations forward their measurements. These data are processed to determine the dynamic error value in the satellites' positions. The dynamic correction value calculated by the ground stations have two key attributes, range and integrity. The range provides the precise location of the GPS satellites as measured by each of the multiple ground stations. The integrity of the correction is provided through numerous repeated measurements conducted independently at each ground station and the INMCCs. It is these corrections that a GAGAN receiver uses to compensate for the GPS errors.

- GSAT-8 and GSAT-10: The corrections calculated by the INMCC based on the data received from the reference stations are transmitted to GSAT-8 and GSAT-10 for retransmission back to Earth for GAGAN receivers. The two INMCCs in Bengaluru are connected to the Indian uplink stations, two in Bengaluru and one in New Delhi, using two fibre optic links operating at 2 MB, as well as two 128 kb VSAT terminals.

- The GAGAN infrastructure relies on other elements to minimise the positional errors. The radio signals between the Earth and the satellites must pass through the ionosphere, which is subject to variability from space weather arising primarily from the variability in the solar wind, charged particles emitted by the Sun. On arrival, the charged particles interact with the atoms and molecules in the Earth's upper atmosphere and strip electrons from them. The Total Electron Content (TEC) is a measure of the density of electrons in the Earth's ionosphere. This electron density in the ionosphere has a direct influence on the quality of signal reception from the navigation satellites in orbit to receivers on Earth. Monitoring TEC helps determine the magnitude of the interference satnav signals are subjected to.

As with all other aspects of the Earth's atmosphere, the TEC value is subject to random changes and is particularly influenced by solar activity. India is especially prone to TEC variations because the Equatorial Ionization Anomaly (EIA - a concentrated region of ionisation near the magnetic equator) and the EEJ happen to lie directly over India. A network of TEC stations around India continuously measure these variations and provide corrections for the positions of the navigation satellites to maintain the accuracy of the SBAS system. The TEC measurements are also used by the INMCC to process corrections, along with the data from the reference stations.

Facility	Number	Provision
Master Control Centre (INMCC)	2	Bengaluru × 2
Land Uplink Station (INLUS)	3	Bengaluru × 2 Delhi
Indian Reference Station (INRES)	15	Airports in the following cities Ahmedabad, Bengaluru, Bhubaneswar, Kolkata, Delhi, Dibrugarh, Gaya, Goa, Guwahati, Jaisalmer, Jammu, Nagpur, Porbandar, Port Blair and Trivandrum
Total Electron Content (TEC) Stations	18	TEC stations to monitor the total electron content over the Indian subcontinent.[756]
Data Communication Network (DCN)	4	Two optical fibre cable networks Two VSATs
Geosynchronous Satellites	3	GSAT-8, GSAT-10 and GSAT-15

Table 13-2 Overview of the GAGAN Architecture

SBAS can help replace the function of a traditional Instrument Landing Systems (ILS) used by pilots, particularly in complex terrains, difficult weather conditions or at night. Unlike ILS that has to be built around airports and individual runways, SBAS does not make a similar demand on an airport infrastructure. In addition to the benefits of fuel saving, reduction in delayed landings and take-offs, efficient use of airspace capacity, reduction in environmental impact and enhanced navigation, GAGAN improves safety in the aviation sector. Any aircraft equipped with a GAGAN receiver can provide a satellite-based location in real time to ground-based air traffic controllers.[757]

In addition to India's GAGAN , other SBAS implementations include:

- Japan – Multi-Functional Satellite Augmentation System (MSAS)

- US – Wide Area Augmentation System (WAAS)

- Canada – Canadian Wide Area Augmentation System (CWAAS)

- Europe – European Geostationary Navigation Overlay Service (EGNOS)

- Russia – System for Differential Correction and Monitoring (SDCM)

- China – Satellite Navigation Augmentation System based on the BeiDou constellation

Airport Authority of India is widening the scope of GAGAN beyond aviation and looking to offer the service to additional sectors, such as vehicle tracking, train tracking, disaster management, marine and farming applications, as well as to surrounding nations that happen to fall in GAGAN's footprint.[758]

SBAS relies on GEO satellites, which maintain their positions over the same points on Earth, and associated ground stations. Therefore, SBAS systems have evolved regionally. There are several SBAS networks around the world, but only three (WAAS, MSAS and EGNOS) are fully operational. Since most international flights operate globally, SBAS systems have been designed to be compatible with each other, and providers collaborate to ensure that all SBAS systems integrate into a seamless global navigation system. GAGAN is designed to interoperate with WAAS, EGNOS and MSAS. SBAS cooperation is coordinated via the Interoperability Working Groups EGNOS/MSAS and EGNOS/WAAS, and interoperability tests are regularly conducted. Countries that provide the SBAS service do not charge for it.

Global Navigation Satellite Systems

The US developed the first GPS in 1973, and it is the first of what is now called Global Navigation Satellite Systems (GNSS). GPS has been fully operational since the mid-1990s. A satnav receiver is able to determine its position, speed and time from a minimum of four navigation satellites in Earth orbit. Local topography, such as trees, high-rise buildings or mountains could reduce the field of view. With a constellation of at least 24 satellites in orbit at around 26,000 km altitude, a minimum of eight GPS satellites will be overhead at any one time, which would provide an adequate level of the signal in most inhabited parts of the world.

In the interest of maintaining control over critical national services independent of the US's GPS and, to some measure, securing sovereignty, other nations are developing their own systems. Russia, China, Japan and the European Union (EU) have or are developing their own GNSS infrastructures, GLONASS, BeiDou, Quasi-Zenith Satellite System (QZSS) and Galileo, respectively.[759] India's development of IRNSS, despite its not being global, is part of this growing international infrastructure. The Japanese QZSS has been under development for several years. Once complete, by 2018, the final configuration will consist of four satellites, three of which will be in a very unusual orbit. The QZSS satellites will be at GSO (35,786 km) inclined at 43°. The fourth satellite is for redundancy, an in-orbit spare. This unique orbit will allow at least one satellite to be overhead (elevation 60° or more) over Japan at any one time providing a navigational signal to receivers on Earth directly below.[760]

	GPS	GLONASS	GALILEO	COMPASS/Bei Dou-2
Number of satellites	32	27	24 (6 spares) by 2020	35 (by 2020) + 5 GEO
Number of orbital planes	6	3	3	3
Semi-major axis (km)	26,560	25, 508	29, 601	21,500
Orbital revolution period	11:58 H	11:15 H	14:07 H	12:35 H
Inclination	55°	64.8°	56°	55°
Satellite mass	1,100 kg (IIR)	1,400 kg	700 kg	2,200 kg

Table 13-3- Summary of GNSS systems in operation and in development

By the end of the second decade of the 21st century, over a hundred navigational satellites (32 GPS, 27 GLONASS, 30 Galileo, 11 IRNSS, 4 QZSS and 35 BeiDou) will be orbiting the Earth. Larger constellations offer higher positional accuracy and inherently more reliable service, suggesting the need for interoperability and cooperation. Just as there is a requirement for interoperability between the SBASs for use predominantly in civil aviation, there is also a desire for interoperability between the satellite navigation (regional and global) systems. The larger nations with larger investment now have operational GNSS programmes (if not yet fully mature) for global navigational systems.

Lack of cooperation between the US, Russia and China has hindered interoperability for reasons of political interest, the pursuit of commercial advantage and prestige of maintaining national self-sufficiency.

System	Status
The US's GNSS GPS (Global Positioning System)	The US military established the GPS system in 1978, and in 1996, it was made available globally for civilian use but with limited accuracy. The restrictions were removed to its present form in 2000. GPS consists of 24 operational satellites in six orbital planes inclined at 55° at an altitude of 26,560 km and has two master control stations (primary and backup), eleven command and control antennas and fifteen monitoring stations. All ground stations are located around the world.
Russia's GNSS GLONASS (Globalnaya Navigazionnaya Sputnikovaya Sistema)	By 2015, GLONASS consisted of 27 of the eventual 29 satellites in 3 orbital planes inclined at 64.8° at 19,000 km orbit. It was established during the Soviet Era and became operational in 1995 but was not maintained for several years. It has been developed by the Russian Federation in two phases 2002–2011 and New GLONASS 2012–2020. GLONASS has been operating as planned since 2011. It has 60 ground-based monitoring stations, with at least 25 located outside the Russian Federation.
China's GNSS BeiDou Navigation Satellite System (BDS)	Currently, BeiDou consists of 20 satellites providing primarily regional coverage. The number of satellites in orbit is set to increase from the current 20 to 35 by 2020 making BeiDou a GNSS. Eight satellites in the BeiDou constellation are in GSO or GEO orbits providing IRNSS-like service over China. The remaining will be at medium Earth orbit of 21,000 km in three orbital planes inclined at 55°.
EU's GNSS Galileo	Galileo (named after the 17th-century Italian astronomer) is a joint ESA and EU programme for a GNSS with 18 satellites already in orbit of an eventual 30. The first satellite was launched in 2005, and Galileo became partially operational in 2016. Galileo is due to become fully operational by 2020 with all 30 satellites (24 operational and 6 in-orbit spares) at 23,000 km orbit in 3 orbital planes. The ground segment consists of 16 sensor stations, 6 tracking and command facilities and 5 uplink stations.
India's regional navigation system IRNSS (Indian Regional National Satellite System)	IRNSS is a regional (not global) satellite system consisting of 7 satellites (three in GEO and four in GSO orbits) with a footprint over India and 1,500 km beyond. The final number of satellites in the constellation is not defined but will increase beyond 7. It is designed to have an operational lifetime of 12 years.

Table 13-4 Summary of National Satellite Navigation Systems

But this is changing. International cooperation, through the UN, EU and more recently BRICS, has demonstrated remarkable benefits of space-based search and rescue service (COSPAS-SARSAT), SBAS systems in the US and Europe and collaboration in human spaceflight making the ISS viable. The UN-sponsored International Committee on Global Navigation Satellite Systems has been promoting cooperation for mutual benefit between nations on matters of civilian use of GNSS resources since 2005. Although it has not yet achieved its vision of the "best satellite-based positioning, navigation and timing for peaceful uses for everybody, anywhere, anytime", there has been progress.[761]

Interoperability between nations is to the benefit of all the nations that participate. These benefits include minimisation of the risk of in-orbit collisions between satellites and between satellites and space debris, the prevention of radio interference and maximisation of the density of satellites in constellations. Collaboration also simplifies the receiver design that can work with multiple constellations, increase in signal strength and improvement in location accuracy for end users.

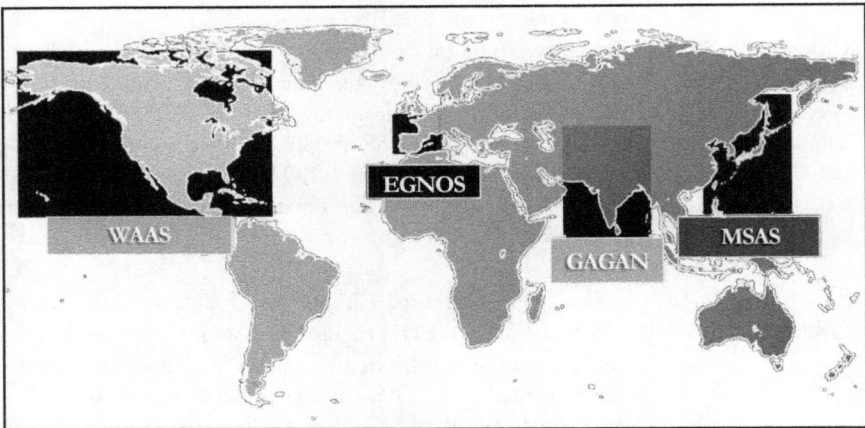

Figure 13-7 Satellite-Based Augmentation Systems. Credit ESA

The US established bilateral agreements with Russia and ESA in 2004, with India in 2005, with Japan in 2008 and with China in 2014 to support interoperability with the US's GPS. To help deliver a global, compatible and interoperable GNSS system, cooperative agreements are facilitated through international organisations, such as the ICAO, UN's International Telecommunications Union, the Radio Technical Commission Aeronautics and the European Organisation for Civil Aviation Equipment.

While India is exploiting both IRNSS and GAGAN for national programmes, such as farming, fishing, aviation, town planning and new business opportunities, it has also been in consultation with the US on

GPS-related cooperation since 2005. India's IRNSS functionality, except the RS, fits naturally with its space programme's raison d'être to support India's social and economic development. To facilitate the development of commercial applications and international cooperation, ISRO released the Signal in Space Interface Control Document (ICD) in June 2014. The document details the technical parameters that developers require to build systems and applications using the navigation signals from the IRNSS constellation.[762] ICDs have been released by Russia for GLONASS (version 5.1 in 2008), by the US for GPS (version H in 2013), by China for BeiDou (v2.0 in 2013), by India for IRNSS (v1.1 in 2017[763]) and by the EU for Galileo (v1.2 in 2015).

As a result of international collaboration, it is now possible to buy a single handset that can receive and process signals from multiple (BeiDou, Galileo and GLONASS) constellations. Traditionally, a minimum of four satellites from a constellation of about 30 was necessary to get a prompt and accurate position. In the not-too-distant future, there may be 30 satellites overhead at any one time from a constellation of over a 100.

◀ ◇ ▶

Chapter Fourteen
Human Space Flight

W ithout Gagarin, there would have been no Armstrong, asserted the Russian rocket engineer Boris Chertok (1912–2011).[764] Yuri Gagarin from the USSR was the first man in space in 1961. Eight years later, Neil Armstrong from the US became the first man to walk on the surface of the Moon. Just as the US's space programme was driven by the success of the USSR's programme, the success of the Chinese Human Spaceflight (HSF) programme has been the impetus for that of India's. China space program is operated by the China National Space Administration (CNSA) and started in the mid-1950s. The US has barred NASA from cooperating with CNSA but despite that, China has the most advanced space program after the US and Russia.

HSF is a natural aspiration for any national space programme and India is no different but there are suggesting that China's HSF has been a motivation for India's. China's first spaceflight with an astronaut was completed in 2003, and by the following decade, China had flown ten astronauts, including a woman, in four flights.[765] India has not launched any Indian's to space and does not have an active Human Spaceflight program. ISRO presented plans for its HSF programme to the Prime Minister of India in October 2006 to secure funds and initiate the programme in April 2007. This timeline suggests that this was most likely triggered by China's success of Shenzhou 5 in 2003 and Shenzhou 6 in 2005. However, a decade after presenting those plans, ISRO still does not have a formal approval from the government to proceed with its HSF programme.

From the outset, ISRO has been clear that its main objective is to use space to address the social and economic needs of the nation. HSF does not naturally fit into this vision. Apart from the enormous cost of HSF, any funds diverted to such a programme would not be available for ISRO's earth observation and communication capabilities responsible for delivering those social and economic benefits. Besides, the vision of its founder Vikram Sarabhai explicitly excluded manned-spaceflight "There are some who question the relevance of space activities in a developing nation. To us, there is no ambiguity of purpose. We do not have the fantasy of competing with the economically advanced nations in the exploration of the moon or the planets or manned space-flight. But we are convinced that if we are to play a meaningful role nationally, and in the community of nations, we must be second to none in the application of advanced technologies to the real problems of man and society."[766]

Financial and political commitment aside, India lacks one of the key elements for a HSF programme, a heavy-lift launch vehicle. Even if a spacecraft qualified for a human crew and a fully trained human crew were available, India does not have a reliable human-rated launch vehicle to deliver the spacecraft to Earth orbit. However, the development of some critical elements associated with HSF are in progress. For example, a spacecraft re-entry experiment was conducted in 2007, a mock-up of a crew capsule was flown and recovered after a sub-orbital flight in 2014, and work is progressing on astronaut training facilities, space food and launch escape system.

India's First and Only Astronaut Rakesh Sharma

Of India's 1.3 billion nationals, only one has had the first-hand experience of spaceflight with direct support of the Indian government. Two astronauts with connections to India, Kalpana Chawla and Sunita Williams have flown as part of the American space programme. Four Indian astronauts (two from the Indian Air Force and two ISRO) have trained for a specific mission but only one went to space. In April 1984, Wing commander Rakesh Sharma (born 1949) from the IAF spent eight days aboard the Soviet space station Salyut 7 in LEO. India did not solicit the spaceflight but accepted after it was offered a second time by the USSR. The first offer was made by Leonid Brezhnev (1906–1982), leader of the USSR, in 1978 to the then Indian Prime Minister Morarji Desai (1896–1995). Following a general election in India, Indira Gandhi was elected, and Brezhnev successfully repeated the offer to the new Prime Minister in 1980.[767]

The offer to fly an Indian astronaut onboard the USSR's space station Salyut 7 was part of the USSR's wider Interkosmos programme and not an offer exclusively made to India. The Interkosmos programme was a part of the IGY objective to build international cooperation but also a strategy to extend the USSR's political influence. In addition to India, nationals of 15 other countries, including Hungary, Vietnam, Poland, East Germany,

Afghanistan and France, experienced spaceflight between 1978 and 1988 through the Interkosmos programme. The US pursued a similar dual pronged policy. India already had a strong relationship with the USSR going back to the early 1960s when the latter had played a key role in India's very first rocket launch from Thumb and assisted in the launch of India's first satellite in the 1970s. It was not just in space technology that India received offers of assistance but in other sectors, too, during the period of the Cold War. Through such offers of collaboration, both the US and the USSR attempted to persuade non-aligned nations, including India, to side with their respective geopolitical world view.

When the offer came from the USSR, ISRO was making progress with its INSAT satellite programme and developing the infrastructure on the ground to support it. Launch vehicle technology was still at an early stage. ISRO had achieved initial success with the SLV-3, but the PSLV was still more than a decade away. HSF did not feature in its plans, and ISRO could not justify investing time and resources to gain the experience of eight days of human spaceflight that had no foreseeable value once the mission was over. So, when the Prime Minister passed on the offer of a free spaceflight for an Indian astronaut, ISRO declined. The Prime Minister then approached the Indian Air Force (IAF). Upon receiving a positive response, she took the political decision in 1980 to proceed. She instructed ISRO to provide full support and, if possible, to design meaningful scientific experiments that could be conducted during the mission.

The search for a suitable candidate then commenced. Reminiscent of the secret plan under which the USSR had selected Yuri Gagarin for his flight in April 1961, a secret programme (named Pawan 'the wind' in Hindi) was initiated in India. Around 200 IAF test pilots volunteered for "something extraordinary" under this programme.[768] It was only when the selection process reduced the number of candidates to a handful, and the medical tests started that the true nature of the mission, HSF, was disclosed to the applicants. All candidates were put through a series of medical, physical and psychological test at the Institute of Aviation Medicine (now the Institute of Aerospace Medicine). Four candidates were shortlisted, and they travelled to Moscow for further medical tests. Eventually, Ravish Malhotra (born 1943) and Rakesh Sharma were selected as the final two candidates. Although only one would undertake the spaceflight, both underwent identical training. The decision for Sharma to be part of the primary crew and Malhotra in the backup, was made prior to their arrival at Star City in Russia for training in 1982. Sharma was born in the town of Patiala in Punjab in 1949; two years after the country had gained independence.[769] He had joined IAF in 1970, and with 50 hours of training in a Russian MIG 21 fighter, he was on the front line. By his 23rd birthday, he had completed 21 operational missions in defensive and offensive roles during the 1971 Bangladesh Liberation War.

After the war, he became a test pilot. Prior to being selected for the Interkosmos mission, Sharma and Malhotra had known each other and both trained at the National Defence Academy but their time there did not overlap. Malhotra had left before Sharma arrived. The National Defence Academy is a joint services academy of the Indian Armed Forces, where cadets of the three services (army, air force and navy) train together. In the 1970s, both had served at India's Air Force Test Pilot School within the Aircraft Systems Testing Establishment in Bangalore. In 1974, Malhotra was selected for the USAF Test Pilots Course. He attended the class 74A at Edwards Air Force Base and graduated as an Experimental Test Pilot.[770]

Figure 14-1 Soviet-Indian prime and backup crew from left- Yuri Malyshev, Ravish Malhotra, Rakesh Sharma, Georgi Grechko, Anatoly Berezov and Gennady Strekalov. Credit Sputnik

Following their selection and medical assessments, Sharma and Malhotra arrived in Moscow with their families on 20 September 1982 to begin their training. They were eventually rostered as Research Cosmonauts, Sharma as part of the primary crew and Malhotra in the backup.[771] The primary crew (Commander: Yury Vasilyevich Malyshev (1941–1999), Flight Engineer: Nikolai Nikolayevich Rukavishnikov (1932–2002 – later Rukavishnikov was replaced by Gennady Mikhailovich Strekalov (1940-2004)) and Research Cosmonaut: Rakesh Sharma) and the backup crew (Commander: Anatoli Nikolayevich Berezovoy (1942–2014), Flight Engineer: Georgi

Mikhaylovich Grechko (1931–2017) and Research Cosmonaut: Ravish Malhotra) underwent identical training. Should an illness, injury or any other issue prevent any one of the primary crew from flying, the backup crew would step in.

Figure 14-2 Soyuz Capsule Abort System that Helped Gennady Strekalov Survive a Launch Failure Six Months before His Flight with Sharma. Credit NASA[772]

The 18-month training included assessments, training in a centrifuge and weightless environment, simulations for launch and landing in the Soyuz launch vehicle and survival training to cater for the eventuality that during return the capsule may land in a remote location resulting in a delay between landing and recovery.[773] Since this was a Russian programme, learning the Russian language was a key requirement. Malhotra had some experience with the Russian language. While at the National Defence Academy, he had selected Russian as his foreign language, but it was entirely new for Sharma, who found it to be the "most difficult part of the training."[774] On 26 September 1983, six months before their flight on Soyuz T-11, the primary and backup crews, including Sharma and Malhotra, watched the launch of Soyuz T-10-1. With a two-member crew, T-10-1 was to fly to Salyut 7 with a mission to augment its solar arrays. The launch did not go to plan. A fuel leak and fire resulted in the destruction of the launch vehicle and launch pad. The Capsule Abort System engaged removing the crew capsule from the top of the rocket a few seconds before a massive explosion engulfed the launch site. As an experienced fighter and test pilot, Sharma was familiar with such life-and-death situations. He did not tell his wife about the launch failure "as had been my practice right through my flying and testing career."[775] Gennady Strekalov, one of the two who survived that day, returned to the launch pad six months after his death-defying experience and sat alongside Sharma and Malyshev for another attempt to launch to Salyut 7 when Nikolai Nikolayevich Rukavishnikov (1932–2002) assigned as the flight engineer on Sharma's flight fell ill.

Figure 14-3 Crew of Soyuz T-11 in Star City. (Right to left) Gennady Strekalov, Yuri Malyshev, Rakesh Sharma, Ravish Malhotra and Sharma's wife, Madhu. 15 April 1984. Credit Sputnik

On Tuesday, 3 April 1984, at 10:38, Rakesh Sharma with Commander Yuri Malyshev and Gennady Strekalov blasted off from the Baikonur

Cosmodrome aboard the Soyuz T-11 spacecraft. Ten minutes later, Soyuz T-11 was in a 224 kilometre LEO. There was no repeat of the dramatic events of six months earlier. Sharma, a career pilot, used to looking out of a window when flying, found the absence of windows at launch unnerving. The navigation, cabin pressurisation and warning system displays were placed in front of the research cosmonaut's seat and it was his duty to monitor the health of these systems during flight. As the only one of the crew without previous spaceflight experience, his role during launch was to monitor instruments and participate only if an emergency arose. None did. Twenty-five hours after launch, Soyuz-T-11 gradually caught up and docked with Salut 7, which had the crew of Salyut T-10 on-board since February. For the next eight days, the six cosmonauts lived and worked together onboard Soyuz-7.[776]

Sharma adapted quickly to microgravity without any ill effects. A day after arrival, he spoke to the Indian Prime Minister from orbit. Reminiscent of President Nixon's (Richard Milhous Nixon, 1913–1994) speaking to Neil Armstrong and Buzz Aldrin during their historic Moonwalk in July 1969, this was the political highlight of Sharma's mission. The brief conversation in Hindi with Prime Minister Indira Gandhi has come to define Sharma's spaceflight. She asked him "Upur se Bharat kaise dikhta hae (how does India look from above)?" He replied "Saare jahan se acha (best in the world)". His natural, spontaneous and unrehearsed response were words from the lyrics of a song he was very familiar with since his student days.[777]

Sharma's mission made a substantial contribution to a sophisticated science experiment called Terra. During April, the huge Indian land mass is clear and cloud free, unlike Europe and the northern latitudes. The Terra Experiment was the collective name for the study of India's natural resources using data from a variety of sources, including photographic surveys of Indian territory with a multi-zonal MKF-6M apparatus and KATE-140 camera onboard Salyut 7. The images produced formed part of a larger collection, including visual observations and photographic surveys using hand-held cameras, aerial surveys and surface measurements of experimental sections of Indian territory by Indian specialists.[778] The original plan involved Sharma taking images from orbit during the nine passes over India, which was increased to 11 in support of Terra. Around 2,000 images were taken from space during the mission and later added to ISRO's remote sensing database. The database had been initiated almost a year earlier at the launch of India's second remote sensing satellite INSAT-1B.

During the eight days in space, Sharma recorded biomedical data from his own body with equipment designed and built by ISRO scientists. He conducted experiments on the phenomenon of undercooling and microgravity and its effects on semiconductor alloys using silver and germanium. Sharma's yoga experiment was the "most curious of the medical experiments."[779] It was designed to study the potential of yoga techniques to mitigate the harmful effects of weightlessness on the human body; bones and muscles tend to deteriorate in extended periods of microgravity. Collectively, the crew conducted 43 experiments over the eight days of the mission.

Sharma went to space aboard Soyuz T-11 on 3 April as the 138th person to enter space and returned to Earth aboard Soyuz T-10 on 11 April, landing as planned in the USSR 46 km to the east of the city of Arkalyk. The return to Earth, potentially as hazardous as the launch, was also incident free, although Sharma recalls that the sound of the parachute cables chafing against the brackets was very unnerving "I was certain that the parachute was going to part company."[780] After his successful flight, he was awarded two medals by the USSR, Hero of the Soviet Union and The Order of Lenin. This was an integral part of the Interkosmos programme, a package deal that all foreign cosmonauts received. Sharma is the only Indian recipient of these awards and with the demise of the USSR on 26 December 1991, will remain so. From the Indian government, he received the Ashok Chakra, and in the interest of diplomatic consistency, the two Russian crew were also awarded the Ashok Chakra.

Speaking to the media about his experience, Sharma endorsed passionately the role of international collaboration. Not the collaboration of Interkosmos, which was driven by narrow political objectives of the Cold War, but one borne out of a grander recognition that future space exploration should be conducted by people from planet Earth and not just by representatives of a few of its nations. He advocates the idea that not every nation should have to reinvent the wheel, nor should they have to develop from scratch their HSF programme. It would not only prevent plundering the meagre resources of planet Earth but also not undermine the collective enterprise that human space exploration ought to be. Sharma is not keen on being defined by his spaceflight, nor is he preoccupied with it but is bemused that others are "It was just an event. It was given to me. I did it, and I want to move on". It was no "cakewalk" he insists, but it was not as fulfilling as it would have been had he been a career astronaut'.[781]

As a backup pilot, Ravish Malhotra was never called up. Following the joint celebratory tours around India, he returned to his career as a test pilot. Malhotra declined an offer to return to the USSR as an air attaché and turned instead to the private sector and worked for an aerospace company. Since his return in 1984, he has never been back to Russia. Following his spaceflight, Sharma returned to the IAF as a Flight Commander with an operational squadron before joining Hindustan Aeronautics Ltd. as chief test pilot. Later he headed the aerospace and defence division of an American Software Company in India.[782]

Figure 14-4 Rakesh Sharma with Hero of the Soviet Union and The Order of Lenin Awarded by the USSR. August 2013. Credit Author

From ISRO's perspective, the flurry of media interest upon Sharma's return to India was followed by nothing; just as ISRO had anticipated. Even though the eight-day-long spaceflight was a great morale boost for the nation, Sharma's flight, in the absence of a coherent Indian plan for HSF, has remained one of the "isolated artefacts of curiosity."[783] However, even before Sharma had completed his mission, a demand arose again for another Indian astronaut. This time, it was not from the USSR, but the US.

Still-born Astronaut

By the time Sharma's mission ended, the American Space Shuttle had been flying successfully for three years. To recover some of the huge costs, a commercial objective was built into the Space Shuttle programme from the outset. NASA was looking for ways to promote the Shuttle's commercial potential and offered to fly a payload specialist as part of the commercial deal to launch a satellite from the Space Shuttle.[784] This also served as NASA's attempt to counter, compete with or simply emulate the USSR's Interkosmos programme.

India was encouraged by the 1975 SITE programme to have its own communication satellite programme, resulting in the Indian Satellite (INSAT) series. ISRO commissioned four satellites of the first INSAT series (INSAT-1A through 1D) from the US-based Ford Aerospace and Communication Corporation. INSAT-1A was launched by a Delta launch vehicle in April 1982, and INSAT-1B was launched by the Shuttle in August 1983. INSAT-1C was assigned to mission STS-61-I for launch by the Space Shuttle Columbia on 25 June 1986. It would be launched along with three other payloads and a crew of five (including an Indian astronaut) for a seven-day mission.[785] Once the agreement was signed, the search for an Indian payload specialist to join the crew was initiated. Since INSAT was an ISRO project, the proposed astronaut was to be from ISRO, not the IAF.

Initially, 400 volunteers with science and engineering backgrounds applied from various ISRO centres. Forty were shortlisted for further assessment, including a week-long screening consisting of medical, stress and psychological tests. In the summer of 1985, seven candidates were invited to a final selection board headed by Professor U.R. Rao and included Rakesh Sharma and NASA astronaut Paul Joseph Weitz (born 1932). The panel selected two candidates, Paramaswaren Radhakrishnan (born 1943), a scientist from VSSC in Trivandrum, and Nagapathi Chidambar Bhat (born 1948), an engineer from ISRO Satellite Centre in Bangalore. The candidates had to complete additional medical tests determined by NASA at the Johnson Space Centre. ISRO chose to make the public announcement on the candidate selection only after the tests in the US were complete. Until such announcement, both were prohibited from sharing the decision

330

publicly. In July 1985, both candidates flew out to Huston, Texas, and successfully completed the medical evaluation.

Radhakrishnan, as a 22-year-old electronics engineer, was recruited in 1966 to INCOSPAR, which in 1969 became ISRO. He had established a strong connection with the American space programme from the outset. As a new recruit, Radhakrishnan's first role was to visit schools, colleges and universities as part of a three-member team teaching students about spaceflight. He crisscrossed India for almost a year using a NASA-supplied Chevrolet truck loaded with models of rockets, satellites, Apollo modules and cine film of rocket launches. In 1966, the US was still developing the Apollo programme (the Saturn 5 launch vehicle and Command, Service and Lunar Modules) that would take Americans to a return journey to the surface of the Moon. Later Radhakrishnan served in ISRO in numerous roles including the design and development of the power system for India's first satellite, Aryabhata

Figure 14-5 N.C. Bhat and P. Radhakrishnan. Credit Bert Viz

Bhat joined ISRO in 1973 just as India's satellite programme was getting underway. Based at the ISRO Satellite Centre in Bangalore, he contributed to several of ISRO's key space missions. He was involved in fabricating several elements for ISRO's first satellite Aryabhata, the solar panel unfurling mechanism for IRS-1A, the antenna used on-board India's first Moon mission Chandrayaan-1 and some of the experiments on SRE-1 in 2007, India's first spacecraft that returned to Earth after 12 days in space.

Radhakrishnan was concerned that his age, 41, would work against him. He remembers a series of progressively tougher tests during the selection process, which included "countless questionnaires to fill in, and endless interviews, which mostly consisted of a monologue from the candidate, punctuated occasionally by a prompting question from the other side. The purpose was to probe into the deepest crevices of the candidate's mind."[786] Bhat recalls the use of eye drops to dilate the pupil so that pictures could be taken of the retina. He unnecessarily suffered from blurred vision for three days before his sight returned to normal "Later, I came to know that they had forgotten to put another eye drops which restores the dilated pupil to normal."[787]

Figure 14-6 P. Radhakrishnan (back row third from left) and N. C. Bhat (back row third from right) during High-altitude Training with the Indian Airforce. 1985.
Credit N.C. Bhat

Preparation for the mission started following the successful medical tests conducted by NASA. The training included learning about the technical details of the INSAT satellites, how they were built, readied for launch and operated once in orbit. Both candidates visited the ISRO Satellite Centre in Bangalore, National Remote Sensing Agency (now National Remote Sensing Centre) in Hyderabad and SAC in Ahmedabad. Their training included gaining familiarity with high G forces in a centrifuge and altitude flights with the IAF.

The launch of INSAT-1C was scheduled for November 1986. During a five-day mission, the payload specialist would help prepare and launch INSAT-1C from the Space Shuttle cargo bay while in LEO. As with all space missions, there were a primary and a backup crew. The decision for primary/backup crew was to be made in September, that is, two months before launch. Until then, both would undergo identical training in India and the US. The Institute for Aerospace Medicine familiarised them with the space food that was being prepared by the Defence Food Research Laboratory (DFRL) in Mysore. The DFRL had provided food for the series of Indian Antarctic missions, as well as Sharma's visit to Salyut 7[788]. In addition to the NASA supplied food aboard the Space Shuttle (72 different items and 20 beverages), the Indian astronauts would take Indian space food, too. The preliminary group of 12 foods suggested by the DFRL included peas pulav, chicken pulav, lemon rice, chicken masala, peas paneer, kheer rice pudding, pineapple juice, mango juice, grape juice, sooji halwa, mango fruit bar and chapattis.[789]

Then, on 28 January 1986, Space Shuttle Challenger exploded shortly after launch. Radhakrishnan and Bhat were at Ford Aerospace in the US for familiarisation with the INSAT spacecraft. They did not see it live but an hour afterwards on TV. They described the horror of seeing the death of seven astronauts as "very tragic and touching."[790] The incident sent shock waves across the US and space agencies around the world. The spectacular destruction during the launch of the most sophisticated spaceship ever built and the largest loss of life in a single US space incident brought NASA to an existential crisis. NASA's initial statement indicated that the Space Shuttle mission would be rescheduled within six months, and ISRO's provisional response was to proceed on that basis. However, the magnitude of the incident dictated that a longer period of review would be necessary.[791] Radhakrishnan concluded that his dream of spaceflight "went up in smoke in that moment."[792]

It wasn't just the Indian satellite and the astronaut flight that was impacted. Investigation of the Challenger accident was completed and the report published on 29 October 1986. Before the end of the year, on 27 December 1986, President Reagan (Ronald Wilson Reagan, 1911–2004) issued a directive asserting that the Space Shuttle "shall no longer provide launch services for commercial and foreign payload unless those spacecraft have unique, specific reasons to be launched aboard the space shuttle."[793] Before the accident, two dozen Shuttle flights (there were four shuttles in total) were scheduled per year. Afterwards, it was just 14 flights per year.[794] In practice, it was even lower. Between the first and last Shuttle flights (12 April 1981 and 16 May 2011), 135 space missions were completed averaging to about 12 missions per year.[795] The mission to launch Indian communication satellites from the Space Shuttle ended abruptly, and with it, the hopes of two Indian engineers who had suddenly acquired and then lost the opportunity of spaceflight. On that news, Radhakrishnan concluded, "so here I am, a still-born astronaut."[796]

Roadmap for Human Spaceflight

ISRO's HSF programme has had several false starts. In 2006, over 80 senior scientists met at ISRO headquarters in Bangalore concluding unanimously that time was right for ISRO to proceed with its HSF.[797] India's Eleventh Five Year Plan[798] (2007–20012), stated "The development of a manned mission would take about 10 to 12 years, and it is planned to focus on developing critical technologies required during 11th plan period and achieve substantial progress towards realising a manned mission during 12th plan period".

A mission with specific objectives related to a human crewed spacecraft was completed in the following year when ISRO demonstrated its ability to launch and recover a 550-kilogram module in January 2007. SRE-1was ISRO's first attempt to recover a spacecraft it had launched. Following the launch aboard a PSLV and twelve days in Earth orbit, the capsule was remotely commanded to de-orbit and was recovered by the Indian coast guard after splashdown in the Indian Ocean. In addition to basic on-board microgravity experiments, ISRO tested the capsule's navigation, guidance and control systems. The recovered SRE-1module is now a key exhibit in ISRO's space museum in St. Mary Magdalene Church, the cradle of India's space programme located within the grounds of VSSC in Kerala.

SRE-1was an important step towards the HSF programme. Its success helped ISRO secure a $2.1 million (Rs.8.5 crore) feasibility study on HSF. Six months after SRE-1, in June 2007, ISRO chairman constituted a steering committee on Human in Space Programme. The committee produced a four-volume report by early 2008 consisting of a project report, cost estimate, executive summary and a prologue. This report was submitted to the government in February 2008. So, by mid-2008, ISRO had a clear vision on how its HSF programme would develop based on detailed research, feasibility studies and a successful re-entry mission. By the end of the year, an opportunity arose for potential international collaboration that could hasten the programme timeline compared to ISRO's doing it alone. A MoU was signed following Russian President Dmitry Anatolyevich Medvedev's (born 1965) visit to India in December 2008. The Russians were probably influenced by India's success with SRE-1in 2007 and ISRO's first and highly successful mission to the Moon, Chandrayaan-1, which had been launched three months before Medvedev's visit. Under this agreement, Russia would fly an Indian astronaut aboard a

Russian spacecraft by 2013 and help India build its own spacecraft capable of carrying a human crew based on the Soyuz design for launch by GSLV-Mk3 by 2016.

Buoyed by the success of SRE-1, Chandrayaan-1 and the prospects of rapid progress through the Russian MoU, the ISRO chairman Kasturirangan unveiled plans for two- and three-member crew modules designed for a week-long space mission during the 96th Indian Science Congress in January 2009.[799] K. Radhakrishnan, the VSSC director at the time but became ISRO chairman about nine months later, was also present. He also spoke about an Indian space mission by Indian astronauts in an Indian spacecraft, along with missions to the Moon and Mars. One spectacular media headline that followed was "India plans to hoist tricolour on the Moon by 2020."[800] In 2009, the Government of India, too, appeared to be in favour not only of HSF but also of a human mission to land on the Moon by 2020.

Figure 14-7 Space Recovery Experiment Capsule on display at VSSC. Credit ISRO

India's plans were very likely motivated by the September 2008 success of the Chinese Shenzhou 7 mission, which carried for the first time a crew of three. By 2016, China successfully completed 6 missions carrying 15

335

astronauts, 13 men and two women (Shenzhou 5 on 15 October 2003 with one astronaut; Shenzhou 6 on 16 October 2006 with two; Shenzhou 7 on 25 September 2008 with three; Shenzhou 7 on 25 November 2008 with three; Shenzhou 9 on16 June 2012 with three; Shenzhou 10 on 11 June 2013 with three and Shenzhou 11 on 17 October 2016 with three). Ostensibly, India is not in a space race with China, but in 2014, ISRO revealed that it prepared its mission to Mars in haste only after the Chinese mission to Mars, Yinghuo-1, aboard a Russian rocket failed to leave Earth orbit in November 2011.[801]

Over the next couple of years, the HSF programme made no real headway. The following year, 2010 was not a good year for ISRO. The MoU with Russia had not been productive and was terminated in October 2010.[802] Why that happened is unclear, but probably because India and Russia could not agree on the commercial arrangements underpinning the technology transfer of Soyuz from Russia to India.[803] Two GSLV launches, GSLV D3 in April and GSLV F06 in December, failed during launch.

The Twelfth FYP (2012-2017), published in 2011, merely restated the Eleventh FYP objective to "develop the critical technology and subsystems related to Human Space Flight programme", confirming that, between the 11th and 12th Plans, India's HSF programme made no major progress. Unexpected media statements in 2012 from the IAF, claiming that the IAF was proceeding with the astronaut crew selection process, added to the confusion.[804] The source of the confusion was to some extent competitive posturing between the IAF and ISRO. The IAF has had a strong connection with the HSF in the past. Astronaut training is centred around the Institute of Aerospace Medicine which is an IAF body, and the first astronaut, Rakesh Sharma, was a wing commander in the IAF.[805]

At the end of 2013, ISRO issued a press release insisting that media reports on Manned Mission to Moon were unfounded.[806] Despite the absence of an essential element, the heavy launch vehicle, ISRO has been quietly making progress towards human spaceflight capability. In March 2012, the Minister of State confirmed in the Indian parliament that Rs.145 crore ($22.5 million) had been allocated for ISRO to pursue the development of critical technologies for the HSF programme.[807] These funds were split between various tasks: Rs.61 crore ($9 million) for the crew module, Rs.27 crore ($4 million) for qualifying the launch vehicles for a human crew, Rs.36 crore ($5.5 million) for contracts with institutions and Rs.21 crore ($4 million) for other tasks, such as crew module aerodynamics, space suits and life support systems. A mock-up of the crew capsule was tested in a sub-orbital flight on 18 December 2014 as the LVM3-X/CARE mission, a sub-orbital version of the SRE-1 but using the LVM3 with a non-active cryogenic third stage.

The HSF programme is not one of ISRO's priorities, but with the allocated funds, it has been quietly working on the development of an astronaut training programme and an astronaut crew capsule under the guidance of a dedicated project director for the HSF programme.[808] Progress is being made in many key prerequisites, including space suits, environmental and life support systems, and emergency Capsule Abort System at the launch pad and during the early phase of launch. Since March 2009, ISRO has had a MoU with the IAF's Institute of Aerospace Medicine to conduct basic research on the physical and psychological requirements for the human spaceflight crew and expand the institute's existing facilities to cater for the HSF programme as a pre-project R&D activity.

Figure 14-8 Crew Module in the Andaman Sea after Splashdown. 18 December 2014. Credit ISRO

ISRO has also entered into agreements with a Bangalore-based third party to initiate the development of spacesuits and a Mysore-based company to develop a space food menu for Indian astronauts.[809] The initial plan envisaged a crew module designed for two, but this has since been expanded to include a 3-astronaut variant. Preliminary designs indicate that the crew module will be a twin-walled structure with a sealed, all-welded internal shell.

A service module with a re-entry engine is to support the crew module.

Initial plans call for a mission duration of a few hours for the first mission that could be extended up to seven days. The orbit has been designed to allow the crew capsule to re-enter the Earth's atmosphere automatically in the event of a service module engine failure.[810] The capsule would be launched from Sriharikota using GSLV-Mk3/LVM3 return for a splashdown in the Bay of Bengal or the Indian Ocean. The crew module is designed with some capability to manoeuvre in the atmosphere along both down and across ranges to support navigation to a predetermined splashdown location.

Once the launch vehicle (LVM3) is ready, integrated tests with the crew module, including the Emergency Capsule Abort System (CAS), can commence. The CAS is a small rocket at the top of the main launch vehicle. In an emergency, it fires like an ejector seat (but is located above the crew) and pulls the crew capsule away from the launch vehicle for a parachute landing a few kilometres away from the launch pad.[811] Prior to the first human orbital flight, multiple test-flights of the crew module in orbit with dummy passengers and non-human occupants (the US had used a chimpanzee, the USSR used 4 dogs and the Chinese had used rats and guinea pigs. These initial missions test various subsystems, including life support, guidance, navigation and re-entry.[812] The USSR had conducted seven flights without human occupants using varying sophistication in the onboard subsystems before Yuri Gagarin's historic flight on 12 April 1961.

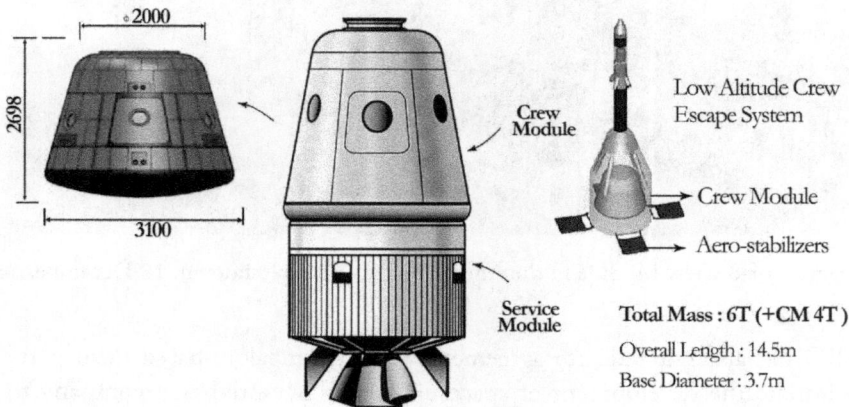

Figure 14-9 Proposed Crew Vehicle (left), Crew Module Attached to the Service Module (centre) and the Emergency Capsule Abort System (right). Credit ISRO

The crew capsule of each version was an improvement on the previous. Only three of the seven flights were completely successful before Gagarin made his attempt. Between 1959 and 1961, the US too conducted over a dozen tests on the ground and sub-orbital launch on their Mercury spacecraft before America's first astronaut Alan Shepard's (1923–1998) flight in May 1961. China conducted four test flights of their Shenzhou spacecraft before their first human spaceflight with Yang Liwei (born 1965) in 2003. ISRO had planned to follow-up SRE-1 with SRE-2, but following a series of delays, SRE-2 has been quietly withdrawn.[813] ISRO's 2014 LVM3-X/Crew Module Atmospheric Re-entry Experiment test flight was sub-orbital, and the crew module was of an initial rudimentary design. Since then, ISRO has not scheduled any additional test flights of the crew module, so a date for India's first flight with a human crew is not imminent and remains unknown. The success of India's HSF programme will be measured more in terms of political and national prestige than results in science or technology.

The HSF programme will require a profound shift in ISRO's capability and capacity, but success will not translate into meaningful national economic benefits for which ISRO was established. Further, the high cost of HSF may well delay or prevent its other objectives. Kasturirangan, who became ISRO chairman in 1994, understood the scale of the challenges inherent in the HSF programme and noted "People give you a wrong impression about the type of resources needed for human space mission" and the "returns will never be big."[814] Ten years ago, India committed to a desire for HSF but ISRO still has no formal government approval to proceed. In its absence, ISRO is undertaking work on some aspects of HSF in the background.

The story of Indian astronauts is a short one. Rakesh Sharma continues to speak and write about his experience and the future for India's HSF programme. A recent announcement indicated that Bollywood film about his experiences is underway.[815] Kalpana Chawla, was killed when Space Shuttle Colombia disintegrated during re-entry over Texas on 1 February 2003. Announcements about a film about her career have also been published but here former husband, Jean-Pierre Harrison has insisted in the past that he does not support the project.[816].

Ravish Malhotra retired from his work in the private sector and is not actively involved in any space projects. P. Radhakrishnan continues to live in Kerala where he frequently participates in the media as a science communicator in both his native Malayalam and English. N.C. Bhat retired in 2011 from ISRO but continues to share his experience in mechanical design as a consultant. Since 2015, he has been involved with Team Indus in support of their lander and rover mission to the Moon. Sunita Williams completed two spaceflights to the ISS and clocked up a total of over 300 days in space and over 50 hours of spacewalk. She is now part of a team of astronauts preparing to fly the next generation of US's commercially built human-rated spacecraft from Boeing CST-100 and Space-X Dragon.

Each national government that has launched humans into space started with men and then included women. The Soviets put the first woman in space in 1963, two years after the first man. The gap between the first male astronaut and its first female astronaut for the US was 22 years. The Chinese took just nine years. In early 2016, the IAF Chief Arun Raha announced that women in the IAF could qualify as fighter pilots. Perhaps, India's first man in space could be a woman.

◄ ◊ ►

Chapter Fifteen
Moon, Mars and Science

I n October 2008, ISRO launched Chandrayaan-1ndia's first mission beyond Earth orbit, marking a dramatic shift from its founding principles. When the US and USSR governments were making unprecedented financial and political commitments to their space programmes for human spaceflight, science and the exploration of the solar system, Vikram Sarabhai had stated in 1966 "in India the immediate goal of our space research is modest. We do not expect to send a man to the Moon or put white, pink or black elephants into orbit around the Earth."[817] He had ruled India out of competing with "economically advanced nations in the exploration of the Moon or the planets or manned space-flight" and insisted that India focused only on the "application of advanced technologies to the real problems of man and society". Four decades later, a spacecraft made in and launched from India arrived in lunar orbit. Was this a break from the founding principles or a "course correction" to reflect the ambitions of a new generation in a new century?

India's space programme was intended from the outset to help improve the quality of life of ordinary Indians. To achieve societal development, Sarabhai and others had not only to build the technological infrastructure from scratch but also to embark on crystallising in the fabric of Indian society the scientific temper that Nehru had imbued in the Indian Constitution. Sarabhai was a scientist and understood the transformative power of science and technology to change society. The scale of the benefit of the space programme is not always immediately obvious or tangible. Changing the fabric of society to permeate scientific temper is a complex and slow process.

Scientifically literate populations, industrial centres of excellence, R&D laboratories and educational institutions are some of the foundations of modern developed societies. Momentous and technologically stunning missions of exploration, like Sputnik or Apollo, inspire the younger generations to take up careers in science and engineering. Of all the scientific fields (i.e., astronomy, physics, genetics, geophysics or organic chemistry), it is space science that tends to hold the most fascination for young students contemplating a future career in science and technology. A large-scale space programme can also secure political and financial commitment because of its capacity to deliver national prestige, security, as well as economic development. An Indian spacecraft in orbit around the Moon could be a powerful catalyst to inspire a new generation of scientists and engineers to guide India through the next cycle of national development.

Destination Moon

By the turn of the millennium, ISRO had established a reliable track-record of over a decade in launching and operating satellites in Earth orbit. It operated one of the most sophisticated national constellations of remote sensing and communication satellites in the world. It had a launch vehicle (PSLV) with a proven track record and had successfully launched its next generation of launch vehicle GSLV-D1. Sarabhai's original goals for India's space programme had been largely met. ISRO had matured as an organisation and was optimistic in its technical competence. It was time for a new generation of ISRO leaders to contemplate new ambitious goals based on its new-found confidence.

India's mission to the Moon emerged from a paper presented in 1999 under the guidance of the then ISRO chairman K. Kasturirangan. Even though going to the Moon offered no obvious national development outcomes, it was a mission beyond Earth orbit, uncharted territory for ISRO. Operating a satellite around the Moon would be just like operating any other satellite in orbit; only that it would be further away. While ISRO could use much of the existing infrastructure, it would still need to build new deep space communication infrastructure and develop the navigation and guidance systems required for a journey from the Indian coastline to the lunar orbit.

Once initial calculations confirmed that a PSLV could be used for a mission to the nearest celestial body, planning for such a mission began. Getting government approval was, however, not a trivial process. In March 1998, a new nationalist government by Bharatiya Janata Party had come to power, and a couple of months later, India conducted a second series of nuclear tests, known as Pokhran-2. International sanctions followed as they had done following Pokhran-1 nuclear tests in May 1974. For the Indian government, an Indian mission to the Moon would not only help recover national prestige but announce India's arrival as a space power on the international stage.[818] This was almost a repeat of what the US and USSR governments had done during the Cold War in pursuit of national reputation. Political and financial commitment from the Indian government for space projects that would raise the national profile on an international stage was not difficult in principle to acquire. It did however involve a long and involved process. Kasturirangan recalls "we had to go through an elaborate process of consultation and justification with the scientific community, academics, the political system, and the public media before this mission was given the go-ahead."[819] It took over four years to convince the government that there was value to a purely scientific mission to the Moon.[820]

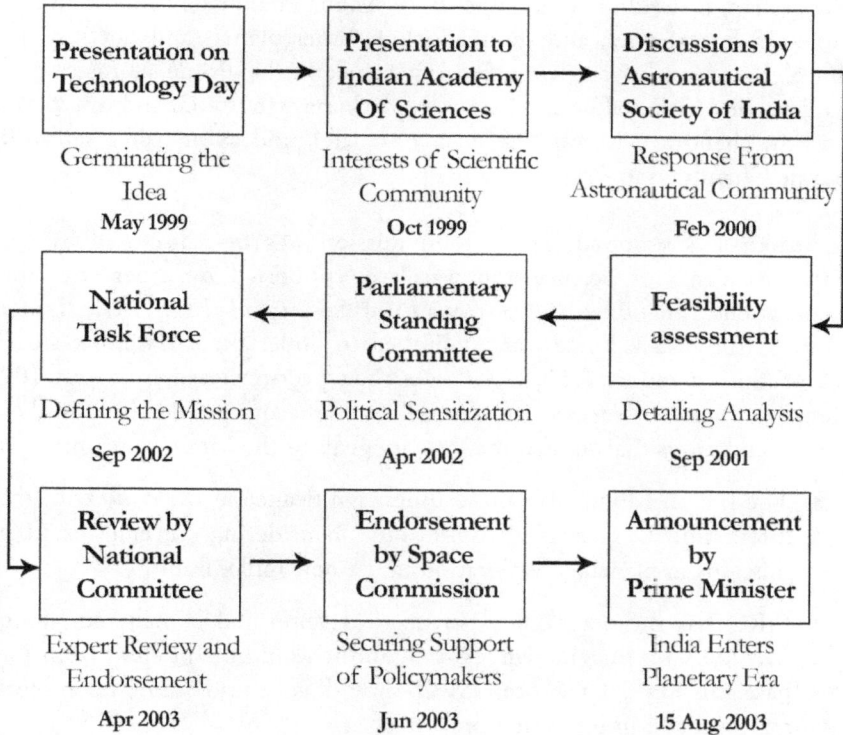

Presentation on Technology Day	Presentation to Indian Academy Of Sciences	Discussions by Astronautical Society of India
Germinating the Idea	Interests of Scientific Community	Response From Astronautical Community
May 1999	Oct 1999	Feb 2000

National Task Force	Parliamentary Standing Committee	Feasibility assessment
Defining the Mission	Political Sensitization	Detailing Analysis
Sep 2002	Apr 2002	Sep 2001

Review by National Committee	Endorsement by Space Commission	Announcement by Prime Minister
Expert Review and Endorsement	Securing Support of Policymakers	India Enters Planetary Era
Apr 2003	Jun 2003	15 Aug 2003

Figure 15-1 Steps Leading up to India's First Moon Mission. Credit K. Kasturirangan

On 15 August 2003, the Prime Minister of India announced to the world India's intention to launch a mission to the Moon, Chandrayaan-1. Though publicly unstated, there was probably another motivation behind India's announcement. On 23 January 2003, China had announced its plan for a mission to the Moon.[821] In September 2003, the ESA's SMART-1 arrived in lunar orbit, and two months later, in October 2003, China achieved spectacular success with its first orbital HSF mission. India would have known about China's imminent HSF plans. In the absence of an Indian mission to the Moon, India risked its space programme being perceived internationally as immature and second to that of China. The Indian space programme needed to set itself a bold new objective. ISRO was already familiar with designing, building, launching and operating satellites in Earth orbit. If India could get to the Moon before China, then it would be the fifth space agency to do so, after the USSR (1959), US (1959), Japan (1990) and ESA (2003).[822] Kasturirangan had originally planned for a single mission to the Moon called Somayaan, but the Prime Minister was of the view that if ISRO was to engage in an exploration of the solar system, it should have a longer-term vision. The mission should not be a one-off, but a series, and should include other planets and not just the Moon. Based on advice from Sanskrit scholars, the Prime Minister also changed the name of the mission from Somayaan to Chandrayaan (in Sanskrit, chandra = Moon and yaan = vehicle), and as first of a series, it became Chandrayaan-1.

Kasturirangan's proposal for a Moon mission was first discussed by the Indian Academy of Science and was later supported by other scientific organisations, including the Astronautical Society of India. The Lunar Mission Study Task Force was established to undertake a feasibility study and produce a report. This report was then peer-reviewed by about 100 scientists from a variety of disciplines, and it provided four key recommendations that framed the primary goals of the Moon mission:[823]

- The Indian Moon Mission assumes significance in the context of the international scientific community considering several exciting missions in planetary exploration in the new millennium.

- ISRO has the necessary expertise to develop and launch the Moon Mission with imaginative features, and it would be different from the past missions. Therefore, ISRO should go ahead with the project approval and implementation.

- Apart from technological and scientific gains, it would provide the needed thrust to basic science and engineering research in the country. The project would help the return of young talent to the arena of fundamental research.

- The academia, in particular, university scientists, would find participation in such a project intellectually rewarding. In this context, the scientific objectives would need further refinement to include innovative ideas from a broader scientific community through Announcement of Opportunity, etc.

The reference to academia and Announcement of Opportunities in the final recommendation would subsequently transform the mission into an international, award-winning, and most scientifically successful mission that ISRO had ever initiated.[824]

Building Chandrayaan-1

Geopolitics and national prestige motivated the mission, but Chandrayaan-1's primary objectives were scientific. A detailed configuration of the spacecraft and launcher evolved over time. Eventually, it was concluded that Chandrayaan-1 would be a 1.5 m cuboid with a mass of 440 kg (524 kg with propellant) powered by a solar panel and lithium-ion batteries for use during a solar eclipse and designed to operate for two years. It would be launched by a PSLV and placed in a 100-km circular polar orbit around the Moon. It would use S-band uplink for remote command, S-band downlink for telemetry and X-band for data transmission from the Moon to Earth. Based on the tried and tested IRS series design (also used for Kalpana-1 and MetSat-1), the single solar array would be canted at 30° to provide 750 W peak power. The spacecraft would be fitted with three separate Solid State Recorders (SSR). SSR1 with a capacity of 32 GB to store scientific data, SSR2 with a capacity of 8 GB for science and spacecraft telemetry data and SSR3 with a capacity of 10 GB exclusively for science data gathered by the one specific instrument, Moon Mineralogy Mapper (M3) instrument.

The payload that actually flew was significantly revised twice from that initially planned. After accommodating all ISRO's six instruments, a calculation determined that an additional 10 kg mass and 10 W power consumption were still available. ISRO issued an Announcement of Opportunity, and the unexpected response to this prompted the first design change. The Announcement of Opportunity was an invitation to scientists, engineers or institutions beyond ISRO (national and international) to make use of this excess available capacity. It received an unexpectedly large international response. Two instruments proposed by NASA M3 and Miniature Synthetic Aperture Radar (Mini-SAR), looked promising but together they exceeded the 10-kg allowance.

To accommodate both, ISRO redesigned the spacecraft and increased the available capacity from 10 kg to 25 kg. Now, instead of two additional instruments, ISRO could accommodate even more. Eventually, six instruments were incorporated from international partners. M3 from NASA; Mini-SAR also from the US and UK; Near-Infrared Spectrometer (SIR-2) from ESA; Sub keV Atom Reflecting Analyser (SARA) also from

ESA, India, Japan and Sweden; Radiation Dose Monitor (RADOM) from the Bulgarian Space Science Institute; and Chandrayaan-1 X-ray Spectrometer (C1XS), a collaboration funded by ESA and produced by the Rutherford Appleton Laboratory and University of Wales, UK.

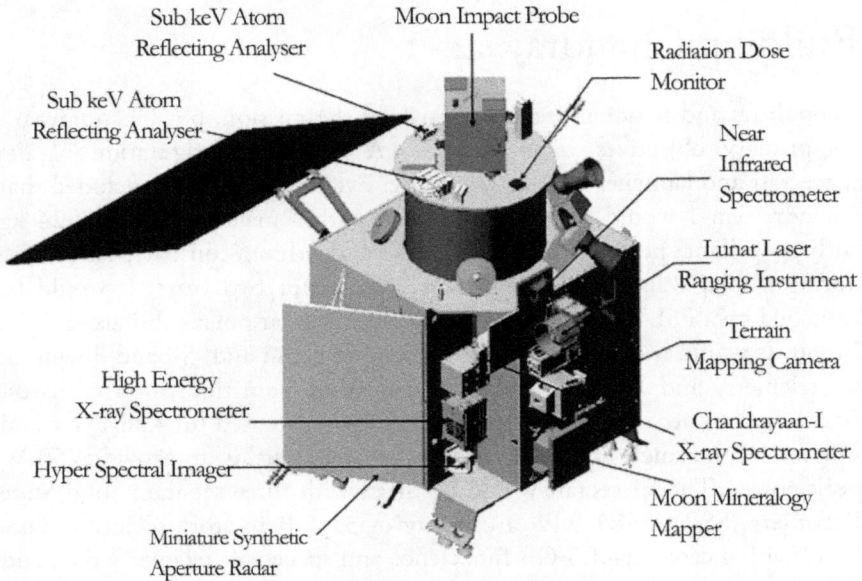

Figure 15-2 Chandrayaan-1 and Its Science Payload. Credit Adapted from ISRO

The second significant design change came when a Moon Impact Probe (MIP) was added. The MIP was designed to detach from Chandrayaan-1 and make an impact on the surface of the Moon delivering the Indian Tricolour to Earth's nearest neighbour. Abdul Kalam, the then President of India and a highly respected former ISRO engineer who had helped develop India's first satellite launcher SLV-3, had advocated the MIP. His elevated political position ensured his proposal was taken seriously. Further, the delivery of the Indian flag to the surface of the Moon was a source of national pride and would attract additional government commitment.

The MIP, at 35 kg, called for a major project redesign, including optimising subsystems shared by payloads, such as power and data storage, and reducing built-in redundancies. The number of star sensors was reduced from four to two; the twelve Reaction Control Systems were reduced to eight and two tanks of 35 litres capacity each used to pressurise fuel were replaced with one of 67 litres capacity. To minimise cost overrun and reduce the delay these major redesigns caused, ISRO chose to dispense with the traditional practice of building three models (structural, engineering and flight) and instead went straight to building a single flight model.[825]

The Moon mission also required building additional ground infrastructure. The Moon at 384,400 km was more than ten times farther than the most distant assets ISRO operated at the time, the satellites in GSO at 36,000 km. Getting a spacecraft to the Moon, entering precise lunar orbit and then operating the spacecraft from Earth were a fresh set of challenges for ISRO. Additional infrastructure required to support Chandrayaan-1 included a new deep space communication facility, a Mission Operations Complex in Bangalore and an Indian Space Science Data Centre to house the large and complex data-set Chandrayaan-1 would return.[826] Multiple fibre optic communication links (2 × 2 Mbps, 14 Mbps and 512 Mbps) were established between the Mission Operations Complex and the Data Centre to support redundancy and real-time backup. ISRO also initiated a programme to build large steerable antennae to meet Chandrayaan-1's communication requirements. An 18-m dish was completed in 2006 and a 32-m antenna by the end of 2007. Initially, the launch of Chandrayaan-1 was scheduled for 9 April 2008. To meet the schedule, ISRO purchased the 18-m antenna from Germany. This antenna alone could have provided the communication link for Chandrayaan-1.[827] However, a 32-m antenna was built jointly by Electronics Corporation of India Limited (ECIL) in Hyderabad, Bhabha Atomic Research Centre (BARC) in Mumbai, various ISRO labs and private Indian industries. Since then, another 11-m antenna has also been added. All three are located at ISRO's Byalalu site, 40 km from Bangalore. ISRO's ISTRAC facility in Bangalore operates all antennae in Byalalu and elsewhere.

The collection of antennae contributed to building IDSN. To ensure international interoperability, IDSN complies with the international standards of CCSDS.[828] The IDSN would be essential to support future ISRO science missions, such as Astrosat, Aditya-L1 and missions to Mars, which at the time were years away. From the outset, the 32-m antenna was built and operated to international standards for communication with spacecraft from other nations in deep space. Communication antennae around the world allow nations to communicate with their spacecraft, even when they are not in the sky over them. These international collaborative events have either been mutual, where no money changed hands or subject to a formal commercial arrangement. The IDSN alone, including the 18-m and 32-m antennae, cost Rs.100 crore ($15 million). The total cost of Chandrayaan-1 including the IDSN, was Rs.386 crore (about $60 million).[829] A WikiLeaks document reveals the US's estimate that the mission cost "USD 89 million (Rs. 514 crore) not including partner country instruments."[830] ISRO is able to recoup some of this investment through commercial use of its 32-m antenna by international agencies.

Journey to the Moon

Of the total 126 spacecraft that have gone from the Earth to the Moon by mid-2016, most have used the Lunar Transfer Trajectory.[831] While US's Saturn 5 could deliver 35,000 kg to the lunar orbit, Chandrayaan-1was launched by a PSLV with a capacity to deliver only 675 kg. PSLV's limited power restricted Chandrayaan-1's maximum science payload to 105 kg and defined the route it would take from the Earth to the Moon. The most fuel-efficient approach involves first an Earth orbit immediately after launch and then gradually stretching that orbit to a highly elliptical one that eventually spans the distance of the Moon. When close enough to be within the gravitational influence of the Moon, the spacecraft would fire its engine and be captured in a lunar orbit.

Date	Event	Delta V (m/s)	Orbit (km)
22/10/08	Launch		254 x 22,932
23/10/08	Earth Bound Orbit 1	344	299 x 37,908
25/10/08	Earth Bound Orbit 2	328	336 x 74,715
26/10/08	Earth Bound Orbit 3	221	347 x 165,015
29/10/08	Earth Bound Orbit 4	77	459 x 266,613
03/11/08	Earth Bound Orbit 5 & Lunar Transfer Trajectory	60	972 x 379,860
05/11/08	TCM while en-route to the Moon	0.8	815 x 378,515
08/11/08	Lunar Orbit Insertion	817	507 x 7,510
09/11/08	Lunar Bound Orbit 1	57	200 x 7,502
10/11/08	Lunar Bound Orbit 2	868	183 x 255
11/11/08	Lunar Bound Orbit 3	31	101 x 255
12/11/08	Lunar Bound Orbit 4	59	101 x 102.8

Table 15-1 Chandrayaan-1Trajectory from the Earth to the Moon

Key research undertaken by R.V. Ramanan in orbital mechanics defined Chandrayaan-1's trajectory from the moment it left Indian soil until lunar orbit insertion. The most fuel-efficient trajectory that Ramanan calculated for the PSLV required a series of five Earth orbits, each with an increasing apogee (going further from the Earth) using Chandrayaan-1's onboard LAM. The initial lunar orbit was highly elliptical, and a series of four LAM burns in lunar orbit were required to achieve the 100-km circular lunar orbit.[832] A circular pole-to-pole 100-km orbit was critical to achieving the spatial resolutions over the lunar poles that the science instruments were designed to capture. Even after the final orbit was achieved, Chandrayaan-1 would not remain precisely in it. Variations in the Moon's density would change Chandrayaan-1's orbit, requiring an orbital correction manoeuvre from ISRO engineers every four weeks or so.[833]

Chandrayaan-1 was launched on PSLV-C11 at 10:19 UT from Sriharikota on 22 October 2008 and released twenty minutes later into a transfer orbit around the Earth at 254 km x 22,932 km with an inclination of 17.9°. Following five Earth orbits, Chandrayaan-1 arrived in lunar orbit just over two weeks later. During its first four days in lunar orbit, Chandrayaan-1 circularised its orbit from 507 km x 7,510 km to 101 km x 102 km through a series of four LAM burns.

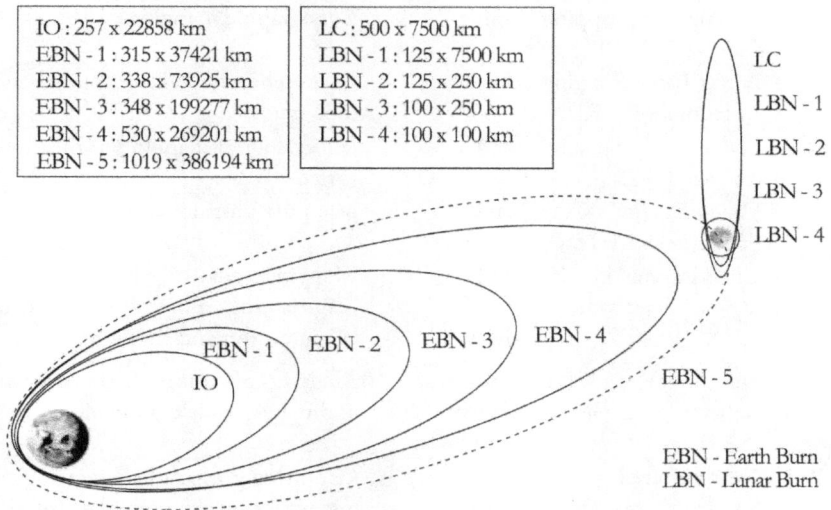

IO : 257 x 22858 km
EBN - 1 : 315 x 37421 km
EBN - 2 : 338 x 73925 km
EBN - 3 : 348 x 199277 km
EBN - 4 : 530 x 269201 km
EBN - 5 : 1019 x 386194 km

LC : 500 x 7500 km
LBN - 1 : 125 x 7500 km
LBN - 2 : 125 x 250 km
LBN - 3 : 100 x 250 km
LBN - 4 : 100 x 100 km

LC
LBN - 1
LBN - 2
LBN - 3
LBN - 4

EBN - 1   EBN - 2   EBN - 3   EBN - 4
IO
EBN - 5

EBN - Earth Burn
LBN - Lunar Burn

Figure 15-3 Chandrayaan-1 Orbit profile. Credit ISRO

Chandrayaan-1's position throughout this period was regularly and frequently monitored by IDSN, as well as NASA's Deep Space Network (DSN). These data sets were used by the Precise Orbit Computation using Accelerometer Data (PROCAD) software that was developed especially for Chandrayaan-1.[834] The spacecraft used four high precision accelerometers to autonomously measure its speed during a LAM burn and to terminate the burn dynamically when the required velocity was reached. Usually, the precise position of a spacecraft is calculated by measurements from tracking stations, which can be many hours after an engine burn. Using PROCAD, ISRO was able to determine Chandrayaan-1's orbit with high precision dynamically at any time, including during an engine burn.[335]

Many nations, one spacecraft

The unmanned lunar landers and rovers of the USSR and the nine Apollo missions of the US had already undertaken extensive scientific work from lunar orbit and on the surface.

	Payload	Origin	Objective
1	Moon Impact Probe (MIP)	India	Vertical analysis of lunar atmosphere
2	Moon Mineralogy Mapper (M3)	US	High-resolution assessment of lunar chemical and mineralogical resources
3	Terrain Mapping Camera (TMC)	India	Topographic mapping
4	Lunar Laser Ranging Instrument (LLRI)	India	Comprehensive 3D mapping and topography
5	Hyper Spectral Imager (HySI)	India	Mineralogical mapping of the lunar surface
6	High Energy X-ray Spectrometer (HEX)	India	Mapping thorium and lead
7	Chandrayaan-1 X-ray Spectrometer (C1XS)	UK but funded by ESA	Mapping relative abundances of magnesium, aluminium, silicon, calcium, iron and titanium
8	Sub keV Atom Reflecting Analyser (SARA)	India, Japan, ESA and Sweden	Analysis of atmospheric content and magnetic field anomalies
9	Near-Infrared Spectrometer (SIR-2)	Germany, ESA, Norway and Poland	Capturing geological, mineralogical and space weathering data of the lunar surface
10	Miniature Synthetic Aperture Radar (Mini-SAR)	US and UK	Measure low energy particles directly from the Sun and those reflected by the lunar surface
11	Radiation Dose Monitor (RADOM)	Bulgaria	Radiation sensor to measure radiation dose at all points along Chandrayaan-1's journey

Table 15-2 Chandrayaan-1 Payload Overview

The Apollo missions had taken twenty-four men to the Moon, twelve of whom visited the surface in six missions. Both the US and the USSR brought back samples of rocks and soil, 382 kg by the US during the six Apollo missions and 326 grams by the USSR with three unmanned sample return missions. In addition, both nations had conducted extensive experiments from lunar orbit. What science could be done by an Indian mission to the Moon that had not already been done by the numerous missions already completed? Chandrayaan-1 was equipped with an array of eleven instruments. ISRO scientists set three key goals for Chandrayaan-1. Deploying scientific instruments not used during the Moon missions of the 1960s and 1970s, Chandrayaan-1 was to (i) prepare a three-dimensional atlas of the entire surface of the Moon with a resolution of 5–10 m., (ii) identify the location and distribution of lunar minerals, light and heavy elements and (iii) search for and map the distribution of water, if it existed.

Two of ISRO's instruments, the TMC and LLRI, were designed to capture data to meet the primary objective of the mission, to prepare a high-resolution three-dimensional map.[836] Although 98% of the lunar surface had already been imaged in the past, only specific areas, such as potential landing sites, had been targeted for high-resolution imaging. TMC was designed to cover the entire lunar surface.[837] Data from three separate instruments, HySI, M3 and SIR-2, were compiled to meet the second objective of the mission, to produce a high-resolution map of the Moon's mineral and chemical composition. HEX and M3 were to detect the presence of water. RADOM, along with TMC, was activated while Chandrayaan-1 was still in Earth orbit. The remaining nine instruments were commissioned gradually between 16 November and 9 December 2008.

The Moon was known to be extremely dry. The Earth's axis of rotation is inclined at 23° to the plane of the orbit, so both poles get some sunlight over a year. The Moon's orbit is inclined at 1.5°, so craters at its poles get almost none. If water existed, it would be trapped and frozen in the deep craters at the lunar poles which are never illuminated by sunlight. Chandrayaan-1's unusual pole-to-pole orbit allowed it to look directly down into these areas during each orbit. If water was present, Chandrayaan-1 had a good chance of detecting it.

Moon Impact Probe

The MIP, at 35 kg, is best considered a separate spacecraft designed to detach from Chandrayaan-1 when it arrived in lunar orbit, descend to, and impact on the surface planting the Indian flag in a predefined location. It was the result of a significant mission redesign introduced at short notice and produced by ISRO at VSSC, Trivandrum. MIP had three key objectives, close-up observation of the Moon, testing technology for navigating to a predefined point on the lunar surface and preparing for future soft-landing missions.

MIP carried three payloads, a Moon imaging system (a camera), a radar altimeter and a neutral mass spectrometer to detect and quantify the chemical constituents in the tenuous lunar atmosphere during the journey from orbit to the surface. The radar altimeter used radio signals 100 times a second to measure altitude during the descent. Trial runs on Earth from aircraft demonstrated an accuracy of between 2 and 4 m. Chandrayaan-1 arrived in lunar orbit on 8 November, and the orbit was circularised by 12 November. MIP separated from Chandrayaan-1 two days later, on 14 November at 20:06 IST; the date was selected to coincide with the birth anniversary of India's first Prime Minister Jawaharlal Nehru. MIP then started its journey to the surface, impacting at 1.75 km/s near the lunar south pole (latitude 89° south), close to crater Shackleton at 20:31 IST. The impact site was named Jawahar Sthal (Jawahar site).

Figure 15-4 Moon Impact Probe. Credit ISRO

During the 25-minute descent, the onboard instruments relayed data to Earth via Chandrayaan-1. There was an inherent risk of damage to Chandrayaan-1's instruments from the exhaust of MIP's engine when it detached and left Chandrayaan-1. To mitigate this risk, a special non-degassing propellant based on Viton rubber was deployed for the tiny (8 g) de-orbit motor and for the two spin motors.[838] During its journey from 100-km orbit to the lunar surface, MIP's neutral mass spectrometer called Chandra's Altitudinal Composition Explorer (CHACE) detected evidence of water and carbon dioxide, along with other trace molecules.

Moon Mineralogy Mapper

The M3 was a state-of-the-art high-resolution imaging spectrometer jointly designed and developed by NASA's Jet Propulsion Laboratory (JPL) and Brown University. A key instrument with a mass of 8.2 kg, it was designed to help map properties of the Moon from polar orbit to improve the understanding of the evolutionary processes responsible for the Moon's current state.

M3 operated in two modes, target or global. In the target mode, M3 used its onboard telescope to look in detail at narrow strips of the Moon from its 100-km orbit.[839] In the global mode, M3 did not engage the telescope and imaged the whole Moon, albeit at a lower resolution. M3 split moonlight (that is, sunlight reflected by the Moon) with a spectrometer and examined the light at various wavelengths using a solid-state sensor. M3 data measurements of the absorption features of rocks and minerals were designed to help understand lunar geological processes and facilitate a high-resolution assessment of lunar chemical and minerals present.

A payload operations centre was developed by the M3 team to process M3 data. As agreed in the NASA/ISRO MoU for the M3 instrument, NASA delivered a replica of the payload operations centre to the Indian Space Science Data Centre to help process raw science data.[840] M3 was scheduled to make four two-month observational slots over the planned two-year mission. During these active slots, Chandrayaan-1 was to capture and transmit M3 data for two hours each day. However, issues of thermal control and loss of a star tracker resulted in operational challenges that required modifications to the initial schedule. M3's commissioning phase was extended to 2009. On 31 January 2009, M3 was operated in its global mode for several discrete periods.[841] M3 made observations periodically, rather than the planned 48-minute period, during each of the twelve orbits.[842] Most of the observations were carried out under less-than-favourable viewing conditions in low resolution using global mode. In mid-May 2009, the spacecraft lost the second-star tracker, and ISRO decided to raise the orbit to 200 km from the existing 100 km.[843] Between 18 November 2008 and 16 August 2009, 958 images were recorded by M3 during five observational periods. M3's discovery of unusual rock types (Mg spinel anorthosite) and a kilometre-sized structure (a crystalline anorthosite exposure) provided enough evidence for the existence of a global magma ocean in the Moon's early history.[844] M3's primary discovery was evidence of water (and Hydroxyl) molecules on surface of the Moon. The evidence was stronger at higher latitudes especially at the poles where the temperature was lower compared to that at the lunar equator. This water is not in pools of water on the surface but traces of water molecular and hydroxyl molecules. However, water that is present deep inside a crater could be present as ice frozen since formation. This detection has been confirmed with data from CHACE and other spacecraft.

Terrain Mapping Camera

TMC was developed by ISRO at the SAC in Ahmedabad. At almost 7 kg, the stereo colour camera was designed to capture high-resolution colour images at 5 m resolution. TMC was equipped with three viewing windows, looking straight down from orbit, forward in the direction of travel (by an angle of 25°) or behind (by an angle of 25°). It used a digital linear Active Pixel Sensor (APS), which an enhanced form of CCD producing data directly in a digital output. Images from the TMC were used to compile a 3-D atlas of the Moon's surface, both its near and far sides. The first images taken by the TMC were not of the Moon but the Earth. A week after launch, while still in Earth orbit, the TMC was commanded to image the Earth as part of the testing and commissioning activity.[845]

TMC had imaged 50% of the Moon, mostly around the key Polar Regions, when the mission came to a premature end. The high-resolution images from the TMC recorded detailed views of the central peak of the crater Tycho and combined with HYSI data confirmed the existence of crystalline feldspars in the lunar highlands. TMC also recorded images of a well-preserved lava tube near the landing site of Apollo 15. Data from the TMC supported the theory that the surface of the Moon was covered by an ocean of magma for the first billion years of its 4.5 billion-year-long history.

Lunar Laser Ranging Instrument

LLRI was built by ISRO's Laboratory for Electro-Optics Systems in Bangalore. At about 10 kg, LLRI was one of the heavier instruments in the science payload. It was used to measure the lunar terrain (heights of mountains and depths of valleys) to a resolution of 5 m. Activated for the first time on 16 November; it was designed to systematically cover the day and night side of the Moon. LLRI directed pulses of infrared laser light down to the lunar surface ten times a second. The faint reflections were collected by its 7-cm optical telescope and directed onto a sensitive photodetector. LLRI data was combined with TMC data to produce a topographical map of the lunar polar regions. LLRI allowed scientists to study the morphology of the lunar surface features and to help provide a density distribution of the crust.

Hyper Spectral Imager

HySI, an optical camera using a solid-state image sensor, was designed to acquire stereoscopic data to produce a mineralogical map of the lunar surface. It had a mass of 4 kg and was developed by ISRO's SAC, Ahmedabad. The Active Pixel Sensor used in HySI was of the same type used in TMC. With a spatial resolution of 80 m, it had the ability to image the Moon in 20 km wide strips from an orbit of 100 km.

HySI targeted deep craters and their central peaks where the material from the lower crust or upper mantle could be observed and analysed. It produced images in a narrow range of wavelengths in 32 contiguous bands that helped produce a mineralogical map of the Moon by combining its data with that from other instruments, especially C1XS, M3 and SIR-2. HySI data was also used to confirm the discovery of a new spinel-rich rock type on the lunar far side composed of magnesium and aluminium minerals.

High Energy X-Ray Spectrometer (HEX)

HEX was developed by ISRO's SAC in Ahmedabad, with contributions from ESA. Although referred to as an X-ray Spectrometer, the higher energies of the detected radiation were more consistent with gamma-rays. HEX was the first instrument designed to capture data on high-energy X-rays in the lunar environment, and at 16kg, it was also the heaviest instrument on-board.

Unlike the Earth, the Moon does not have a significant magnetic field or an atmosphere. Consequently, high-energy radiation and cosmic rays from the Sun and the cosmos can reach the lunar surface and penetrate up to a meter below.[846] This radiation interacts with some constituents that make up the rocks and soil, raising their energy level (of the constituent atoms and molecules) to an excited state. Such excited states are not stable and return to their ground state by emitting X-rays or gamma-rays. It is this radiation that HEX was designed to detect. The precise energies of these X-rays and gamma-rays indicate the chemical composition of the lunar surface. On Earth, when the Sun heats up the equator, volatile material, such as water, moves towards higher latitudes. HEX was designed to detect and measure the transport of such volatile material from the high temperature lunar equatorial region (more than 100°C) towards the lunar poles (less than 0°C).

Radioactive elements, such as uranium, thorium, radium and lead, as well as potassium, which have a very long half-life, provide the Moon with an internal heat source. By detecting the gamma-rays from the decay of these heavy elements and combining it with data from TMC and HySI, HEX was to map the distribution of these elements across the lunar landscape. However, immediately after arrival in lunar orbit, Chandrayaan-1 experienced thermal control problems that limited the operations of HEX and other instruments. HEX was not able to operate between noon and midnight (lunar time) because its sensitive detector could not be adequately cooled.[847] However, it did collect data on the high-energy X-ray background from the Moon, averaging over a significant portion of the lunar surface.[848]

Chandrayaan-1 X-ray Spectrometer

C1XS, at 5.2 kg, was a sophisticated version of an instrument that ISRO had initially designed in-house. Funded by ESA, C1XS was designed and developed at the Rutherford Appleton Laboratory in the UK. Apart from ISRO, additional support came from Brunel University in the UK and the French National Centre for Scientific Research in Toulouse. C1XS was similar to HEX but sensitive to lower energy X-rays. It used a new, enhanced version of the CCD, a swept charge device that offered a higher resolution and could operate at room temperature. Data from C1XS was designed to help quantify the relative abundance and distribution of elements, such as magnesium, aluminium, silicon, calcium, iron and titanium. With this data, an estimate of the chemical composition of the Moon could be mapped to a resolution of around 25 km. Sodium was detected on the Moon for the first time by C1XS.[849]

Every 11 years, the Sun goes through a cycle of maximum and minimum solar activity, including sunspots, solar flares and coronal mass ejections. Though random, the likelihood of solar flares increases and decreases within a solar cycle.[850] Chandrayaan-1was launched in 2008 when the cycle was at a minimum. However, two months after it arrived in lunar orbit, a weak solar flare initiated a solar wind, and on 12 December 2008, C1XS picked up X-rays generated on the surface of the Moon by this solar wind. It was a weak signal, but Chandrayaan-1 was there to see it.[851]

Sub keV Atom Reflecting Analyser

SARA was also a product of international collaboration. It was jointly designed and developed for the ESA by the Institute of Space Physics, Kiruna, Sweden; ISRO's VSSC; Japanese Aerospace Exploration Agency (JAXA) in Tokyo and the University of Bern, Switzerland. At 3.5 kg, SARA consisted of three distinct sensors that between them detected and measured low energy neutral atoms generated by particles emanating from the Sun impacting the top-most layer of the lunar surface. In the absence of a magnetic field and an atmosphere, solar radiation can reach and interact with the material of the lunar surface completely unhindered. SARA could measure the protons reaching the Moon from the Sun and those that were reflected. It determined that 20% of the protons absorbed an electron from the lunar surface to create a hydrogen atom. In places, where oxygen atoms were present, they combined to form water. In October 2009, SARA repeated an observation made by CHACE a month earlier that water was present on the Moon.[852]

Near-Infrared Spectrometer (SIR-2)

SIR-2 was an instrument based on SIR-1 carried by the European SMART-1 spacecraft to the Moon in 2003. It was an enhanced version of SIR-1 and jointly developed and funded by the ESA; Max Planck Institute for Solar System Research in Göttingen, Germany; the University of Bergen (UiB), Norway; and the Polish Academy of Science, Warsaw. SIR-2 was one of the smallest and lightest (2.3 kg) of the Chandrayaan-1 payload.

SIR-2 was designed to collect and analyse sunlight reflected by the Moon's surface. The collected light was directed through an optical fibre to a spectroscope to understand geological and mineralogical details, space weathering and the vertical distribution of crustal material on the lunar surface. By combining SIR-2 data with that from HySI, it was possible to identify and locate minerals, such as olivine and pyroxene, for potential future exploitation of lunar resources.

Miniature Synthetic Aperture Radar

Mini-SAR was also a product of international collaboration. Interestingly, it had a non-civilian contributor. The instrument was led by the US's Naval Air Warfare Center with support from NASA and Johns Hopkins University in the US. Mini-SAR was built in the UK by three companies, Raytheon, BAE Systems and Surrey Satellite Technology. At 8.77 kg, Mini-SAR targeted one of the mission's key objectives of discovering the presence of water on the Moon. Measurements in 1994 by the NASA Clementine mission had raised the possibility of water within the Shackleton crater at the southern lunar pole. Mini-SAR was to help provide evidence to confirm that observation.

By transmitting and subsequently receiving reflected radio waves, Mini-SAR was to map both lunar poles for water ice by measuring properties of reflectance, roughness and polarisation. In the absence of sunlight in the craters at the poles, deposits of water ice could survive for millions of years. During each sweep over the poles, Mini-SAR collected data in 8-km wide swaths with a resolution of 150 m. It was switched on for ten-minute intervals every orbit as it passed over the poles, above 80° latitudes. It was switched off at other times.

The NASA spacecraft Lunar Reconnaissance Orbiter (LRO) in lunar orbit at the time also carried a Mini-SAR instrument. A joint experiment was scheduled for 20 August 2009 to help detect surface (frozen) water. LRO Mini-SAR was configured to receive radio signals transmitted by Chandrayaan-1 Mini-SAR.[853] It was a novel and technically challenging experiment as both spacecraft were to target the crater Erlanger, only 10 km in diameter while moving at 1.6 km/s. The experiment was conducted as planned with both spacecraft looking at crater Erlanger for at least 35 seconds, but subsequent analysis indicated that Chandrayaan-1 had lost the capability to point with the precision required due to hardware failure. A further attempt was being planned a week later when further hardware deterioration ended the Chandrayaan-1 mission.

Radiation Dose Monitor

RADOM was developed by the Solar Terrestrial Influences Laboratory of the Bulgarian Academy of Sciences in Sofia, Bulgaria. At about 0.1 kg, it was the smallest and lightest of all the eleven science instruments onboard Chandrayaan-1. RADOM was a miniature dosimeter-spectrometer designed to measure energetic radiation present in the environment through which Chandrayaan-1 travelled and operated. It was the first instrument to be switched on, just two hours after launch, and it collected data while in Earth orbit, en-route to the Moon, during the initial series of lunar orbits and the designated 100-km and subsequent 200-km lunar orbits.

RADOM collected data throughout the Chandrayaan-1mission, from 22 October 2008 to 31 August 2009. Measurements of radiation intensity around the Earth were in line with the measurements of similar experiments on-board the ISS. The radiation that RADOM detected came from various sources, cosmic radiation from the cosmos reflected by the Earth or Moon, the steady stream of radiation from the Sun, as well as instances of increased radiation from high solar activity, such as solar flares. On 13 March 2009, RADOM recorded increased radiation resulting from a small magnetic storm on the Sun.[854]

Science from Chandrayaan-1

The primary mission objectives of Chandrayaan-1were to create a high-resolution three-dimensional atlas of both the near and far side of the Moon, including the Polar Regions, and to conduct a chemical and mineralogical mapping of the entire lunar surface. Before the termination of the mission, the M3 instruments had covered 90% of the lunar surface, but the TMC had managed only around 50%, though most of that included the important Polar Regions not covered in detail before.[855] The relative concentrations and locations of aluminium, magnesium and silicon, along with iron-bearing minerals, such as pyroxene, were integrated into a 3D map by combining data from TMC, LLRI and C1XS.

Chandrayaan-1's data revised three key assumptions about the Moon:

- The Moon was not dry. Data, primarily from M3 but also MIP and Mini-SAR, recorded evidence of water and hydroxyl molecules with quantities increasing towards the poles. On 14 November at 20:31 IST, MIP impacted in the lunar south pole. During its journey from the 100-km orbit to the surface, MIP collected images and data. The data from the CHACE instrument independently verified M3's detection of water in the region of lunar South Pole through which it descended.

- Solar wind protons were not absorbed by the Moon. The SARA instrument measured that around 20% of the solar wind protons arriving on the Moon were reflected back into space as neutral energetic hydrogen atoms. These protons were integral to the process of making water on the Moon.

- The Moon was geologically active. Hydrated magma indicated episodic explosive volcanic events. M3 data revealed spectra of magma flows around Lowell and Bullialdus craters. The TMC and HySI instruments identified uncollapsed lava tubes, which could be used in the future for human habitation.

The planned two-year mission was declared prematurely over at 01:30 IST on 29 August 2009 after 312 days of operation when Chandrayaan-1 finally lost its ability to transmit. The early termination of the mission was caused by the spacecraft's inability to regulate its internal temperature. On arrival in lunar orbit, Chandrayaan-1 found itself in a temperature regime of +100°C on the day side and -100°C on the night side of the Moon. Designed to operate at 40°C or lower, Chandrayaan-1 was frequently in excess of 50°C, and at one stage, 80°C.

The excessive radiant heat from the lunar surface was higher than what the ISRO engineers had anticipated. A series of innovative workarounds were employed that included re-orientating the spacecraft, operating some of the instruments only on the night side of the orbit or at times switching them off altogether to reduce internally generated heat. To move to a lower temperature environment, on 21 May 2009, ISRO decided to increase the orbit from a 100-km to a 200-km altitude.[856] This decision reduced the spatial resolution of the data collected by some of the instruments.

Figure 15-5 Polar Region of the Lunar Surface Captured by Chandrayaan-1. 15 November 2008. Credit ISRO

In addition to resolving thermal problems, the engineers also came up with innovative solutions using the onboard gyroscopes to maintain spacecraft attitude control when initially one, and then both, star sensors failed. Intervention and monitoring on a 24/7 basis by ISRO engineers was required to allow all onboard instruments to operate and continue to collect data. The loss of the mission was eventually identified to be a hardware failure of five individual DC/DC converter components used to convert and send telemetry data from the onboard systems and instruments to Earth. The manufacturer of the component "expected operating lives that exceed ten years" on unmanned spacecraft in a space environment.[857] One report suggested that ISRO had "underestimated temperatures around the Moon, so the probe had been overheating for months."[858] Despite the premature end, a probe committee reviewing the mission formally concluded in its report that the Chandrayaan-1mission was "quite successful". During the 3,400 orbits, Chandrayaan-1 collected over 7,000 images, including images of the Earth from an Earth orbit of 7,000 km and a full Earth image from lunar orbit (400,000 km) using TMC. TMC and HySI had also covered 50% of the lunar surface and M3 90%.

ISRO concluded that Chandrayaan-1 had completed 95% of its mission objectives.[859] The mission's success was also recognised internationally with high profile awards from the International Lunar Exploration Working Group, the American Institute of Aeronautics and the Astronautics and National Space Society. Apart from meeting its chief scientific objectives, Chandrayaan-1 will be remembered for three other key achievements. The strong international collaborative spirit it engendered, the wealth of experience many young ISRO engineers gained in managing and operating a spacecraft in lunar orbit and the recognition that ISRO as an organisation had the capability to successfully undertake complex space science missions beyond Earth orbit.

ISRO engineers calculated that the Moon's tenuous atmosphere would decay Chandrayaan-1's orbit and it, like the MIP, would crash land on the Moon by around 2012. But it did not. Surprisingly, Chandrayaan-1 no longer operational was discovered using Earth based radar, still orbiting the Moon in 2017.[860]

Chandrayaan-2: Journey to the Lunar Surface

On 12 November 2007, a year before Chandrayaan-1 arrived at the Moon, ISRO signed an agreement with Russia for Chandrayaan-2, a joint mission to the Moon in 2011 or 2012.[861] It would have three elements: an orbiter, lander and rover. Russia had agreed to contribute the lander and rover. India would supply the PSLV launch vehicle and the orbiter. Subsequently, the launch date slipped to 2013, and India agreed to provide the rover, as well as the orbiter. PSLV was replaced by GSLV. Russia was expected to provide the lander which would deliver the rover to the lunar surface.

However, in November 2011, a Russian rocket Zenit-2SB carrying two spacecraft to Mars, one Russian and the other Chinese, failed to leave Earth orbit, with the loss of both spacecraft. The resulting agency-wide review forced Roscosmos to first delay and then withdraw from the planned joint mission with India to the Moon.[862] Initially, ISRO sought to find alternative partners in NASA or ESA but then decided to go solo making it a wholly Indian mission.

The provisional mission design consists of a lander, at 1.25 tonnes with four instruments and a 20-kg rover with two instruments. Although small, the rover will have about twice the mass of NASA's rover Sojourner, which operated on the surface of Mars for about three months in 1997. The power and range of the transmitters on the rover and lander are not sufficiently strong to communicate directly with the Earth, so the orbiter is used as a relay. The orbiter is configured to carry 5 instruments. Each of the three components, orbiter, lander and rover, has a distinct collection of science packages designed to make the best use of its local environment. Unlike Chandrayaan-1, Chandrayaan-2 does not involve international collaboration.

Chandrayaan-2's journey from the Earth to the Moon will be similar to that of Chandrayaan-1. Following launch, Chandrayaan-2 will first orbit the Earth in increasingly larger orbits through eight Earth bound orbits extending the orbit each time. The ninth burn (Trans Lunar Injection) will extend its orbit to intersect with that of the Moon's orbit. Once in lunar orbit, a series of four lunar bound manoeuvres will result in the desired 100 x 100 km lunar orbit. It is from here that the lander containing the rover will separate from the orbiter on its journey for a soft landing on the surface of the Moon. The Chandrayaan-2 orbiter is designed for a mission of one year, and the lander/rover are expected to operate for about two weeks. The temperature on the lunar surface ranges from a maximum of about 100°C at midday to a minimum of -180°C at night. A day on the Moon lasts 28 earth days. Chandrayaan-2 is not designed to survive the extreme range of temperatures it will experience in one lunar day.

There was a proposal to provide Chandrayaan-2 with a nuclear power source to sustain it through the 14 days of a lunar night when the solar panels receive no sunlight, but ISRO is no longer pursuing that option.[863] Testing on several elements of the Chandrayaan-2 mission, including soft landing on the surface, the lunar rover exiting the lander and the rover traversing the lunar surface has been completed. These tests were conducted in the open air within the perimeter of Bangalore's HAL airport and ISRO Satellite Integration and Test Establishment, which is a part of ISAC also in

Bangalore. The lander descent engine has also undergone testing.

Orbiter	Origin	Purpose
TMC-2	SAC, Ahmedabad	Generate a map of the lunar geology and mineralogy. Terrain Mapping Camera-2
Imaging Infra-Red Spectrometer	SAC, Ahmedabad	Locate, identify and study the minerals and water molecules on the lunar surface from orbit.
Synthetic Aperture Radar	SAC, Ahmedabad	Locate and identify the chemical constituents in the first few metres under the lunar surface. This will include water and water ice including in areas that are not in permanent shadow.
CHACE-2	Space Physics Laboratory, Thiruvananthapuram	Study the lunar exosphere (100 km). Chandra's Altitudinal Composition Explorer (CHACE-2, a neutral mass spectrometer)
X-Ray monitor	Spectrometer from ISAC and X-ray Monitor from PRL	Large-scale mapping of specific elements across the lunar surface. Chandrayaan-2 Large Area Soft X-ray Spectrometer and Solar X-ray Monitor
Lander	Origin	Purpose
Seismometer	LEOS	Monitor moonquakes. Laboratory for Electro Optics Systems, Bangalore
Thermal Probe	Space Physics Laboratory, Thiruvananthapuram	Measure temperature at the landing site for the duration of the mission.
Langmuir Probe	VSSC	Measure electrons released by the interaction between high energy solar radiation and lunar surface particles.
Rover	Origin	Purpose
Laser Induced Breakdown Spectroscopy	LEOS	Qualitative and quantitative elemental analysis of lunar surface material.
Alpha Particle Induced X-ray Spectroscope	PRL, Ahmedabad	Identify the elemental composition of the lunar surface material. The spectroscope carries a small source of alpha particles to which the source material is exposed. A measurement of the resulting X-rays helps to identify the source.

Table 15-3 Overview of Chandrayaan-2 Science Payload. Credit ISRO

Chandrayaan-1 with a mass of 1.3 tonnes was launched using a PSLV-XL. Chandrayaan-2's mass is expected to be twice that so will be launched using a GSLV Mk2. By March 2015, ISRO had used a third of the allocated Rs. 603 crore ($90 million) for the mission,[864] and in February 2016, the ISRO chairman asserted "ISRO is ready for Chandrayaan-2" for launch in 2018[865] or early 2018.[866]

Orbiter

The orbiter is designed to operate for a year in a 100-km lunar orbit. It has five instruments, all of which have been produced by ISRO. TMC-2 will continue the work of TMC-1 and fill in the gaps from the Chandrayaan-1 mission to provide a full 3-D map of the lunar surface. The Imaging Infra-Red Spectrometer will scan the lunar surface using a wider wavelength range (extended from 3 microns in Chandrayaan-1 to 5 microns in Chandrayaan-2) to once again map the distribution of minerals, water molecules and hydroxyl present on the lunar surface.[867]

A SAR operating in a dual frequency is expected to detect further evidence in support of water ice located in the first few tens of metres below the lunar surface and in the regions permanently in shadow. CHACE-2 will look at the composition of the tenuous lunar atmosphere during descent. This is similar to the instrument carried aboard the MIP that detached from Chandrayaan-1and captured data as it descended to its lunar surface impact.

Data from a joint instrument, Chandrayaan-2 Large Area Soft X-ray Spectrometer and Solar X-ray Monitor, will be used to map the distribution and relative concentration of elements, such as aluminium, silicon, calcium and magnesium, on the lunar surface. Combining this data with that collected by C1XS on Chandrayaan-1 will continue building a complete picture of the chemical constituents of lunar surface material.

Lander

The lander will be ISRO's first attempt to soft land on an extra-terrestrial body. It will be equipped with a high-resolution descent camera, a throttleable liquid fuel main engine with attitude control thrusters and a high degree of automation to ensure a soft landing in an obstacle-free environment. Given the time delay between Moon-Earth communication, the lander is designed with built-in autonomy to land using real-time data from the altimeter and descent camera without intervention from Earth. Following separation, an initial de-orbit burn will bring the lander down from 100 km to a 15-km altitude. A further series of engine burns will bring the lander above the landing site with zero-horizontal velocity. From this point, the lander will descend vertically to the surface through another series of engine burns until it is at an altitude of 4 m above the surface. At 4 m, the engine will cut-off, and the lander will free fall to soft landing. In the 1/6 lunar gravity this is not a hard landing as it would be from 4 m on Earth.[868] Since the Moon is devoid of any substantial atmosphere, parachutes do not work. The engine has been subjected to high-altitude tests lasting over 500 seconds. Following a successful landing, the rover will exit from the lander and explore the immediate area of the landing zone. The lander will remain stationary for the duration of the two-week mission.

The lander's onboard science payload consists of a seismometer for studying moonquakes, a thermal probe (thermometer) to measure the temperature at the landing site over the duration of the mission and two experiments (Langmuir Probe and Radio Occultation) to detect electrons emanating from a few centimetres above the lunar surface. On Earth, the high-energy radiation from the Sun is deflected by the Earth's magnetic field or absorbed by its atmosphere. In the absence of both on the Moon, this high-energy radiation from the Sun reaches its surface, and the interaction of the radiation with particles of dust causes atoms to release electrons resulting in a thin layer of plasma over the sunlit side of the Moon.

Two experiments on the lander will measure the Total Electron Count (TEC) and its variation in density and temperature. The Langmuir probe on the lander will directly collect these data and transmit it from the lander to the orbiter. The radio communication link itself can also be used to infer TEC. The Orbiter will orbit the Moon once every hour. As it rises and sets over the horizon, the radio signals will travel through this layer of plasma. The variation in the intensity of the signals received by the orbiter when it is close to the lunar horizon will help to measure the density of this plasma.[869]

Rover

At 20 kg, the rover has a limited payload. The onboard cameras, antennae and communication systems are powered by a battery, which is charged by a single solar panel. It communicates its data to Earth via the orbiter a 100 km above and not the lander a few metres away. It has six wheels controlled by 10 motors guided by a pair of stereoscopic cameras providing a 3D view of its environment. To test the traction of its wheels and driving performance on Earth, a helium balloon is attached to the rover to simulate the reduced lunar gravity. The rover's science payload consists of two instruments. A Laser Induced Breakdown Spectroscope will look at the same plasma that the Langmuir probe on the lander will examine.

The second instrument, Alpha Particle Induced X-ray Spectroscope, will also identify chemical compositions but use a radioactive sample, typically Curium 244 that emits Alpha particles, X-rays and photons. The Alpha particles emitted are deflected by the atomic nuclei in the lunar dust. By measuring this deflection, the mass of the deflecting nuclei and its chemical composition can be determined. These Alpha Particle Induced X-Ray Spectroscopes are small, lightweight instruments that have been used to collect data on the Moon, Mars and a comet.

Figure 15-6 Chandrayaan-2 Rover. Credit ISRO

Both the rover and the lander will communicate with the Earth via the orbiter, which will be contactable above the horizon for about 30 minutes during each orbit. Chandrayaan-2 is scheduled for a launch sometime in early 2018.

Why India Went to Mars

India went to Mars in 2013 because both Japan and China had tried and failed. If India could succeed where Japan and China had not, it would be next after the US, USSR/Russia and ESA to orbit Mars. India had published its intensions for the exploration of Mars, the solar system and human spaceflight but the decision to go to Mars in 2013 was in response to external events. While India's journey to the Moon in 2008 was planned and matured over time, the mission to Mars was more opportunistic and finalised in haste. India always had plans for a mission to Mars, the decision to go in 2013 was in response to the failure of China's mission to Mars in

2011. Publicly, ISRO had declared that there "is no race with anybody" but later conceded that the decision to go to Mars was taken following the failure of the Chinese mission.[870]

Orbital mechanics dictate that the opportunity for a journey to Mars comes around only once every two years. After 2011, the next opportunity would be the autumn of 2013. Neither Japan or China could attempt a Mars mission for 2013 but India could. Japan had unsuccessfully attempted to fly its spacecraft Naomi to Mars in 1999, and although eventually, it got to Mars in 2003, Nozomi failed to enter the Martian orbit and flew past instead. Exactly when ISRO began seriously contemplating a mission to Mars is unclear. Ajey Lele, a Senior Fellow at the Institute for Defence Studies and Analyses in New Delhi, considers that this could have been an outcome of the 2006 meeting of leading Indian scientists to discuss the future of India's space programme.[871] That was the time when India's first mission to the Moon was taking shape. Through this scientific collaboration, India wanted the US to start to ease its sanctions on high technology. In March 2006, India and the US signed a deal that paved the way for scientific instruments from the US to be installed on Chandrayaan-1. In media reports of that agreement, the NASA chief stated, "We are also looking at sharing of data for Earth sciences and Earth resources and broader scientific cooperation in exploring beyond the Earth."[872]

By November 2008, ISRO's first Moon mission Chandrayaan-1 had been in lunar orbit for two weeks and had successfully deployed its MIP to the surface. Riding high on this success, ISRO's then chairman Madhavan Nair publicly announced that "the study for Mars exploration has already started."[873] During the 2010 Indian Science Congress, Nair confirmed India's plans for a Mars mission to a visiting journalist from the UK.[874] During the same Congress, Abdul Kalam, who had been instrumental in incorporating the MIP on Chandrayaan-1, raised the idea of Mars as a mission objective.[875] By the time the Chandrayaan-1 mission came to a premature end in 2009, India may not have had a specific timeline, but it did have a firm conviction that Mars was the next destination.

Prior to the failure of the Chinese space mission to Mars in November 2011, China had chalked up a series of successes in space. It successfully completed three human spaceflights, including one with three astronauts; delivered a module for its space station to orbit and placed its Chang'e 1 spacecraft in lunar orbit in October 2007, a year before Chandrayaan-1. Although its first Mars mission failed in 2011, China was not culpable for the failure. Its spacecraft, Yinghuo-1, was designed to enter Martian orbit after hitching a ride on an ambitious Russian space mission called Phobos-

Grunt. This Russian spacecraft was to land on the Martian moon Phobos, collect a sample of material from the surface of Phobos and return it to Earth. Russia signed an agreement with China that Yinghuo-1 would be launched on the same rocket as Phobos-Grunt. However, neither Phobos-Grunt nor Yinghuo-1 left Earth orbit following a launch vehicle failure. The fourth stage of the launch vehicle, along with both spacecraft, disintegrated during Earth re-entry several weeks after launch.[876] Following this failure in 2011, China could not have readied another Mars mission for November 2013, but India could.

Further, while India's PSLV-XL launcher could just about get a spacecraft to Mars in 2013, for the next two opportunities in 2016 and 2018, an additional velocity of at least 380 m/s would be required compared to 2013. That was just outside the reach of the PSLV-XL.[877] So, India had to go in 2013 or wait until 2020. India saw 2013 as the best opportunity to get to Mars before China and took it.

From Sriharikota to Mars

India's first Mars mission called Mars Orbiter Mission (MOM) or Mangalyaan (in Sanskrit, Mangala = Mars and Yaan = vehicle) was launched at 14:38 IST on 5 November 2013, not towards Mars, but a 50-km cubic volume of space near where Mars would be at 07:27 IST on 24 September 2014.[878] Not only was Mars a target moving around the Sun at 24 km/s, but MOM would be launched from Earth moving at 30 km/s around the Sun. In addition, MOM's launch pad at Sriharikota) was moving at about 0.5 km/s (Earth's rotation). MOM would have to hit that small target after a journey of 667 million kilometres 300 days after launch. Timing was critical. If MOM arrived a little earlier or later, it would either crash into Mars or pass it by. Once it arrived at the precise time and place, there was one more critical step; MOM had to fire its engine to slow down to enter orbit around Mars, a key manoeuvre that ISRO had not conducted before.

On 24 September, everything went to plan. MOM arrived at precisely the right spot, its engine fired, and it entered the planned highly elliptical orbit of 421 by 77,000 km. With that, India became the fourth space agency to arrive at Mars, and it had done it on its first attempt. The Indian Prime Minister was present at Sriharikota watching MOM arrive and embraced the ISRO chairman on live TV in celebration. In some ways, India's arrival at Mars was tantamount to a rebirth of the Indian Space Programme. With the widespread national and international press coverage that followed, many within India, as well as outside, realised for the first time that India had an active and effective space programme. As part of the planning process for MOM, ISRO had studied historical Mars missions, particularly the 29 missions that had failed of the total 55 until then. Each of the 29 failures were "carefully addressed during the preparation for MOM."[879] ISRO had been designing, building, launching and operating satellites for over three decades. On the face of it, MOM was yet another mission to put a satellite in orbit, further away but conceptually the same. In practice, building and sending a satellite to Mars introduced a host of new engineering challenges.

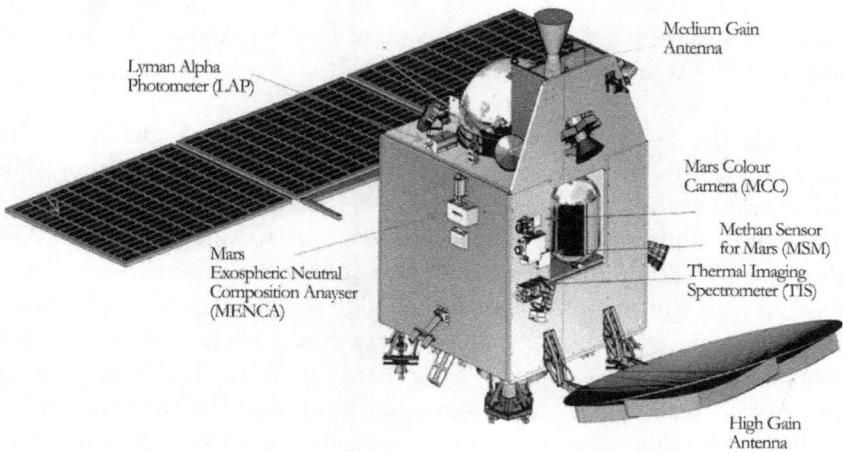

Figure 15-7 Mars Orbiter Mission and Its Science Payload. Credit Adapted from ISRO

Although launched on 5 November, MOM remained in Earth orbit until 1 December before embarking on a nine-month journey to Mars. During this extended period, MOM was exposed to potential harm from the intense radiation environment of space as it crossed the radiation fields around the Earth (Van Allen belt) 39 times. Japan's attempt to reach Mars in 1999 with its Nozomi spacecraft had failed in part because of radiation damage to some of the control systems.[880] MOM's radiation shielding was designed for protection against an accumulated dose of 9 krads for packages inside the cuboid and 15 krads for packages mounted on the outside.[881]

The vast Earth-Mars distance generated two additional challenges. While ISRO was familiar with operating satellites in Earth orbit at 36,000 km, Mars-Earth distance varied between about 50 million km and 400 million km. The round-trip communication interval between Earth and MOM could take up to 42 minutes (21 minutes each way at 400 million km or 5 minutes at 50 million km). ISRO not only had to develop the infrastructure to communicate over this vast distance, but it also had to build in additional level of autonomy, given the time lag, MOM needed to "think and do" by itself.[882]

The solar panel was MOM's only source of electrical power. Mars orbits the Sun at a distance that is about 1.5 times that of the Earth, where the intensity of the Sun is only 42% compared to that at the Earth-Sun distance.[883] The battery and efficiency of its solar panel had to be designed to operate in that context. While the battery used was the same as in Chandrayaan-1, a single 36 Ah lithium-ion battery, MOM was fitted with a three-panel solar array (each panel 1.8 m by 1.4 m) generating 840 W instead of a single panel solar array required for Chandrayaan-1 which had generated 750 W in Lunar orbit.

To catch the 2013 Earth-to-Mars launch window (21 October – 19 November), ISRO worked to a hectic schedule with no contingency. The launch window is the period when a journey from Earth to Mars can take place with minimum energy. This window appears once every 780 days when the orbits of Mars and Earth line up so that the angle between Earth, Mars and the Sun is 40°. This configuration is necessary for a Hohmann Transfer Orbit, and all spacecraft that have ever gone to Mars from Earth have used it. MOM's journey from Earth to Mars had three phases, Geocentric, Heliocentric and Martian, determined by which body is producing the dominant gravitational influence, the Earth, Sun or Mars. The Geocentric phase, where MOM was in the Earth's gravitational field, lasted until it had travelled 918,347 km from the Earth. It entered the Martian phase once it got closer than 573,473 km to Mars. In between, MOM was predominantly under the Sun's gravitational force, and thus in the heliocentric phase. The MOM spacecraft design was a modified version of that used for Chandrayaan-1 which in turn was based on the IRS/INSAT bus. ISRO chose a design with proven reliability given the tight timeline. On 21 September 2012, just a month after the formal announcement, ISAC in Bangalore took delivery of the structure of the MOM spacecraft built by Hindustan Aeronautics Limited (HAL). MOM's sub-systems, some of which were already being developed, were then transferred to the structure, and the spacecraft began to take shape. Installation of science instruments started in April 2013, testing from August onwards, and on 3 October 2013, the spacecraft was delivered to Sriharikota to begin preparation for launch.

The PSLV-C25 launch from the FLP was initially scheduled for 28 October but then pushed to 5 November 2013. This was the first time Sriharikota conducted a launch in the month of November is avoided as the rainy season typically spans October and November. MOM was launched by the most powerful variant of ISRO's reliable and in practice only operational launch vehicle, the PSLV-XL. Despite this, there was a velocity shortfall of around 1.5 km/s. This shortcoming in PSLV-XL's power dictated the overall mission profile, the smaller science payload of 15 kg, the six Earth orbits (known as Gravity Assist) and the highly elliptical Martian orbit of 418 by 76,872 km when it got there. ISRO turned this highly elliptical orbit into an opportunity by configuring experiments that exploited such an orbit.[884] During the same launch window, NASA launched its spacecraft called Maven to Mars. With a launch vehicle that can put seven tonnes into GTO compared to PSLV's 1.3 tonnes, Maven was launched two weeks after MOM and was on its way to Mars an hour after launch compared to MOM's three weeks and arrived at Mars three days before MOM.

PSLV's unusual destination called for a unique launch trajectory. Typically, a PSLV delivers its payload to Earth orbit in about twenty minutes, from where the payload proceeds under its own power to the final destination. The fourth stage ignition and payload separation is monitored from ground stations in Port Blair in Brunei, Biak in Indonesia and Canberra in Australia. For MOM's eventual trajectory to Mars, an interval of 24 minutes was required between the end of stage three and the ignition of stage four. By that time, the ground track of the fourth stage would have travelled beyond Australia. The nine-minute fourth stage burn would take place over the south Pacific, well beyond the reach of any of ISRO's or its international partners' ground stations.[885] To ensure that the telemetry of the fourth stage burn and MOM separation was recorded, ISRO commissioned two ships from the Shipping Corporation of India, Nalanda with a 4.6 m antenna and Yamuna with a 1.8 m antenna. Poor weather conditions at sea delayed these ships from getting to the operational sites as scheduled, resulting in the launch date moving from 28 October to 5 November. They arrived on location about three days before launch and positioned themselves approximately 2,000 km apart near Fiji.[886] One would monitor the fourth stage ignition and burn and the other the release of MOM from the spent fourth stage.

During the 300-day cruise from Earth to Mars, ISRO monitored MOM using its ground stations during the 12 hours a day that it was visible in the sky from India and with assistance from NASA's DSN at other times.[887] The first picture taken by the Mars Colour Camera was of the Earth two weeks after launch. The other four instruments, too, were switched on for short periods for testing and calibration. To ensure that MOM arrived at the right place at the right time with high precision, four Trajectory Control Manoeuvers (TCM) were scheduled, of which only three were carried out, TCM-1 on 11 December 2013, TCM-2 on 11 June 2014, TCM-3 scheduled for August 2014 was omitted as no correction was deemed required and TCM-4 on 22 September 2014. The TCMs were designed to cater for small course variations resulting from navigation errors and effects of external forces, such as gravity or solar radiation pressure.

The first two TCMs required engine burns for forty seconds and nine seconds, collectively changing MOM's velocity by 9.33 m/s.[888] The third TCM was skipped in favour of revised TCM-4. Two days before arrival, a joint LAM and eight-thruster burn lasting for four seconds reduced MOM's velocity by 2.1 m/s. Although a very short burn resulted a tiny change in velocity, this burn was significant as it verified that the LAM operated as expected. For additional assurance, ISRO had been mirroring on the ground the LAM burns using an identical LAM at the VSSC. Had this test on 22 September not gone according to plan, ISRO would have had time to engage plan B for Mars Orbit Insertion. Plan B involved using only the eight small thrusters to perform the breaking manoeuvre to enter Martian orbit. The lower capacity of the smaller thrusters would have required a longer burn of ninety minutes. The mid-point of the burn would have remained the same, but it had to start earlier. That is why a test was conducted on 22 September. Plan B would have resulted in an even more elliptical orbit (an apoapsis of 0.27 million km) and only about 3 kg of propellant would have remained making mission success a "touch and go."[889]

Date	IST	Event	Engine Burn duration (s)	Apogee (km)
05/11/13	14:38	Launch	935	23,550
07/11/13	01:17	1st orbit increase	416	28,825
08/11/13	02:18	2 ND orbit increase	570.6	40,186
09/11/13	02:10	3rd orbit increase	707	71,636
11/11/13	02:06	4th orbit increase	incomplete	78,276
12/11/13	05:03	4th orbit increase (supplemental)	303.8	118,642
16/11/13	01:27	5th orbit increase	243.5	192,874
01/12/13	00:49	Trans Mars Injection	1328.89	

Table 15-4 Series of Earth orbits prior to departure for Mars

Navigation and guidance had brought MOM to within 1,847 km of the surface of Mars on 24 September 2014. Travelling at 6.5 km/s and accelerating by virtue of its approach to Mars, MOM had to slow down by 1099 m/s to enter orbit.[890] Had the braking manoeuvre failed, the mission would have become a flyby and not an orbiter. To decelerate, MOM had to point in the opposite direction of travel and fire its LAM and the eight small thrusters for 24 minutes. Two key events coincided with this breaking manoeuvre. Five minutes after firing the LAM and the eight small thrusters, MOM entered Mars' shadow and lost sight of the Sun for the first time since leaving Earth. A few minutes later, still during the engine burn, MOM disappeared behind Mars and lost contact with Earth. Both events were expected and planned for. The required commands for reorientation, start engine firing, stop engine firing and re-point to Earth had been uploaded to MOM a few days earlier. Given the 42-minute lag in round trip communication, real-time communication was not practical, and all spacecraft far from Earth are designed to operate autonomously. Following the engine burn, which consumed 250 kg of propellant, MOM decelerated and entered Martian orbit.

A few minutes after MOM came out of the eclipse, it reoriented its high gain antenna to point to Earth and then sent signals to Earth that the engine burn and orbit insertion went to plan. About 12.5 minutes after transmission, NASA's DSN station in Canberra, Australia, received the signals from MOM and forwarded them to ISRO, confirming that MOM had entered orbit around Mars and could finally live up to its name, Mars Orbiter Mission. MOM achieved an orbit of 421 x 76,993 km with a period of 73 hours and had 37 kg of usable propellant left. On 24 September 2014, MOM joined four other Mars orbiters, ESA's Mars Express and three NASA orbiters, Mars Reconnaissance Orbiter, Mars Odyssey and MAVEN (Mars Atmosphere and Volatile EvolutioN), which had arrived just three days earlier. In addition, NASA was operating two rovers, Opportunity and Curiosity, on the Martian surface.

Science from Martian Orbit

The MOM was designed with six objectives, three technology and three scientific.[891]

- Design and develop a Mars orbiter with the capability to perform earth bound manoeuvres, Martian Transfer and Mars Orbit Insertion after nearly 300 days of travel.

- Incorporation of autonomous features in spacecraft.

- Design, plan and operate deep space communication with the orbiter (ca 400 million km).

- Exploration of Mars surface features, morphology, topography, mineralogy.

- Study of constituents of Martian atmosphere and dynamics of the upper atmosphere.

- Detect emanations of gaseous constituents from the surface/subsurface looking for clues for geological and biological activities.

- Unlike Chandrayaan-1, the haste at which the Mars mission was developed did not allow time to engage international partners. MOM is carrying a package of five scientific instruments selected from a shortlist of twelve, all of which were designed and built in-house by ISRO.

Instrument	Origin	Purpose
Lyman Alpha Photometer (LAP)	Laboratory for Electro Optics Systems, Bangalore	Help understand the processes responsible for the current Martian atmosphere
Methane Sensor for Mars (MSM)	SAC, Ahmedabad	Detect, measure and understand the presence of methane on Mars
Mars Colour Camera	SAC, Ahmedabad	Image Martian surface and atmospheric features as they change during the day and seasons
Thermal Imaging Spectrometer (TIS)	SAC, Ahmedabad	Image Mars using infrared wavelength. It can operate during the day or at night
Mars Exospheric Neutral Composition Analyser (MENCA)	VSSC, Thiruvananthapuram	Measure and analyse the constituents of the Martian exosphere

Table 15-5 Overview of Mars Orbiter Mission's Science Payload

Two instruments are designed to look at the Martian atmosphere, two at the surface and one sampling Mars' upper atmosphere. Between them, the instruments measure the vertical composition of the Martian atmosphere; investigate the rate at which the atmosphere has been dissipating into space; image the Martian surface, thermally during day and night and visually during day time; and attempt to detect and measure emissions of methane.

Lyman Alpha Photometer

Mars today has no permanent deposits of surface liquid water on its surface and only a very thin atmosphere, typically around 1% that of the Earth. Astronomers estimate that in the past Mars had a much denser atmosphere and enough water to cover the Martian surface in a 500 m deep ocean.

Lyman Alpha Photometer (LAP) is an instrument designed to collect data that will eventually lead scientists to understand what happened to the original Martian water and atmosphere. It was built by ISRO's Laboratory for Electro-Optics Systems in Bangalore. LAP has three objectives:

- Generate spatial (whole of Mars) and temporal (through all seasons) profiles of hydrogen and deuterium Lyman alpha intensities

- Characterise deuterium-enrichment in the upper atmosphere as the light hydrogen atoms escape, and

- Determine with the help of (i) and (ii) the rate at which Mars is losing water as hydrogen/deuterium atoms have their origins in water molecules.

Hydrogen exists in two stable forms known as isotopes. Hydrogen (H) with a single proton in the nucleus and Deuterium (D) with a proton and a neutron in the nucleus. LAP measures the relative abundances of these two types of hydrogen. LAP is an optical instrument that looks at the Martian atmosphere and, using an absorption cell technique, measures first the incoming hydrogen and then the deuterium Lyman alpha radiation. On Earth, 99.98% of all hydrogen is H. A series of measurements can be used to generate an intensity ratio for D/H isotopes.

Mars is about half the size of the Earth but only about 10% of its mass. This low mass, and thus weaker gravity, allows the very light hydrogen molecules to escape easily from Mars. The absence of a Martian magnetic field allows the UV radiation from the Sun to interact directly with the upper atmosphere. The Ozone layer that acts as a barrier on Earth is not present on Mars. This UV radiation breaks the water molecules in the Martian atmosphere into O and H2. Mars has been experiencing this slow evaporation of its atmosphere for millions of years. As the rates of evaporation of the heavier deuterium and lighter hydrogen, forms are different, over time more H escapes than D increasing the D/H ratio.

LAP was first switched on and tested on 6 February 2014 16 million km from Earth when still end route to Mars. The basic health checks to verify the instrument's operation were conducted over twenty minutes. It now collects data from orbit for about 30 minutes on either side of a close approach of each orbit when MOM is closer than about 3,000 km to Mars. The data LAP has collected is currently being processed but not yet published, and it continues to function as planned.[892]

Methane Sensor for Mars

Produced by the SAC in Ahmedabad, MSM weighs 2.54 kg and is attracting the most attention. On Earth, methane is associated with life. When it was detected on Mars in 2003, there was speculation that this was the long sought after evidence of life on Mars. Since then, methane on Mars has been detected by telescopes on Earth, from Martian orbit and by rovers on the surface of Mars.[893] Although the methane detections have been repeated and confirmed, the instances of detection have been sporadic, and the concentrations detected extremely tenuous. Adding to the mystery is the sudden disappearance of the detected methane. Once produced, Methane vanishes in several decades so the detected Methane is not ancient but new. Why and how Methane is produced is a puzzle. Methane can have multiple sources other than undiscovered Martian life form, geochemical, geological and perhaps local processes acting on organic chemicals brought to the surface of Mars by meteorites.

MSM is designed to detect methane with a sensitivity of 38–60 parts per billion (ppb) during a single ten-second observation. The typical background level of methane on Mars is 0.7 ppb. The 2003 measurement detected 250 ppb, but all other observations have been in the range of 10–100 ppb. NASA's rover Curiosity, also known as Mars Science Laboratory, has been on the surface of Mars with a sophisticated collection of scientific instruments since August 2012. It sampled the Martian atmosphere six times between October 2012 and June 2013 and found no methane. However, a few months later it detected significant levels of methane in short bursts. The levels were tiny, but still an order of magnitude above the background levels. Subsequent measurements confirmed that normal background levels had returned. Detection of methane is potentially the first step in the discovery of evidence of ancient or existing life on Mars but what has been observed on Mars to date does not offer a coherent picture.[894]

It is designed to measure a total column of methane in the Martian atmosphere from orbit. During its orbit, about twice each week, MOM can observe specific areas of Mars from around 400 km and the full disc from 77,000 km. It uses a Fabry-Perot Etalon sensor, the first time one has been used in space. It detects and measures methane variations over place and time. As the instrument uses reflected light, only the day side of the atmosphere can be sampled. MSM was tested observing the Earth before it arrived at Mars.[895] MSM was first activated after launch, while MOM was in Earth orbit and observed the Sahara Desert. It was reactivated seven times in total before arriving at Mars. The observations of the Earth and the darkness of space were used for initial calibration. MSM's initial observations of Mars were recorded from near apogee. At a distance of nearly 77,000 km where the relative motion between MOM and Mars is less than 100 m/s, the whole disc of Mars can be seen but only at low resolution. MSM has not yet recorded close-up observations when MOM is at perigee. This will be a challenge because, although the images will be of a higher resolution, at perigee, MOM will have a relative motion of 4,000 m/s. Over time, these observations could help identify the geographical site associated with methane detections on Mars.

MOM is expected to make repeated measurements over many years from which seasonal patterns may emerge. Collectively, these data will help determine the dynamic nature of the methane cycle on Mars. In March 2015, ISRO published MSM's observations of Mars for the first-time reporting that it had not recorded any significant observations of methane until then, but the observations were sufficient to indicate that the MSM instrument was functioning as expected.[896]

Mars Colour Camera

The Mars Colour Camera, too, was built by SAC in Ahmedabad. At 1.27 kg, it is the lightest instrument in the science package. It was constructed from components from disparate sources, including in-house, commercial off-the-shelf (COTS) components. For example, the primary lens (105 mm with f4.0) in the camera was a COTS product weighing 620 g. It was customised and qualified for use in space by engineers at SAC. The modified version now in Martian orbit weighs 310 g. The Mars Colour Camera uses a single commercial high-speed snapshot colour CMOS sensor with a RGB Bayer filter as used in domestic cameras and camcorders. From 77,000 km, Mars subtends an angle of 4°, which the camera can easily capture in its field of view of 5.7°. From its highly elliptical orbit, MOM can take images of Mars at 19 m resolution from 370 km or 4 km resolution from 77,000 km.[897] The Mars Colour Camera was developed using the three-model philosophy. A verification model was developed for demonstrating the proof of concept. A flight model-like and an identical flight model were developed and were subjected to qualification and acceptance level tests, respectively. The development of these models ran almost parallel with feedback from one model incorporated into the other and verified quickly to meet the challenging timeline.

Mars Colour Camera is designed to image the Martian dust storms, polar ice cap and atmospheric phenomena, as well as surface features, such as volcanoes, valleys and mountains. MOM's highly elliptical orbit will also allow it to image Deimos, one of Mars' two moons, from a unique view point. The MCM was activated while in Earth orbit and took images of the

Earth in three imaging sessions. Two sessions were conducted on 19 September 2013 and one four days later. The very first image shows India, as well as parts of Asia and Africa. These were taken on 19 September from 7,240 km. The camera's images comply with the internationally recognised Planetary Data System standard (PDS). To verify the characteristics of the images, ISRO arranged to capture equivalent images of the Earth from INSAT 3A at the same time as the Mars Colour Camera to match the illumination. This exercise helped ISRO scientists to calibrate the Mars Colour Camera.

By August 2016, MCM had returned more than 540 images, and ISRO has made these images available for scientific investigation and publication by academics and researchers in India. Mars has been imaged by several spacecraft (mostly from NASA) from orbit and the surface using higher quality instruments for many years. The camera's contribution is not expected to reveal anything dramatic. For a mission that was primarily a technology demonstrator, images from the Mars Colour Camera are contributing to scientific publications. There is a remote possibility that in time high-resolution images from the camera combined with data from the MSM may identify geographical locations on Mars where the first extra-terrestrial life may reside.

Thermal Imaging Spectrometer

ISRO has been working with remote sensing Thermal Infrared Imaging Systems for many years. They are used in EO satellites, including Kalpana and INSAT-3D. They have also been used by NASA for exploring Mars: 1976 Viking1 & 2 Infrared Thermal Mapper, 1997 Mars Global Surveyor – Thermal Emission Spectrometer and 2001 Odyssey – Thermal Emission Imaging System. The TIS in MOM, weighing 3 kg, was produced by SAC in Ahmedabad. Like the Mars Colour Camera, this instrument was constructed from in-house and customised COTS products.

TIS is sensitive to the infrared part of the spectrum (7,000 to 13,000 nanometers) rather than the visible spectrum (400 to 700 nanometers) where the Mars Colour Camera operates. TIS is a spectrometer used to analyse the light rather than a camera that uses light to produce an image. It uses a slit and a grating to split the individual components of infrared light to discover the chemical and mineralogical composition of the source. TIS optics consists of f/1.4 lens assembly with a focal length of 75 mm and field of view + or – 3.18° that directs light to a grating where it is dispersed into constituent wavelengths.[898] Another set of optics then refocuses the dispersed spectrum on a 160 by the 120-pixel sensor.

Like all MOM's instruments, TIS was operated before it arrived in the Martian orbit. It was first activated while in Earth orbit on 23 November 2013 to conduct basic health checks. Key system operations were conducted by using TIS to first look at "dark space" (a blank area of space) to establish the instrument's zero signal baseline, then a 10-minute observation of the Sahara Desert region of the Earth followed by another dark space observation. The measurements recorded during these observations were compared to those from a laboratory and used to calibrate TIS. Five additional dark counts were conducted on 6 February, 13 March, 5 May, 26 June and 18 August 2014 during the Earth-Mars cruise. From the unique characteristic spectra of minerals and soil types, TIS can eventually generate a chemical and mineralogical map of the entire Martian surface. The team that developed TIS has also developed a spectral signal library of specific minerals found on Earth, such as olivine and serpentine. If similar signals are present in Martian observations, identification of minerals on Mars will not be a complex process.

Mars Exospheric Neutral Composition Analyser

MENCA was built at the Space Physics Laboratory located within the VSSC complex. At 3.56 kg, it is the heaviest of the five instruments and based on the CHACE (fitted to the MIP) used to analyse the contents of the tenuous lunar atmosphere as the MIP descended from orbit to surface impact. MENCA is designed to collect data on how far the outer Martian atmosphere (known as the exosphere) extended and what it was made of. It is unique in that it does not analyse electromagnetic emissions or reflections from Mars but particles that make up the exosphere in-situ. Taking advantage of the highly elliptical orbit, MENCA samples its environment five times along each orbit and build up a radial profile of the composition of the exosphere between 400 km and 77,000 km. Over time, MENCA will build a profile of Mars' exosphere in altitude, daily and seasonal variations. It will also be used to study the environment of Phobos during its encounters.

During the development stage, MENCA was repeatedly tested and calibrated. The vacuum chamber used to simulate space-like environment has a built-in facility through which known gases can be introduced at a controlled rate. Also, a reference mass spectrometer is used to ensure that MENCA calibration is consistent with a known source. MENCA was calibrated on three separate occasions at the VSSC where it was developed. The first instance was before testing and evaluation, second after testing and evaluation and a third just before it was transported from VSSC to ISAC in Bangalore for integration into the spacecraft.

MENCA was activated to test functionality while MOM was still in Earth orbit and later during the cruise to Mars. It was first operated in Martian orbit on 29 September 2014 just five days after arrival. MOM's orbit was temporarily lowered to 260 km to collect data from four orbits during late December 2014. These measurements allow scientists to understand why Mars' atmosphere is so tenuous, about 1% that of Earth's, and how the current carbon dioxide rich constitution came about. They also allow scientists to model the mechanism and the rate at which these molecules escape from Mars' upper atmosphere.[899] Data from MENCA is formatted to Planetary Data System (PDS) standard and archived at the ISRO Space Science Data Centre.

Mission Status

India's first mission to Mars, which was developed on a very short time scale, has succeeded in every respect. It entered orbit as planned, all onboard instruments have been returning data. MOM has survived a communication blackout and a whiteout when planetary alignment forces MOM to fend for itself without any assistance from ISRO. A blackout (Earth–Sun-Mars), is a period when the Sun and Earth appear in the same direction and ISRO cannot discern the weak radio signals from Mars are drowned out by the closer Sun. A little like listening to a phone call in a noisy room. The maximum duration of the blackout in June 2015 was 17 days (between 6 – 22 June). A whiteout (Sun–Earth–Mars) is a period when the Sun and the Earth appear in the same point in the Martian sky, and MOM cannot discern the weak radio signals from Earth because of the interference from the Sun. The maximum duration of a whiteout was been around 14 days during 16 – 29 May 2016.

As the first mission to Mars, ISRO had designated MOM as a technology demonstrator with an operational period of 6 months. Once the initial six months of operations were complete, ISRO declared this primary mission a success. Since then, MOM has been operating in an open-ended extended mission phase. MOM completed its 100th orbit on 22 June 2015. With ample reserves of fuel, the data that the onboard instruments it continues to return data. By August 2016, MOM had generated six peer-reviewed publications, and more results will be published following the Announcement of Opportunity making MOM data available to researchers.[900] Data from MOM's instruments can be used as an independent source to support scientific observations and conclusions made by other spacecraft.

MOM remains healthy as a spacecraft, and each of the five onboard instruments continues to observe Mars and return data from orbit. When MOM was launched, its mere five instruments were not expected to make any ground-breaking scientific discoveries. After all, what could MOM achieve with its collection of meagre instruments that has eluded the American and European missions that have been scrutinising Mars for decades with higher specification instruments?

Figure 15-8 Mars full disc captured by Mars Colour Camera from an altitude of 66543 km in October 2014. Credit ISRO

A possible, although not probable, achievement would be if a detection of methane by the methane sensor were linked to a surface feature imaged by the colour or thermal camera helping to identify a specific geographic location as a methane source.[901] Those sites could then be scrutinised by future missions for the potential Martian microbes on the surface of Mars. The mission, spacecraft and all onboard instruments have enjoyed remarkable success. However, a risky and hazardous scenario could have ended MOM mission in February 2017. In its then orbit, MOM would have entered a series (in consecutive orbits) of solar eclipses that would have lasted for up to eight hours.

These eclipses would have coincided with MOM's orbit at apogee, where it moves the slowest and thus would have stayed in the shadow longer. With the solar panel as the only power source, the battery would have drained completely, raising for the first time the prospect of MOM shutting down since leaving Earth, even though it still had about 30 kg of propellant, sufficient for several years of operation.

On 17 January 2017, ISRO manoeuvred MOM into a new orbit that not only avoided the February 2017 eclipses altogether but also extended its life. In the words of ISRO chairman Kiran Kumar "Because of the crucial orbital change, the MOM now gets three additional years' life. We are expecting it to transmit data till 2020."[902] The new orbit will reduce the time spent in the shadow by about half for these and future eclipses. Following this manoeuvre, MOM still has 13 kg of fuel, which will allow it to continue to gather data into the next decade. In June 2017, MOM completed 1000 days in Martian orbit, well beyond the intended 3 months.

Astrosat - Astronomy from Orbit

Observational astronomy using telescopes has been conducted from India since the 17th century by gifted observers, such as Father Eugène Lafont and C.V. Raman. They observed from places, such as Calcutta and Trivandrum, and official observatories, such as Kodaikanal and Madras. However, some types of astronomical observations cannot be made from India or anywhere else on the surface of the Earth. The history of science illustrates that new instruments or vantage point give rise to the tantalising possibility of new scientific discovery.

For ISRO, Astrosat would be a fresh new vantage point. Although we can see stars in the night sky, most of the electromagnetic radiation coming to Earth from space is blocked by the atmosphere. Only visible light and some ultraviolet, infrared, and short-wave radio can make it from space to the surface of the Earth. When scientists want to study other wavelengths including X-rays, they have to get their instruments into space.

The first X-rays from the sun were detected by a V2 missile launched in the US in 1948.[903] This was one of about 100 V2s that were recovered from Germany by the US in the few weeks after the end of World War II. In

1962, a sounding rocket accidentally detected the first X-ray source beyond the solar system.[904] India's first satellite Aryabhata, launched in 1975, carried a science payload that collected X-ray data from several cosmic sources, including a black hole. About a decade later, an experiment (Anuradha) to study low energy cosmic rays was designed, built and tested in India. It operated successfully between 29 April and 6 May 1985 onboard Spacelab-3 within the Space Shuttle Challenger.

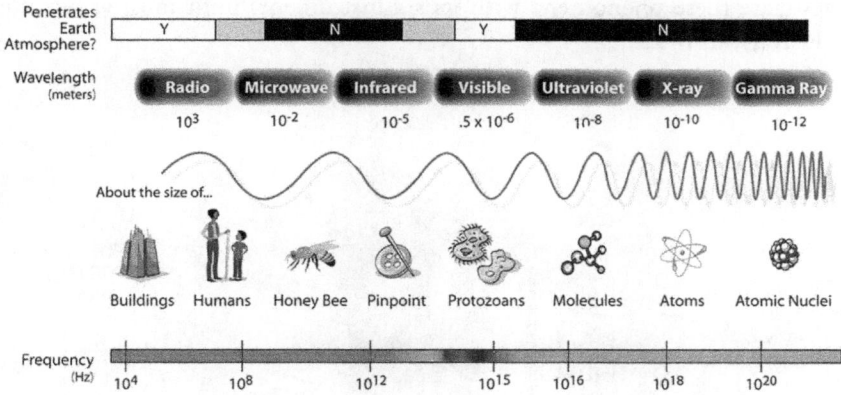

Figure 15-9 Electromagnetic Spectrum. Credit NASA

During this time, the experiment recorded 10,000 alpha particle events, a similar number of heavier ions and about 15,000 galactic cosmic rays.[905] An X-ray experiment dubbed Indian X-ray Astronomy Experiment as part of Indian Remote Sensing Satellite (IRS-P3) payload and operated between 1996 and 2004. It collected data to generate a light curve of variation in X-ray sources. It also observed binary and intense X-ray sources. On 28 September 2015, India launched its first dedicated astronomy satellite to observe the cosmos using wavelengths that are otherwise blocked by the atmosphere.

Astrosat carries six instruments looking at a range of frequencies accessible from space. Canada and the UK contributed, but Astrosat is predominantly an Indian mission with input from six Indian institutions.[906] With a total mass of 1,513 kg at launch with 42 kg of propellant, Astrosat is designed to operate for a minimum of five years in a 650-km equatorial orbit inclined by 6°. The orbit and inclination were selected to allow Astrosat to avoid an unusual phenomenon known as the South Atlantic Anomaly (SAA). It is a section of an inner Van Allen radiation belt that surrounds the Earth but dips to lower altitudes over the South Atlantic. The SAA is a high radiation hazard for all spacecraft in LEO that must travel through it. The Hubble

Space Telescope observations are paused during this time amounting to a loss of up to 15% of observing time. The ISS crew also take precautions, and spacewalks are not scheduled during these times. High energy radiation (UV, X-rays and gamma-rays) to which Astrosat instruments are tuned are produced during some of the most energetic events in the cosmos. During its five years, Astrosat will look at individual hot, bright stars, very small but extremely dense neutron stars, rotating neutron stars, supernovae, black holes and supermassive black holes at the centres of galaxies.[907] Astrosat will investigate these phenomena with its six instruments built mostly, but not exclusively, in India.

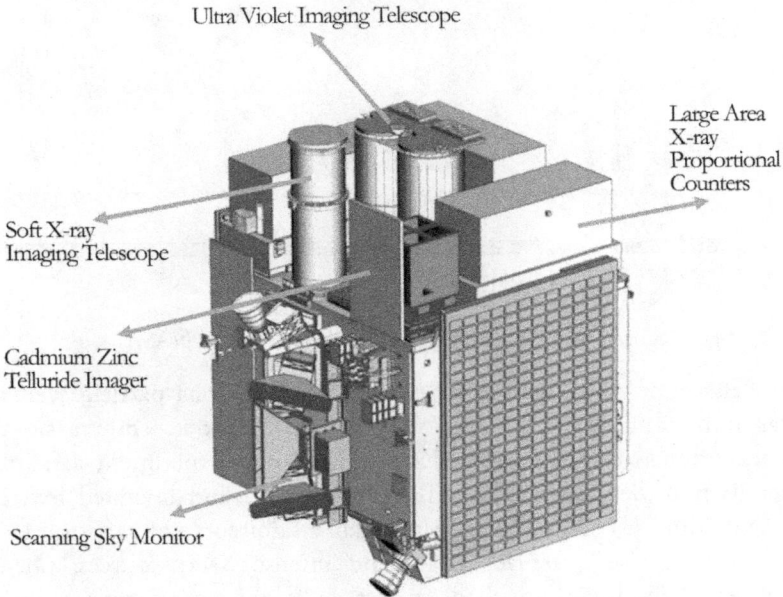

Figure 15-10 Astrosat and its Science Payload. Credit Adapted from ISRO

Four of the instruments UVIT, CZTI, LAXPC and SX are located on the same side, so they point in the same direction. The SSM on a rotating platform scans the sky looking for sporadic X-ray sources. The role of the last instrument, the CPM, is to monitor the radiation in the environment where Astrosat operates to help ensure that Astrosat subsystems and instruments avoid damage from charged particles, predominantly in the SSA. In this sense, it is not a science instrument like the others.

Instrument	Origin	Purpose

Charged Particle Monitor (CPM)	TIFR	Not strictly a science instrument. Designed to prevent damage to Astrosat's sensitive sensors by intense radiation, the CPM is constantly monitoring the radiation in Astrosat's environment. When it detects excessive radiation from the Sun, Moon and Earth, it places other instruments in a safe mode.
Scanning Sky Monitor (SSM)	ISAC and Inter University Centre for Astronomy and Astrophysics	Detect intergalactic X-rays (2.5–10 keV) and map their sources in the cosmos.
Large Area X-ray Proportional Counters (LAXPC)	TIFR	Similar to SSM but LAXPC detects X-rays of higher energy (3–80 keV). It is located on a platform that allows it to track an object for greater detail
Ultraviolet Imaging Telescope (UVIT)	ISRO, TIFR, Indian Institute of Astrophysics, and Canadian Space Agency.	This instrument consists of two telescopes, one sensitive to the visible spectrum (visible 320–550 nm) and the other Near Ultraviolet (NUV 200–300 nm).
Soft X-ray Telescope	TIFR and University of Leicester, the UK	Designed to observe low energy (0.3–8 keV) X-ray sources.
Cadmium Zinc Telluride Imager (CZTI)	TIFR and IUACC	A special sensor designed to detect hard (10–100 keV) X-rays and gamma-rays

Table 15-6 Overview of Astrosat's Science Payloads

Charged Particle Monitor

Astrosat will avoid the SAA completely in about a third of its orbits. During other orbits, it will spend about 15 to 20 minutes of its 97-minute orbit in the SSA. The CPM detects this and other radiation hazards and shuts down or suspends instruments to prevent damage. The intensity and distribution of the charged particles in the SAA are well known but can dynamically change because of solar flares or coronal mass ejection events. These changes are unpredictable, so too is the precise time Astrosat enters and leaves the SAA. The CPM is designed to detect the actual radiation levels in real-time to allow Astrosat to enter a temporary safe mode if required. The CPM was the first sensor to be switched on, a day after launch. Although its primary function is to ensure that Astrosat runs efficiently and safely, its data will help refine the SAA model, making it scientifically valuable. Unlike the other instruments, CPM was not a proposal-based instrument. It was developed by TIFR, and it runs at all times.

Scanning Sky Monitor

The SSM is located on a rotating platform looking for transient X-ray sources in the energy range of 2.5–10 keV. It was developed by ISAC and the Inter University Centre for Astronomy and Astrophysics and weighs 75.5 kg. The high energy of X-rays cannot be focused using lenses or mirrors, so a different detection mechanism is used. The SSM instrument consists of three almost identical separate gas-filled containers with a total area of 180 cm². An incoming X-ray generates a voltage within the container by ionising the gas (mostly argon, some xenon and a tiny bit of methane).[908] While the detection principle is similar to that used in the LAXPC, the SSM is designed to record directional information of incoming X-rays. The voltage pulse, along with the time and position within the container, and the orientation of the rotating platform at the time of detection are used to identify in the source in the sky generating the X-rays. In addition to the scanning mode, the SSM can operate in a stare mode to examine specific objects of interest. For example, on 26 October 2015, the day it was first switched on, first it detected during the scanning mode and then 'stared' at the neutron star binary pulsar 4U0115+63. The SSM is highly sensitive. It has a restricted field of view and must avoid bright objects, including the Earth, Moon, Sun and zodiacal lights (a collection of dust within the ecliptic that is lit up by the Sun). With a field of view of 22° by 100° at any one time, the SSM can monitor half of the sky facing away from the Sun about four times a day.

Large Area X-ray Proportional Counters

The Large Area Proportional Counters (LAXPC) instrument is similar to the SSM, but instead of operating on a rotating platform and scanning the sky for transient and random X-ray events, LAXPC has a field of view of 0.9° by 0.9° and targets individual objects. It is collocated on a platform with three other instruments that can track a specific object in the sky. LAXPC detects X-rays of energies (3–80 keV) higher than SSM (2.5–10 keV) and also uses containers of gas where X-rays trigger a voltage pulse, its amplitude proportional to the energy and number of X-rays detected.[909] The three identical LAXPC containers are rectangular (120 cm x 50 cm x 70 cm) in shape. At a total area of 6,000 cm² these are currently the largest LAXPCs operating in space. LAXPC can detect X-rays at a very short timescale, in the order of a few milliseconds.

At 419 kg, the LAXPC is the heaviest instrument on Astrosat. Despite the large collecting area, LAXPC requires a day or two of collecting data to complete an observation of a single source. The high energy X-rays LAXPC is tuned to detect are not emitted by the Earth or the Moon, so it has only to avoid the Sun (by at least 30°) when observing. The higher energy X-rays in LAXPC's range allow it to observe some of the most energetic events in the cosmos. It will search for black holes with mass in the order of magnitude as the Sun, as well as supermassive galaxies that are nine orders of magnitude more massive. It will also study pulsars, quasars, neutron stars and violent events deep in the nucleus of active galaxies.

The unique feature of Astrosat is the ability for several instruments to observe a single object at the same time. Observation by one instrument can initiate a more detailed observation by another. For example, LAXPC could observe in detail bursts and flares that were initially detected by the SSM. On 19 October 2015, when it was first switched on, it observed a dark area of the sky before moving to a known stable source used for calibration, the Crab Nebula. By chance, these initial observations helped to identify a small drift of the satellite (0.25° per hour), which the engineers were able to fix easily.[910] The LAXPC counters were activated on 20 October 2015 with the voltage gradually increased to the high voltage required for operation. On 24 October 2015, LAXPC observed another well-known source GRS 1915+105 which further aided calibration.

Soft X-ray Telescope

Unlike light, X-rays are not reflected by mirrors, so an X-ray telescope uses a different technique to construct an image. Most X-rays are either absorbed by or penetrate glass, but if the angle is shallow (a grazing angle), then X-rays are reflected but only at a shallow angle. Astrosat's SXT is based on this technique and uses two reflections down the side of the telescope tube lined with conical reflecting surfaces (0.2 mm aluminium plated with gold) rather than a mirror. This approach only works for lower energy (0.3–8 keV) X-rays, otherwise known as soft X-rays. The SXT has a diameter of 38 cm and a length of nearly 250 cm. The reflected X-rays are captured on a sensitive CCD at the prime focus operating at a very low temperature of -80°C to minimise background noise. SXT avoids the Sun (45°) and the Earth (6° the upper atmosphere can generate X-rays as a result of the solar wind) but is not concerned with avoiding the Moon since it is not a source of X-rays. TIFR developed the telescope. The CCD detector was provided by the University of Leicester in the UK.

SXT was activated on 26 October 2015 and calibrated with an onboard source. On the same day, it looked at its first astronomical object, an active galaxy well known as a strong X-ray source. SXT also observed Tycho, a supernova remnant that was recorded as a naked eye observation from the 14th century before the telescope was invented. Despite the sensitivity of the CCD, SXT requires several hours (at least half a day, accumulated over several orbits) to complete a typical observation.

Ultraviolet Imaging Telescope

The UVIT is one instrument made up of two almost identical telescopes that can operate in three parts of the spectrum. One telescope looks at the visible spectrum (visible 320–550 nm) and Near Ultraviolet (NUV 200–300 nm) using a beam splitter. The second telescope is designed to observe in the range known as the Far Ultraviolet (FUV 130–180 nm). In both UV ranges, a built-in grating is used to disperse the incoming light to conduct a spectroscopic analysis. Both telescopes have an aperture of 375 mm with a focal length of 4750 mm. The imaging sensors used are semiconductor devices that detect individual photons originating from either visible, NUV or FUV.[911] Individual photons are amplified, counted and integrated over the period of the exposure. UVIT is a collaboration between ISRO, TIFR, Inter University Centre for Astronomy and Astrophysics, Indian Institute of Astrophysics and the Canadian Space Agency. The UV photon counting detector was provided by Canada.[912] The lead institution for UVIT is the Indian Institute of Astrophysics.

UVIT's first target was NGC 18 (object number 18 listed in the New General Catalogue used by astronomers to identify deep sky astronomical objects), an open cluster of about 1,500 stars in the constellation of Cepheus. It was the first object to be observed by UVIT on 30 November 2015, 62 days after launch. Although not an object used for calibration as the Crab Nebula, NGC 18 has been observed from space in the past, so an assessment of UVIT could be made by comparing observations. Since astronomical objects are fainter in UV than visible, the UV photons are first amplified before they can contribute to the detected signal. To avoid interference from background photons and in the interest of sensor safety, UVIT is operated only at night. It has a Moon and Sun avoidance angle of 45° and 12° from the limb of the Earth and can typically make an observation of an optical or UV source of a maximum of 30 minutes at a time.

Cadmium Zinc Telluride Imager

CZTI is not a telescope in the traditional sense but a single solid-state detector divided into four identical and independent quadrants with 16 individual CZTI modules. CZT is an alloy of cadmium, zinc and tellurium that can run as a radiation detector at room temperature. The 64 modules of 256 pixels each produce an image with a resolution 16,384 pixels. All the CZTI modules have built-in collimators providing a field of view of 4.6° by 4.6°. A large radiator plate is used to dissipate the heat produced by the instrument during operation to maintain the nominal operating temperature of 0 to 15°C.

CZTI is located on the same platform as the telescopes so can observe the same object at the same time as the other instruments. It operates in the 10–100 keV range, also known as hard X-ray range. These new energies extend Astrosat's capacity to study X-ray binaries stars, active galactic nuclei and gamma-ray bursts. It is also expected to reveal polarisation in the hard X-ray part of the spectrum. The CZTI only needs to avoid the Sun when observing, as neither the Moon nor the Earth emits hard X-rays. Once Astrosat arrived in orbit, the CZTI was the first primary scientific payload to be activated on 15 October 2015. The Crab Nebula was used to calibrate the timing capability of the instrument.

Operational Status

Unlike Chandrayaan-1 and MOM, Astrosat did not have far to travel. Consequently, the mass, electrical power and data transmission capacities of its instruments are an order of magnitude larger. The spacecraft and all the instruments are operating in orbit as planned. Two solar panels provide 1,600 watts via two 36 Ah lithium-ion batteries. It has two 120 GB solid state storage devices enough to store data collected in four orbits. Astrosat can generate up to 420 GB of data daily. Typically, Astrosat has a line of sight of ISTRAC in Bangalore for about 10 minutes during 10 or 11 of its 14 daily orbits. Using two X-band carriers, Astrosat can transmit data to Earth at 105 Mb per second. The first six months after launch were used for calibration and performance evaluation, and then, the fully operational mode was activated. Astrosat is reserved for use by Indian institutions only for the second six-month interval. After the first year in orbit, Canada will have 5% and the UK 3% observing time throughout the mission lifetime. In the second year, 10% of the time is set aside for international proposals from nations that have not contributed to Astrosat, which will increase to 20% in the third year.[913]

Scientific experiments make a tangible contribution to a national space programme. They help develop key technical infrastructure and allow scientists and engineers to acquire operational experience through developing and operating leading edge experiments. They also boost morale and enhances job satisfaction while gaining scientific credibility for those involved. Further, the sense of discovery emanating from pure research attracts curious and talented individuals for careers in space. An invigorated and growing space programme can deliver long-term national economic and societal development, national prestige at home and international recognition beyond.

Future Science and Interplanetary Missions

At the present ISRO has ruled out HSF but continues to focus on primarily on development in launch vehicles including reusable technology, infrastructure including a second VAB, potentially another launch pad or even another launch site, return to the Moon, Mars and other goals in the Solar system. It is doing this in the backdrop of the maintaining and enhancing its Earth observation, navigation, weather and communication satellite constellations. Missions in the pipeline include:

Return to Mars

As MOM approached its second year in orbit, ISRO issued a call for proposals to the scientific community and academics for India's second mission to Mars: MOM-2. It is seeking proposals for scientific objectives to target and the instruments that can deliver them. ISRO's initial design of MOM-2, also an orbiter, includes a payload of 100 kg (MOM was just 15 kg) designated for the much lower orbit of 5,000 km, instead of MOM's 77,000 km. One report suggests that the orbit will be 200 by 2,000 km.[914] The date of launch is not established but it is unlikely to be before the 2020 launch window.[915]

Aditya-L1

India's first science mission in Earth orbit was Astrosat launched in 2015, the second called Aditya-L1 is due for launch in 2019/20 to examine the Sun. Aditya, from the Sanskrit name for the Sun God, was originally designed to observe the Sun's outer atmosphere, known as the Corona, from an SSPO. The mission was initially proposed in 2008 when Chandrayaan-1 was in final stages of preparation. The initial designs published in 2011, Aditya-1 consisted of a single instrument (a Visible Emission Line Chronograph) on a 400kg spacecraft in an 800 km SSPO.

A polar orbit is not the most convenient for solar observations. The Earth itself is "in the way" and obscures the Sun for half of every orbit. The stability of the viewing platform is also harder to maintain from a spacecraft moving at about 7 km second. The Aditya-1 mission has since been revised to include a larger science payload and a destined for the Lagrangian L1 orbit instead of SSPO and renamed as Aditya-L1. The Earth has 5 gravitationally "neutral" points in space known as Lagrangian points (L1 thru to L5). Here the force of gravity between the Sun and the Earth is equal. L1 is a special point on a line drawn from the centre of the Earth to the Centre of the Sun. The L1 point is 1.5 million km from the Earth towards the Sun. Once the decision was taken to use L1 for Aditya, Aditya was renamed to Aditya-L. Being further away than the 800 km, it will require additional fuel to get to the new L1 orbit. Once a spacecraft arrives in a Lagrangian point, it has a minimal force of gravity acting on it and stays put with little or no station keeping requirements. From L1 Aditya-L1 will have uninterrupted view of the Sun at all times. It will carry a total of 7 instruments to observe the Sun.

- The original Visible Emission Line Chronograph from the Indian Institute of Astrophysics (IIA). It is designed to study the solar corona and monitor Coronal Mass Ejections.

- A Solar Ultraviolet Imaging Telescope (SUIT) from the Inter-University Centre for Astronomy & Astrophysics designed to image Solar Photosphere (the visible "surface of the Sun).

- Aditya Solar Wind Particle Experiment (ASPEX) from the PRL to study the particles emanating from the Sun.

- Plasma Analyser Package for Aditya (PAPA) from VSSC to analyse the composition of the solar wind.

- Solar Low Energy X-ray Spectrometer (SoLEXS) from ISAC to monitor solar X-ray flares

- High Energy L1 Orbiting X-ray Spectrometer (HEL1OS) from PRL, ISAC and Udaipur Solar Observatory observe and monitor dynamic solar events that can give rise to disruptive space weather effects on Earth

- Magnetometer from LEOS and ISAC to measure the variations in Interplanetary Magnetic Field

Venus Orbiter Mission

Reports of an Indian mission to Venus first surfaced in 2012 with a timeline of an orbiter to arrive at Venus in three years later. This announcement was at about the same time that the Mars Orbiter Mission acquired the formal approval. The mission to Venus was then shelved.

Like MOM, the Venus mission would also be launched by a PSLV and enter a highly elliptical orbit (500 x 60,000 km). Details available are tentative and likely to change. It is expected to carry 5 instruments, use a single solar panel providing at least 500 W and have a total mass of around 175 kg. Over time this apogee would be reduced offering closer views of Venus and its atmosphere. In Mid-2017, ISRO formally made an "Opportunity Announcement" for science instruments for a mission to Venus.[916] The opportunity is open only to scientists in India. The mission plan is still at a very early stage what detail is available is tentative. A launch date has not yet been established but expected sometime in the middle of the next decade.

Team Indus

An Indian private sector entity, Team Indus is planning to send a rover to the surface of the Moon. Team Indus is entirely independent of ISRO but has commercially engaged ISRO to launch its rover to the Moon on one of its PSLV. It is participating in the international Google Lunar Xprize competition. The Google Lunar XPrize is a competition designed to foster innovation in low cost development of robotic exploration of the Moon with a prize of $20 million. The $30M Google Lunar XPRIZE is a global competition to challenge and inspire engineers and entrepreneurs to develop low-cost methods of robotic space exploration. To win, a privately funded team must successfully place a robot on the surface of the Moon, travel at least 500 meters and transmits high-definition video and images back to Earth. Competing with Team Indus are 4 other teams from Japan, two from the US and one from Israel.

All the contenders have signed contracts with a launch service provider to transport the rovers from the Earth to the Moon. Following a ride share agreement, one of ISRO's PSLV will launch Team Indus and the Japanese rovers to the Moon in December 2017. Team Indus is entirely independent of ISRO. The launch to space using ISRO's PSLV is a commercial arrangement. As part of Chandrayaan-2 mission, ISRO is planning to send a lunar rover to the surface of the Moon. Team Indus is private initiative, and may succeed in that goal ahead of ISRO.

❮ ◈ ❯

Chapter Sixteen
Space and National Security

I ndia established a nuclear programme long before it embarked on a space programme. INCOSPAR, which later became ISRO, was founded in 1962 under the Atomic Energy Commission, binding India's space programme to its nuclear programme from inception. The Department of Space did not exist until 1972. The AEC was established in 1948, rather hastily after independence, by Homi Bhabha, a gifted and internationally accomplished physicist. The emergence of the AEC was a product of the unique phase India was passing through when the democratic institutions and the now infamous bureaucratic procedures had not yet been firmly established. India's space programme has a social and economic development agenda at its heart but military objectives have gradually crept in the space programme has matured.

Space Infrastructure

One of the earliest external threat came from China when it conducted its first nuclear test in 1964. A decade later India responded. On 18 May 1974, a small fission device with a yield of 8 kt (although a larger yield was initially claimed) was detonated underground at the Pokhran Test Range located in the isolated Thar desert region in the state of Rajasthan. With Pokhran-I, India became the sixth nation to join the nuclear weapons club. The detonation was categorised as a Peaceful Nuclear Explosion, an internationally accepted form of nuclear testing.

Pokhran-II came two decades later. On 11 May 1998, India conducted three further nuclear tests, one was a fusion device and the other two, fission. Two days later, two more fission devices were detonated, all underground at Pokhran. Just as India had responded to China's nuclear test, Pakistan responded to India's. Between 28 and 30 May, Pakistan tested six devices, also underground, in the Chagai district of the province of Baluchistan. The exclusive club of six nations with the capability to deploy nuclear weapons since 1974 became seven amidst widespread international condemnation. The diplomatic fury was accompanied by economic sanctions formalised by the UN resolution 117 in June 1998. In the resolution, the UN recognised the new threat to international peace and raised its concerns about a potential arms race in South Asia. In 2016, India had around 110 nuclear devices and Pakistan about 140.[917] Neither India or Pakistan has conducted further tests since May 1998.

During the 1990s, as India developed its domestic space launch capability (the SLV-3, ASLV-3, and the PSLV), it was inevitably seen as a potential delivery mechanism for nuclear weapons. The dual use nature of missile technology, as well as the close connection between the civil nuclear power and nuclear weapons programmes, has helped and hindered India's space programme. As a nuclear power, India acquired greater political and diplomatic influence on the international stage. The sanctions after Pokhran II held back ISRO's cryogenic engine programme from which it is still recovering. India's first explicit use of space for national defence started with the Technology Experiment Satellite (TES). In 1999, the Indian Army supported by the IAF engaged in military action (the Kargil War) to regain positions that Pakistani soldiers and Kashmiri militants had occupied on the Indian side of the Line of Control. The Indian forces would have benefited from satellite imagery, but ISRO at the time did not have such capability. The US had satellites with sufficient resolution capable of assisting the Indian military but refused India's request for satellite images. The Kargil Review Committee Report highlighted India's deficiency in space-based high-resolution reconnaissance,[918] and TES was ISRO's prompt response to this recognition of a shortfall in India's strategic infrastructure at a critical time.

TES was designed and built in record time and placed in a 568 km SSO on 22 October 2001. With a panchromatic camera designed to provide a resolution of 1 m, it was arguably India's first instrument with military photo reconnaissance capability.[919] It incorporated a step and stare mode that provided high resolution in a wide field of view.[920] TES was classified, and unlike data from other Indian satellites, TES data was not made commercially available.[921] Before TES became operational, India had purchased images from the IKONOS satellite that made 1-m resolution images commercially available from 1 January 2000. While TES provided high-resolution imagery sufficient for strategic purposes, ISRO was also using it as an experimental test bed for numerous technologies. TES was designed to operate for only three years but returned data for over a decade. On 11 January 2007, China launched a ballistic missile from the Xichang Satellite Launch Centre that intercepted and destroyed one of its non-operational weather satellites, Fengyun-IC, in Earth orbit at an altitude of

863 km. The international community was surprised and alarmed by the demonstration of this military capability. The Indian military and political leaders recognised that they would have to respond. As a signatory to the Outer Space Treaty, India had maintained a strict stance on not weaponising space. Another incident in the following year further highlighted India's military shortcomings. The sophisticated terrorist attack in Mumbai in November 2008 motivated India to initiate its first dedicated photo reconnaissance satellite called RISAT-2 (Radar Imaging Satellite-2). Since the first INSAT series, RISAT-2 is the only satellite that ISRO has launched and operates but had not built.

RISAT-2 was acquired from Israel Aerospace Industries (IAI) in exchange for providing satellite launch services at a very short notice. It has a mass of 300 kg and is based on the TecSAR mini satellite designed and built by IAI. It was launched on 20 April 2009, six months after the Mumbai attacks, and placed into orbit by PSLV-C12, along with a 40-kg student-built microsatellite Anusat. RISAT-2 was designed to monitor hostile incursions across Indian borders, suspicious vessels at sea and for search and rescue.[922] It uses radar rather than visible or infrared light for imaging and incorporates Synthetic Aperture Radar (SAR) technology that can operate in spot, mosaic and strip imaging modes. The combination of a 3-m communication dish, radar and SAR allows RISAT-2 to provide imaging capability at a resolution of 1 m at any time (day or night) and in any weather (radar can penetrate clouds). Three years after RISAT-2, ISRO completed and launched RISAT-1. It was placed in a 536 km SSPO at an inclination of 41°. RISAT-1 is equipped with a larger data storage capacity (300 GB), higher capacity data transmitter and a battery to cater for the longer periods of an eclipse in the SSPO.[923] Instead of a traditional dish antenna, ISRO produced an unusual 6 m (along-track) x 2 m (cross-track) antenna customised for the SAR function. The orbits of RISAT-1 and RISAT-2 take them over parts of the globe not just India. Most of the data they collect cover countries other than India. Antrix and the National Remote Sensing Centre make this data commercially available on the international market, along with data from other ISRO remote sensing satellites. RISAT-1's polar orbit brings it over Norway's Svalbard ground station (SvalSat). Because of its unique geographical position, SvalSat is the only commercial station that can offer images from anywhere in the world an hour after they were taken. In 2015, Antrix entered into a commercial agreement with SvalSat that allows it to directly download data and generate images from RISAT-1.[924]

On 30 August 2013, ISRO launched GSAT-7, the first spacecraft dedicated to the Indian defence forces, specifically to provide communication services for the Indian Navy. Until then, the Navy had been using the British Inmarsat-C service. Also known as INSAT-4F and Rukmini, GSAT-7 was designed to replace the VSAT satellite communication service available to the military at the time. VSAT is a commercially available satellite-based data communication service of up to a few Mbit/s that requires a dish, typically about a meter in diameter. It is a service used by most commercial shipping and others who operate in remote locations and available from several international providers. GSAT-7 provides the Indian Navy's warships, including the Rajput-class destroyers, Brahmaputra-class frigates, Sukanya-class patrol vessels and submarines, with secure encrypted digital voice and data communication using four Ku, one S-band and three Ultra High Frequency and C-band transponders.[925]

Two years after the Navy's GSAT-7, the Indian Army received its communication satellite, GSAT-6 (also known as INSAT-4E), with five C-band and five S-band transponders in GEO. GSAT-6 is equipped with an unusually large 5 m diameter antenna, which was unfurled after launch. While the C-band communication footprint covers the whole of India and beyond, S-band transmissions use the large antenna to concentrate signals in five separate spot beams over India. This not only increases signal strength that facilitates small handheld communication devices but also allows a frequency to be reused dynamically, as each spot beam is geographically separated from the other. Radio frequency spectrum is a limited resource and is allocated by the International Telecommunications Union. Satellite operators are required to strictly adhere to frequencies allocated to prevent unintended radio interference and impairing operations of other satellites.

ISRO has built but not launched the dedicated satellite for the IAF, GSAT-7A, due to the non-availability of a frequency in the radio spectrum required for the satellite.[926] Unmanned Aerial Vehicles, airborne radar and autonomous missiles of the IAF are increasing the already high demand for a dedicated communication service.

On 2 October 2016, the ISRO chairman asserted that the "Indian space agency will not be found lacking in helping secure India's national interests now and in future.[927] This was the first time ISRO had publicly declared the active role of its satellites in supporting Indian military operations. Before India operationalised its overt military space assets, "the satellite requirements of its armed services were being met from existing facilities."[928] Having to use shared commercial communication satellites for secure military communications illustrated India's limited appetite and capacity to exploit space for military use. Further, in the absence of a reliable heavy launch vehicle, India must turn to a foreign launch service provider for launching its exclusively military satellites. In the interest of national security, nations usually avoid foreign nationals in the chain of design, build, launch and operations of strategic national space capability. ISRO, like NASA, is a public agency and a product of a democratic nation. When required for reasons of national security, its assets can be requisitioned for military use. Any of ISRO's satellites with a communication payload or the capability to image surface features at a high resolution of about 1 m or less can be used by the military for surveillance, intelligence gathering and offensive or defensive communication. At the beginning of 2014, four of ISRO's 25 operational satellites were suitable for military use.[929]

By the end of 2016, India's military presence in space broadly consisted of two dedicated satellites (GSAT-6 and GSAT-7) for use by its armed forces, the Cartosat and RISAT series and the NavIC constellation.[930] RISAT-1 and RISAT-2, though motivated by and designed for strategic requirements, are formally classified as Earth Observation satellites.[931] When required to meet national security demands, this capacity is shared between the Indian Navy, Army and Air Force. The military services ISRO provides include high-resolution real-time images for surveillance, navigation for military assets and secure voice, video and data communication the use of small mobile handsets.

Even though space was militarised with the advent of communication satellites, it has not been systematically weaponised as land, air and sea have been. By mid-2016, there were 1,419 operational satellites in Earth orbit, of which only 146 were categorised as military.[932] Further, none of the spacecraft currently in Earth orbit is designed and deployed with a capability and sole intention to destroy other spacecraft in orbit.[933] China, Russia and especially the US have developed a significant space-related military capability. India's military investment and presence in space is minimal.[934] Even though India's space programme has been uniquely civilian in its origin, ISRO is now active in pure scientific research and commercial space services and provides support to the Indian Armed Forces and intelligence services.

For All Mankind

The UN was constructed by peoples who had experienced first-hand the brutality of World War II. It was charged with the responsibility for developing a progressive international law and promoting international cooperation. The unstated hope of the member nations was that the UN would prevent another world war. The sudden arrival of Sputnik in 1957, introduced a new realm of space. It was extraordinarily different from anything humans had experienced before. In the midst of distrust that accompanied the Cold War, the UN was in the right place at the right time to facilitate international discussion and cooperation and develop the conventions, principles and treaties under international law on how this new realm should be accessed as a shared resource. A year after the launch of Sputnik, the United Nations General Assembly (UNGA) created Committee for the Peaceful Uses of Outer Space (COPUOS), a committee that encapsulated its remit in its name. Four years after Sputnik, in 1961, Yuri Gagarin demonstrated that space was an environment open to exploration by humans.

Figure 16-1 Soviet, UK and USA Ambassadors signing the Outer Space Treaty observed by US President Johnson. Washington DC January 27, 1967. Credit UNOOSA

The first step in this framework was the adoption of the Declaration of Legal Principles Governing the Activities of States in the Exploration and Use of Outer Space by the United Nations General Assembly (UNGA) in 1963. The five legal principles led to five UN resolutions and then five treaties. These five treaties lay out the framework for international law to protect space as an environment and ensure it is open to peoples of all nations for exploration for the mutual benefit of all mankind. Over the subsequent half century, through a number of conferences, committees, sub committees and working groups, most UN member nations have agreed to adhere to them. The UNGA's five Legal Principles that determine how member nations operate in space are:

- Declaration of Legal Principles Governing the Activities of States in the Exploration and Use of Outer Space. Adopted on 13 December 1963.

- Principles Governing the Use by States of Artificial Earth Satellites for International Direct Television Broadcasting. Adopted on 10 December 1982.

- Principles Relating to Remote Sensing of the Earth from Outer Space. Adopted on 3 December 1986.

- Principles Relevant to the Use of Nuclear Power Sources in Outer Space. Adopted on 14 December 1992.

- Declaration on International Cooperation in the Exploration and Use of Outer Space for the Benefit and in the Interest of All States, Taking into Particular Account the Needs of Developing Countries. Adopted on 13 December 1996.

Throughout human history, the discovery of a new resource has usually been followed by exploitation, contamination, conflict over ownership and denudation of the very attributes that made the resource unique. The South American rain forests, coral reefs off the coast of Australia, barren landscape of Antarctica and the slopes of Mount Everest are no longer as pristine nor plentiful as they once were. Shared common environments where individual actors focus only on their own specific needs inevitably lead to a premature loss for everyone. To protect space from such an eventuality and in keeping with the key elements of the UN charter, the pursuit of international peace, collaboration, friendship and harmony, the UNGA set up the COPUOS very early into the space age.

On 14 December 1972, the race to the Moon ended with the departure of Apollo 17 from the surface of the Moon. That was followed by the Apollo-Soyuz Test Program and the symbolic handshake between the American and Russian astronauts three hours after their spacecraft docked in Earth orbit in July 1976. Since then, the number and types of spacecraft entering space have increased, along with the number of countries that have developed the capability to do so. While space programmes in the past were shaped by political rivalries, today, they are more pragmatic, targeting commercial, economic and national developmental goals. The services delivered from space have evolved and become an essential component of a modern national infrastructure. The spacecraft in space delivering have become assets requiring military protection.

India has been a member of COPUOS from the outset in 1959. Throughout its development, the Indian space programme has nurtured close connections with the UN. In addition to financial support, the UN has provided a platform to support ISRO's international collaborative goals and provided a global window for its achievements. It was the COPUOS Scientific and Technical Subcommittee that in 1962 advocated the building of the equatorial sounding rocket launch station at Thumba. It was under the UN flag in Thumba that the US and USSR scientists, along with scientists from other nations, worked side-by-side, an unexpected COPUOS outcome in the midst of the Cold War. Since its inception, COPUOS has convened three international conferences, UNISPACE-I in 1968, UNISPACE-II in 1982 and UNISPACE-III in 1999. They shared a single remit, to examine the practical benefits of space exploration and the opportunities available to non-space-faring nations with special relevance to the needs of developing nations. The next one, known as UNISPACE+50, is scheduled for 2018.

In February 1968, Sarabhai led a formal ceremony at TERLS where Thumba was dedicated as a UN facility. Six months later, Sarabhai was in Vienna as the Vice-President and Scientific Chairman of the first UN Conference (UNISPACE-I) on the Peaceful Uses of Outer Space. Yash Pal from ISRO was the Secretary-General of UNISPACE-II in Vienna in August 1982. Four years after he stood down as the Chairman of ISRO, U.R. Rao held the post of COPUOS chairman during UNISPACE-III. M.Y.S. Prasad, a former director of Sriharikota, contributed to the Inter-Agency Debris Coordination Committee (IADC) Debris Mitigation Guidelines.

Until recently, only a "few nation-states controlled access to space, owned the most space assets and considered it their domain."[935] The COPUOS has become a forum for the development of international space law to prevent those 'few nations' from becoming the gatekeepers to space. Over a period of 12 years since 1967, five treaties have been adopted by the UN.

- The Outer Space Treaty. Treaty on Principles Governing the Activities of States in the Exploration and Use of Outer Space, including the Moon and Other Celestial Bodies. Came into force on 10 October 1967.

- The Rescue Agreement. Agreement on the Rescue of Astronauts, the Return of Astronauts and the Return of Objects Launched into Outer Space. Came into force on 3 December 1968.

- The Liability Convention. Convention on International Liability for

Damage Caused by Space Objects. Came into force on 1 September 1972.

- The Registration Convention. Convention on Registration of Objects Launched into Outer Space. Came into force on 15 September 1976.

- The Moon Agreement. Agreement Governing the Activities of States on the Moon and Other Celestial Bodies. Came into force on 11 July 1984.

It is these treaties that individual nations can choose to sign and ratify and thus commit to international law. Although not all member states sign and ratify all the treaties, most do adhere to them. So far, space activities in India have been the exclusive domain of the state-run ISRO. India has complied with Outer Space Treaty in the absence of domestic legislation. However, as the private space sector takes root in India as it has done elsewhere, domestic legislation will be required to ensure that treaty obligations continue to be met.

War and Space

The Outer Space Treaty focused on the peaceful uses of outer space and preventing the use of weapons of mass destruction in space. In the 1960s, with the growing nuclear arsenals of the two superpowers, the presence of nuclear weapons in space seemed inevitable. The Outer Space Treaty does not explicitly prevent militarisation of space assets, only weapons of mass destruction. To address that shortcoming, two additional treaties have attempted to stop space from becoming a future war zone, but with minimal engagement from key protagonists, both have been ineffective. In 2000, the UNGA passed the Prevention of an Arms Race in Outer Space (PAROS) Treaty. In 2008, the Treaty on the Prevention of the Placement of Weapons in Outer Space and the Threat or Use of Force Against Outer Space Objects (fortunately abbreviated to PPWT) was presented at the UN Conference on Disarmament (CD). The CD is the international platform for defence and military negotiations.[936] It was established in 1979 and has a membership of 65 nations. It meets annually presided by its members on a rotating basis with a specific remit of disarmament.[937] The nature and scope of disarmament include the cessation of the nuclear arms race, nuclear disarmament and prevention of a nuclear war and an arms race in outer space. PAROS and PPWT attempt to diminish the prospect of using space for war arising from the new capabilities based on technological innovations in lasers, communication and space transport.

With the more powerful and rich nations attempting to gain the upper hand, neither PAROS nor PPWT has made any substantial progress. The patterns of history prevail in the present. When one nation invests in a programme for national defence, its neighbours see it as a threat, initiating an arms race. The nuclear arms race of the 1960s is being replayed but with space as the new high ground. In 1949, the USSR developed its nuclear arsenal because of the US, China in response to the USSR, India because of China and Pakistan because of India. PAROS is in part a response to this build-up of nuclear arsenals on Earth and the fear of this race continuing into outer space. PPWT, primarily a Chinese and Russian initiative, was introduced because PAROS had not made much headway. With "its large missile defence program, technical superiority and huge military budget the United States has consistently refused to negotiate PAROS in the CD."[938] The US considers the PAROS principles too restrictive for its future ambitions and regards the Russian proposal as a tactic to prohibit or limit its access to space. It considers the PPWT ambiguous and its restrictions incompatible with its plans particularly regarding commercial opportunities in space.

The CD has already negotiated treaties, such as Non-Proliferation of Nuclear Weapons, the Convention on the Prohibition of Military or Any Other Hostile Use of Environmental Modification Techniques (ENMOD) and the Treaty on the Prohibition of the Emplacement of Nuclear Weapons and Other Weapons of Mass Destruction on the Sea-Bed and the Ocean Floor and in the Subsoil thereof (also known as the Seabed Treaty). By June 2015, there were 77 parties to ENMOD and 89 nations signed the Seabed Treaty.[939] In 1979, the Agreement Governing the Activities of States on the Moon and Other Celestial Bodies (also known as the Moon Treaty or Moon Agreement) came into effect, and there was the expectation that it would be supported by member nations in the same way that they supported ENMOD. After all, in 1979, conceptually the realm of seabed was perhaps just as unique and challenging an environment as the surface of the Moon.

The United Nations Office for Outer Space Affairs (UNOOSA maintains a record of the status of international agreements relating to activities in outer space. As on 1 January 2016, its records show that while most nations have ratified the Outer Space Treaty, Rescue Agreement, Liability Convention and the Registration Convention, most nations with a developed space programme, including the US, Russia and China, have refused to sign the Moon Treaty. India is among the four nations that have signed the Moon Treaty but have not yet ratified it. In its 2005 publication Vision for 2020, the US's Space Command expressed its vision for "Full Spectrum

Dominance". In this vision, the "medium of space is the fourth medium of warfare, along with land, sea, and air."[940] One assessment of this vision concludes, "nearly every country in the world but the US supports the preservation of space from weaponisation."[941] The US appears unable to endorse a treaty that will prevent implementing its principle of Full Force Integration (the integration of space forces with air, land and sea forces, enabling war fighters to take full advantage of space capabilities as an integral part of special, joint and combined warfare). Modern warfare, as with other aspects of 21st-century societies, has moved from a national to a global context. The US, USSR and China are consolidating their armed forces (land, sea and air) with their space capabilities under an aerospace or space command.

In 1982, the US's Air Force Space Command (AFSPC) headquartered at Peterson Air Force Base, Colorado, was established with a remit to provide global war-fighting capabilities. The AFSPC was responsible for space-based capabilities, including secure real-time audio and video communication, navigation and surveillance delivered by an established network of space-based assets. In 1985, acknowledging the increasing value of space for military applications, the US Joint Chiefs of Staff created a new unified command, the US Space Command. Its contribution was credited for the remarkably expeditious military success of the US-led coalition in the Gulf War in 1991. All wars since then have relied heavily on the insights of C4ISR (command, control, communications, computers, intelligence, surveillance and reconnaissance) that modern space technology offers.

Each of the US's military services Army, Navy and Airforce had its own Space Command but merged into a single Strategic Command in 2002. The US Space Command was disbanded, and its responsibilities were passed on to the US Strategic Command. It is one of nine unified commands in the US Department of Defence providing intelligence and cyber support and monitoring spacecraft and debris. It manages the orbit of ISS and triggers spacecraft (including ISS) orbital manoeuvre to avoid space debris when required. The targeted destruction of the satellite USA-193 satellite in orbit was also conducted by the US Strategic Command. Combining "military and space" has been the central, consistent theme as the US military evolved since the first Gulf War. Speaking in 2016, the commander of the US Strategic Command Cecil Eugene Diggs Haney asserted "my mission-space goes from under the sea all the way up to geosynchronous orbit."[942]

As the only superpower and with the largest economy, the US can set the international agenda for strategic use of space. It is probably the vision of

405

the US armed forces to extend operations to space that prevents it from making progress with international treaties, such as PPWT and PAROS. The military in China, Russia and India will be pressured to follow, but their economies are not sufficiently large to support this expensive endeavour. One assessment of "Vision 2020" paints a dark picture of the profound loss to all of humanity if war were executed in space. It would include the loss of a global space economy of over $300 billion, end of international cooperation that has taken decades to develop, huge loss of operational satellites and the societal services they provide and undermining the use of LEO for future generations because of the debris generated.

The Prevention of an Arms Race in Outer Space continues to be a reoccurring CD agenda each year. An amended draft of the PPWT was presented by China and Russia in 2014 making it the basis for further discussions at CD. However, adoption of the amended draft still looks unlikely because most states do not see an acceptable verification mechanism. What constitutes a weapon in space is also a matter of debate. An existing satellite in orbit could act as a weapon. A state, should it wish to, could manoeuvre its own satellite close to another state's satellite and command it to self-destruct. Russia and China, however, have been pressing hard to get a formal agreement. During the 2016 CD, Venezuela and the Russian Federation declared in writing that they "will not be the first to place weapons of any kind in outer space."[943] Support for the agreement is also channelled through the increasingly influential BRICS organization. A relatively new phenomenon that brings together India, China and Russia on one side against the US. Apart from the real possibility of challenging the US's ambitions of militarising space, a more trusting and cooperative relationship between Russia, India and China could be an unexpected outcome of BRICS.

The BRICS conference in the Russian City of Ufa in 2015 declared "negotiations for the conclusion of an international agreement or agreements to prevent an arms race in outer space are a priority task of the Conference on Disarmament, and support the efforts to start substantive work, inter alia, based on the updated draft treaty on the prevention of the placement of weapons in outer space and of the threat or use of force against outer space objects submitted by China and the Russian Federation."[944] The CD 2015 report commented on the US's responses to the 2014 joint draft from China and Russia.[945] A Russian view concludes, "Experts have noticed a positive reaction to the Russian proposal by India, and no objections in principle by other BRICS nations.[946]

In 2015, India publicly declared its position that "Outer space and celestial bodies were the common heritage of humankind and had to be used for the benefit and interest of all humankind in the spirit of cooperation. The prevention of an arms race in outer space and doing so urgently would avert a grave danger for international peace and security."[947] While it had ratified ENMOD and accessioned the Seabed Treaty, when the PPWT was established in 2008, India did no sign concerned that it may curtail its future space ambitions.

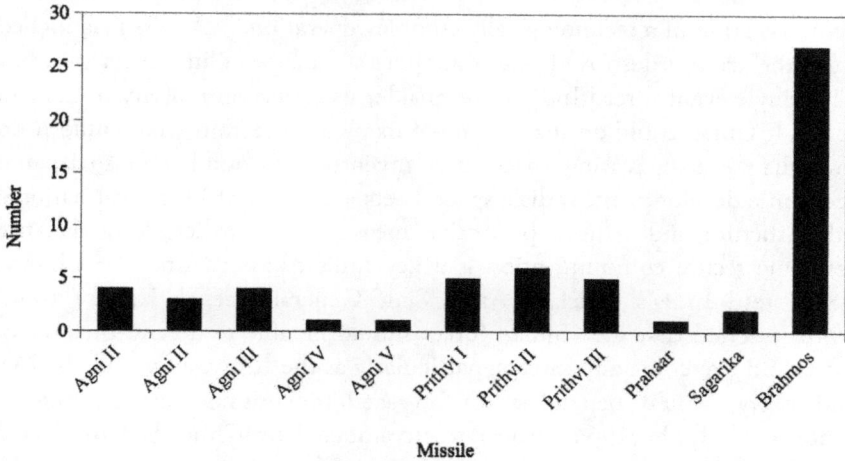

Figure 16-2 Number of tests in India by missile type. Credit Carnegie Endowment for International Peace

In the meantime, India is developing its conventional military capacity informed by its achievements in space. ISRO is a civilian entity and DRDO is not. Since both are funded by central government, it is inevitable that there is a flow of technology, know-how and personnel between them. Abdul Kalam moved to DRDO to develop India's ballistic missiles after leading the development of the SLV-3 in ISRO. Several missiles have been in development since the 1980s. Agni V has demonstrated a range 5,000 km. Brahmos is India's most tested missile developed with collaboration with Russia.[948]

Russia, China and now through the BRICS membership, India are challenging the US to come to an agreement that will cut back on its ambitions for military superiority in space. In the absence of an agreement, Russia, China and India will be forced to invest in defence infrastructure to keep up with the US. Even with their combined resources, China and Russia cannot financially compete with the US. Just as with the race to the Moon, the US can bring to bear a magnitude of resources that other nations cannot match. Large scale expenditure on defence has been a considerable hurdle for BRICS nations, but historically not for the US. A successful collaboration between the BRICS nations could potentially have an impact on keeping a check on the US's military expansion in space.

Anti-Satellite Weapons

China's 2007 anti-satellite test (ASAT) was a highly successful demonstration of a technologically complex operation.[949] A missile launched from the ground destroyed one of China's defunct satellites in Earth orbit. This single event forced India to reconsider its stance on military options in space. If China could do that to one of its own spacecraft, what could it do to India's ever-increasing space asset inventory? Grounded in social and economic development, India's space assets are a part of its critical national infrastructure and require protective measures.[950] Further, long distance real-time secure communication is a key military asset. After the Chinese ASAT test, India's Chief of Army Staff General Deepak Kapoor (born 1948) asserted that the "Indian forces should be able to defend their space assets and overcome any threats, particularly as the Army, as well as the IAF and Navy, would depend a lot on satellite communications even in peacetime."[951] The 1964 nuclear test conducted by China had motivated India's nuclear weapons programme; that cycle may now be repeated for ASAT weapons.

ASAT weapons are designed to attack spacecraft for strategic and tactical military advantage. Traditional weapons, such as guns, machine guns and cannons, have been attached to all military vehicles, including ships, helicopters and aircraft. Doing that on a spacecraft in orbit is problematic. The recoil from their use is tantamount to a thrust changing the spacecraft's attitude or orbit.[952,953] Several designs of ASAT weapons have been tested since the late 1950s, including high energy ground-based lasers, charged particle beams and the increasingly effective cyber technology that can electronically jam or interfere with a target spacecraft rendering it inoperable without destroying it.

Most ASATs are technologically advanced versions of missiles originally developed as anti-aircraft missiles that can be launched from land, air or sea. To target a spacecraft in Earth orbit, the ASAT weapon or missile must travel faster, further and higher while navigating to a small target that itself is moving at around 8 km/s. Such a sub-orbital missile does not need to carry an explosive warhead because it can destroy its target by the energy resulting from the high-speed collision, making it a kinetic kill weapon. An ASAT weapon can also be co-orbital, a small explosive carrying satellite delivered to orbit where it approaches its target slowly and explodes when close by.

China, having publicly and spectacularly made its point, provided assurances to the US that it "will not conduct future ASAT tests in space."[954] A year after China's ASAT test, however, the US demonstrated its ASAT capability by destroying its own failed photo reconnaissance satellite, USA-193. It was destroyed in February 2008 using a missile launched from the warship Lake Erie stationed near Hawaii. The satellite USA-193 launched in December 2006 had failed a few months later. It was in a low (around 250 km) and decaying orbit and would have re-entered within a few weeks anyway. Unlike the Chinese ASAT test, the US had announced their intentions before the strike and insisted it was not an ASAT test. It claimed that the action was necessitated by the potential risk from a possible impact should its large fuel tank survive re-entry. Most debris resulting from the destruction of USA-193 burnt up during re-entry within a few weeks and all within a year because of the lower orbit.

The US, Russia and China have successfully conducted ASAT tests in the past. India is under pressure to do so, too. India's armed forces have expressed their desire for military capability in space. As part of its Integrated Guided Missile Development Programme (IGMDP), India has been developing ballistic missiles since 1983, when IGMDP was established. It was headed initially by Abdul Kalam following his work as the project scientist on SLV-3. The primary objective of IGMDP is to counter the threat from hostile neighbours. The programme has developed a series of missiles, including Akash, Prithvi and Agni, which can be used for a variety of military tactical scenarios. These missiles use rocket technology, including one, two or three-stage rocket motors mostly using solid propellant with a short range. Variants that are more recent have evolved to long-range multi stage missiles using a liquid engine and Ramjet technology.

India's long-range missiles, Agni-5 and Prithvi-3, have the capability for sub-orbital spaceflight. If India were to develop ASAT capability, it would probably incorporate Agni and Prithvi missile technologies. Following the successful test flight of Agni-V in 2012, the head of DRDO, V.K. Saraswat (born 1949), declared that Agni-V could be modified to deliver "defence satellites into a low Earth orbit during an emergency."[955] However, as a military programme, the status of India's ASAT weapon development has not been made public.

Before the Chinese ASAT test of 2007, formal announcements by the foreign and defence ministers of the government of India had kept its plans for military use of space ambiguous. They reiterated India's stance on the peaceful use of outer space while keeping the option open for military use of space in the future.[956] As a civilian organisation, ISRO has not historically been involved in developing overt military technology. Given the changing geopolitical landscape and the ISRO chairman's determination to support India's strategic objectives, it is inevitable that ISRO will pivot to a more military posture in the future. In 2015, the then director of Sriharikota highlighted ISRO's existing capability that could be re-tasked in the interest of national security. He pointed out "any country that has the capability to precisely launch a satellite to a precise orbit also has co-orbital ASAT or a Kinetic Kill capability to destroy a satellite."[957]

India has also belatedly recognised the significance of cyber security threats to its armed forces and space assets. It has been slow in responding to cyber threats to its terrestrial IT systems, including banks, railways and online services.[958] ISRO may already have been impacted by cybersecurity breaches of its terrestrial and space assets.[959] Recognising the urgency, in 2012, the Indian government committed to training around half a million cyber security specialists dubbing them "cyber warriors."[960]

Each of the three services Army, Navy and IAF, has constituted its own space and cyber cell, as well as dedicated special forces, including Para-SF battalions (Army), Marine commandos (Navy) and Garuds (IAF). Their defence-specific space requirements are managed by a small tri-services unit, the Integrated Space Cell, operating under the Integrated Defence Headquarters. Progress on developing a new joint services command for space, cyberspace and clandestine warfare in the form of special forces has been slow and is still awaiting government approval.[961] Most developed nations rely at least in part on space based infrastructure for the security services. As in other countries, a nation turns to its space agency to fulfil its national security needs.

In 2015, a decade after it was first proposed, the defence minister Manohar Parrikar (born 1955) directed the integrated defence staff to "work out and fine-tune" the "basic structures" for the Cyber, Aerospace and Special Operations Commands.[962] As with its potential ASAT response, India has not formally announced plans for a large-scale space command capability. ISRO operates its space programme under domestic policies and laws, primarily the Satellite Communication Policy published in 2007 and Remote Sensing Data Policy published in 2011. India is in the early stages of developing its policies for space security and military. Dr Rajeswari Pillai Rajagopalan, Senior Fellow and Head of the Nuclear and Space Policy

Initiative at the Observer Research Foundation, asserts "India must take steps to declare a space policy, or at least its key aspects."[963] When the organisational structures are finalised, ISRO will have to reconcile its primary objective of providing space-based services for national development with its commercial, scientific and strategic capabilities.

Space Debris

The Chinese ASAT carried out in 2007 generated one of the largest pools of space debris ever recorded, including 3,000 trackable fragments (10 cm or larger). Space debris is defined as any non-functional man-made object that could pose the risk of unintended collision to operational spacecraft in Earth orbit or those transiting that region to or from an interplanetary mission.

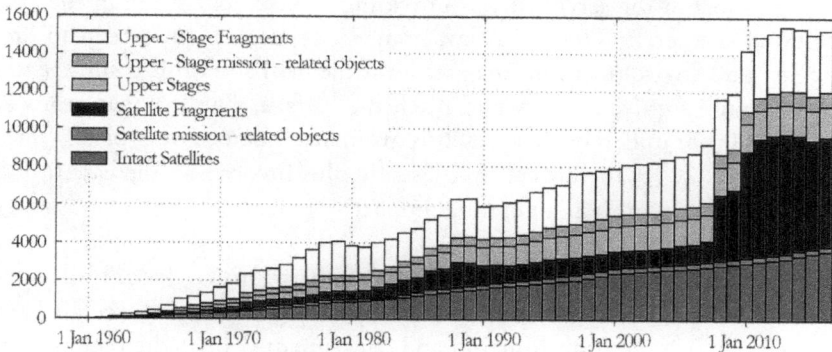

Figure 16-3 Accumulation of space debris Jan 1960 – Jan 2017 in all orbits. Credit ESA

The high-velocity impact between Fengyun-1C at 954 kg and the ground-launched missile (catalogued as SC-19 by the US military) at 600 kg took place within a popular 863 km SSPO. Over time the 1.5 tonne of debris, including fragments of solar panels, antennae, batteries, sensors and transmitters, formed a vertical, pole to pole ring around the Earth. Only 10% of the around 30,000 fragments are sufficiently large to be tracked.[964] During a statement at the UNGA, one scientific assessment concluded that the number of catalogued objects larger than 1 cm "which accounts for more than 25 percent of all catalogued objects in low Earth orbit, will stay in orbit for decades, and some for more than a century."[965] Around 80% of the debris from the Chinese ASAT will remain in Earth orbit for the next century. Had the collision taken place at a lower altitude, the lifetime of the debris would have been significantly shorter.

Two years later, on 10 February 2009, an inactive 900 kg Russian Cosmos 2251 satellite and an active 700 kg Iridium 33 satellite unintentionally collided generating around 2,000 fragments larger than 10 cm and many smaller will stay in orbit for decades. This was the first unintended collision between spacecraft in Earth orbit and took place at an altitude of 800 km at a speed of over 10 km per second at almost 90° to each other; one was in an equatorial and the other in a polar orbit. Spacecraft have been leaving Earth for space for nearly 60 years, but most of the space debris is a product of the last decade. The two events of 2007 and 2009 generated two-thirds of the space debris that is regularly monitored to mitigate the potential risk posed to operational spacecraft in orbit.

Space debris, especially that is small and untraceable, is a hazard for all users of space, no matter who was responsible for creating it. The speed in LEO of even small objects is around 7.5 km per second a resulting collision is a very energetic event. Such an impact could damage or destroy a spacecraft leading to loss of the service it was providing or even loss of life in the case of crewed spacecraft. The risk is not limited to space. It can extend to life, property and environment on the ground if the debris in orbit is sufficiently large, it could survive re-entry and reach the surface. The ISS that orbits at around 450 km undergoes space debris avoidance manoeuvre typically once a year, but in 2015, there were five. Usually, this involves an increase in ISS orbital altitude to prevent a potential impact.

Space debris is not just a danger to spacecraft in orbit. Uncontrolled re-entry is a hazard to life and property on Earth. Historically spacecraft have not had a built in de-orbit function and consequently, once the fuel ran out a spacecraft would simply be abandoned in orbit. New private operators such as OneWeb with plans to deliver broadband services from space with large constellations of around 900 satellites build in a de-orbit function at the outset. That will not only ensure that LEO is cleared of non-function spacecraft, but a controlled re-entry will take place over uninhabited parts of the Earth.[966] Space debris has been increasing ever since the launch of Sputnik in 1957 as the number of spacecraft and nations that rely on them have increased. Humans have continuously been in space since October 2000 with a maximum of 13 (in 2009) at any one time.

The number of operational spacecraft in orbit is around 1,500. More countries are demanding greater space-based services including navigation, communication and meteorology and more countries are developing the capability to build, launch and operate spacecraft. As the number and density of spacecraft in orbit increase so does the risk of collision. The runaway effect of debris from one collision causing another, generating more debris and further collisions, the Kessler Syndrome is now considered as a serious threat.

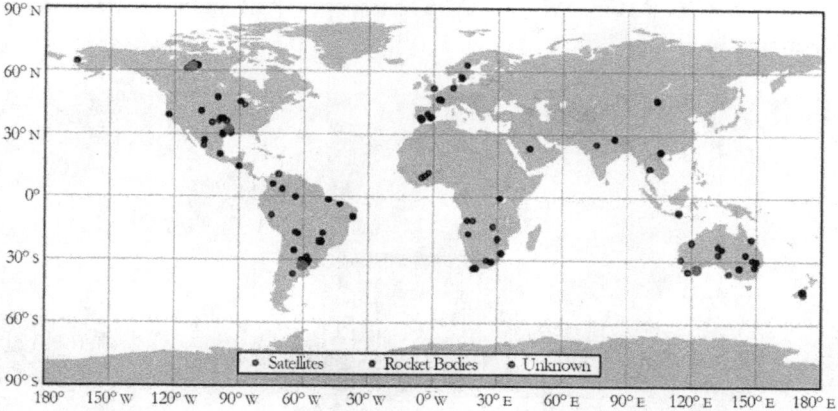

Figure 16-4 Location of debris objects that have been recovered after re-entry.
Credit ESA

In 2010, the British Parliament highlighted the risk stating that the Kessler Syndrome "could make it impossible to operate satellites safely in the future."[967] Debris in LEO of about 400 km is not a serious concern, as there is sufficient atmospheric drag to force it to re-enter within about two years or sooner. With increasing altitude, the rate of orbital decay is smaller where it will remain in orbit for decades. Objects of about 10 cm in LEO are routinely tracked using Earth-based radar, while objects 1 m or larger are monitored using Earth-based optical telescopes. Space debris in GEO is too far for radar and monitored only by optical telescopes.

There are over 20,000 large objects over 10 cm in LEO that is, up to 2,000 km, with a majority concentrated around 750 km. Of the 20,000 large objects in orbit, approximately 1,500 are active spacecraft. Space debris is tracked by a global network of observatories coordinated primarily by two organisations, the International Scientific Optical Network (ISON) managed by the Keldysh Institute of Applied Mathematics of the Russian Academy of Sciences and the Space Surveillance Network (SSN) managed by the US military. ISON is an international project with about 30 optical telescopes in 20 observatories located in ten countries, while the SSN consists of ground-based radars and optical sensors at 25 sites worldwide.

ISRO has now joined the global effort to track space debris with two installations. A Multi-Object Tracking Radar consisting of large phased-array-antenna-based tracking radar recently established at Sriharikota. It will have the capability to locate and track debris of 25 cm x 25 cm at 1,000 km orbit. Two 1-m optical telescopes fitted with 4000 x 4000 CCDs are being installed 2,000 km away on Mount Abu and will be remotely controlled by ISRO from its MCF. Mount Abu telescopes will help to track debris fragments in GEO.

Figure 16-5 Multi-Object Tracking Radar. Credit ISRO

The most significant contribution to the growth of space debris has been from accidental collisions between spacecraft and deliberate ASAT weapon tests. However, unintended explosions from unused fuel, pressure vessels and pyrotechnics of spacecraft in LEO have also contributed to increasing the number of objects in space well beyond the number of spacecraft launched. On 22 December 2007, the UNGA endorsed the Space Debris Mitigation Guidelines that derived from the work done by IADC. The IADC is an international forum for exchanging information between member space agencies on space debris research activities designed to identify debris mitigation guidelines.

The guidelines reflect the practices that have been developed and are being practiced by some national and international organisations. Space debris can arise from each of the three phases that constitute a space mission - launch, operation and end-of-life. The UN invited the IADC member states to voluntarily incorporate three principles in their national space management procedures to mitigate space debris: (a) limit debris released during normal operations, (b) minimise the potential for in-orbit break-up and collision and (c) remove non-operational objects from populated regions.[968]
The IADC guidelines include the following:

- Space debris mitigation should be considered at the mission design phase. Launch vehicles and spacecraft should be designed to minimise debris release, such as connectors between stages, spacecraft release mechanisms, sensor covers and so on, during normal operations. The design phase should also consider minimizing probable failure modes leading to accidental breakups.

- Pyrotechnics built into launch vehicles and spacecraft for stage separation or as self-destruction mechanism should be made safe once a spacecraft is in orbit.

- At the end of the mission, the risk of accidental break-up or explosion should be minimised by releasing residual propellants, controlled discharge of high-pressure vessels, dampening flywheels and momentum wheels and disconnecting and discharging batteries.

- Spacecraft terminating their operational phases in LEO region should be directed to leave the operational orbit and re-enter through a de-orbit trajectory.

- Spacecraft in GEO orbits should be moved 235 km or more above their allocated orbit to free the slot for future spacecraft. Orbital slots in GEO is a "scarce natural resource, whose importance and value increase rapidly with the development of space technology", with capacity for about 2000 satellites, as calculated in the 1980s.[969] GEO slots are allocated to individual countries by the UN's International Telecommunications Union. By carefully and dynamically managing their positions, ISRO operates more satellites in GEO than GEO slots allocated to it.[970]

India has ratified two treaties that have come out of IADC, the treaty banning nuclear weapons tests in the atmosphere, outer space and under water and the convention on the prohibition of military or any other hostile use of environmental modification techniques. Space debris is increasingly seen as the "greatest risk to the space environment" that can only be mitigated through international collaboration.[971] As the number of spacecraft in orbit increases, abiding by the IADC guidelines will become increasingly more significant. While an international consensus on mitigating the risk of debris proliferation is becoming established, it is taking place against a backdrop where space is the next theatre for war (after land, sea and air) and a centre commercial competition.

◁ ◇ ▷

Chapter Seventeen
The Road Ahead

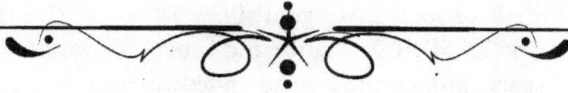

In 1970, Vikram Sarabhai noted that a holistic approach was necessary for national development. He wrote "there is a totality of the process of development which involves not only advanced technology and hardware but the imaginative planning of supply and consumption centres, of social organisation and management, to leapfrog from a state of backwardness and poverty."[972] India's space programme is one essential element of a larger national undertaking that is necessary for that leap. Even before becoming India's first prime minister, Jawaharlal Nehru had initiated policies and the building of institutions recognising the profound role of science in fulfilling his vision of a self-sufficient India "it is science alone that can solve the problem of hunger and poverty, of insanitation and illiteracy."[973] Since its inception, India's space programme has always received strong political support from every Indian government irrespective of its political colour. Relative to other parts of the Indian government behemoth, the DOS under which ISRO operates is seen as a beacon of efficiency and technological excellence.

Independent India had emerged into a world immersed in optimism, uncertainty and the beginnings of a Cold War. In 1947, it was not just India that awoke "to life and freedom" but most of Europe after the horrors of World War II. The tumultuous events of two world wars in the same century had redrawn the map of Europe, forced people apart in some places and together in others, introduced the most powerful weapons ever devised and triggered a Cold War that prevailed almost for the rest of the century. Partly in response to this emerging Cold War, President Harry S. Truman's (1884–1972) inaugural address in 1949 directed the US to "embark on a bold new programme for making the benefits of our scientific advances and industrial progress available for the improvement and growth of underdeveloped areas."[974] The targets of this policy were the Asian and African countries that had gained or were pursuing independence. India was a top target. A division emerged based on western democracy and the USSR's communism. Countries that aligned with the US and democratic ideals came to be known as the "west" and those that aligned with the

USSR and communist ideals became the Warsaw Pact. Nehru refused to join either bloc, rejecting both "communist teachings... and anti-communist teachings" insisting that "If I join any of these big groups I lose my identity."[975]

Nehru and his successors have guided India to receive technological assistance from the USSR, US, France and others for various programmes, including the space programme, since independence. That support was modulated by the changing geopolitical landscape.[976] The US quietly supported India's desire for a nuclear weapon but only after China had successfully tested its first nuclear weapon in 1964. A decade later, when India conducted its first nuclear test, Pokhran-1, the US introduced sanctions. The Nixon government, with an eye on air bases in Pakistan to monitor the USSR, always considered Pakistan a natural ally. The US supported Pakistan in the 1971 Bangladesh Liberation War and imposed on India one of the earliest sanctions regimes in its history, leading to an almost three-decade long technological apartheid.[977]

Why should a country faced with problems of illiteracy and basic healthcare invest large sums of its national income on a spacecraft to Mars? How can India justify receiving international aid while choosing to invest in a space telescope in Earth orbit for pure scientific research? Investment in space programme for nations struggling to address basic needs of employment, healthcare and education for large proportions of its populations appears wrong and inappropriate. The effectiveness of space based solution to address them are not tangible nor obvious. Science and technology were key contributors to creating the developed world. Since independence, a series of India's elected leaders have navigated India's continued development based on science and technology.

Rockets or Rotis

Has the Indian space programme had a substantial impact on reducing poverty? Some observers conclude that the answer is mixed, pointing out that India's space assets have cost more than the revenue they generate.[978] India could have acquired its space assets commercially, and substantial capital expenditure could have been saved. The infrastructure at Sriharikota, VSSC, SAC, ISAC, MCF and PRL for designing, building and launching spacecraft; communication; spacecraft management and R&D would not be required. However, ISRO was not designed to be a commercial entity despite its deepening commercial activities. It provides services, including healthcare, education, fishing advisories and weather services to farmers that make a daily impact on the lives of millions of its citizens in a myriad of ways. In a delightful and detailed audit of the social impact of ISRO's services, S.K. Das records one example of profound but unmeasurable outcome, "you see that old lady with a bent back? It is for people like her that ISRO would have made the journey safe with its landslide maps."[979]

Further, there is no universally fixed definition or mechanism to measure poverty. What poverty is and how it is measured changes with time and place.[980] Despite an expensive and targeted programme, a 'War on Poverty,' initiated by the Lyndon Johnson administration in 1964, the US has been unable to eradicate poverty.[981] Since the remarkable technological success of its Mercury, Gemini and Apollo space programmes in the 1960s, the poverty rate in the US has hovered between 12% and 15%.[982] In India, during the last quarter of the 20th century, poverty fell from about 50% to about 25%.[983] As a consequence of the economic reforms of 1991, the Indian economy has grown at a remarkable pace pulling around 10 million people a year above the poverty threshold in the last decade.[984] The growth of the Indian economy has mirrored the reduction of poverty.[985]

India's economic transformation over the last two decades has quietly turned it from a net recipient to a donor of international aid.[986] While India still receives and accepts international aid to help address the needs of its 300 million poor, the UK is winding down its contribution of $300 million (Rs.2037.5 crore) per year, along with other western nations. At the same time, India's contribution to international assistance has been growing. India has (a) pledged $10 billion (Rs.66,800 crore) in line of credit towards development projects in Africa, (b) committed $1 billion (Rs.6,700 crore) to Afghanistan for its fight against terrorism,[987] (c) contributed $825 million (Rs.5,500 crore) in grants for students from 156 developing countries to study in Indian universities via the Indian Technical and Economic Cooperation Programme and (d) become the largest donor to neighbouring countries, including Bhutan, Bangladesh, Nepal and Sri Lanka.

By 2020, India is expected to be the youngest country in the world with half of India's 1.3 billion population under the age of 26. With more people in the workforce, this demographic dividend will add to the boom in consumer spending and fuel future growth. There will be greater demand for existing space based services and new services such as mobile telephony and satellite based internet access. For many these have become a routine expectation, such that, speaking at UNISPACE-II, Yash Pal, a highly respected scientist who has held several senior posts within ISRO, posed the question, "should mankind not begin to conceive of new "minimal human rights" which would include the right to communicate...?"[988] ISRO's achievement in nation building are substantial and growing. Perhaps its most significant contribution is in fostering a sense of national pride, prestige and self-sufficiency as envisaged by its founders. In the absence of the domestic capability, including space-based services India would be forced to do without them or continue to depend on other nations at high cost. Both options are costly and would perpetuate, not diminish poverty.

Satellite TV, Demand and Supply

The unqualified success of SITE, which demonstrated the benefit of DTH broadcasting from space in 1975–76, inspired a nationwide appetite for DTH services in the 21st century India. Between 2010 and 2015, the DTH market grew by 25% annually.[989] Only 53 million of the 250 million Indian households can receive DTH services. The other 200 million are approximately equally split between those that have cable TV and those that do not have a TV at all. The popularity of large high-definition (HD) flat screen TV is adding to HD television content further increasing the demand for data capacity. By 2016, India had seven DTH suppliers (Videocon d2h, Sun Direct, Reliance Digital, Dish TV, Tata Sky, Aortal DD Direct Plus and Digital TV) providing 900 TV channels, of which 50 carried HD signal. This is expected to grow to 1,300 with 150 HD channels in 2017.[990]

The growing economy and middle class has triggered a surge in demand for DTH services, and ISRO is struggling to meet. By 2015, Indian DTH services were using 78 Ku-band DTH transponders, of which ISRO was supplying only 19.[991] The remaining 59 were purchased from foreign satellite providers by Indian DTH suppliers through ISRO in an administratively complex and costly process.

Each satellite has only a limited number of transponders; INSAT-4B, for example, has just 24. Further, only a fraction of the transponder capacity owned by ISRO is available for use by domestic DTH suppliers. Most are commercially leased or used for national services, such as COSPAS-SARSAT, GAGAN , meteorology and remote sensing. At the end of the 11th FYP in 2011, only 198 transponders of the planned 500 were available from the INSAT/GSAT satellites. The loss of two communication satellites GSAT-4 and GSAT-5P and 40 transponders between them in 2010 further exacerbated the problem.[992] In the 12th FYP (2012–2017), the projected demand was for 794 transponders, but ISRO planned to increase to only 398 transponders.[993] These transponders were to be delivered through the launch of 14 additional communication satellites over the period of the

12th FYP.[994] As conservative as it was, ISRO still has not been able to meet this target. Despite its many successes, ISRO is failing to meet the domestic demand for transponders and is looking to lease them commercially from international satellite operators.[995]

Private Sector

The global space industry estimated to be worth $323 billion (Rs. 2,041,038 crore) in 2015. A tiny amount of this revenue accounts for satellite manufacturing (0.04%) and launch services (0.01%) around the world. The US dominates the commercial launch market and ISRO engaged in just 1 commercially procured launch in 2014 and two in 2015.[996] In India, space activities remain primarily within the purview of the state-owned ISRO. In 1992, ISRO established Antrix as an independent entity to conduct its commercial activities. Through Antrix, ISRO has been generating income by providing services, including the launch of foreign satellites, the sale of data from its EO satellites and the leasing of transponders on its communication satellites. However, a nascent commercial space sector is beginning to emerge in India. Some of the reasons for weak private sector participation can be traced back to the early economic policies. For the first four decades of Independence, Indian economy was guided largely by the socialist ideals of its government. Burdensome red tape hindered entrepreneurship and industrial innovation, and the private sector in India was virtually non-existent until economic controls were relaxed. This pre-liberalisation period between 1947 and 1991 is cynically known within India as the 'Licence-Permit Raj.'[997] In 1991, India's economic regulations were relaxed, from when the economy has seen steady growth.

Over the last two decades, a $100 billion (about Rs. 670,000 crore) information technology industry has developed in India, and it has come a long way since the initial low-cost outsourced call centres and first line software support for the US and European multinationals. Since then it has moved to higher value services in software development, cloud services and data centre hosting. Other private sector industries, such as pharmaceuticals, materials and machine tools, have also been flourishing.[998] Despite progress in some sectors, such as IT, engineering, semiconductors and telecommunications, India is still largely procuring Western technology. For example, commercial aircraft, nuclear power, specialised medical scanners and defence systems. While internationally, the commercial space sector has been growing led by the private sector, entrenched Indian bureaucracy and government regulations continue to block private sector opportunities in India.

With over two decades of existence, Antrix has demonstrated the value of commercial opportunities in space. Indian technology start-ups are now participating actively in the resurgence in private sector space companies. This is an international phenomenon known as New Space 2.0. Over the last two decades, governments around the world have turned to IT systems to deliver their services. Now they want to exploit Space based services too. Recently the DOS identified 160 projects across 58 ministries and departments where the Indian private sector could help deliver government obligations using geospatial data.[999] The regulatory and legal framework required for New Space 2.0 is not yet in place. The government is beginning to address this along with structures needed for governance, investment, collaborative working between government and non-government (including academia) participants as well as international partners.

Private sector start-ups have emerged in the Indian space sector. Earth2Orbit, founded in 2007 and headquartered in Mumbai, offers services, including the launch of foreign satellites by ISRO. Satsearch is an online repository for the small but growing supply chain of private companies producing space hardware. Based in India, the companies listed are located around the globe. Mysore based Bellatrix Aerospace was established in 2015. It conducts research and development for orbital launch vehicles and electric propulsion for satellites. Dhruva Space, based in Bengaluru, is planning to build small satellites in India for amateur radio operators and, with ISRO's help, place them in orbit. Axiom Research Labs is participating in the Google Lunar X-Prize (GLXP) through Team Indus. The GLXP is an international competition for the private sector to land a rover on the Moon, travel at least 500 m and send back high-resolution images before the end of 2017. In 2014, Team Indus was one of three teams to win the milestone prize of $1 million (Rs.62 crore). It has signed a contract with ISRO to launch its lunar rover to the Moon current scheduling means the first rover to land on the Moon from India may not be ISRO's.[1000]

While no private company in India is engaged in large-scale projects for space, many provide space-related components, subsystems or services. ISRO planned 50 launches during the 12th FYP (2012–2017), which would require 200 rocket engines, but it did not have the necessary manufacturing capacity.[1001] It relied on private companies, including Godrej and Boyce to make engines for PSLV and GSLV and parts for the cryogenic engines; Larsen and Tourbo for rocket motor cases, rocket nozzles, solar arrays and wind tunnels; Walchandnagar Industries for antennae, satellite components and rocket motor casings; Venketeswara Industries for hardware and Tata Advanced Material Limited for advanced composite materials. Currently, only 40% of the materials required for the cryogenic stage are manufactured domestically. The quantities required for the rest are not sufficiently large to justify establishing an Indian manufacturing facility.[1002] A total of 130 organisations contributed to the making of ISRO's MOM and the PSLV-XL launcher, of which only nine were from the public sector.[1003] Around 500 companies in India supply products or services to help ISRO's programmes. Instead of a few large private companies being involved in large-scale space projects, many Indian private companies make a small contribution to ISRO's space missions.

Worldwide, the number of companies offering products or services, such as satellite launch, spacecraft subsystems or components, sub-orbital human spaceflight. There is vigorous growth in several areas of space technology, and many private companies are already engaged to varying degrees in a variety of space ventures. They include human spaceflights, cargo delivery to the ISS, asteroid mining, designing and building space ports, exploration of the Moon, hotel in space and the development of single-stage-to-orbit launch vehicles.[1004] India's private sector is not yet technologically matured to compete in this highly competitive and growing international market dominated by companies, such as Space-X, Lockheed Martin, Blue Origin and Airbus. The US private sector is already responsible for the delivery of materials to and from the ISS, and soon, it will provide human spaceflight capability, too.

Despite the 1991 economic liberalisation, India's economy is still not sufficiently free to provide the support required by new start-ups. The barrier for the private sector making to make a substantial contribution is still largely structural and a product of the bureaucratic mentality that values administrative power over innovation and technical progress. Speaking in October 2014, ISRO chairman Radhakrishnan stated "Indian space agency will be creating a single entity to undertake launch missions, the entity will be a mix of public, private sectors, and ISRO itself."[1005] If such an entity is created and it includes ISRO, there is a high probability that its success will be undermined by the very bureaucratic hurdles that have prevented its establishment to date. What the Indian space programme requires is an entirely private space sector modelled on the remarkably successful Indian IT sector. Private space companies require an environment free of excessive regulations where young Indian entrepreneurs are free to innovate. ISRO has established and grown its intellectual property portfolio to 10 trademarks, 45 copyrights and 270 patents. It has listed 29 spin-offs that have come out of its R&D work. They include fire extinguishing powders, search and rescue beacons, ground penetrating radar, adhesives and GPS software for mobile computers.[1006] The Vikram 1601 microprocessor used by all ISRO's launch vehicles for guidance, navigation and trajectory control, was developed in-house by ISRO at its Semi-Conductor Laboratory in northern India.[1007]

As part of ISRO's technology transfer programme, over 300 technologies associated with electronics, computer-based systems, polymer chemicals, electro-optical instrumentation, satellite communications and broadcasting have been passed on to Indian industries.[1008] However, ISRO remains the intellectual property owner of the key technologies used by the private sector to manufacture launch vehicles and spacecraft components. The growth of these private companies is inherently tied to that of ISRO's. If ISRO were to allow its intellectual property to be used by the Indian private sector internationally for commercial purposes, the economies of scale might allow industry to compete on the world stage. ISRO itself would benefit from the resulting lower cost, shorter lead times and higher availability.

After national governments driving space operations for over half a decade, a tipping point has now arrived where the public demand for space services is at a level that makes commercial sense for private industry to step in and take a more prominent role. ISRO recognises that it can only deliver by boosting its engagement with the private sector. International treaties, such as Article VI of the Outer Space Treaty, require governments to assume liability for the space activities of private companies within their national borders. A new "risk v reward quotient" is attracting private investment in innovative public-private partnerships.[1009] In early 2016, the ISRO chairman announced the establishment of an industrial consortium managed by Antrix that will integrate and launch PSLV from 2020.[1010] Another consortium, Alpha Design, overseen by ISRO, is building two IRNSS standby satellites. The first satellite was built with the active participation of ISRO and the second will be built by the consortium alone. ISRO is also looking at the option of setting up a special economic zone (SEZ) close to SDSC-SHAR, where independent companies could be set up to do international and domestic space business. In this way, ISRO hopes to foster and grow a private sector space capability within India.

On its 50th anniversary, The Outer Space Treaty is being reviewed to reassess its applicability in the 21st century. Particularly in respect of its growing commercial opportunities in space. During a hearing of the space subcommittee of the Senate Commerce Committee in the US in June

2017, the value of US's continued commitment to the OST was debated. One view highlighted the international destabilising consequence of the US pulling out of the OST would be to create " confusion and uncertainty, hindering new commercial developments as well as established private sector space activities."[1011],[1012]

The number of private space companies has now exceeded government agencies in the US. It was the US that led the content of the original OST in 1967. Through the publication of the draft bill entitled the American Space Commerce Free Enterprise Act (FEA) of 2017, the US is set to shape the regulation of private space activity. Once approved, the FEA bill will determine if the US will exit the OST, amend it or how its obligations of international laws will be implemented in domestic law. The US has approved two private space ventures (the Bigelow inflatable habitat attached to the International Space Station, and Moon Express lunar lander mission) using a cumbersome process involving three separate agencies. The Federal Communications Commission (FCC) for telecommunications, National Oceanic and Atmospheric Association (NOAA) for remote sensing and the Federal Aviation Administration (FAA) for launches and re-entry. Ideally, the new bill will identify a single one-stop-shop that can process private sector space projects. The implementation model US settles on will most likely inform how other nations will interpret international law.

ISRO's ecosystem has evolved such that Indian space-related private sector companies work only with ISRO, rather than the international space market.[1013] Perhaps, as ISRO engages with government initiatives, such as Make in India and Digital India, private sector investment could see a sharp growth, with Indian companies involved in space service emerging on the international stage, just as the Indian IT companies did in the 1990s. Further, to encourage foreign investment, the limit for Foreign Direct Investment (FDI) was raised from 26% to 49% in 2014. The Make in India initiative of the government has a particular focus on space, allowing 100% FDI.[1014] Cooperation between India and Russia in the space sector, where the two nations have had a long and deep tradition, is also expected to be one of the areas of growth.[1015] Between 2004 and 2013, only 3% of the global launches took place in India, and only 3% of the satellites were manufactured in India.[1016] India is sixth on the list of nations with indigenous space capability and has a vast potential for growth.

Research and Development

Since independence, India has underinvested and lacks in the number and quality of R&D institutions.[1017] In 1970, ISRO established the Sponsored Research (RESPOND) programme to exploit the wider potential of Indian universities, industry and institutions to conduct leading edge research in space technology and space applications.[1018] The main objective was to establish strong links with academic and professional institutions throughout India to carry out quality research to support ISRO's programmes. Since the 1970s, ISRO has allocated 1% of its annual space budget to the RESPOND programme for sponsored research.[1019]

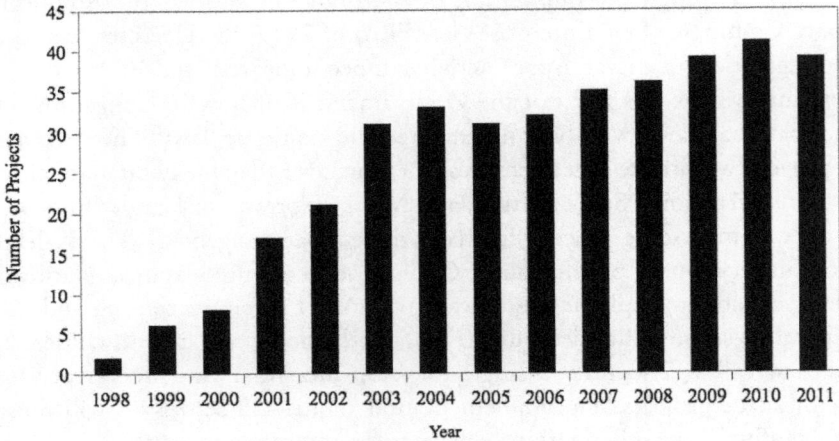

Figure 17-1 Number of Projects in the RESPOND Programme between 1998 and 2011. Credit Adapted from Vikas Patel and Ankita Patel[1020]

By the 1980s, RESPOND had matured creating stronger links with some institutions that became the base for 'ISRO cells,' including the IISc in Bengaluru and the Indian Institutes of Technology in Chennai, Mumbai and Kanpur. Over time, through this initial connection, many promising young graduates have found their way into full-time employment within ISRO. One study listed 239 projects between 1998 and 2011. The number of RESPOND programmes initiated and completed each year varied. A total of 21 RESPOND projects were completed between 18 different institutions and universities in the period 2010–11. Between them, they generated eight PhDs, two M. Tech thesis, produced 51 staff (principle investigators, fellows and associates) and published 78 papers in India and internationally.[1021] A more recent ISRO study of 71 completed RESPOND projects between 1998 and March 2014 concluded that 15 (21%) were good, 17 (24%) were excellent and 39 (55%) were very good. The 71 projects were distributed across 60 institutions at a cost of 732.06 lakh (1.1 million USD) and generated 18 PhDs.[1022]

During the 1950s, it was the industrial development in the private sector along the US east-coast, the Boston-Harvard-MIT route (also known as Route-128), that provided much of the technical and engineering innovation that allowed the US to become the world leader in military technology and helped NASA's Apollo programme to land on the Moon.[1023] In the 21st century, Silicon Valley on the US's west coast became the cradle of the information revolution and continues to be the primary source of fresh ideas that shape the technology revolution worldwide.

In 1984, NASA established the Office of Commercial Programs as a culmination of the strategy initiated by the Reagan administration's National Policy on the Commercial Use of Space. In the US, the private sector continues to lead the technological innovation for government-sponsored space projects. Perhaps, ISRO hoped that RESPOND would be the spark to inspire a Route-128-like phenomenon in India. However, in India, this golden triangle of "lab-academia-industry" has not fully emerged.[1024] In the absence of a Route-128 equivalent in India, ISRO's ability to grow remains limited.

During the late 1960s, Vikram Sarabhai visited MIT several times and recognised that R&D companies on Route 128 were stimulating innovation and industrial growth.[1025] Although ISRO has a long way to go to match the US's route 128 model, it has taken the first steps. Now, ISRO is looking beyond the RESPOND programme for new ways to engage industry. Under a new initiative starting in 2017, ISRO will offer larger sponsorships in the region of Rs.20–30 lakhs ($30,000–40,000) over longer periods extending between 2 to 3 years to any parties, not just academic and government agencies. The R&D projects will be more advanced, sophisticated and riskier designed to create materials, components and systems for future launch systems.[1026]

Global Space Market

Determining a precise quantitative value of ISRO's economic output is difficult as it is for any large complex organisation supported by the public sector. This is true for NASA, JAXA and other national space programmes around the world. Development resulting from access to education and medicine and economic growth enabled by the national space programme are mostly intangible and inherently difficult to measure. Satellite-based early warning systems save lives, livestock, crops and urban infrastructure, the value of which is impossible to quantify. The space race of the 1960s not only enhanced national prestige but also accelerated the development of technology, generating wider social and economic benefits and enhanced quality of life in Russian and American societies. As developed and developing nations increase their dependency for mobile communications, navigation systems and online access, the commercial market for building and launching space-based services will continue to grow.

Modern economies are increasingly reliant on space based services (such as traffic lights, transport system, banking services, freight etc.) where most service users are usually unaware of the dependency on orbiting spacecraft. This increasing demand is manifest in the observed growth of the global space market, and the trend is set to continue. In India, the number of landline phones, around 40 million, was surpassed quickly by an estimated 850 million mobile phones. During 2010–2011, India saw a growth of 15 million mobile phone subscriptions, every month.[1027] The number of space-based services, including telephone communication, satellite TV and Satnav services, is set to increase in the coming decade.

The number of personal computers, including laptops and tablets, is increasing by around 15% per year. By 2020, India will have the largest working age population in the world, which will further increase the demand for space-based services. At the beginning of the Space Age, developments in the space industry helped stimulate the consumer market; today, it is the other way around. Innovation in hardware and software, especially in digital consumer products, such as high-speed network switches, computer memory, digital cameras and particularly smart phones, has helped to develop the small satellite market. The hardware used for modern tablets and smart phones is smaller, cheaper, faster and demands less power than equivalent hardware only a few years ago. Modifications required to get this hardware to work in the extreme conditions of space are not onerous.

Today, small satellites with powerful functionality can be developed quickly at low cost by assembling subsystems using COTS components. Small satellites are low in weight and consequently cheaper to deliver to orbit. Commercial organisations that in the past found the cost of space assets prohibitive are now helping to build a new customer base for low-cost space services. This new market offering a range of services, including communication, search and rescue, disaster management and basic research, is attracting clients, such as academics, amateur radio enthusiasts and small countries that otherwise would not be able to justify the cost. Cubesat, a 10-cm cube, is a popular nano satellite used predominantly for testing and R&D by institutions, universities and non-profit organisations.

The use of small satellites has seen a growth from about 25 in 2013 to over

80 in each of the subsequent 3 years. They typically cost between $10,000 and $20,000 (Rs.6 to 13 lakhs) but a new suit of launchers dedicated to small satellites is expected to dramatically reduce the costs to around $30,000 (Rs. 3 lakhs).[1028] However, the income they generate is minimal, and occasionally for student/academic research projects, they are launched without charge.

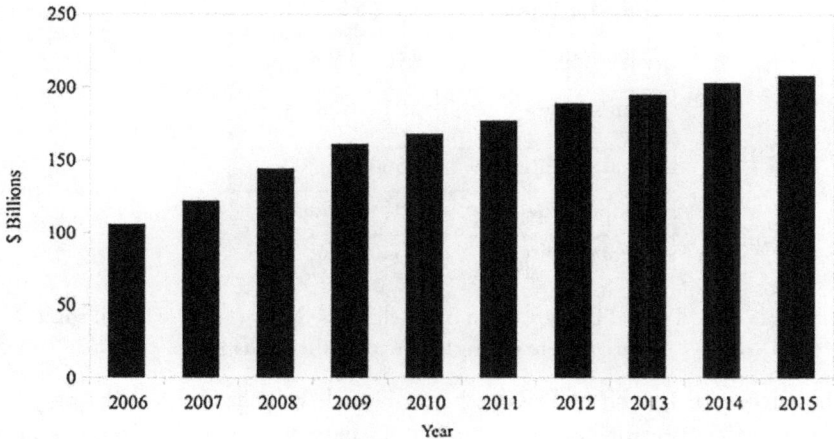

Figure 17-2 Global Satellite Industry Revenues ($ Billions). Credit Adapted from The Satellite Industry Association Report 2016

Of the 92 space launches conducted worldwide in 2014, 23 were subject to a commercial agreement. In 2014, ISRO conducted four launches, one of which was commercial.[1029] In June 2014, the launch of SPOT 7 and four other smaller satellites using a PSLV generated $15 million (Rs.100 crore) for ISRO. With that launch, ISRO clocked 40 satellites launched on a commercial or bilateral (inter-governmental arrangements where services rather than money are exchanged) basis. Most of these were nano satellites, with less than 10 kg mass, and only two were sufficiently large to carry a payload with remote sensing capabilities.

ISRO is generating sufficient income from launch and other services to claim financial self-sufficiency, but it has a long way to go to exploit its potential in the launch services market. In 2017, Google sold its satellite business, Terra Bella, to Planet Labs Inc. a San Francisco start-up that operated a fleet of 60 small satellites providing live high-resolution images from orbit. Terra Bella provides applications and services that exploit data in real-time for commercial application. In 2017, PSLV-C37 launched Cartosat-2D but made the headlines for the 104 satellites launched in one go. Most of the satellites, 88, were for a private US company Planet Labs.

ISRO and PSLV is well placed for this small satellite market but the income they generate is low.

Satellite Class	Mass Range*
Femto satellite	10 – 100 g
Pico satellite	< 1 kg
Nano satellite	1 – 10 kg
Micro satellite	11 – 100 kg
Mini satellite	101 – 500 kg
Small satellite	<500 kg
Medium satellite	501 – 2,500 kg
Large satellite	> 2,500 kg

Table 17-1 Typical Classification of Satellites by Mass (* There is still some variation in definitions within the industry)

The successful launch of ISRO's heavy launch vehicle GSLV-Mk3 in June 2017 is a significant achievement for its launch capability. It must now operationalise the GSLV-Mk3. Until that happens, not only will ISRO continue to bear the huge costs of commercial launches of its communication satellites by ESA's Ariane, but its portfolio of commercial services will remain limited. More significantly, it will not achieve one of its key objectives, self-sufficiency.

Regional Space Power

Most of ISRO's commercial clients are geographically distant from India. Singapore, Europe and Canada. Near neighbours, such as Sri Lanka, Bangladesh, Pakistan and Afghanistan, have typically procured space services from the US, Europe or China but not India. In November 2014, only a few months after coming to office, Prime Minister Modi (Narendra Damodardas Modi, born 1950) initiated a programme to change that. He expressed his willingness for Indian space assets originally designed exclusively for Indian use, be available to the region around India. Addressing the 18th South Asian Association for Regional Cooperation (SAARC) summit in Nepal in November 2014, he committed ISRO to design, build and launch a communication satellite for use by SAARC member nations. In addition to providing each SAARC member the use of the communication satellite, he offered to extend the range of the IRNSS footprint for Satnav capability to all SAARC countries.[1030] The inclusion of the foreign minister as a member of the Space Commission also points to the strategic shift that India wants to assert its regional influence.[1031]

The satellite was launched in May1027 using a GSLV-Mk2 at the cost of $35 million (Rs.200 crore) and making one Ku-band transponder available to each of the participating SAARC nations without charge. Each nation will be responsible for building the supporting ground infrastructure and deciding how the communication capacity is used (i.e. for DTH, VSAT, tele-education, telemedicine, disaster management support or other).[1032] The plans for an almost 2-ton SAARC satellite with 12 Ku-band transponders did not progress smoothly. Pakistan withdrew from the project citing concerns about the security of its communications going through an Indian satellite. Initially, it had supported the idea and offered financial and technical assistance with the satellite construction. Fearing that Pakistan would want a stake in the project management, India declined. This was as close the two neighbouring space agencies (ISRO and SUPARCO) have ever come to a collaborative space venture.[1033] The interest from Bangladesh and Afghanistan had also waned as they progress with plans for their own communication satellites. In the absence of unanimous agreement in SAARC, India renamed the satellite the South Asia Satellite (also known as GSAT-9) and launched on a GSLV-Mk2 in May 2017. Maldives, Bangladesh, Sri Lanka, Bhutan, Nepal, and India will use it for communication.

International Collaboration

India has benefitted from its cooperation with many nations along its journey to becoming a successful space faring country. During the Cold War, in the 1960s and 70s, the USSR and US kick-helped to start India's space programme with equipment, training, rocket motors and satellite launch services. In the 1980s, France helped with liquid engine technology know-how, and in the 1990s, the USSR assisted at least temporarily with cryogenic engine technology. Other nations that have played a key role in India's space programme include Japan, Germany, the UK and Israel.[1034] In addition to his intellect, Bhabha had the contacts essential for international collaboration, and Nehru, who was committed to developing a science-based self-reliant India, considered such collaboration to be critical. It was also made possible by the deep personal connection between Bhabha and India's first prime minister, Jawaharlal Nehru that stretched back to Bhabha's boyhood with the socially connected Tata family.[1035]

Some space projects rely on international collaboration others (for example, the ISS, Search and Rescue, SBAS and sharing of deep space communication networks when operating interplanetary missions, debris mitigation, climate change monitoring and human missions to Mars) require it. Since the Apollo missions of the early 1970s, no humans have returned. When they do, that too could be a collaborative effort. The International Lunar Decade, an initiative designed to provide a framework for strategically directed international cooperation for a permanent return to the Moon, will be launched on 20 July 2019 – the 50th anniversary of the Apollo 11 landing.[1036]

Since 2000, the ISS has maintained a continuous human presence in space made possible by international collaboration facilitated by a complex set of agreements between 15 sovereign nations[1037] Apart from the engineering achievement in building and operating it and the value of the science conducted in orbit, it is the trust and fostering of deeper relationships between the peoples of member nations on Earth is probably the ISS's greatest accomplishment. The more nations compete and collaborate in sports, arts, music, business and science, the less likely they are to fight on the battlefield.

India's space assets have allowed it to participate in and contribute to global collaborative ventures. These include international search and rescue (COSPAS-SARSAT), high precision air traffic management (GAGAN), satellite Navigation (NavIC), space debris avoidance (Multi Object Tracking Radar) and soon climate change.[1038] International collaboration is essential to achieve human civilisation's grand ambitions in space. Human missions to establish and inhabit a permanent, sustainable Moon base, to walk on Mars and to visit other bodies of the solar system are complex and expensive undertakings. They will only be possible through international collaboration grounded in science and technology.

Science and technology underpin human civilisation allowing 7 billion people to live longer, healthier and more productive lives. It was also the scientific and technological capability that facilitated the vast destruction and enormous loss of life and destruction of brutal wars, bloody revolution and ruthless conquest that is also part of human civilisation. It was from the final years of World War II that the UN emerged as an instrument to prevent its repetition. The pressing need to "learn to live together in peace and harmony" preoccupied the minds of world leaders in the immediate aftermath.[1039] The UN was founded with noble aims, including "cooperate in solving international problems of an economic, social, cultural or humanitarian character." Despite its many repeated failures, the UN is still the most powerful instrument the planet has ever seen for uniting all peoples from all nations.

The progress of human civilisation in the 21st century hinges on cooperation in space. India, representing such a large part of it, is playing an increasingly significant role. The European Enlightenment and Industrial Revolution emerged from national programmes engaging science and technology on a large scale. Developing nations have arrived at that point in their journey to modernise where they, too, are now investing, economically and intellectually, in science and technology and are poised to enter the first world. Economic growth, increased investment and achievements in space are critical if India is to complete its journey from the third world to the first.

The longstanding tensions between India and China could softened as both nations find themselves on the same side through their BRICS membership. The US and Russian collaboration on the ISS in space has contributed to reducing the likelihood of and perhaps preventing conflict between them on Earth. On Cosmonauts Day 12 April 2016, President Putin in a live call to the ISS, asserted that "in spite of any difficulties that we are encountering on Earth, people are working in space shoulder to shoulder, hand in hand.[1040]

Humans have not left Earth orbit since 1972. Al Worden, the command module pilot of Apollo 15, was one of only two dozen humans who did and holds the view that "we will go to Mars in the 2030s time frame, and if we are smart we will include the Chinese and Indians not only for the technical expertise but the economics."[1041] Rakesh Sharma, the first Indian to go to space concluded that the next time humans leave Earth, they should do so not as Americans, Indians or Chinese but as people from planet Earth."[1042]

Value of Space

In 1948, British scientist Fred Hoyle (1915–2001) predicted that "once a photograph of the Earth, taken from the outside, is available... a new idea as powerful as any in history will be let loose."[1043] That was a decade before Sputnik and two decades before an iconic image of Earth was recorded from the lunar orbit. After what had been a particularly violent year, 1968 concluded with the optimism of successfully completing one of humanity's oldest dreams. On the first human journey from the Earth to the Moon, the Apollo 8 commander Bill Anders concluded that they went to the Moon but "what we really discovered was the planet Earth."[1044] The event is widely considered to have initiated the international environmental movement. The ideas that led to Yuri Gagarin's journey to Earth orbit or Neil Armstrong's walk on the Moon were at the outset fanciful, outlandish and reckless. The veteran ISRO scientist Yash Pal was referring to the same audacious and risky ambition when he said in Hindi "pagalpan hai hum logon meh," which loosely translates to "there is madness in us all."[1045] The same apparently irrational logic drives a developing nation to invest in a space programme to address the challenges of poverty. In the past it was slavery, child labour and profound inequality of sex and race that that shamed and undermined the notion of "human civilisation". Today it is inequality and widespread poverty. Space technology is an integral part of the solution.

Figure 17-3 Earth from Apollo 8 in lunar orbit. Christmas Eve 1968. Credit NASA

Human civilisation throughout its long history has repeatedly demonstrated that science-based societies provide higher quality and more fulfilling lives for their citizens than those that are not. Benefits of space-borne resources offer a greater potential to transform large developing nations, such as India, than the developed countries from where spaceflight first emerged. Although the Indian and Japanese space agencies, ISRO and JAXA, were established in 1969, the role of Japan's space programme has been very different to that of India's.[1046] From the outset, India's space programme has been deeply intertwined with its ambitious country-wide developmental goals. The bold decisions taken by its founders, Bhabha, Sarabhai, Dhawan and others, demonstrated that the basic needs of undeveloped communities could be met with space technology with a modest financial commitment from the central government. Space-based assets provide services that exploit the unique global perspective available only from Earth orbit, not only services, such as early warning systems for tsunami, search and rescue and weather, but also communication, navigation and satellite television that have become critical to the quality of life in 21st-century societies. They also offer something more profound but less tangible, a unique perspective of our place in the universe.

Chasing a Chimera?

India's remarkably successful space program is matched by the audacity in the ambitions of its founders. R. Aravamudan was one of the first engineers selected by Vikram Sarabhai to join the Indian Space Program in 1962. He describes the bewilderment of his colleagues when he chose to take up this position. "When they heard of my decision to volunteer, most of my colleagues thought I was mad. Why on earth would I want to sacrifice a career with steady growth prospects to chase a Chimera" (a fire breathing hybrid creature from Greek Mythology - something that does not really exist).[1047]

Pursuing India's first nuclear power station in the 1960s, Homi Bhabha had a clear vision of India's trajectory to development. Developing nations should not repeat the technological steps of developed nations "that is one thing we must not do if we are ever to catch up."[1048] In the past telephone connections, TV reception and electricity supply came into private houses via vast ground-based networks. Many parts of India and other developing nations have chosen to skip this traditional approach of physical cables stretching across the nation. Homes today have a mobile telephone, a satellite dish for TV and electricity from solar panels on the roof. China surpassed one billion mobile phone users in 2012 and India in 2016. Traditional ground-based infrastructure does not have the agility to scale to meet the increasing demand.[1049] By the start of the next decade, high-speed global Internet access from space will transform communication services, especially for rural areas not served today.[1050] As countries around the world increase their dependency on space-based services, India through ISRO is better placed than most to face the challenges of the coming decades.

Writing in 2005, Abdul Kalam expressed his vision for India's objectives between 2005 and 2030. He wanted to see India progress to manned missions to the Moon and Mars, reusable spacecraft, harnessing energy for use in Earth, interplanetary exploration, integrated disaster management and navigation satellites.[1051] About half in Kalam's list have been largely met. At the outset, they were just as ambitious as ambitious as those that remain. With over a decade to go other developing nations can find India's progress an inspiration.

The formidable challenges of improving the quality of lives on a national scale are not unique to India. India's space programme serves as a model for other developing nations faced with similar social problems. Apart from being a role model, India's space programme has a tangible impact on its neighbouring countries. The footprint of its communication, weather, search and rescue and navigation satellites spills over its national borders. Solutions in the 21st century are guided by the pressing global need to sustain the limited natural resources, minimise the impact of global warming and responsibly use land and sea for food production. In a world that has never been more interconnected and interdependent, the solutions for social development must also be global. Irreversible erosion of natural resources can only be replaced by sustainable development once disparities among nations are addressed.[1052] Around 1.5 billion people of South Asia (Afghanistan, Bangladesh, Bhutan, India, Maldives, Nepal, Pakistan and Sri Lanka) live a predominantly agrarian lifestyle making a particularly heavy demand on the environment. It is in these regions where high population density places tremendous pressure on the environment that strong strategies for sustainable management are required.[1053]

Only space-based observation can provide the coverage required for large scale land management, irrigation, drought mitigation, harvest maximisation and monitoring of natural resources, such as fishing zones, forests, mines and coastal erosion. Long term strategic decisions based on information from space-based assets lead to better decision-making and interventions.[1054] Only space-based assets can bring cost-effective education to large, remote and dispersed populations. The cost of knowledge is high, but the cost of ignorance is higher. Through its space programme, India is fulfilling its tryst with destiny. Throughout human history, progress has relied on exploiting the latest technological innovations. In the past, tools made from stone, iron and steel shaped human progress. Civilisation would not have been possible in the absence of the plough, printing press and the internal combustion engine. Today, it is space technology. From a standing start in 1963, India has demonstrated the power of space-based technologies to transform a nation. Developing countries will remain as developing countries unless they engage with modern space technologies.

◄ ◊ ►

Appendices

Abbreviations

AEC	Atomic Energy Commission
AAI	Airport Authorities of India
ALTIKA	Altimeter in Ka band
APPLE	Ariane Passenger Payload Experiment
ARIES	Aryabhatta Research Institute for Observational Sciences
ASAT	Anti-satellite
ASI	Astronomical Society of India
ASS	Aligarh Scientific Society
ATM	Automatic Teller Machine
BARC	Bhabha Atomic Research Centre
BNCSR	British National Committee for Space Research
C1XS	Chandrayaan-1 X-ray Spectrometer
C4ISR	Command, Control, Communications, Computers, Intelligence, Surveillance and Reconnaissance
Cartosat	Satellite for Cartographic Applications
CCD	Charge-Coupled Device
CCSDS	Consultative Committee for Space Data Systems
CCSRC	Commonwealth Consultative Space Research Committee
CD	Conference on Disarmament
CE	Cryogenic Engine
CERN	European Centre for Nuclear Research
CES	Crew Escape System
CHACE	Chandra's Altitudinal Composition Explorer (a neutral mass spectrometer)
CLGS	Closed Loop Guidance System
CNES	Centre national d'études spatiales (National Centre for Space Studies, France)
CNSA	China National Space Administration
COPUOS	Committee on the Peaceful Uses of Outer Space
COSPAR	Committee on Space Research
COTS	Commercial off-the-shelf
CPM	Charge Particle Monitor
CSIR	Council of Scientific and Industrial Research
CUS	Cryogenic Upper Stage
CZTI	Cadmium Zinc Telluride Imager
DAE	Department of Atomic Energy
DFRL	Defence Food Research Laboratory
DOS	Department of Space
DMR	Dual Mode Ramjet

DRDO	Defence Research and Development Organisation
DSN	Deep Space Network
DTH	Direct-to-Home
ECIL	The Electronics Corporation of India Ltd
EDUSAT	Educational Satellite
EEJ	Equatorial ElectroJet
EGC	Engine Gimbaling Control
ELDO	European Launcher Development Organisation
ENMOD	Convention on the Prohibition of Military or Any Other Hostile Use of Environmental Modification Techniques
EO	Earth Observation
EOSAT	Earth Observation Satellite Company
EPS	Electric Propulsion System
ESA	European Space Agency
ESCES	Experimental Satellite Communication Earth Station
EU	European Union
FAA	Federal Aviation Administration
FAC	Failure Analysis Committee
FCC	Filling Control Centre
FEA	Free Enterprise Act
FLP	First Launch Pad
FUV/NUV	Far Ultra Violet / Near Ultra Violet
FYP	Five Year Plan
GAGAN	GPS Aided Geo Augmented Navigation
GEO	Geostationary Earth Orbit
GEOSAR	Geostationary Search and Rescue
GNSS	Global Navigation Satellite System
GPS	Global Positioning System
GSAT	Geosynchronous Satellite
GSLV	Geosynchronous Launch Vehicle
GSO	Geosynchronous Orbit
GTO	Geosynchronous Transfer Orbit
GTS	Great Trigonometrical Survey
HAL	Hindustan Aeronautics Limited
HEF-20	High Energy Fuel
HLV	Heavy-lift Launch Vehicle
HSF	Human Spaceflight
HTPB	Hydroxyl-terminated Polybutadiene
HySI	Hyper Spectral Imager
IACS	Indian Association for the Cultivation of Science
IADC	Inter-Agency Debris Coordination Committee
IAF	Indian Air Force
ICAO	International Civil Aviation Organisation

ICSU	International Council for Scientific Unions
ICT	Information and Communication Technology
IDSA	Institute for Defence Studies and Analyses
IDSN	Indian Deep Space Network
IGMDP	Integrated Guided Missile Development Programme
IGY	International Geophysical Year
IIRS	Indian Institute of Remote Sensing
IISc	Indian Institute of Science
IIST	Indian Institute of Space Science and Technology
IIT	Indian Institute of Technology
INC	ISRO Navigation Centre
INCOSPAR	Indian National Committee for Space Research
INMCC	Indian Master Control Centres
INSAT	Indian National Satellite System
IPRC	ISRO Propulsion Complex
IQSY	International Quiet Sun Years
IRNSS	Indian Regional Navigational Satellite System
ISAC	ISRO Satellite Centre
ISITE	ISRO Satellite Integration and Test Establishment
ISON	International Scientific Optical Network
ISRO	Indian Space Research Organisation
ISS	International Space Station
IST	Indian Standard Time
ISTRAC	ISRO Telemetry, Tracking and Command Network
ITAR	International Trade in Arms Regulation
IYAS	International Years of the Active Sun
JAXA	Japanese Aerospace Exploration Agency
JPL	Jet Propulsion Laboratory
LAM	Liquid Apogee Motor
LAP	Lyman Alpha Photometer
LAXPC	Large Area Xenon Proportional Counter
LEO	Low Earth Orbit
LEOS	Laboratory for Electro-Optic Systems
LEOSAR	Low Earth Orbit Search and Rescue
LH2	Liquid Hydrogen
LIDAR	Light Detection and Ranging
Li-Ion	Lithium-Ion
LISS	Linear Imaging Self Scanner
LLRI	Lunar Laser Ranging Instrument
LOX	Liquid Oxygen
LPSC	Liquid Propulsion Systems Centre
LUT	Local User Terminal
LVM3-X/CARE	Crew Module Atmospheric Re-entry Experiment

M3	Moon Mineralogy Mapper
MCC	Mission Control Centre
MCF	Master Control Facility
MENCA	Mars Exospheric Neutral Composition Analyser
Mini-SAR	Miniature Synthetic Aperture Radar
MIP	Moon Impact Probe
MIT	Massachusetts Institute of Technology
MLP	Mobile Launch Pedestal
MMH	Monomethyl Hydrazine
MOM	Mars Orbiter Mission
MON	Mixed Oxides of Nitrogen
MoU	Memorandum of Understanding
MSM	Methane Sensor for Mars
MSS	Mobile Satellite Service
MST	Mobile Service Tower
MTCR	Missile Technology Control Regime
NARL	National Atmospheric Research Laboratory
NASA	National Aeronautics and Space Administration
NGC	New General Catalogue
NGO	Non-Governmental Organisation
NNRMS	National Natural Resources Management System
NOAA	National Oceanic and Atmospheric Administration
NPL	National Physical Laboratory
NRSA	National Remote Sensing Agency
NRSC	National Remote Sensing Centre
OST	Outer Space Treaty
PAROS	Prevention of an Arms Race in Outer Space
PBAN	Polybutadiene Acrylonitrile
PSDS	Planetary Data Standard
PPWT	Treaty on the Prevention of the Placement of Weapons in Outer Space and the Threat or Use of Force Against Outer Space Objects
PRL	Physical Research Laboratory
PROCAD	Precise Orbit Computation using Accelerometer Data
PSLV	Polar Satellite Launch Vehicle
R&D	Research and Development
RADOM	Radiation Dose Monitor
RCS	Reaction Control System
Resourcesat	Satellite for Resources Management
RESPOND	Sponsored Research
RFNA	Red Fuming Nitric Acid
RISAT	Radar Imaging Satellite
RLV	Reusable Launch Vehicle
RLV-TD	Reusable Launch Vehicle Technology Demonstrator

RRI	Raman Research Institute
RS	Restricted Service
SAA	South Atlantic Anomaly
SAARC	South Asian Association for Regional Cooperation
SAC	Space Applications Centre
SAC	Space Application Centre
SAR	Synthetic Aperture Radar
SARA	Sub keV Atom Reflecting Analyser
SARAL	Satellite with ARGOS and ALTIKA
SATNAV	Satellite Navigation
SBAS	Satellite-Based Augmentation System
SDSC	Satish Dhawan Space Centre
SEP	Société Européenne de Propulsion
SIR-2	Near-Infrared Spectrometer
SITE	Satellite Instructional Television Experiment
SITVC	Secondary Injection Thrust Vector Control
SLP	Second Launch Pad
SLR	Satellite Laser Ranging
SNEPP	Subterranean Nuclear Explosion for Peaceful Purposes
SPROB	Solid Propellant Booster Plant
SPT	Stationary Plasma Thrusters
SPS	Standard Positioning Service
SRE	Space Capsule Recovery Experiment
SROSS	Stretched Rohini Satellite Series
SSAB	Solid Stage Assembly Building
SSM	Scanning Sky Monitor
SSN	Space Surveillance Network
SSO	Sun-synchronous Orbit
SSPO	Sun-synchronous Polar Orbit
SSTO	Single Stage to Orbit
SSR	Solid State Recorder
STEP	Satellite Telecommunications Experiments Project
SUPARCO	Space and Upper Atmosphere Research Commission
SVAB	Second Vehicle Assembly Building
SXT	Soft X-Ray Telescope
TCM	Trajectory Control Manoeuvres
TEC	Total Electron Content
TERLS	Thumba Equatorial Rocket Launching Station
TES	Technology Experiment Satellite
TIFR	Tata Institute of Fundamental Research
TIS	Thermal Imaging Spectrometer
TMC	Terrain Mapping Camera

TSTO	Two-Stage-to-Orbit
TT&C	Telemetry, Tracking and Command
UDMH	Unsymmetrical Dimethylhydrazine
UHF	Ultra-High Frequency
ULV	Unified Launch Vehicle
UN	United Nations
UNGA	United Nations General Assembly
US/USA	United States of America
USSR	Union of Soviet Socialist Republics
UTC	Coordinated Universal Time
UVIT	Ultraviolet Imaging Telescope
VAB	Vehicle Assembly Building
VHF	Very High Frequency
VHRR	Very High-Resolution Radiometer
VRC	Village Resource Centre
VSAT	Very Small Aperture Terminal
VSSC	Vikram Sarabhai Space Centre

List of Interviews

Dr M.Y.S. Prasad. Director of Sriharikota. Sriharikota 15/01/2015. Interview at Sriharikota on operations at Satish Dhawan Space Centre. He was a member of the UN's Interagency Space Debris Coordination committee.

Professor Mustansir Barma. Former Director of TIFR Mumbai 5/01/2016. Interview at TIFR in Mumbai on the history of TIFR, Homi Bhabha and the deep connections between ISRO and TIFR.

S.K. Shivakumar. Director of ISAC. Bangalore 26/03/2014. Interview at ISAC in Bangalore on the evolution of Indian Deep Space Network.

Dr Mylswamy Annadurai. Mars Orbiter Mission Programme Director. Bangalore 26/03/2014. Interview at ISAC in Bangalore on Chandrayaan-1and the Mars Orbiter Mission.

Sundaram Ramakrishnan. Director of VSSC. Thiruvananthapuram, Kerala 21/08/2013. Interview at VSSC in Thiruvananthapuram on the development of ISRO's launch vehicles.

Professor U.R. Rao. Former Chairman of ISRO. Bangalore 6/03/2013. Interview at ISRO headquarters in Bangalore on Professor Rao's role in the development of India's first satellite and ISRO satellite programme.

Amrita Shah. Author and Vikram Sarabhai's biographer. IISc Bangalore 24/08/2013. Interview at IISc in Bangalore on the life and legacy of Vikram Sarabhai.

Reg Turnill. Former BBC Journalist. Kent, UK 3/11/2011. Telephone interview in London, UK, on the German rocket scientist Wernher von Braun.

Al Worden Apollo15 Command Module Pilot. Michigan, USA 3/9/2014. Telephone interview in Michigan, USA, on the space programmes of China and India and Worden's recollections of photographing the crater on the Moon called Sarabhai.

P. Radhakrishnan. ISRO Scientist. Thiruvananthapuram, Kerala 9/9/2013. Telephone interview in Thiruvananthapuram on P. Radhakrishnan's recollections as one of the two ISRO scientists selected to fly aboard the Space Shuttle.

N.C. Bhat. ISRO Scientist. Bangalore 8/2/2014. Telephone interview in Bangalore on N.C. Bhat's recollections as one of the two ISRO astronauts selected to fly aboard the Space Shuttle.

Ravish Malhotra. Former Indian Air Force. Test Pilot and Astronaut. Bangalore 10/09/2013.Telephone interview in Bangalore on Ravish Malhotra's experiences during the astronaut selection and the Indo-Soviet flight training for flight aboard the Soyuz to the USSR Salyut space station.

Rakesh Sharma. Former Indian Air Force. Test Pilot and Astronaut. Coonoor, Tamil Nadu 12/08/2014. Interview in Coonoor on Rakesh Sharma's experiences during the astronaut selection, training, flight aboard the Soyuz and eight-day on-board the USSR Salyut space station in 1984.

Professor Roddam Narasimha. Aerospace Scientist. Bangalore 6/2/2015. Telephone interview in Bangalore on Professor Narasimha's recollections of Satish Dhawan and his role as a member of the Space Commission in the Department of Space.

Gloria Morris and Lucy Morris. London 5/7/2016. Stephen Smith's granddaughter and great granddaughter. They provided personal details of Stephen Smith's life via email and telephone.

V.S. Hegde. Director at Antrix. Bangalore. Interview in Bangalore on ISRO's commercial operations.

Mrinalini Sarabhai. Vikram Sarabhai's widow. Ahmedabad 3/11/2012. Telephone interview to discuss Vikram Sarabhai's legacy.

Monica Sarabhai. Vikram Sarabhai's daughter. Ahmedabad 26/6/2015. Monica Sarabhai provided consent to use one of the family pictures. Contact via Facebook.

Professor Praful Bhavsar. Former Director of Space Applications Centre. Ahmedabad 18/12/2012. Professor Bhavsar held several roles, including that of the project scientist for the first rocket launch from Thumba. He shared via telephone and email personal recollections and some images of the first launch event on 21 November 1963.

Professor Jacques Blamont. Astrophysicist and pioneer of the French space programme. Paris. 19/04/2013. Telephone interview in Paris, France, on Professor Blamont's recollections of the sodium payload of the first rocket launch from India on 21 November 1963, his personal recollections of Vikram Sarabhai and the wider collaboration between the Indian and French space programmes.

Professor Rajesh Kochhar. London. 23/06/2016. Meeting with Professor Kochhar during his trip to London, UK. He provided assistance that led to establishing contact with Stephen Smith's family in London.

Dr Rajinder Singh. Historian and Physicist. Oldenburg, Germany 31/08/2015. Correspondence by email from University of Oldenburg, Germany. Dr Singh has written extensively on the scientific contributions of Indian scientists, especially in the pre-independence period.

Melvyn Brown. Kolkata. 2/4/2014. Interview in Kolkata, India, on Melvyn Brown's research on Stephen Smith, his meeting with Stephen Smith's son Hector and the wider experiences of the Anglo-Indian community in India.

Indian Currency

The Indian rupee, managed by the Reserve Bank of India, is the currency used across India's 29 states and 7 union territories. It is usually abbreviated to Rs. The International Organization for Standardisation's currency code for the Indian rupee is INR. The rupee is made up of 100 paise. The name rupee is derived from the rupiya, a silver coin first issued by Sultan Sher Shah Suri in the 16th century. The nomenclature used to express large currency (and large numbers in general) uses the Hindi words, lakh and crore. The location of the comma in a series of zeros is also unique.

Hindi	International Scientific Notation	Comma Placement	USD ($) 2015 Rate
1 lakh	100 thousand (10^5)	1,00,000	1,576
1 crore	10 million (10^7)	1,00,00,000	157,600

As a British colony, the value of the Indian rupee was pegged to British pounds. From 1927 to 1966, it was 13 rupees to the British pound. In 1966, the rupee was devalued and pegged to the US dollar at a rate of 7.5 rupees to the US dollar. This value lasted until the US dollar devalued in 1971. Since then, the value of the Indian rupee has fluctuated along with all other international currencies.

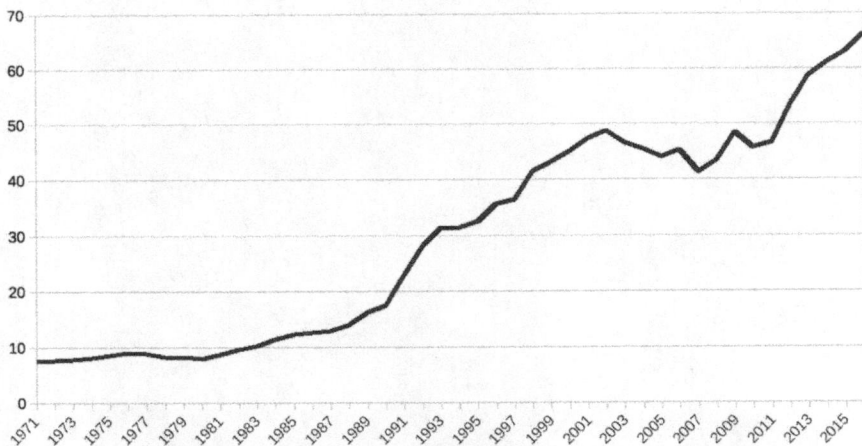

The Exchange Rate of the US Dollar and Indian Rupee between 1971 and 2015

Through the 1980s, the value remained less than ten rupees to the US dollar and saw a sharp increase in 1991 following the relaxation of financial controls and increase in foreign investment. By the end of 1993, the exchange rate was around 30 rupees to the US dollar.

I have attempted to present all monetary values in both Indian rupee and the US dollar. The approximate value of the alternate currency, either Indian rupee or the US dollar, is given in brackets. The values presented throughout should be considered approximate given the rounding errors and fluctuating exchange rates. In addition, there are inconsistencies in the values documented in publically available sources.

Types of Orbits

Orbits are classified on their altitude (height from the Earth's surface), inclination (offset from the equator) and eccentricity (extent of deviation from a circular orbit). There are three primary classifications of orbits. Low Earth Orbit (LEO) between 160 to 2,000 km, Medium Earth Orbit (MEO) between 2,000 to 35,786 km and Geostationary Earth Orbit (GEO), almost circular orbit 35,786 km above the Earth's equator (or 42,164 km from the centre of the Earth). An orbit of 35,768 km but not over the equator is a Geosynchronous Orbit (GSO). All GEO and GSO orbits have a period of almost 24 hours (or specifically sidereal day (23.934461223 hours). With increasing altitude, the spacecraft's speed reduces and the time taken to go around the Earth once (also known as the period) increases. Almost all satellites in GEO orbit the Earth in the same direction as the Earth's rotation, that is, anti-clockwise if looked down upon from the North Pole.

Altitude above Sea Level (km)	Speed km/s	Period (minutes)	Period (Hours)	Re-entry after*
100	7.84	86	1.4	87 days
200	7.78	88	1.5	130 days
300	7.73	90	1.5	139 days
400	7.67	92	1.5	174 days
500	7.62	94	1.6	218 days
1,000	7.35	104	1.7	1.2 years
2,000 (Start of MEO)	3.89	127	2.1	24 years
35,786 (GEO)	3.07	1435	23.9	43 years
50,000	2.66	2219	37.0	60.5 years
200,000	1.39	15550	259.2	242 years

The values in the re-entry column are approximate. These are periods by when atmospheric drag would naturally cause a satellite orbit to decay and re-enter.

In practice, these periods are influenced by several factors in addition to drag including mass and surface area (1000 kg/1 m² is assumed for the table above). Further, the Earth's magnetic field and solar radiation contribute to orbital decay, and both are subject to variation. The re-entry figures assume the that the spacecraft's orbit is not actively managed by the satellite operators.

There are many orbital configurations. Five of the most common are:

- Low Earth Orbit: This is a circular or elliptical orbit parallel or inclined to the equator with a wide altitude range.

- Polar Orbit: An orbit that takes the spacecraft over the North and South Poles and is thus tilted around 90° to the equator. Typically, they have a period shorter than two hours. An Earth observation satellite in a polar orbit can cover the entire surface in one day.

- Sun-synchronous Polar Orbit (SSPO): An orbit where the altitude and inclination are uniquely tuned so that a spacecraft crosses the equator at a same time of the day on every orbit. This is a very useful feature for EO satellites that require consistent illumination.

- GSO and GEO: Both orbit the Earth at an altitude of 35,786 km and have a period of 24 hours, that is, they circle the Earth once every day. A GEO orbit is parallel to the Earth's equator, while GSO is not. A spacecraft in GEO orbit appears to be stationary in the sky. This orbit is ideal for satellite TV. Once the receiving dish is set up to point to a specific satellite, it does not need to be adjusted. A spacecraft in GSO also takes 24 hours to orbit the Earth, but it will appear to move up and down in the sky every day.

- Highly Elliptical Orbit: Communication satellites in GEO cannot be accessed by users living in higher latitudes on Earth. More than about 55° above or below the equator, a satellite in GEO orbit is close to or below the local horizon. A satellite in HEO (instead of GEO) is a solution for such users. A Molniya orbits is a specific HEO of 63.4°.

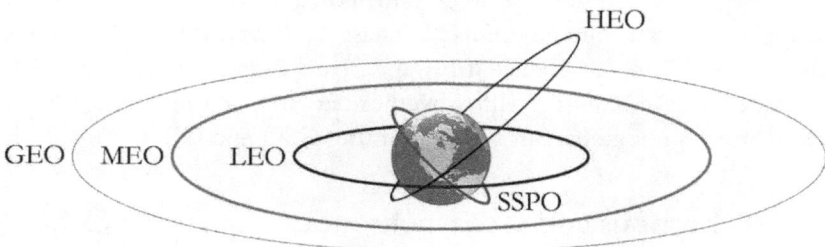

Typical Orbital Configurations

Low Earth Orbit

There is no well-defined point at which the Earth's atmosphere ends, and space begins. Like a twilight sky where day fades into night, the Earth's atmosphere gradually fades into space. Internationally, 100 km above sea level is regarded as the official boundary, where space begins. This boundary is known as the Karman Line, named after Theodore von Karman, a

Hungarian-American scientist. At that altitude and speed, a spacecraft would orbit the Earth once in 86 minutes.

An altitude of 160 km is the lowest with sufficient vacuum for a spacecraft travelling at 7.8 km/s to sustain an orbit to fulfil a function. However, the orbit is not stable, atmospheric drag will cause the orbit to decay in a matter of days. At higher altitudes, there is less drag, speed of the spacecraft in orbit is lower, the period longer and the decay of orbit slower. LEO is used mostly by spacecraft for EO, photo reconnaissance and science and sometimes by communication satellites. The Hubble Space Telescope, Astrosat and the ISS are in LEO.

Polar Orbit

A polar orbit is one where a satellite's orbit is inclined at around 90° to the equator (pole to pole). The Sun Synchronous Polar Orbit SSPO is a special type of polar orbit that combines the altitude and inclination, such that a satellite in SSPO crosses the equator at the same time each day. For example, ISRO's Cartosat-2 orbits the Earth at 635 km with an inclination 97.87° and has a period of 94.7 minutes crossing the equator at 09:30 hours every day. Most EO satellites are in Polar Orbit.

Medium Earth Orbit

A spacecraft in orbit between 2,000 km and 37,786 km is said to be in MEO with an orbital period between 2 and up to 15 hours. Most global positioning constellations use MEO, including those of the US, China, Europe and Russia, but not India (Glonass 11 h 15 m, GPS 11 h 58 m, Beidou 12 h 35 m and Galileo 14 h and 7 m). This configuration ensures that there are at least four satellites overhead at any one time. India's NavIC constellation is not global but regional. It uses GSO and GEO.

Geosynchronous and Geosynchronous Equatorial Orbit

GSO and GEO orbit the Earth once every 24 hours and are predominantly used for communication satellites. Because of their unique properties, GEO orbits have limited availability. Their allocation is determined by the UN's International Telecommunications Union. India has 11 GEO slots. India's current occupancy in GEO/GSO consists of

- 32.6E   IRNSS-1F
- 47.9E   INSAT 4CR
- 55.0E   IRNSS-1A and IRNSS-1B (both inclined at 29°)
- 55.0E   GSAT-8, GSAT-16

- 74.0E Kalpana-1, GSAT-7, GSAT-14, GSAT-18, INSAT-3DR
- 83.0E GSAT-6, GSAT-10, GSAT-12, GSAT-19E, IRNSS-R1C, INSAT 4A, INSAT 3D at 82. °1 in reserve
- 93.5E INSAT 4B, GSAT-15, GSAT-17
- 97.2E SOUTH ASIA SAT
- 111.8E IRNSS-1D, IRNSS-1E (both inclined at 29°)
- 129.5E IRNSS-1G

GEO slots are typically divided into 2° longitude slots, which at 35,786 km amounts to an arc about 1,500 km long. When operational, each satellite must be confined to a box of a minimum 0.1° (about 70 km). By high precision control, satellite operators can co-locate multiple satellites in a single slot provided the radio frequencies do not interfere. From GEO, satellites have the whole Earth in their field of view and thus are used for planet-wide meteorological and communication services.

Satellite Communication

Satellite communication makes use of the Electromagnetic Spectrum between 3 GHz to 30 GHz. However, this is not exclusive. While some satellite communication takes place outside this range, the 3–30 GHz range is also used by some terrestrial applications. The table below indicates the typical use of a particular band of the spectrum.

Band	Very Low Freq. (VLF)	Low Freq. (LF)	Medium Freq. (MF)	High Freq. (HF)	Very High Freq. VHF	Ultra-High Freq. (UHF)	Super High Freq. (SHF)	Extremely High Freq. (EHF)
Wave	100km	10 km	1 km	100 m	10 m	1 m	10 cm	1 cm
Freq.	3 KHz	30 KHz	300 KHz	3 MHz	30 MHz	300 MHz	3 GHz	30 GHz
Typical Use	Maritime radio and navigation beacons	AM Radio	Maritime and aviation radio communication	Shortwave radio communication and Satellite-based Search and Rescue	Terrestrial FM radio, TV and navigation aids	Cell phone, GPS and TV	3 to 30 GHz band for satellite communication	Radio Astronomy

Ground stations with a large dish for high-speed communication are required for most satellite services. An uplink station sends a signal from the ground to the satellite, and a downlink station receives a signal from the satellite.

The uplink and downlink stations are typically located within a single installation. To prevent interference, up and downlink channels operate at different frequencies.

Band and Frequency Range (GHz)	Description

L-Band 1 – 2	Used for GPS signals and satellite mobile phones by Iridium and Inmarsat for remote users on land, air and sea. The impact of weather on this frequency is minimal. Hardware is not expensive. The Safety-of-Life signal used by civilian aviation Satellite-Based Augmentation Systems, such as GAGAN, is in a nearby L5 GHz.
S-Band 2 – 4	Used terrestrially for weather, navigation radar and maritime communication. It was also used for in-space communication between Space Shuttle and ISS. Terrestrial use includes home broadband and cordless telephones in some parts of the world.
C-Band 4 – 8	Used by satellite networks (uplink 5.925-6.425 and downlink 3.625-4.200) data, communication, TV and amateur satellite radio. Typically, it requires a large dish (3 m or larger). These frequencies are not sensitive to "rain fade", that is, the quality of the signal is not affected by rain or moisture. It is commonly used in areas where tropical rain is common. The first transatlantic TV signal in 1962 by Telstar used C-band. ISRO's Synthetic Aperture Radar (RISAT) satellites use the 5.35 GHz frequency to produce radar images. Terrestrial use includes home broadband. The newer multiband routers operate at 5 GHz.
X-Band 8 – 12	Primarily used by the military for communication, air defence, weather monitoring, maritime and air traffic control. It is also used by spacecraft to transmit data by spacecraft beyond Earth orbit, on the surface of Mars, in orbit around Saturn and on the edge of the solar system. NASA's Deep Space Network operates in this band. A slice of this frequency is dedicated for use by Amateur Radio Satellite AMSAT.
Ku Band 12 – 18	Used for satellite communication and DTH satellite TV transmissions (uplink 14.00-14.50 and downlink 10.95-12.75). It is also used by NASA's Tracking Data Relay Satellite (TDRS). TDRS was first used in 1975 during the Apollo-Soyuz mission for communication between two crewed spacecraft in space. Since then, it has been used by the space shuttle. Today, it is used by the ISS for routine communications. Two-way Internet connection, including VSATs, use Ku band. These frequencies are almost exclusively used for satellite communication. Ku-band offers higher signal strength and higher data throughput. Although a smaller dish suffices, this frequency band is susceptible to signal deterioration from rain and snow.
Ka-Band (26 – 40 GHz)	Used for higher bandwidth satellite communication. Currently, used by Inmarsat and planned for the Iridium Next satellite series using an uplink frequency of 27.5 GHz and 31 GHz. This band will also be used by the James Webb Space Telescope to be launched in 2018. The Ka band is more susceptible to rain attenuation than the Ku and the C bands. Terrestrially, this band is used by military aircraft for high resolution targeting radar.

Common Frequency bands used by satellites

The device on a satellite that receives signals at one frequency and retransmits at another is called a transponder. Typically, satellites have around two dozen transponders. For example, INSAT-3A is equipped with 12 C–band, 6 upper extended C-band, 6 Ku-band and 1 Satellite Aided Search & Rescue transponders. Some satellites can have over a hundred transponders. A communication satellite is predominantly a vehicle for a collection of transponders. Typical transponders perform the following functions:

- Receive an uplinked signal from earth stations

- Change the frequency of the received signal

- Amplify the signal

- Retransmit the signal back to Earth for end users or another earth station.

Modern satellites have Regenerative Transponders. These are transponders that provide more complex processing and offer multiplexing, compression and multiple audio and video channels.

The UN's International Telecommunications Union allocates not only GEO slots but also radio frequencies for use by private suppliers and countries. Because the demand for frequencies is extremely high, two technical solutions are commonly implemented for frequency reuse:

- Polarisation: A communication signal at one frequency can have vertical or horizontal polarisation. By using the polarisation attribute, a single frequency can be used to carry two sets of signals (for example, two TV stations at the same time) in the same geographic area without generating interference.

- Spot beams: A satellite in orbit can use the same frequency to carry different information at the same time by focussing its transmissions to small areas on Earth (for example, Kerala and Maharashtra). As the two footprints do not overlap, there is no interference. This technology was a major step in the development of cellular mobile phone technology.

International Treaties

International laws that describe how nations access, use and explore space are defined predominantly by five treaties that have emerged from the United Nations General Assembly over the last fifty years. UN's 1967 Treaty on Principles Governing the Activities of States in the Exploration and Use of Outer Space, including the Moon and Other Celestial Bodies. It is the primary treaty that forms a general legal framework for developing the law of outer space. The four other treaties deal specifically with certain concepts included in the 1967 Treaty. Each treaty is preceded within the UN by a formal resolution(s) voted on by member states and then a declaration of legal principle(s). Once the treaty is established it then up to individual states to formally sign, ratify and establish local laws to implement it.

Treaty	Date	No. of Articles	India's Status
Treaty on Principles Governing the Activities of States in the Exploration and Use of Outer Space, including the Moon and Other Celestial Bodies. Sometimes abbreviated to the Outer Space Treaty.	10/10/67	17	Ratified
Agreement on the Rescue of Astronauts, the Return of Astronauts and the Return of Objects Launched into Outer Space. Sometimes abbreviated to the Rescue Agreement.	3/12/68	10	Ratified
Convention on International Liability for Damage Caused by Space Objects. Sometimes abbreviated to The Liability agreement.	1/9/72	28	Ratified
Convention on Registration of Objects Launched into Outer Space.	1/9/76	12	Ratified
Agreement Governing the Activities of States on the Moon and Other Celestial Bodies. Sometimes abbreviated to the Moon Treaty	11/7/84	21	Signed but not ratified

The Outer Space Treaty is 50 years old in 2017 and the subject of fresh debate. As the private sector engages in commercial operations such as space launch service, asteroid mining, space tourism and exploitation of the lunar resources, governments will enact national laws to implement the OST. Commercial entities are preparing to invest in commercial ventures but will want legal clarity before they proceed. It is not clear how, if at all, the treaty will be amended, which nations will choose to abide by it or declare that they will not. Of particular interest and areas of contention, include:

Outer Space Treaty

- Article 6 of the OST requires individual states to "authorise and supervise" all space activities conducted by "non-governmental entities" in their jurisdiction. This could become be a huge burden for governments and businesses.

- Article 7 commits a launching state to international liabilities for physical damages caused by its space objects. The complex international agreements behind space missions today (spacecraft built by one nation, launched by another and operated by yet another) could result in complex and unclear legal obligations.

- Article 12 requires that "installations, equipment and space vehicle on the Moon" belonging to one nation shall be open to visits by others on the basis of reciprocity. Space is becoming a source of interest for national security for many nations. This article may be a source of contention as nations innovate and develop new technologies with potential military use not envisaged at the time the Moon Treaty was written. For example, the treaty of Versailles that ended the First World War placed limits on weapons that Germany could have and deploy. These restrictions did not extend to rockets (i.e. V1 and V2) for military use because they had not been invented when the treaty was signed.

Moon Treaty

- Article 3 of the Moon Treaty explicitly prohibits nuclear weapons in space (on the Moon or in lunar orbit) but does allow for the use of military personnel participating in scientific research. This potential ambiguity could be exploited by some states as for example commercial whaling in pursuit of "scientific research" has been on Earth.

- Article 7 requires that States operating on the Moon shall maintain the "existing balance in its environment" avoid "harmful contamination". What constitutes a "balance" and a "harmful contamination" will probably be open to interpretation by each member state.

- Article 8 Asserts that States that launch spacecraft and personnel shall "retain jurisdiction and control over any personnel". This requirement could undermine UN's Human Rights Act which for example, protects the rights of individuals from one nation to seek asylum in another.

- Article 11 states that "Neither the surface nor the subsurface of the Moon, nor any part thereof or natural resources in place, shall become property of any State". This could mean that state run organisations cannot engage in commercial mining operations. Private business investors too may seek legal assurances prior to investing in missions to exploit minerals on asteroids and the lunar surface.

Individual nations can formally choose to sign and ratify these treaties. Although not all member states sign and ratify all treaties, most do adhere to them. The distinction between terms, such as signed, ratified and accessioned, is described here: http://ask.un.org/faq/14594. A status of which nations have signed which treaty is maintained by the UN here http://www.unoosa.org/oosa/en/ourwork/spacelaw/treaties/status/index.htm l.

The following section summarises some of the contents of the five treaties. Full text is available here http://www.unoosa.org/pdf/publications/STSPACE11E.pdf.

Outer Space Treaty

Article No.	Treaty on Principles Governing the Activities of States in the Exploration and Use of Outer Space, including the Moon and Other Celestial Bodies. Also known as the Outer Space Treaty.
1	The exploration and use of outer space, including the Moon and other celestial bodies, shall be carried out for the benefit and in the interests of all countries, irrespective of their degree of economic or scientific development, and shall be the province of all mankind.
2	Outer space, including the Moon and other celestial bodies, is not subject to national appropriation by claim of sovereignty, by means of use or occupation, or by any other means.
3	States Parties to the Treaty shall carry on activities in the exploration and use of outer space, including the Moon and other celestial bodies, in accordance with international law, including the Charter of the United Nations, in the interest of maintaining international peace and security and promoting international cooperation and understanding.
4	States Parties to the Treaty undertake not to place in orbit around the Earth any objects carrying nuclear weapons or any other kinds of

	weapons of mass destruction, install such weapons on celestial bodies, or station such weapons in outer space in any other manner.
5	States Parties to the Treaty shall regard astronauts as envoys of mankind in outer space and shall render to them all possible assistance in the event of accident, distress, or emergency landing on the territory of another State Party or on the high seas. When astronauts make such a landing, they shall be safely and promptly returned to the State of registry of their space vehicle.
6	States Parties to the Treaty shall bear international responsibility for national activities in outer space, including the Moon and other celestial bodies, whether such activities are carried on by governmental agencies or by non-governmental entities, and for assuring that national activities are carried out in conformity with the provisions set forth in the present Treaty.
7	Each State Party to the Treaty that launches or procures the launching of an object into outer space, including the Moon and other celestial bodies, and each State Party from whose territory or facility an object is launched, is internationally liable for damage to another State Party to the Treaty or to its natural or juridical persons by such object or its component parts on the Earth, in airspace or in outer space, including the Moon and other celestial bodies.
8	A State Party to the Treaty on whose registry an object launched into outer space is carried shall retain jurisdiction and control over such object, and over any personnel thereof, while in outer space or on a celestial body.
9	In the exploration and use of outer space, including the Moon and other celestial bodies, States Parties to the Treaty shall be guided by the principle of cooperation and mutual assistance and shall conduct all their activities in outer space, including the Moon and other celestial bodies, with due regard to the corresponding interests of all other States Parties to the Treaty.
10	In order to promote international cooperation in the exploration and use of outer space, including the Moon and other celestial bodies, in conformity with the purposes of this Treaty, the States Parties to the Treaty shall consider on a basis of equality any requests by other States Parties to the Treaty to be afforded an opportunity to observe the flight of space objects launched by those States. The nature of such an opportunity for observation and the conditions
11	In order to promote international cooperation in the peaceful exploration and use of outer space, States Parties to the Treaty conducting activities in outer space, including the Moon and other celestial bodies, agree to inform the Secretary-General of the United Nations as well as the public and the international scientific community, to the greatest extent feasible and practicable, of the nature, conduct, locations and results of such activities.
12	All stations, installations, equipment and space vehicles on the Moon

	and other celestial bodies shall be open to representatives of other States Parties to the Treaty on a basis of reciprocity. Such representatives shall give reasonable advance notice of a projected visit, in order that appropriate consultations may be held and that maximum precautions may be taken to assure safety and to avoid interference with normal operations in the facility to be visited.
13	The provisions of this Treaty shall apply to the activities of States Parties to the Treaty in the exploration and use of outer space, including the Moon and other celestial bodies, whether such activities are carried on by a single State Party to the Treaty or jointly with other States, including cases where they are
14	This article describes the 6 steps on how States manage the administrative process associated with the treaty.
15	Any State Party to the Treaty may propose amendments to this Treaty.
16	Any State Party to the Treaty may give notice of its withdrawal from the Treaty one year after its entry into force by written notification to the Depositary Governments. Such withdrawal shall take effect one year from the date of receipt of this notification.
17	This Treaty, of which the Chinese, English, French, Russian and Spanish texts are equally authentic, shall be deposited in the archives of the Depositary Governments. Duly certified copies of this Treaty shall be transmitted by the Depositary Governments to the Governments of the signatory and acceding States.

Rescue Agreement

Article No.	Agreement on the Rescue of Astronauts, the Return of Astronauts and the Return of Objects Launched into Outer Space Also known as the Rescue Agreement
1	Each Contracting Party which receives information or discovers that the personnel of a spacecraft have suffered accident or are experiencing conditions of distress or have made an emergency or unintended landing in territory under its jurisdiction or on the high seas or in any other place not under the jurisdiction of any State shall immediately: (a) Notify the launching authority (b) Notify the Secretary-General of the United Nations.
2	If, owing to accident, distress, emergency or unintended landing, the personnel of a spacecraft land in territory under the jurisdiction of a Contracting Party, it shall immediately take all possible steps to rescue them and render them all necessary assistance.
3	If information is received or it is discovered that the personnel of a spacecraft have alighted on the high seas or in any other place not under the jurisdiction of any State, those Contracting Parties which are in a position to do so shall, if necessary, extend assistance in

	search and rescue operations for such personnel to assure their speedy rescue.
4	If, owing to accident, distress, emergency or unintended landing, the personnel of a spacecraft land in territory under the jurisdiction of a Contracting Party or have been found on the high seas or in any other place not under the jurisdiction of any State, they shall be safely and promptly returned to representatives of the launching authority.
5	This article covers 5 steps that States should follow when they become aware of space objects have returned to their territory.
6	For the purposes of this Agreement, the term "launching authority" shall refer to the State responsible for launching, or, where an international intergovernmental organisation is responsible for launching, that organisation, provided that that organisation declares its acceptance of the rights and obligations provided for in this Agreement and a majority of the States members of that organisation are Contracting Parties to this Agreement and to the Treaty on Principles Governing the Activities of States in the Exploration and Use of Outer Space, including the Moon and Other Celestial Bodies.
7 - 10	These articles describe steps on how States manage the administrative process associated with the treaty.

Liability Agreement

Article No.	Convention on International Liability for Damage Caused by Space Objects. Also known as the Space Liability Agreement
1	This article provides the formal definitions for damage, Launching, Launching State and Space Object.
2	A launching State shall be absolutely liable to pay compensation for damage caused by its space object on the surface of the Earth or to aircraft in flight.
3	In the event of damage being caused elsewhere than on the surface of the Earth to a space object of one launching State or to persons or property on-board such a space object by a space object of another launching State, the latter shall be liable only if the damage is due to its fault or the fault of persons for whom it is responsible.
4	In the event of damage being caused elsewhere than on the surface of the Earth to a space object of one launching State or to persons or property on-board such a space object by a space object of another launching State, and of damage thereby being caused to a third State or to its natural or juridical persons, the first two States shall be jointly and severally liable to the third State.
5	Whenever two or more States jointly launch a space object, they shall be jointly and severally liable for any damage caused.
6 -20	This article describes how culpability of damage is established,

	liability may be limited and claims for compensation may be processed.
21 -24	This article describes how large-scale damage should be addressed along with existing international law.
25-28	These articles describe steps on how States manage the administrative process associated with the treaty.

Registration Convention

Article No.	Convention on Registration of Objects Launched into Outer Space. Also known as the Registration Convention
1	This article defines the terms Launching state, Space object, State of registry.
2	This article provides details on the registering the space object and responsibilities where two or more States work together.
3	The Secretary-General of the United Nations shall maintain a Register in which the information furnished in accordance with article IV shall be recorded. There shall be full and open access to the information in this Register.
4	This article defines the details of the space object that is to be recorded in the register. That includes name of the launching state, date, number of objects and details of the orbits.
5-12	These articles describe steps on how States manage the administrative process associated with the treaty.

Moon Treaty

Article No.	Agreement Governing the Activities of States on the Moon and Other Celestial Bodies. Also known as the Moon Agreement.
1	1. The provisions of this Agreement relating to the Moon shall also apply to other celestial bodies within the solar system, other than the Earth, except insofar as specific legal norms enter into force with respect to any of these celestial bodies. 2. For the purposes of this Agreement reference to the Moon shall include orbits around or other trajectories to or around it. 3. This Agreement does not apply to extra-terrestrial materials which reach the surface of the Earth by natural means.
2	All activities on the Moon, including its exploration and use, shall be carried out in accordance with international law, in particular the Charter of the United Nations, and taking into account the Declaration on Principles of International Law concerning Friendly Relations and Cooperation among States in accordance with the

	Charter of the United Nations,5 adopted by the General Assembly on 24 October 1970, in the interest of maintaining international peace and security and promoting international cooperation and mutual understanding, and with due regard to the corresponding interests of all other States Parties.
3	1. The Moon shall be used by all States Parties exclusively for peaceful purposes. 2. Any threat or use of force or any other hostile act or threat of hostile act on the Moon is prohibited. It is likewise prohibited to use the Moon in order to commit any such act or to engage in any such threat in relation to the Earth, the Moon, spacecraft, the personnel of spacecraft or manmade space objects. 3. States Parties shall not place in orbit around or other trajectory to or around the Moon objects carrying nuclear weapons or any other kinds of weapons of mass destruction or place or use such weapons on or in the Moon. 4. The establishment of military bases, installations and fortifications, the testing of any type of weapons and the conduct of military manoeuvres on the Moon shall be forbidden. The use of military personnel for scientific research or for any other peaceful purposes shall not be prohibited. The use of any equipment or facility necessary for peaceful exploration and use of the Moon shall also not be prohibited.
4	1. The exploration and use of the Moon shall be the province of all mankind and shall be carried out for the benefit and in the interests of all countries, irrespective of their degree of economic or scientific development. Due regard shall be paid to the interests of present and future generations as well as to the need to promote higher standards of living and conditions of economic and social progress and development in accordance with the Charter of the United Nations. 2. States Parties shall be guided by the principle of cooperation and mutual assistance in all their activities concerning the exploration and use of the Moon. International cooperation in pursuance of this Agreement should be as wide as possible and may take place on a multilateral basis, on a bilateral basis or through international intergovernmental organisations.
5	1. States Parties shall inform the Secretary-General of the United Nations as well as the public and the international scientific community, to the greatest extent feasible and practicable, of their activities concerned with the exploration and use of the Moon. Information on the time, purposes, locations, orbital parameters and duration shall be given in respect of each mission to the Moon as soon as possible after launching, while information on the results of each mission, including scientific results, shall be furnished upon completion of the mission. In the case of a mission lasting more than sixty days, information on conduct of the mission, including any scientific results, shall be given periodically, at thirty-day intervals.

	For missions lasting more than six months, only significant additions to such information need be reported thereafter. 2. If a State Party becomes aware that another State Party plans to operate simultaneously in the same area of or in the same orbit around or trajectory to or around the Moon, it shall promptly inform the other State of the timing of and plans for its own operations. 3. In carrying out activities under this Agreement, States Parties shall promptly inform the Secretary-General, as well as the public and the international scientific community, of any phenomena they discover in outer space, including the Moon, which could endanger human life or health, as well as of any indication of organic life.
6	1. There shall be freedom of scientific investigation on the Moon by all States Parties without discrimination of any kind, on the basis of equality and in accordance with international law. 2. In carrying out scientific investigations and in furtherance of the provisions of this Agreement, the States Parties shall have the right to collect on and remove from the Moon samples of its mineral and other substances. Such samples shall remain at the disposal of those States Parties which caused them to be collected and may be used by them for scientific purposes. States Parties shall have regard to the desirability of making a portion of such samples available to other interested States Parties and the international scientific community for scientific investigation. States Parties may in the course of scientific investigations also use mineral and other substances of the Moon in quantities appropriate for the support of their missions. 3. States Parties agree on the desirability of exchanging scientific and other personnel on expeditions to or installations on the Moon to the greatest extent feasible and practicable.
7	1. In exploring and using the Moon, States Parties shall take measures to prevent the disruption of the existing balance of its environment, whether by introducing adverse changes in that environment, by its harmful contamination through the introduction of extra-environmental matter or otherwise. States Parties shall also take measures to avoid harmfully affecting the environment of the Earth through the introduction of extra-terrestrial matter or otherwise. 2. States Parties shall inform the Secretary-General of the United Nations of the measures being adopted by them in accordance with paragraph 1 of this article and shall also, to the maximum extent feasible, notify him in advance of all placements by them of radioactive materials on the Moon and of the purposes of such placements. 3. States Parties shall report to other States Parties and to the Secretary- General concerning areas of the Moon having special scientific interest in order that, without prejudice to the rights of other States Parties, consideration may be given to the designation of such areas as international scientific preserves for which special protective arrangements are to be agreed upon in consultation with

	the competent bodies of the United Nations.
8	1. States Parties may pursue their activities in the exploration and use of the Moon anywhere on or below its surface, subject to the provisions of this Agreement. 2. For these purposes States Parties may, in particular: (a) Land their space objects on the Moon and launch them from the Moon; (b) Place their personnel, space vehicles, equipment, facilities, stations and installations anywhere on or below the surface of the Moon. Personnel, space vehicles, equipment, facilities, stations and installations may move or be moved freely over or below the surface of the Moon. 3. Activities of States Parties in accordance with paragraphs 1 and 2 of this article shall not interfere with the activities of other States Parties on the Moon. Where such interference may occur, the States Parties concerned shall undertake consultations in accordance
9	1. States Parties may establish manned and unmanned stations on the Moon. A State Party establishing a station shall use only that area which is required for the needs of the station and shall immediately inform the Secretary-General of the United Nations of the location and purposes of that station. Subsequently, at annual intervals that State shall likewise inform the Secretary-General whether the station continues in use and whether its purposes have changed. 2. Stations shall be installed in such a manner that they do not impede the free access to all areas of the Moon of personnel, vehicles and equipment of other States Parties conducting activities on the Moon in accordance with the provisions of this Agreement or of article I of the Treaty on Principles Governing the Activities of States in the Exploration and Use of Outer Space, including the Moon and Other Celestial Bodies.
10	1. States Parties shall adopt all practicable measures to safeguard the life and health of persons on the Moon. For this purpose, they shall regard any person on the Moon as an astronaut within the meaning of article V of the Treaty on Principles Governing the Activities of States in the Exploration and Use of Outer Space, including the Moon and Other Celestial Bodies and as part of the personnel of a spacecraft within the meaning of the Agreement on the Rescue of Astronauts, the Return of Astronauts and the Return of Objects Launched into Outer Space. 2. States Parties shall offer shelter in their stations, installations,
11	1. The Moon and its natural resources are the common heritage of mankind, which finds its expression in the provisions of this Agreement, in particular in paragraph 5 of this article. 2. The Moon is not subject to national appropriation by any claim of sovereignty, by means of use or occupation, or by any other means. 3. Neither the surface nor the subsurface of the Moon, nor any part thereof or natural resources in place, shall become property of any State, international intergovernmental or non-governmental

organisation, national organisation or non-governmental entity or of any natural person. The placement of personnel, space vehicles, equipment, facilities, stations and installations on or below the surface of the Moon, including structures connected with its surface or subsurface, shall not create a right of ownership over the surface or the subsurface of the Moon or any areas thereof. The foregoing provisions are without prejudice to the international regime referred to in paragraph 5 of this article.

4. States Parties have the right to exploration and use of the Moon without discrimination of any kind, on the basis of equality and in accordance with international law and the terms of this Agreement.

5. States Parties to this Agreement hereby undertake to establish an international regime, including appropriate procedures, to govern the exploitation of the natural resources of the Moon as such exploitation is about to become feasible. This provision shall be implemented in accordance with article 18 of this Agreement.

6. In order to facilitate the establishment of the international regime referred to in paragraph 5 of this article, States Parties shall inform the Secretary- General of the United Nations as well as the public and the international scientific community, to the greatest extent feasible and practicable, of any natural resources they may discover on the Moon.

7. The main purposes of the international regime to be established shall include: (a) The orderly and safe development of the natural resources of the Moon; (b) The rational management of those resources; (c) The expansion of opportunities in the use of those resources; (d) An equitable sharing by all States Parties in the benefits derived from those resources, whereby the interests and needs of the developing countries, as well as the efforts of those countries which have contributed either directly or indirectly to the exploration of the Moon, shall be given special consideration.

8. All the activities with respect to the natural resources of the Moon shall be carried out in a manner compatible with the purposes specified in paragraph 7 of this article and the provisions of article 6, paragraph 2, of this Agreement.

| 12 | 1. States Parties shall retain jurisdiction and control over their personnel, vehicles, equipment, facilities, stations and installations on the Moon. The ownership of space vehicles, equipment, facilities, stations and installations shall not be affected by their presence on the Moon. |

2. Vehicles, installations and equipment or their component parts found in places other than their intended location shall be dealt with in accordance with article 5 of the Agreement on the Rescue of Astronauts, the Return of Astronauts and the Return of Objects Launched into Outer Space.

3. In the event of an emergency involving a threat to human life, States Parties may use the equipment, vehicles, installations, facilities

	or supplies of other States Parties on the Moon. Prompt notification of such use shall be made to the Secretary-General of the United Nations or the State Party concerned.
13	A State Party which learns of the crash landing, forced landing or other unintended landing on the Moon of a space object, or its component parts, that were not launched by it, shall promptly inform the launching State Party and the Secretary-General of the United Nations.
14	1. States Parties to this Agreement shall bear international responsibility for national activities on the Moon, whether such activities are carried on by governmental agencies or by non-governmental entities, and for assuring that national activities are carried out in conformity with the provisions set forth in this Agreement. States Parties shall ensure that non-governmental entities under their jurisdiction shall engage in activities on the Moon only under the authority and continuing supervision of the appropriate State Party. 2. States Parties recognise that detailed arrangements concerning liability for damage caused on the Moon, in addition to the provisions of the Treaty on Principles Governing the Activities of States in the Exploration and Use of Outer Space, including the Moon and Other Celestial Bodies and the Convention on International Liability for Damage Caused by Space Objects, may become necessary as a result of more extensive activities on the Moon. Any such arrangements shall be elaborated in accordance with the procedure provided for in article 18 of this Agreement.
15	1. Each State Party may assure itself that the activities of other States Parties in the exploration and use of the Moon are compatible with the provisions of this Agreement. To this end, all space vehicles, equipment, facilities, stations and installations on the Moon shall be open to other States Parties. Such States Parties shall give reasonable advance notice of a projected visit, in order that appropriate consultations may be held and that maximum precautions may be taken to assure safety and to avoid interference with normal operations in the facility to be visited. In pursuance of this article, any State Party may act on its own behalf or with the full or partial assistance of any other State Party or through appropriate international procedures within the framework of the United Nations and in accordance with the Charter. 2. A State Party which has reason to believe that another State Party is not fulfilling the obligations incumbent upon it pursuant to this Agreement or that another State Party is interfering with the rights which the former State has under this Agreement may request consultations with that State Party. A State Party receiving such a request shall enter into such consultations without delay. Any other State Party which requests to do so shall be entitled to take part in the consultations. Each State Party participating in such consultations

	shall seek a mutually acceptable resolution of any controversy and shall bear in mind the rights and interests of all States Parties. The Secretary-General of the United Nations shall be informed of the results of the consultations and shall transmit the information received to all States Parties concerned. 3. If the consultations do not lead to a mutually acceptable settlement which has due regard for the rights and interests of all States Parties, the parties concerned shall take all measures to settle the dispute by other peaceful means of their choice appropriate to the circumstances and the nature of the dispute. If difficulties arise in connection with the opening of consultations or if consultations do not lead to a mutually acceptable settlement, any State Party may seek the assistance of the Secretary-General, without seeking the consent of any other State Party concerned, in order to resolve the controversy. A State Party which does not maintain diplomatic relations with another State Party concerned shall participate in such consultations, at its choice, either itself or through another State Party or the Secretary-General as intermediary.
16	With the exception of articles 17 to 21, references in this Agreement to States shall be deemed to apply to any international intergovernmental organisation which conducts space activities if the organisation declares its acceptance of the rights and obligations provided for in this Agreement and if a majority of the States members of the organisation are States Parties to this Agreement and to the Treaty on Principles Governing the Activities of States in the Exploration and Use of Outer Space, including the Moon and Other Celestial Bodies. States members of any such organisation which are States Parties to this Agreement shall take all appropriate steps to ensure that the organisation makes a declaration in accordance with the foregoing.
17	Any State Party to this Agreement may propose amendments to the Agreement. Amendments shall enter into force for each State Party to the Agreement accepting the amendments upon their acceptance by a majority of the States Parties to the Agreement and thereafter for each remaining State Party to the Agreement on the date of acceptance by it.
18	Ten years after the entry into force of this Agreement, the question of the review of the Agreement shall be included in the provisional agenda of the General Assembly of the United Nations in order to consider, in the light of past application of the Agreement, whether it requires revision. However, at any time after the Agreement has been in force for five years, the Secretary-General of the United Nations, as depositary, shall, at the request of one third of the States Parties to the Agreement and with the concurrence of the majority of the States Parties, convene a conference of the States Parties to review this Agreement. A review conference shall also consider the question of the implementation of the provisions of article 11, paragraph 5, on

	the basis of the principle referred to in paragraph 1 of that article and taking into account in particular any relevant technological developments.
19	1. This Agreement shall be open for signature by all States at United Nations Headquarters in New York. 2. This Agreement shall be subject to ratification by signatory States. Any State which does not sign this Agreement before its entry into force in accordance with paragraph 3 of this article may accede to it at any time. Instruments of ratification or accession shall be deposited with the Secretary-General of the United Nations. 3. This Agreement shall enter into force on the thirtieth day following the date of deposit of the fifth instrument of ratification. 4. For each State depositing its instrument of ratification or accession after the entry into force of this Agreement, it shall enter into force on the thirtieth day following the date of deposit of any such instrument. 5. The Secretary-General shall promptly inform all signatory and acceding States of the date of each signature, the date of deposit of each instrument of ratification or accession to this Agreement, the date of its entry into force and other notices.
20 - 21	These articles describe steps on how States manage the administrative process associated with the treaty.

ISRO Spaceflight History

No	Mission	Date	Overview
1	Aryabhata	19/04/75	ISRO's first experimental satellite, Aryabhata, had a science payload of three instruments. It had a mass of 360 kg and was placed in an orbit of 563 x 619 km by a Soviet launch vehicle. Only two of Aryabhata's three instruments, the X-ray astronomy and the solar neutron and gamma-ray experiments, returned data. The electricity supply connection to the third, ionosphere experiment, failed. Aryabhata operated for the first four days and, then, partially until March 1981. It re-entered Earth's atmosphere on 10 February 1992 after 3,500 orbits.
2	Bhaskara-I	07/06/79	Bhaskara-1 had two TV cameras operating in the visible and infrared spectra, which collected remote sensing data for hydrology, forestry and geology applications. It gathered data on the Earth's atmosphere, land and oceans. Initially, an electrical fault prevented the optical camera from operating for over a year. Eventually, the camera was activated and returned images. The satellite operated successfully for over a year and re-entered Earth's atmosphere in 1989.
3	Rohini Technology Payload	10/08/79	The 35-kg payload was dedicated to measuring the in-flight performance of the SLV-3 launch vehicle. SLV-3 was the first Indian launch vehicle capable of placing a satellite in orbit. The second stage failed, and the launch vehicle was lost in the Bay of Bengal.
4	Rohini RS-1	18/07/80	The first success for SLV-3. RS-1 was delivered to an orbit of 305 x 919 km with an inclination of 44.7° and remained in orbit for 20 months.
5	RS-D1	31/05/81	Placed in a lower-than-intended orbit of 186 x 418 km, it carried remote sensing payload of 35 kg. The satellite re-entered nine days later.
6	APPLE	19/06/81	India's first experimental communication

			satellite launched for free by ESA's Ariane. It was placed in GEO of 36,000 km and was located at 102° E longitude. With a mass of 670 kg, it was beyond SLV-3 capability. APPLE operated for over two years.
7	Bhaskara-II	20/11/81	Bhaskara-2 was similar to Bhaskara-1 in design, size, mass, instrumentation, launch and orbital characteristics. The high voltage supply to the camera onboard Bhaskara -2 was redesigned based on the experience of Bhaskara-1. It was considered a complete success. Bhaskara-2 ceased operations in 1984 and re-entered on 30 November 1991.
8	INSAT-1A	10/04/82	India's first fully operational communication satellite. It was built by Ford Aerospace and Communication Corporation and launched by the US's Delta. Following a series of system failures, INSAT-1A was deactivated in September 1983, and ISRO recovered its insured payment.
9	RS-D2	17/04/83	The final flight of SLV-3. The 41.5 kg RS-D2 was delivered to an orbit of 371 x 861 km. Regarded as a complete success, RS-D2 re-entered on 19 April 1990 after 17 months of operations. The SLV-3 programme was terminated early to accelerate the development of the next launch vehicle, ASLV.
10	INSAT-1B	30/08/83	Identical to INSAT-1A and ordered at the same time, INSAT-1B was a complete success, unlike INSAT-1A. It was placed in GEO and operated from 74°E for most of its operational life. It was moved to 93°E in 1992 and decommissioned in August 1993 after operating successfully for longer than its planned seven years. As part of decommissioning, it was raised to a graveyard orbit to clear the GEO orbital slot for a future satellite.
11	SROSS-A	24/03/87	The first satellite in the Stretched Rohini Satellite Series (SROSS), India's first series of scientific satellites built and launched by ISRO. The booster stage worked as planned, but the first stage failed to ignite. The launch vehicle achieved a maximum altitude of 10 km before plunging into the Bay of Bengal.

12	IRS-1A	17/03/88	First fully functional remote sensing satellite built by ISRO. It had two high-resolution cameras, LISS-1 (73 m resolution) and LISS-2 (36.25 m resolution). Launched by the USSR into a polar orbit with 99.08° inclination, which brought it over the same point on Earth every 22 days. IRS-1A operated until July 1996.
13	SROSS-B	13/07/88	Boosters failed to separate after the first stage ignited. The launcher lost attitude control and mission failed.
14	INSAT-1C	22/07/88	Third INSAT procured from Ford. A month after launch, a power system fault crippled INSAT-1C's communication capabilities. After operating partially until 1989, it was abandoned in November 1989, following the loss of attitude control.
15	INSAT-1D	12/06/90	Last INSAT to be built outside India. It replaced INSAT-1B and was identical to it except for a larger battery and propellant capacity for a longer life. The launch was delayed by two incidents. It was first damaged during launch preparations and then again in the San Francisco earthquake of 1989. It was repaired on both occasions, and it operated successfully for over a decade until May 2002 when it lost attitude control.
16	IRS-1B	29/08/91	Remote sensing satellite identical to IRS-1A. Built by ISRO, IRS-1B operated successfully until 20 December 2003.
17	SROSS-C	20/05/92	Operated from 25 May 1992 until re-entry on 15 July 1992. SROSS-C had been delivered to a lower orbit of 391 km x 267 km with an inclination of 46° instead of the planned 437 × 938 km due to partial failure of the 5th stage of the launch vehicle.
18	INSAT-2A	10/07/92	Built by ISRO and launched by ESA. INSAT-2A operated successfully until 2002.
19	INSAT-2B	23/07/93	Launched on ESA's Ariane launcher, INSAT-2B reached its GEO location of 93°E by early August. It operated successfully for its designated lifetime of seven years providing communication, meteorology and satellite-based search and rescue services. It was retired from the main service when it ran out of oxidiser in November 2000 and had

			problems with attitude control. It continued to provide limited services until 2004.
20	IRS-1E	20/09/93	Also known as IRS-P1, IRS-1E was launched on a PSLV test flight. During stage 2 and stage 3 separation, the launch vehicle lost control. Even though subsequent stages worked as planned, IRS-1E failed to get into orbit.
21	SROSS-C2	04/05/94	Launched on the fourth and final ASLV, the satellite was placed in the planned orbit of 600 km x 430 km with an inclination of 45°. The payload consisted of two science instruments, and it discovered 12 gamma-ray burst sources. SROSS-C2 was included in the Interplanetary Network (IPN3), a network of gamma-ray burst detectors in space.
22	IRS-P2	15/10/94	Launched to help manage India's agriculture, hydrology, geology, drought and flood monitoring, marine studies, snow studies and land use through the National Natural Resource Management System (NNMRS). The successful three-year mission was completed in 1997.
23	INSAT-2C	07/12/95	Stationed at 93.5°E, along with INSAT-2B until 2002, it carried an exclusively communications-related payload of 12 C-band, six extended C-band, three Ku-band, one Mobile Satellite Service forward and return transponder and one BSS S-band transponder. After operating for nearly eight years, it was moved to a graveyard orbit in 2003.
24	IRS-1C	28/12/95	The first use of USSR's Molniya 8K78M launcher for an Indian satellite. Placed in a SSO, IRS-1C passes over any given point on the Earth's surface at the same local time ensuring that the Sun was at the same angle providing identical illumination for every pass. It had three onboard cameras, with the highest resolution being 5.8 m.
25	IRS-P3	21/03/96	IRS-P3 carried an X-ray astronomy instrument, along with earth observation instruments, a wide-field sensor and a Modular Optoelectronic Scanner. It successfully studied X-ray sources and pulsars for nearly 10 years until January 2006.

26	INSAT-2D	04/06/97	Operated as planned providing communication services but failed in October 1997. A space weather incident is suspected for the loss of attitude control and loss of communication with Earth. To make up the deficit in India's communication capacity, ISRO purchased a transponder from Arabsat 1C in the following year. This provision was renamed INSAT-2DT.
27	IRS-1D	29/09/97	Launched on PSLV with a larger-than-usual stage 1 (S139 instead of S129), IRS-1D carried an identical payload to IRS-1C, a LISS-3 sensor and a wide field sensor. During launch, stage 4 shutdown prematurely resulting in a lower elliptical orbit than intended. IRS-1D's onboard engines were used to reach orbit. It operated successfully for over 12 years.
28	INSAT-2E	03/04/99	Last of the INSAT 2 series, INSAT-2E was a multi-purpose satellite used for telecommunication, television broadcasting and meteorological services. It was the first to carry a CCD for optical imaging to a resolution of 1 meter. A VHRR used to collect data on concentrations of water vapour and aerosols failed a few months after launch. VHRR was used to image land and oceans with 2 km resolution in visible and 8 km resolution in infrared. INSAT-2E operated successfully for its designed lifetime of 12 years.
29	Oceansat-1	26/05/99	First Indian satellite designed to monitor the ocean. Also called IRS-P4, it was launched along with Korean KITSAT-3 and a German DLR-TUBESAT on a PSLV. It studied the ocean using Ocean Colour Monitor (OCM, a solid-state camera) and a Multi-Frequency Scanning Microwave Radiometer (MSMR). It monitored chlorophyll, phytoplankton and atmospheric aerosols. Oceansat-1 operated for over 11 years, twice its planned mission length, ending its mission on 8 August 2010.
30	INSAT-3B	22/03/00	First of the new INSAT 3 series as INSAT-3A was only launched three years later. It served primarily as a communication satellite with 12 extended C-band transponders, 5 Ku-band transponders and MSS. It provided

			services to commercial telecom and mobile service operators. After a lifetime of 12 years, it was moved to a graveyard orbit in March 2011.
31	GSAT-1	18/04/01	Smaller but similar in function to the INSAT series, GSAT-1 was launched on the maiden flight of ISRO's GSLV. It was an experimental communication satellite with the objective to trial compressed digital TV transmission from a GEO orbit of 48°E. GSLV failed to provide enough power to reach its designated orbital slot. GSAT-1's on-board propulsion system too ran out of fuel while attempting to make up the deficit. Although some communication experiments were conducted, GSAT-1 mission was considered a failure.
32	TES	22/10/01	Launched using PSLV, along with PROBA from Belgium and BIRD from Germany. TES was designed to test multiple new technologies, including light weight structure, high-torque reaction wheels, single propellant tank, solid state data recorder and a high resolution (1 m) optical camera. TES's remote sensing data of civilian areas was made commercially available to the private sector. It was designed to operate for three years.
33	INSAT-3C	24/01/02	Launched to replace the services from the ageing INSAT-2DT and INSAT-2C, as well as increase communication capacity to meet increasing demand. INSAT-3C carried 24 C-band, six extended C-band and two S-band transponders. It had an operational lifetime of 15 years.
34	KALPANA-1	12/09/02	Launched as METSAT, but renamed by the Indian Prime Minister as Kalpana-1 on 5 February 2003 following the loss of the Indian-born American astronaut Kalpana Chawla in space during the re-entry of Space Shuttle Columbia on 1 February 2003. Incidentally, this came shortly after the launch centre itself, Sriharikota, was renamed as the Satish Dhawan Space Centre following the demise of the former ISRO Chairman Satish Dhawan. Kalpana-1 was purely a meteorology mission. It carried an Indian-made 3 band VHRR with a resolution of 8

			km to detect water vapour. It had a mission lifetime of seven years.
35	INSAT-3A	10/04/03	Launched by ESA's Ariane-5, it was designed to provide telecommunication, TV broadcasting, meteorological and search and rescue services. The search and rescue payload had global coverage, including relay beacons on sea, land and air. The meteorological payload included a VHRR, CCD camera and a data relay transponder. INSAT-3A was designed to operate for 12 years.
36	GSAT-2	08/05/03	Primarily a communication satellite, its communication payload included 4 C-band and 2 Ku-band transponders and a MSS payload. A scientific payload to investigate the Earth's ionosphere included Solar X-ray Spectrometer and a radiation monitor. It successfully captured data until April 2011.
37	INSAT-3E	28/09/03	A communication satellite with support for television broadcasting and VSAT used for low speed Internet access to remote locations, including ships at sea. It was also used to conduct tele-education and telemedicine experiments. INSAT-3E was tested at 52°E before moving to its operating longitude of 55°E. On 22 September 2012, it lost service for 11 hours before fully recovering. It was moved to a graveyard orbit on 4 April 2014 when it ran low on fuel, even though the planned lifetime was 15 years.
38	ResourceSat -1	17/10/03	Tenth in the IRS series, it is also known as IRS-P6 and was to continue the role of IRS-1C and IRS-1D. It was placed in an 817-km SSPO crossing the equator at 10:30 am local time 14 times every day. It was equipped with a high-resolution LISS-4 camera operating at visible and infrared with a capability for stereoscopic images of 56 m resolution and a built-in solid-state storage capacity of 120 GB, where data that could not be transmitted immediately could be stored until a ground station came into view. ResourceSat-1 had a mission lifetime of five years.
39	GSAT-3 / EDUSAT	20/09/04	First Indian communication satellite designed exclusively to serve the education sector with distant learning facilities. Its communication

			payload included six extended C-band transponders, five Ku-band transponders with regional coverage and one Ku-band transponder with national coverage. Placed in a geostationary slot at 74°E for most of its life, it operated from an inclined orbit in its final year of operation. GSAT-3 was retired in December 2010.
40	Cartosat–1	05/05/05	First Indian satellite (also known as IRS-P5) to provide in-orbit stereoscopic images using two panchromatic cameras sensitive in the visible spectrum. Cartosat-1 was designed to aid cartography in India. Placed in SSPO with an orbital period of 97 minutes, it circles the world like any other SSPO satellite and takes images of the globe, not just India. It was designed to operate for five years.
41	HAMSAT	05/05/05	The first launch from ISRO's Second Launch Pad, HAMSAT was a micro-satellite providing Amateur Radio (HAM) services. It had two transponders, one built by the Indian HAM community and the other by students of the Higher Technical Institute at Venlo in the Netherlands. It was launched along with Cartosat-1. HAMSAT operated for over nine years until a suspected battery failure ended the mission.
42	INSAT-4A	22/12/05	First satellite to bring DTH broadcasting to private homes. At 3,081 kg, INSAT-4A was the heaviest satellite India had built and launched until then. It had 12 Ku-band transponders for DTH and another 12 C-band transponders for communication. It was designed for an operational lifetime of 12 years.
43	INSAT-4C	10/07/06	Commanded to self-destruct as one of the liquid boosters of the launch vehicle failed resulting in a loss of control 55 seconds after launch.
44	Cartosat-2	10/01/07	Also known as IRS-P7, it was launched along with India's SRE, Indonesia's Lap-lan Tubsat and Argentina's Peuensat-1. Cartosat-2 was designed to work with Cartosat-2A (due for launch in 2008) to provide high resolution (1 m) images using a panchromatic camera. It had a 64 GB solid state storage device to store data until transmission and was designed to

			operate for five years.
45	SRE-1	10/01/07	India's first experience of recovering a spacecraft after launch. The 550 kg Space Recovery Experiment capsule was used to conduct microgravity experiments for 10 days in LEO. The recovery phase provided experience in navigation, guidance, thermal protection, parachute deployment and water-based recovery. The SRE-1 capsule was safely recovered from within 15 km of the targeted landing point in the Bay of Bengal. It is currently on display as an exhibit in the museum in VSSC.
46	INSAT-4B	12/03/07	Launched by ESA's Ariane along with the British military satellite Skynet 5A, INSAT-4B is designed to bolster India's satellite communication capacity and carries an exclusive communication payload consisting of 12 Ku and 12 C-band transponders. It was tested at 80.5°E and went operational at 93.5°E in April 2007. On 7 July 2010, one of its two solar panels stopped operating normally. INSAT-4B had a designated lifetime of 12 years.
47	INSAT-4CR	02/09/07	Replacement for INSAT-4C, INSAT-4CR carries 12 transponders for use by DTH television, the Department of Defence, Indian Railways and Oil and Natural Gas Corporation. Even though planned for a lifetime of 12 years, it may not last that long as fuel destined for station keeping was used to get to the designated orbit after the third stage of the GSLV launcher underperformed.
48	Cartosat-2A	28/04/08	Cartosat-2A was launched on PSLV-C9, along with an Indian mini-satellite and eight nano satellites from Japan, Canada, Germany, Denmark and the Netherlands. It has a panchromatic camera with a spatial resolution of less than 1 meter and image resolution sufficiently high for defence purposes and mapping urban and rural infrastructure. Cartosat-2 and Cartosat-2A work as a pair for frequent revisits of the same area to help monitor local situations, such as flooding or earthquake. It was designed for an operational lifetime of five years.
49	IMS-1	28/04/08	Co-passenger with Cartosat-2A, Indian Micro

			Satellite-1 is a low-cost micro-satellite imaging mission previously referred to as Third World Satellite (TWSat). In a polar orbit of 635 km, it can take images with its two medium and low-resolution cameras, and the data is designed to be shared with developing nations.
50	Chandrayaan-1	22/10/08	India's first mission to Moon. Chandrayaan-1carried 11 experiments, six of which came from outside India (the US, UK, Germany, Sweden and Bulgaria). It entered a lunar orbit of 100 x 100 km on 8 November 2008 and successfully released the Moon Impact Probe six days later. In 2009, its orbit was increased to 200 km. It operated for almost a year. All instruments on-board Chandrayaan-1 collected data during the mission. On 29 August, communication was lost with Chandrayaan-1 following a series of technical issues resulting probably from poor shielding. Despite the early termination, the mission was regarded as a success.
51	ANUSAT	20/04/09	A micro-satellite designed, developed and integrated by the Department of Aerospace Engineering at the Madras Institute of Technology, affiliated to Anna University. Anna University Micro-satellite (ANUSAT) was sponsored by ISRO and was launched on PSLV-C12, along with RISAT-2, the primary payload. ANUSAT was placed in an orbit of 550 km with an inclination of 41° and operated until April 2012.
52	RISAT-2	20/04/09	India's first dedicated reconnaissance satellite with day and night all-weather monitoring capability. Acquired in haste from Israel Aerospace Industries following the 2008 terrorist attack in Mumbai and launched by ISRO six months later, RISAT-2 carries a SAR for high resolution (1 meter) 3-D images. The same technology can be used for disaster management and search and rescue. Its launch was not televised live, leading the press to call it a spy satellite.
53	Oceansat-2	23/09/09	Oceansat-2 is designed to provide continuity of service to users of the Ocean Colour Monitor (OCM) instrument on Oceansat-1 launched in 1999. In addition to OCM-2,

			Oceansat-2 carries a scatterometer with a 1 m diameter antenna rotating at 20 rpm for measuring sea surface-level winds and a Radio Occultation Sounder for Atmospheric Studies (ROSA), which can help characterise the Earth's atmosphere and ionosphere by looking at signals from orbiting GPS satellites.
54	ATV-D01	03/03/10	At three tonnes, ATV-D01 is the heaviest sounding rocket ever developed by ISRO until then. It carried a passive scramjet engine combustor module as a test bed for the demonstration of air-breathing propulsion technology. During the flight, the vehicle successfully dwelled for seven seconds in the desired conditions of Mach number (6 + 0.5) and dynamic pressure (80 + 35 kPa). These are conditions required for the stable ignition of the active scramjet engine combustor module planned for future ATV flights.
55	GSAT-4	15/04/10	An experimental satellite launched on GSLV-D3 with an Indian cryogenic third stage. GSAT-4 was intended to test electronic propulsion, miniaturised dynamically tuned gyroscope and a GAGAN payload. 300 seconds after launch, contact was lost, and GSAT-4 did not reach its designated orbit at 82°E. The investigation reported that one second after the ignition of the cryogenic third stage, the fuel turbo pump stopped working.
56	STUDSAT	12/07/10	A PICO satellite (10cm x 10cm x 13.5cm) weighing less than 1 kg, STUDSAT was developed by seven engineering colleges in India. The project was designed to promote space technology in educational institutes. With a lifetime of six months, it played a remote sensing role by taking images with its on-board camera with a resolution of 90 m from its 630 km SSPO.
57	Cartosat–2B	12/07/10	Launched on PSLV-C15, along with an Algerian satellite, a nano-satellite from Canada, another from Switzerland and India's first pico satellite, STUDSAT. Identical to Cartosat-2A with a panchromatic camera capable of imaging a strip of 9.6 km with a resolution of more than 1 metre,

			Cartosat-2B works with Cartosat-2 and Cartosat-2A and has stereoscopic imaging capability. At 630 km SSPO, it can revisit the same location every 4 or 5 days, and with its high-resolution capability, it can satisfy many applications including the military.
58	GSAT-5P	25/12/10	Intended to replace INSAT 3E, GSAT-5 Prime carried 24 C-band and 12 extended C-band transponders. 45 seconds after launch, the communication link between the guidance computer on the third stage and the first stage was lost. After a further 20 seconds, a self-destruct command was initiated and the mission terminated.
59	ResourceSat-2	20/04/11	A remote sensing satellite intended to continue the remote sensing data services provided by ResourceSat-1. Enhancements in LISS-4 include an improved resolution and electronics miniaturisation. ResourceSat-2 was accompanied by two other satellites, Indo-Russian Youthsat and the X-Sat from the Nangyang Technological University of Singapore. It also had two solid state data recorders with a capacity to record up to 200 GB data.
60	YOUTHSAT	20/04/11	Launched in PSLV-C16, along with ResourceSat-2 and Singapore's X-Sat, YOUTHSAT was ISRO's second mini satellite. With an operational lifetime of two years, it was a joint Indo-Russian project to investigate solar activity and Earth's upper atmosphere. Undergraduate and postgraduate students from India and Russia collected data using three instruments: (1) Radio Beacon for Ionospheric Tomography for mapping TEC of the ionosphere, (2) Limb Viewing HySI for measuring airglow emissions of Earth's upper atmosphere (80–600 km) in 450–950 nm and (3) a Russian instrument SOLRAD to study temporal and spectral parameters of solar flares, X- and gamma-ray fluxes, as well as charge particles in the polar cap regions.
61	GSAT-8	21/05/11	Also known as INSAT-4G, GSAT-8 is an advance communication satellite with 24 high powered Ku-band transponders, a two channel GAGAN payload and a lifetime of 12 years. The GAGAN payload in GEO and

			a network of ground stations provide a Satellite-Based Augmentation System that enhances the accuracy and reliability of GPS data for use by the commercial aviation sector. GSAT-8 was tested at 47°E before reaching its operational slot at 55°E.
62	GSAT-12	15/07/11	An exclusively communication satellite designed to meet India's growing communication requirements. GSAT-12 is equipped with 12 extended C-band transponders to facilitate tele-education and telemedicine and for Village Resource Centres. GSAT-12 was launched by an all-women core team and is designed to have an operational lifetime of eight years.
63	Jugnu	12/10/11	A nano-satellite weighing 3 kg, Jugnu was designed and developed by the Indian Institute of Technology, Kanpur, under the guidance of ISRO. It had three objectives: (1) prove the indigenously developed camera system for imaging the Earth in the near infrared region and test image processing algorithms, (2) evaluate GPS receiver for its use in satellite navigation and (3) test the indigenously developed Micro Electromechanical System (MEMS)-based Inertial Measurement Unit in space. It uses a 2 GB data card for on-board storage.
64	Megha-Tropiques	12/10/11	A joint Indo-French mission to study the water cycle and energy exchanges in the tropics. Megha-Tropiques orbits the Earth at 867 km with an inclination of 20° to the equator from where it can image any point in the Intertropical Convergence Zone, six times a day. The main objective of this mission is to understand the life cycle of convective systems that influence the tropical weather and climate and their role in associated energy and moisture budget of the atmosphere in tropical regions. It has four instruments: (1) a Microwave Analysis and Detection of Rain and Atmospheric Structures (MADRAS) imager, (2) an Imaging Radiometer Sounder for Probing Vertical Profiles of Humidity (SAPHIR) developed jointly by CNES and ISRO, (3) a Scanner for Radiation Budget (ScaRaB) and (4) a Radio Occultation Sensor

			for Vertical Profiling of Temperature and Humidity (ROSA), procured from Italy. It has 16 GB on-board solid-state storage. The mission is designed to last for five years.
65	SRMSat	12/10/11	Launched on PSLV C18, along with Megha-Tropiques, VesselSat-1 and Jugnu, SRMSat is a nano-satellite weighing 10.9 kg. It was developed by Sri Ramaswamy Memorial University (SRM), Chennai, as a technology demonstration and EO satellite. Operated by the SRM Institute of Science and Technology, it has been used to monitor greenhouse gases in the atmosphere. It is in LEO of 816 km inclined at 20°. Even though designed to operate only for a year, it remained operational into late 2015.
66	RISAT-1	26/04/12	A remote sensing satellite designed for applications in agriculture, particularly paddy monitoring in the Kharif season, and management of natural disasters, such as flood and cyclone. The key payload of RISAT-1 is the SAR, which enables imaging of surface features during both day and night under all weather conditions. It is in SSPO of 536 km from where it orbits the Earth 14 times every day. It is designed for an operational period of five years.
67	SPOT-6	09/09/12	Launched French earth observation satellite SPOT 6 along with a micro-satellite from Japan in to a 655 km polar orbit inclined at an angle of 98.23° to the equator.
68	GSAT-10	29/09/12	An advance communication satellite similar to GSAT-8 with 30 C-band transponders providing communication services to mainland India and its offshore sites, such as the Andaman and Nicobar Islands. Like GSAT-8, it is also configured to offer navigation support through a GAGAN payload. It operates from 85°E with an expected lifetime of 15 years.
69	SARAL	25/02/13	A joint Indo-French mission, SARAL carries two instruments, a Ka-band Altimeter (ALTIKA) and ARGOS Data Collection System, both built by CNES. The altimetric measurements are designed to study ocean circulation and sea surface elevation and is intended to support many applications,

			including climate monitoring, marine meteorology and sea state forecasting, continental ice studies, environmental monitoring and improvement of maritime security. It is in a SSPO at 781 km inclined at 98.538°. The mission lifetime is determined by its two key instruments, ARGOS is expected to last for 5 years and AltiKa 3 years.
70	IRNSS-1A	01/07/13	First of the seven satellites that comprise the Indian Regional Navigation Satellite System. IRNSS-1A is located at 55°E GSO inclined at 29° and has an expected lifetime of 10 years. IRNSS will provide satellite-based positional information over mainland India and up to 1,500 km beyond.
71	INSAT-3D	26/07/13	An advanced weather satellite incorporating a search and rescue (SARSAT) payload. The mission's primary goals are to provide an operational environmental and storm warning system to protect life and property, to monitor Earth's surface and carry out oceanic observations and to provide data dissemination capabilities. In addition to a six-channel multi-spectral imager, INSAT-3D is equipped with an Atmospheric Sounder. The Atmospheric Sounder provides a vertical profile of temperature, humidity and integrated ozone measurements from the surface to the top of the atmosphere. A Data Relay Transponder receives and relays signals from 1,800 Earth-based Data Collection Platforms (automated weather stations typically located in remote locations) established by the Meteorological Department of India and ISRO. The search and rescue transponder receives and relays distress alerts originating from the distress beacons of maritime, aviation and land based users to the Indian Mission Control Centre located at ISRO's ISTRAC. INSAT-3D is located at 82° E with an expected operational lifetime of 7.7 years.
72	GSAT-7	30/08/13	GSAT-7 is India's first communication satellite for exclusive military use. It is dedicated for use by the Indian Navy to provide independent, secure communication for its fleet of war ships, submarines, aircraft

			and land bases. GSAT-7 was designed and built in India although launched by ESA's Ariane. It is located at 74°E with an expected lifetime of 7 years.
73	MOM	05/11/13	Designed, built and launched by ISRO, MOM successfully entered Martian orbit on 24 September 2014. All five instruments on-board have been activated and returning data. The current orbit of 421 km x 76,993 km takes MOM around Mars once every 72 hours and 51 minutes. Designed for an operational period of only six months, it has already exceeded that target and is expected to operate for several years.
74	GSAT-14	05/01/14	A communication satellite with six Ku-band and six extended C-band transponders intended eventually to replace GSAT-3 capacity. It also has two Ka-band beacons designed to help researchers understand how weather affects Ka-band satellite communications. The first launch attempt on 19 August 2013 had to be abandoned following a second stage fuel leak. The launch was successfully completed on 5 January 2014. GSAT-14 is located at 74°E with a life expectancy of 12 years.
75	IRNSS-1B	04/04/14	Second of the seven satellites that comprise IRNSS. IRNSS-1B is located at 55°E GSO inclined at 29°, identical to but 180° out of phase with IRNSS-1A. It has an expected lifetime of 12 years.
76	Spot-7	30/06/14	Launched SPOT-7 a French earth observation satellite, into a 655 km Sun-Synchronous Orbit (SSO). It is the tenth flight of PSLV in 'core-alone' configuration (without use of solid strap-on motors). There were 4 other co-passengers placed in orbit at the same time.
77	IRNSS-1C	10/16/14	Third of the seven satellites that comprise IRNSS. IRNSS-1C is located at 83°E GEO. It has an expected lifetime of 12 years.
78	GSAT-16	07/12/14	Intended to gradually replace the services provided by INSAT-3E, GSAT-16 has 12 Ku-band, 24 C-band and 12 extended C-band transponders. It provides additional communication capacity to meet India's

			increasing demand for satellite communication. GSAT-16 is located at 55°E with a planned service lifetime of 12 years.
79	LVM3-X/CARE	18/12/14	ISRO's latest generation of launch vehicle, Launch Vehicle Mark 3 (LVM3), took a 3.7-ton early version of ISRO's crew capsule to an altitude of 126 km before a splash down in the Bay of Bengal 19 minutes after launch. It was designated the Crew Module Atmospheric Re-entry Experiment (CARE). The capsule was recovered by the Indian coast guard.
80	IRNSS-1D	28/03/15	Fourth of the seven satellites that comprise IRNSS. IRNSS-1D is located at 111.75°E GSO inclined at 30.5°. It has an expected lifetime of 12 years.
81	GSAT-6	27/08/15	Designed specifically for multi-media mobile applications, GSAT-6 offers multiple audio and video channels using modern encoding and compression technologies. It delivers a variety of services, including interactive services, such as text messaging, weather information and disaster warning. It is India's second satellite with payload specifically for military use. GSAT-6 uses an innovative 6 m antenna that was deployed once in space. It is located at 83°E and expected to operate for 9 years.
82	DMC-3	10/07/15	To date ISRO's heaviest commercial payload delivering of 5 satellites to Sun-synchronous orbit. Total payload mass of 1,440 kg consisted of three identical optical earth observation satellites (DMC3-1, DMC3-2 & DMC3-3 - built in the UK but operated by China). An optical earth observation technology demonstrator micro satellite (CBNT-1) and an experimental nano satellite (De-orbit Sail).
83	TeLEOS-1	16/12/15	Launched six satellites owned by Singapore into a 550 km circular orbit inclined at 15° to the equator. TeLEOS-1 is the primary satellite used for Earth Observation weighing 400 kg. The other five include micro-satellite; VELOX-II (13 kg) 6U-Cubesat technology demonstrator; Athenoxat-1, a technology demonstrator nano-satellite; Kent Ridge-1 (78 kg), a micro-satellite; and Galassia (3.4

			kg) 2U-Cubesat from Singapore Universities.
84	Astrosat	28/09/15	India's first science-only mission in Earth orbit. Astrosat is a multi-wavelength space telescope designed to observe the universe in ultraviolet, X-ray and optical wavelengths. Astrosat is a joint project between several Indian institutions and two international (Canadian Space Agency and the University of Leicester in the UK) institutions, which will collaborate over its five-year mission.
85	GSAT-15	11/11/15	Like GSAT-10, GSAT-15 is intended to help meet the increasing demand for satellite transponder capacity. It has 24 Ku-band transponders and is the third satellite to carry a GAGAN payload, after GSAT-8 and GSAT-10. GSAT-15 carries a Ku-band beacon as well. This allows ground antennae to accurately point towards the satellite. It is located at 93.5°E and has an expected lifetime of 12 years.
86	IRNSS-1E	20/01/16	Fifth of the seven satellites that comprise IRNSS. IRNSS-1E is located at 111.75°E GSO inclined at 28.1°, identical to but 180° out of phase with IRNSS-1D. It has an expected lifetime of 12 years.
87	IRNSS-1F	10/03/16	Sixth of the seven satellites that comprise IRNSS. IRNSS-1F is located at 32.5°E. It has an expected lifetime of 12 years.
88	IRNSS-1G	28/04/16	Last of the seven satellites that comprise IRNSS. IRNSS-1G is located at 129.5°E. It has an expected lifetime of 12 years.
89	RLV-TD	23/05/16	Sub-orbital test flight of ISRO's Reusable Launch Vehicle Launch Technology Demonstrator.
90	Cartosat-2S	22/06/16	Cartosat-2s was accompanied by 19 co-passenger satellites from the US, Canada, Germany and Indonesia, as well as two satellites (SATHYABAMASAT and SWAYAM) from Indian universities/academic institutes. The total weight of all the 20 satellites was about 1,288 kg. Cartosat-2 is designed to operate for five years.
91	ATV-D02	28/08/16	Follow-up flight from the 2010 ATV-D01, but with active twin scramjet engines mounted on the back of the second stage.

			Once the second stage reached the desired conditions for engine start-up, necessary actions were initiated to ignite the scramjet engines, and they functioned for about 5 seconds. After a flight of about 300 seconds, the vehicle splashed down in the Bay of Bengal, approximately 320 km from Sriharikota but was not recovered. It was successfully tracked during its flight from the ground stations at Sriharikota.
92	INSAT-3DR	08/09/16	INSAT-3DR joined other ISRO weather satellites in GEO (INSAT-3A, INSAT-3D and Kalpana). It is equipped with a high resolution (infrared and visible) camera and a SARSAT transponder. A Data Relay Transponder receives meteorological, hydrological and oceanographic signals from Earth-based sensors located in remote uninhabited locations.
93	SCATSAT-1	26/09/16	SCATSAT-1 carries a similar payload as Oceansat-2. A scatterometer (microwave radar) is used to measure wind speed and direction to predict and monitor cyclones. SCATSAT-1 has a solid-state recorder of 32 GB capacity and is designed for an operating life of five years. SCATSAT-1 was one of eight payloads, two of which came from Indian academic institutions and five from other countries. This group of 304 kg in total was placed in the higher orbit of 720 km. It was the first time that ISRO used a single PSLV to deliver satellites to two different orbits.
94	ResourceSat-2A	07/12/16	A remote sensing satellite with similar objectives as its predecessors, Resourcesat-1 and Resourcesat-2. It has three key payloads: a LISS-4 camera with a resolution of 5.8 m operating in the visible and near infrared wavelengths, a lower resolution LISS-3 and an Advanced Wide Field Sensor. Resourcesat-2 orbits the Earth 14 times every day and returns to the same land/sea mass after 24 days.
95	Cartosat-2	15/02/17	Cartosat-2 at 714kg was launched along with 103 smaller satellites. With a total mass of 1378k. The smaller satellites consisted of two ISRO Nano Satellites (INS-1 and INS-2),

			USA (96), The Netherlands (1), Switzerland (1), Israel (1), Kazakhstan (1) and UAE (1). With this single launch, India achieved a record of the most satellites launched at one time.
96	South Asian Satellite	05/05/17	A political initiative by the Indian Prime Minister to engage India's neighbours SAARC nations by providing the free use of one of the on-board 12 transponders to each country.
97	GSAT-19	05/06/17	Unlike the maiden flights of each of its previous launchers (SLV-3, ASLV, PSLV and GSLV Mk1 and GSLV-Mk2) GSLV-Mk3 achieved 100% success. It placed a 3136 kg GSAT-19 communication satellite into a GTO using three-stages - two solid motor strap-ons (S200), a liquid propellant core stage (L110) and a cryogenic stage (C25)
98	Cartosat-2E	23/06/17	This was the 40th flight of the PSLV and 17th in the XL series. The primary payload was the 712 kg Cartosat-2E with a secondary payload of 30 satellites with a combined mass of 243 kg. Cartosat-2E provides high resolution of 0.6 m images produced using its panchromatic and multispectral cameras. Ostensibly Cartosat-2E is an Earth observation satellite, its data used to serve cartographic and remote sensing applications. It is reported to be primarily a military satellite. Once in its designated orbit, ISRO passes operational control to the defence forces. One of the 30 co-passenger satellites were from India, a 15 kg NIUSAT from Nurul Islam University, Tamil Nadu. The other 29 satellites came from Austria (1), Belgium (3), Chile (1), Czech Republic (1), Finland (1), France (1), Germany (1), Italy (3), Japan (1), Latvia (1), Lithuania (1), Slovakia (1). UK (3) and USA (10).

99	IRNSS-1H	31/08/17	IRNSS-1H was intended to replace IRNSS-1A. IRNSS-1A on-board Rubidium clock had failed so was not able to function as designed. However, this mission failed. The payload fairing did not separate and the spacecraft remained inside stage 4.

Index

References

Chapter 1

1. Technically, only V2 was propelled by a rocket engine. V1 was a cruise missile propelled by a pulse jet engine.

2. Perhaps, this 'scramble for India' was in the mind of Otto von Bismarck when he organised the Berlin Conference, aka 'The Scramble for Africa', in 1884. It was designed to agree on the boundaries of the Congo Free State but came to symbolise the division of the whole African continent. The greed, haste and the arbitrariness with which the national boundaries were fixed by Europeans in India, Africa, and the Middle East continues to shape the geopolitics of the 21st century. Michalopouslos, Stelios and Elias Papaioannou. 2011. The Long-Run Effects of the Scramble for Africa. *NBER Working Paper Series. No.17620.* Retrieved from http://www.nber.org/papers/w17620.pdf

3. The British East India Company (also known as the Honourable East India Company) is the largest commercial operation that has ever existed, but it was not alone. Other European nations, including Austria, the Netherlands, Denmark, Portugal and Sweden, too, had East India Companies.

4. Watson, William E. 2003. *Tricolor and Crescent: France and the Islamic World.* Westport, CT: Praeger Publishers. P13.

5. The Governor General was the head of the British administration in India.

6. Haroon, Anwar. 2013. *Kingdom of Hyder Ali and Tipu Sultan.* Bloomington, IN: Xlibris Corporation.

7 Lovell, Sir Bernard. 1979. *In the Centre of Immensities.* First British edition. London: Hutchinson. P130

8. Graves, Donald. 1996. *Sir William Congreve and the Rocket's Red Glare.* Alexandria Bay, NY: Bloomfield. P8.

9. Sir William Congreve (1772–1828). *Jeremy Norman's History of Science.com.* Retrieved from http://www.historyofscience.com/pdf/Congreve-archive.pdf

10. Agrawal, Lion M. G. 2008. *Freedom Fighters of India.* Delhi: Gyan Publishing House. P277.

11. James Dalrymple, a descendant of the acclaimed Indian historian William Dalrymple, was among those captured. Dalrymple, William. 2004. *White Mughals: Love and Betrayal in 18th-Century India.* New edition. Harper Perennialia. P191.

12. Narasimha, Roddam. 1985. Rockets in Mysore and Britain, 1750–1850 A.D. *National Aerospace Laboratories.* P9. Retrieved from: http://www.nal.res.in/pdf/pdfrocket.pdf

13. *Ibid.,* P11

14. Ibid., P11

15. Haroon, Anwar. 2013. *Kingdom of Hyder Ali and Tipu Sultan*. Bloomington, IN: Xlibris Corporation. P300.

16. Agrawal, Lion M. G. 2008. *Freedom Fighters of India*. Delhi: Gyan Publishing House. P279.

17. This was repeated on a much larger scale a century and a half later as the allies (the USSR, UK and the US) scoured Germany for its technical and engineering innovations in aviation and rocketry. Most of the recovered V2 rockets went to the US and USSR.

18. Becklake, E. J. and P. J. Turvey. Congreve at the Rotunda. *Journal of the British Interplanetary Society*. Volume 40. P291.

19. This term is not in common use today but refers to a senior management role, someone with financial and quality responsibilities. Perhaps, the equivalent today would be a Chief Executive Officer.

20. Werrett, S. 2009. William Congreve's rational rockets. *Notes & Records of the Royal Society* 63 (1): 35–56.

21. Sir William Congreve (1772–1828). http://www.historyofscience.com/pdf/Congreve-archive.pdf

22. Narasimha, Roddam. 1985. Rockets in Mysore and Britain 1750-1850 A.D. *National Aerospace Laboratories*. Retrieved from: http://www.nal.res.in/pdf/pdfrocket.pdf.

23. Narasimha, Roddam. 1985. Rockets in Mysore and Britain 1750-1850 A.D. *National Aerospace Laboratories*. Retrieved from: http://www.nal.res.in/pdf/pdfrocket.pdf. Table 4 on page 28 describes these combinations.

24. Gunpowder and black powder are made from similar ingredients (Potassium Nitrate, Sulphur and Carbon) but are not identical. Gunpowder is fine and pure. It explodes when heated or struck. Black powder is coarse and burns when ignited generating smoke.

25. Narasimha, Roddam. 1985. Rockets in Mysore and Britain 1750 -1850 A.D. *National Aerospace Laboratories*. Retrieved from: http://www.nal.res.in/pdf/pdfrocket.pdf. Table 3 lists costs for individual components.

26. Bjerg, Hans Christian. August 2008. "To Copenhagen a Fleet": The British Pre-emptive Seizure of the Danish Norwegian Navy, 1807. *International Journal of Navel History* 7(2). Retrieved from http://www.ijnhonline.org/wp-content/uploads/2012/01/Bjerg.pdf.

27. Graves, Donald. 1996. *Sir William Congreve and the Rocket's Red Glare*. Alexandria Bay, N.Y.; Bloomfield, Ont.: Museum Restoration Service.

28. Braun, Wernher von. 17 July 1969. Pioneers of a New Age. *The New York*

Times, Section C.

29. Ciancone, Michael L. 2010 (Ed.) *History of Rocketry and Astronautics*. American Astronautical Society. P53.

30. During the 1930s, many amateur rocketry groups were established around the world, and they conducted basic rocket test flights. One, for example, was in Manchester, Northwest England. http://astrotalkuk.org/2012/03/27/episode-50-26th-march-2012-manchester-first-rocket-scientists/

31. Gruntman, Mike. 2007. *From Astronautics to Cosmonautics*. Booksurge Publishing.

32. Rao, U. R. 2013. *India's Rise as a Space Power*. Delhi: Cambridge University Press India Pvt Ltd. P47.

33. Sternfeld, Ari. 1937 (Revised edition 1974). Introduction to Cosmonautics. Moscow: ONTI.

34. Gruntman, Mike. 2007. *From Astronautics to Cosmonautics*. Booksurge Publishing.

35. Braun, Wernher von. 17 July 1969. Pioneers of a New Age. *The New York Times*, Section C.

36 Singh, Gurbir. 2011. Yuri Gagarin in London and Manchester: A Smile That Changed the World? Astrotalkuk Publications. P14.

37. *Ibid.*, P14

38. Rogers, Lucy. 2008. It's ONLY Rocket Science: An Introduction in Plain English. New York, NY: Springer. P9.

39. "Novosti nauki i tekhniki: neuzheli ne utopiia?". *Izvestiia VTsIK*. 2 October 1923.

40. Siddiqi, Asif. June 2004. Deep Impact: Robert Goddard and the Soviet 'Space Fad' of the 1920s. *History and Technology* 20(2). Retrieved from http://epizodsspace.no-ip.org/bibl/inostr-yazyki/history-and-technology/2004/2/siddiqi_goddard_space_fad.pdf. Tsiolkovskii's publications contributed to a wider public awareness within the Soviet society of the possibilities of space travel.

41. Ford, Brian. 2013. Secret Weapons: Death Rays, Doodlebugs and Churchill's Golden Goose. Osprey Publishing. P143.

42. The US government had adopted the policy of not permitting 'ardent Nazis' into the country as many were deemed security risks, and at least, some were implicated in war crimes. The applications of German citizens desiring to go to the US were sent by the US Army to the US State Department where they were marked 'Ardent Nazi' and rejected or 'Not Ardent Nazi' when the applicant's skills were deemed useful for the US national interest. The term 'paperclip' came from these applications.

43. Cornwell, John. 2004. *Hitler's Scientists: Science, War, and the Devil's Pact*. New

York, NY: Penguin Books.

44. Burnett, Thom. 2005. Who Really Won the Space Race? Uncovering the Conspiracy That Kept America Second to the Russians. Collins & Brown. P63.

45. Oscar Holderer came to the US along with the second batch of German rocket engineers after the initial group with von Braun. Holderer worked on the wind tunnels used to develop Saturn V for the Apollo programme. Retrieved from http://www.al.com/news/huntsville/index.ssf/2015/05/oscar_holderer_last_of_wer nher.html

46. Siddiqi, Asif A. 2010. *The Red Rockets' Glare: Spaceflight and the Soviet Imagination, 1857-1957*. First edition. Cambridge University Press. P177.

47. Koroleva, Natalya. 2001. *Father*, Volume 2. Moscow, Nauka. P355.

48. Bizony, Piers and Jamie Doran. 2011. *Starman: The Truth Behind the Legend of Yuri Gagarin*. Walker Books. P186.

49 .Conversation between Natalya Koroleva and the author in London. 14 July 2011. http://astrotalkuk.org/2011/07/25/yuri-gagarin-statue-in-london/

50. Neufeld, Michael J. February 2002. Wernher von Braun, the SS, and Concentration Camp Labor: Questions of Moral, Political, and Criminal Responsibility. *German Studies Review* 25 (1): 63.

51. The V2 campaign started in September 1942. By then Germany was on the retreat. Had it been introduced earlier, this new "super" weapon could have had a profound impact in the direction of the war. https://airandspace.si.edu/stories/editorial/remembering-wernher-von-braun-his-100th-birthday

52 Parkinson, B. (Ed.) 2008. *Interplanetary: A History of the British Interplanetary Society*. British Interplanetary Society. P50

53. Space Chronology. (n.d.) Boundary of space breached for the first time. Retrieved from http://www.spacechronology.com/1942.html.

54. Chertok, B. E. 2004. *Rockets and People, Volume 1*. NASA SP-2005-4110. P307.

55. Ford, Brian. 2013. Secret Weapons: Death Rays, Doodlebugs and Churchill's Golden Goose. Osprey Publishing. P143

56. Burnett, Thom. 2005. Who Really Won the Space Race? Uncovering the Conspiracy That Kept America Second to the Russians. Collins & Brown. P77.

57. *Ibid.*, P77

58. Burnett, Thom. 2005. Who Really Won the Space Race? Uncovering the Conspiracy That Kept America Second to the Russians. Collins & Brown. P92.

59. Spangenburg, Ray, and Diane Kit Moser. 2009. Wernher von Braun, Revised Edition. Infobase Publishing. P80.

60. Burnett, Thom. 2005. Who Really Won the Space Race? Uncovering the

Conspiracy That Kept America Second to the Russians. Collins & Brown. P98.

61. Telephone conversation between Reg Turnill and the author. 3 November 2011. Also see http://astrotalkuk.org/2013/02/15/episode-61-reg-turnill-on-wernher-von-braun/

62. Crowley, I. F. and J. R. Trudeau. 2011. *Wernher von Braun: An Ethical Analysis*. Retrieved from https://web.wpi.edu/Pubs/E-project/Available/E-project-121811-161339/unrestricted/von_Braun_IQP_12_20_2011_bw_final.pdf

63. Siddiqi, Asif A. 2010. *The Red Rockets' Glare: Spaceflight and the Soviet Imagination, 1857-1957*. First edition. Cambridge University Press. P224.

64. WikiLeaks. 1973. Von Braun in India 10–23 May. Retrieved from https://www.wikileaks.org/plusd/cables/1973STATE080529_b.html

65. Kalam, A. P. J. Abdul and A. Tiwari. 1999. *Wings of Fire: An Autobiography*. Universities Press. P48.

Chapter 2

66. Kochhar, R. 13 September 2014. Transits of Venus and Modern Astronomy in India. Retrieved from http://www.slideshare.net/rajeshkochhar1/transits-of-venus-and-modern-astronomy-in-india

67. The nation that sent men to the Moon is unquestionably scientifically literate, but it is also a deeply religious one. In a survey on 'average frequency of prayer by country', India came out on top. Some traditions may just be too hard to let go. Many national events and personal choices, including of those who would describe themselves as scientifically literate, are still based on mysticism, superstition and astrology. Pickel, Gert. 2012. *Religion Monitor: Understanding Common Ground – An International Comparison of Religious Belief*. Gütersloh, Germany: Bertelsmann Stiftung. P23. Also, following the launch of PSLV PSLV-C12, ISRO skipped PSLV-C13 and went on to PSLV-C14. http://www.dnaindia.com/scitech/report-superstitions-and-beliefs-of-indian-space-scientists-1915176

68. There is some uncertainty around the exact dates when Indus civilisations prevailed. Traditionally considered to have been around between 2500-1700 BCE, a new study using carbon dating results suggests it may be 2500 years older (5000-1500 BCE). http://www.nature.com/articles/srep26555\

69. Both of the following examples are real! (1) Prior to the launch of India's first mission to Mars, the ISRO chairman visited a temple to make his offering. http://timesofindia.indiatimes.com/india/Isro-chief-seeks-divine-help-for-Mars-mission/articleshow/25238936.cms. (2) Equally bewildering is the online innovation onlineprasad.com, which is apparently very popular. If you are too busy to attend a religious event in a temple in person, you can do it via a proxy online. Services offered include Quick Darshan, Prasad, Devotion and Pooja with the option of prasad, poster and text of the recited prayer delivered to your home by

courier. The latter is a perfect example of how science is used to support and facilitate ancient ritual and is accepted without question.

70. With the help of the USSR, ISRO launched to Earth orbit its first satellite on 19 April 1975.

71. Kochhar, R. K. 1991. The Growth of Modern Astronomy in India, 1651-1960. *Vistas in Astronomy.* Volume 34: 69–105.

72. Eclipses and transits are alignments of the Moon or planets with the Sun. A solar eclipse occurs when the Earth, Moon and Sun line up and the Moon appears for several minutes in exactly the same place in the sky as the Sun. Transits occur when Mercury or Venus line up with the Sun. Over a few hours, the small silhouetted disc of Mercury or Venus moves across the face of the Sun. Transits of Mercury occurred on 9 November 1848, 12 November 1861, 5 November 1868, 6 May 1878, 8 November 1881 and 10 May 1891. Only those in 1861 and 1868 were fully visible from India.

73. Kochhar, Rajesh. 1989. The Transit of Mercury 1651: The Earliest Telescopic Observations in India. *Indian Journal of History of Science* 24 (3): 186–92.

74. Ibid.

75. Deccan Herald. 25 March 1986. Astronomical Discovery Made in India in 1689. http://docrchive.com/document/astronomical-discovery-made-in-india-in-1689-7308221412759902/

76. Astronomical observatories were first established in Europe in the 17th century—the Paris Observatory in 1667, Greenwich Observatory, London, in 1675 and Berlin Observatory in 1688.

77. Even though historian Rajesh Kochhar asserts that the "Madras Observatory had been established as an aid to Trigonometrical Survey of India", it had been established about a decade prior to the start of the GTS. The GTS gave it a new lease of life. Kochhar, R. 13 September 2014. Transits of Venus and Modern Astronomy in India. Retrieved from http://www.slideshare.net/rajeshkochhar1/transits-of-venus-and-modern-astronomy-in-india

78. Markham, Clements. 1871. *A Memoir on the Indian Surveys.* London: W.H. Allen and Co. P238.

79. Markham's memoir is a comprehensive source of information on places, people and events involved in establishing the astronomical observatories in India. Markham, Clements. 1871. *A Memoir on the Indian Surveys.* London: W.H. Allen and Co. P102.

80. Kochhar, R. K. 1991. The Growth of Modern Astronomy in India, 1651-1960. *Vistas in Astronomy.* Volume 34: 69–105.

81. Kochhar, Rajesh. 2014. Madras Observatory - President International Astronomical Union Commission 41: History of Astronomy. http://rajeshkochhar.com/tag/madras-observatory/.

82. Kochhar, Rajesh. 2014. Madras Observatory - President International Astronomical Union Commission 41: History of Astronomy. Retrieved from http://rajeshkochhar.com/tag/madras-observatory/.

83. Proper motion of a star is the actual motion of a star through the cosmos, not just the apparent motion as seen from the Earth as it moves from one end of its orbit to another. This was the first time that technology was sufficiently accurate to confirm that the fixed stars were not really fixed. This technique only works for stars relatively close to the solar system. The vast majority of the stars visible in the night sky are too distant for this technique.

84. Narlikar, Jayant V. 2003. The Scientific Edge: The Indian Scientist from Vedic to Modern Times. Penguin Books.

85. Taylor, Glanville. 1839. Astronomical Observations Made at the Honourable East India Company's Observatory at Madras. For the Years 1836 and 1837. Second edition. IV volumes. Madras, India.

86. With each new instrument comes the prospect of making unexpected revelation. The invention of the telescope in the early 17th century led to the discovery that other planets had moons, atmospheres that changed with the season and surface features, such as mountains and craters. The advent of radio astronomy resulted in one of the most profound discoveries of the 20th century, the Cosmic Microwave Background. Swarup, Govind. 2010. Great Discoveries Made by Radio Astronomers during the Last Six Decades and Key Questions Today. *Pontificiae Academiae Scientiarvm* Acta 21. November. P88. Retrieved from http://www.casinapioiv.va/content/dam/accademia/pdf/acta21/acta21-swarup.pdf

87. Rao, N. K., A. Vagiswari and C. Birdie. 2014. Charles Michie Smith – Founder of the Kodaikanal (Solar Physics) Observatory and Beginnings of Physical Astronomy in India. *Current Science* 106 (3): 447–67. An interesting aside: C.V. Raman's observation of what is now known as the Raman Effect was observed by two Russian scientists a week earlier. Again, speed of publication was key. Kochhar, Rajesh. 18 February 2014. Rise and Decline of Modern Science in India. Oration. Indira Gandhi Prize for Science Popularization of Indian National Science Academy.

88. Rao, N. K., A. Vagiswari and C. Birdie. 2014. Charles Michie Smith – Founder of the Kodaikanal (Solar Physics) Observatory and Beginnings of Physical Astronomy in India. *Current Science* 106 (3): 447–67.

89. Aughton, Peter. 2012. The Transit of Venus: The Brief, Brilliant Life of Jeremiah Horrocks, Father of British Astronomy. Reprint edition. Lancaster: Carnegie Publishing Ltd. P106.

90. The original manuscript with instructions for observing the Transit of Venus on 6 July 1761 sent by the Directors of the East India Company to the Royal Society is available here: https://royalsociety.org/~/media/Royal_Society_Content/exhibitions/transit-venus/manuscripts/MM_10_106_p1.jpg.

91. Hirst, William. 1761. An Account of an Observation of the Transit of Venus over the Sun, on the 6th of June 1761, at Madras. *Philosophical Transactions (1683–1775)*, 52 (1761–1762): 396–398. Retrieved from http://www.jstor.org/stable/pdf/105639.pdf

92. Pasachoff, J. M., G. Schneider and L. Golub. 2004. Explanation of the Black-Drop Effect at Transits of Mercury and the Forthcoming Transit of Venus. *Proceedings IAU Colloquium* 196: 242–53.

93. Sen, Joydeep. 2015. *Astronomy in India, 1784–1876*. Routledge.

94. Kochhar, Rajesh. 1995. *Astronomy in India: A Perspective*. New Delhi: Indian National Science Academy.

95. Ibid.

96. The actual numbers vary and depend on several factors—the relative positions of the Moon and Earth in their respective orbits (the Moon's orbit around the Earth is an ellipse and so is the Earth's orbit around the Sun), the altitude and latitude of the observer and the time of the year.

97 . Rao, N. K., A. Vagiswari and C. Birdie. 2014. Charles Michie Smith – Founder of the Kodaikanal (Solar Physics) Observatory and Beginnings of Physical Astronomy in India. *Current Science* 106 (3): 447–67.

98. All large modern telescopes are built on mountain tops, including Giant Magellan Telescope in Chile, Southern African Large Telescope in South Africa, W. M. Keck Observatory on Mauna Kea in Hawaii and the Gran Telescopio Canarias in Spain. To get above the atmosphere, telescopes are best placed in space, like the Hubble Space Telescope. In India, Kodaikanal was the second, not the first, mountain-top observatory to be sanctioned. Captain W. S. Jacob (1813–1862) had established a small observatory in Poona before serving 11 years as the Director of the Madras Observatory from 1848 onwards. A recognised authority on double stars, Jacob had raised £1,000 from the British Parliament to build a larger observatory at 5,000 feet altitude in Poona. But he died in India eight days after returning from Britain to commence his project. The observatory was never built, giving Kodaikanal Observatory the distinction of being the first mountain-top observatory in India. For more, see Chattopadhyaya, Debi Prasad. 1999. *History of Science, Philosophy and Culture in Indian Civilization: Pt. 1. Science, Technology, Imperialism and War*. New Delhi: Pearson Education India. P364

99. Hassan, S. S., D. C. V. Malik, S. P. Bagare and S. P. Rajaguru. 2010. Solar Physics at the Kodaikanal Observatory: A Historical Perspective. In Hassan, S. S. and R. J. Rutten (Eds). *Magnetic Coupling between the Interior and Atmosphere of the Sun*. Springer. P17.

100. Ibid.

101. Ibid.

102. Anderson, Robert S. 2010. *Nucleus and Nation: Scientists, International Networks, and Power in India*. Chicago: The University of Chicago Press. P37

103. Kochhar, Rajesh. 1995. *Astronomy in India: A Perspective.* New Delhi: Indian National Science Academy. This is a well-researched paper that provides details on these and other Indian observatories.

104. Kochhar, R. K. 1991. The Growth of Modern Astronomy in India, 1651-1960. *Vistas in Astronomy.* Volume 34: 69–105. P92

105. Markham, Clements. 1871. *A Memoir on the Indian Surveys.* London: W.H. Allen and Co. P213.

106. Ibid.

107. Although eclipses (of the Moon and Sun) occur annually somewhere on the Earth, transits of Mercury and especially Venus are rare. Further, these events are localised and are not visible from everywhere on the Earth. For the dates of all the transits of Mercury and Venus from 1000 BC to AD 4000, visit http://www.projectpluto.com/transits.htm

108. Kochhar, Rajesh. 1995. *Astronomy in India: A Perspective.* New Delhi: Indian National Science Academy. P15

109. Vallina, Agust'n Ud'as. 2003. Searching the Heavens and the Earth. The Netherlands: Springer Science & Business Media. P178.

110. Rao, N. K., A. Vagiswari and C. Birdie. 2014. Charles Michie Smith – Founder of the Kodaikanal (Solar Physics) Observatory and Beginnings of Physical Astronomy in India. Current Science 106 (3): 447–67.

111. Vallina, Agust'n Ud'as. 2003. *Searching the Heavens and the Earth.* The Netherlands: Springer Science & Business Media. P179.

112. Times News Network. 11 March 2014. 150-Year-Old St Xavier's College's Observatory Restored. *The Times of India.* Retrieved from http://timesofindia.indiatimes.com/city/kolkata/150-year-old-St-Xaviers-Colleges-observatory-restored/articleshow/31811293.cms

113. Rao, N. K., A. Vagiswari and C. Birdie. 25 May 2011. "Early Pioneers of Telescopic Astronomy in India: G.V. Juggarow and His Observatory". *Current Science* 100(10). Retrieved from http://www.currentscience.ac.in/Volumes/100/10/1575.pdf.

114. Chattopadhyaya, Debi Prasad. 1999. History of Science, Philosophy and Culture in Indian Civilization: Pt. 1. Science, Technology, Imperialism and War. New Delhi: Pearson Education India. P366

115. Interview of the president of the Bangalore Astronomical Society with the author in 2013. It offers an overview of the spread of popular astronomy in India. See http://astrotalkuk.org/2013/11/17/bangalore-astronomical-society/

116. Abbott, B. P. et al. 12 February 2016. Observation of Gravitational Waves from a Binary Black Hole Merger. *Physical Review Letters* 116, 061102. Retrieved from https://physics.aps.org/featured-article-pdf/10.1103/PhysRevLett.116.061102

117. Mehta, Nikita. 3 December 2014. India Joins the Thirty Meter Telescope

Project as a Full Member. *Livemint.* Retrieved from http://www.livemint.com/Politics/ZtRszyDDxMd4BSVFsx7swI/India-joins-the-Thirty-Meter-Telescope-project-as-a-full-mem.html

Chapter 3

118. It has now been replaced with "Harness space technology for national development, while pursuing space science research and planetary exploration."

119. For a review of the updated ISRO website by the author in January 2015, visit: http://astrotalkuk.org/2015/01/04/isro-website-review/

120. Edgerton, David. 1996. The White Heat Revisited: The British Government and Technology in the 1960s. *Twentieth Century British History.* 7 (1): 53–82. P56.

121. Blackett, Patrick. 1953. D-2004-00200-110 (1-6) 31/3/53, Canberra Australia -Text of Presentation. Rich Countries Are Getting Richer. TIFR Archives.

122. A copy of the 12th and previous FYPs for the Department of Science and Technology is available here: http://www.dst.gov.in/about-us/twelveth-five-year

123. Ministry of Science and Technology. 2011. *Working Group Report for the Twelfth Five Year Plan (2012-17).* New Delhi: Government of India, Ministry of Power. http://planningcommission.gov.in/aboutus/committee/wrkgrp12/wg_power1904.pdf

124. Mallick, Sambit. n.d. Building Scientific Institutions in Colonial India: Societies and Associations. NPTEL – Humanities and Social Sciences – Science, Technology and Society. Retrieved from http://www.nptel.ac.in/courses/109103024/pdf/module5/SM%20Lec%2026.pdf

125. 2004, Salam, A. "it is the lack of this contact with others that is the biggest curse of being a scientist in a developing country". A special publication to mark the 40th anniversary of the "Abdus Salam international centre for theoretical research. Retrieved form http://users.ictp.it/~pub_off/books/100_reasons.pdf P30

126. Ibid

127. Out of the 109 members present at the first meeting, 28 were British, 34 were Hindus, and 47 were Muslims. Retrieved from http://aligarhmovement.com/Institutions/scientific_society.

128. Anderson, Robert S. 2010. *Nucleus and Nation: Scientists, International Networks, and Power in India.* Chicago; London: University of Chicago Press. P258.

129. Lourdusamy, J. 2004. *Science and National Consciousness in Bengal, 1870-1930.* New Delhi: Orient Longman. P5.

130. Singh, Rajinder. 2013. The "Forgotten" Astronomical Society of India, Calcutta. *Science and Culture* 79 (9–10). Retrieved from http://www.scienceandculture-isna.org/sept-oct-2013/07%20Art_The_Forgtn_Astrnmcl_Soc._of_Ind._by_Rajinder%20Singh_Pg.369.html/index.html

131. Astronomical Society of India. 1911. *The Journal of the Astronomical Society of India* 1 (7; Session 1910-1911): 160

132. Astronomical Society of India. 1912. *The Journal of the Astronomical Society of India* 2 (7): 222

133. Astronomers using an Earth-based telescope were able to image two planets in the process of formation in the constellation of Taurus. Krol, Charlotte. 19 November 2015. Astronomers Share First Images of New Planet in Formation. *The Telegraph*. Retrieved from http://www.telegraph.co.uk/news/science/space/12004990/Astronomers-share-first-images-of-new-planet-in-formation.html

134. Astronomical Society of India. 1910. The Journal of the Astronomical Society of India. Volume 1 (Issue 6): 131

135. The Lovell Telescope is operated by the University of Manchester. Bernard Lovell had been assigned to work on the British radar programme to detect enemy aircraft during World War II. In a radar system, a radio transmitter sends out a radio signal, and a receiver picks up a faint reflection from a passing aircraft. The strength of the signal and timing characteristics were sufficient to identify the location, speed and direction of the aircraft. While developing this system, Lovell had received spurious radio signals but was unaware of their source. After the War, using surplus equipment, he investigated the mysterious signals and confirmed that the ionised molecules of air reflected his radar signals just as aircraft did. The first time he switched on the experimental equipment in Manchester was on 14 December 1945. This happened to coincide with the annual Geminid meteor shower. Lovell, B. 1968. *The Story of Jodrell Bank*. Oxford University Press. P3.

136. ASI Correspondence. 1 March 1914. Volume 4. P126.

137. Report of the meeting of the society held on Tuesday, 25 June 1912, P223.

138. Report of a meeting of the society held on Tuesday, 28 February 1911, P107.

139. The website of ASI's Public Outreach and Education Committee: http://astron-soc.in/outreach/

140. Nehru, Jawaharlal. 1994. *The Discovery of India*. New Delhi: Oxford University Press. P512. You can also find it here: http://varunkamboj.typepad.com/files/the-discovery-of-india-1.pdf The original copy was published in 1946 and written during Nehru's time in prison. He dedicated this book to "To my colleagues and co-prisoners in the Ahmednagar Fort Prison Camp from 9 August 1942 to 28 March 1945".

141. Rajwi, Tiki. 27 November 2015. To Defend Itself, the Nation has to Become

Self-reliant: U.R. Rao. *Indian Express*. Retrieved from http://www.newindianexpress.com/cities/thiruvananthapuram/To-Defend-Itself-the-Nation-has-to-Become-Self-reliant-U-R-Rao/2015/11/27/article3148482.ece

142. Shah, Amrita. 2007. *Vikram Sarabhai: A Life*. Illustrated edition. India: Viking (India). P63.

143. At 99th position, it only just made it. PTI. 12 November 2015. IISc debuts in top 100 world university ranking. *BusinessLine*. Retrieved from http://www.thehindubusinessline.com/news/education/iisc-debuts-in-top-100-world-university-ranking/article7869064.ece.

144. Dilks, David. 1969. *Curzon in India: Achievement, Vol. 1*. London: Rupert Hart-Davis. P244.

145. A letter dated 3 March 1904 from I.G.R Landes, Barrister-at-Law, who valued the 18 properties at Rs. 125,000. IISc Archives.

146. Ironically, the locals know it better as the Tata Institute rather than the IISc.

147. A letter from D. J. Tata to the Viceroy dated 14 July 1904. IISc Archives.

148. Unknown author. 10 October 1992. In This Issue: Morris Travers (1872-1961). *Current Science* 63 (7). http://www.currentscience.ac.in/Downloads/article_id_063_07_0341_0342_0.pdf. This issue has a graphic showing his Hydrogen liquefier.

149. A letter dated 29 May 1905 from William Ramsay to D. J. Tata providing an update securing a director for the Institute. IISc Archives.

150. Balaram, P. 2008. Morris Travers: Remembering an Institution Builder. Current Science 94(9).

151. Interview with the author on 5 February 2015. Narasimha goes on to share fascinating memories of meeting Satish Dhawan, who was then an associate professor at IISc in 1953. Later, Narasimha moved to Caltech in 1957, from where he recalls seeing Sputnik. He completed his PhD in flow dynamics and returned to India in 1961. He was invited to Thumba to witness the launch of India's first rocket—Nike-Apache. He served as a member of the Space Commission, the government body that approves all ISRO missions, and remembers ISRO's 2008 mission to the Moon, which called for an intense debate prior to approval.

152. Lee, Rachel. September 2012. Constructing a Shared Vision: Otto Koenigsberger and Tata & Sons. *ABE Journal: Architecture Beyond*, https://abe.revues.org/356?lang=en.

153. In a 54 about Satish Dhawan in *A Voyage Through Turbulence*, Roddam Narasimha (Dhawan's student) writes "A postage stamp released in 2009 on the occasion of the centenary of the Institute contains pictures of its several icons: the founder (Jamsetji Tata, Parsi businessman and industrialist from Bombay), his spiritual ally (Swami Vivekananda), his princely supporter (the Maharaja of Mysore), the Institute's first director (British chemist Morris Travers), its greatest scientist (C.V. Raman), its greatest alumnus (the biophysicist G.N. Ramachandran)

and (presumably) its greatest director (Satish Dhawan)". Davidson, Peter A., Yukio Kaneda, Keith Moffatt and Katepalli R. Sreenivasan (Eds) 2011. *A Voyage Through Turbulence*. Cambridge: Cambridge University Press. P374.

154.   For   a   balanced   critique,   see https://desidreaming.wordpress.com/2007/04/30/the-dirty-little-secret-of-the-tatas/. Many other sources also make a similar point but with reduced objectivity.

155. Bulletin of the Atomic Scientists. October 1955. P282

156. Bhabha's thesis was titled *The Theory of Elementary Physical Particles and Their Interactions*. James Clerk Maxwell, Stephen Hawking and Abdus Salam are some of the other recipients of Adams Prize. https://en.wikipedia.org/wiki/Adams_Prize

157. Wadia, S. 2009, Homi Jehangir Bhabha and the Tat Institute for Fundamental Research. Current Science 96 (5): 725–733, 1311.

158. Letter to Sir Sorabji Saklatvala, Chairman, Sir Dorabji Tata Trust. 12 March 1944.

159. Anderson, Robert S. 2010. *Nucleus and Nation: Scientists, International Networks, and Power in India*. Chicago, London: University of Chicago Press. P190.

160. *Ending Extreme Poverty and Sharing Prosperity: Progress and Policies*, a 2015 report from the World Bank noted that a connective infrastructure "is a crucial means of linking the farms and firms where the poor live and work to markets. Electrification of poor areas in South Africa has resulted in a 9 percentage point increase in female labour force participation, consumption, and earnings by allowing reallocation of time use within the household thanks to time-saving appliances'. Policy Research Note, *Ending Extreme Poverty and Sharing Prosperity: Progress   and   Policies.*   P60.   Retrieved   from http://pubdocs.worldbank.org/pubdocs/publicdoc/2015/10/109701443800596288 /PRN03-Oct2015-TwinGoals.pdf

161. Tata Institute of Fundamental Research. 2009. Homi Jehangir Bhabha on Indian Science and the Atomic Energy Programme: A Selection. P143 in section XV11. It is a transcript of an interview titled 'India's developments strategy' with International Science and Technology, October 1963. Ironically, in 2015, India is witnessing a rapid increase in the construction of coal-fired power stations. In August 2011, 554 new power stations were approved with a capacity for 590,000 megawatts of power. Between them they will generate 3.7 billion tons (907.19 billion kg) of carbon dioxide.

162. Rao, P. V. S. 2008. TIFRAC, India's First Computer — A Retrospective. *Resonance*. 13 (5): 420–29.

163. *Ibid.*

164. ISRO's demand for high-performance computing power for calculating numerous complex models and simulations (that is, orbital dynamics, airflows for launch vehicles and nozzle heat dissipation) is high. In 2011, it announced the

completion of India's fastest supercomputer. The new Graphic Processing Unit (GPU)-based supercomputer was named SAGA-220 (Supercomputer for Aerospace with GPU Architecture-220 TeraFLOPS). Retrieved from http://www.isro.gov.in/sites/default/files/flipping_book/64-SI-Apr-Dec2011/files/assets/common/downloads/publication.pdf

165. This event did not attract the level of publicity it should have given the magnitude of the achievement. See http://www.paragonsdc.com/stratex/. Another intriguing discovery was made in 2009 by an ISRO-sponsored high-altitude balloon experiment also launched from Hyderabad. It discovered three previously unknown species of bacteria, which are now named as Janibacter hoylei, after the British astrophysicist Fred Hoyle, Bacillus isronensis after ISRO and Bacillus Aryabhata after India's ancient astronomer Aryabhata and ISRO's first satellite. See http://www.isro.gov.in/update/16-mar-2009/discovery-of-new-microorganisms-stratosphere

166. Nehru deals with the idea of tradition in the chapter titled 'The Burden of Old Tradition'. He acknowledges the deep roots of Indian tradition and the importance of maintaining some of them but posits the case for discarding others. At one point, he states that tradition "has much of good in it, but sometimes it becomes a terrible burden, which makes it difficult for us to move forward". Nehru, Jawaharlal. 2004. *Glimpses of World History*. New edition. New Delhi: Penguin Books India.

167. Jawaharlal, Nehru. 1994. *The Discovery of India*. New Delhi: Oxford University Press. P31.

168. Retrieved from http://www.constitution.org/cons/india/p4a51a.html

169. Kochhar, Rajesh. 2014. Rise and Decline of Modern Science in India. http://www.slideshare.net/rajeshkochhar1/oration-rise-and-decline. Slide 48.

170. As a snapshot of Indian leaders heading multinational companies): Sundar Pichai, CEO, Google; Satya Nadella, CEO, Microsoft; Sanjay Kumar Jha, CEO, Global Foundries; Shantanu Narayen, CEO, Adobe; Francisco D'Souza, CEO, Cognizant; Sanjay Mehrotra, CEO, SanDisk Corporation; Rajeev Suri, President and CEO, Nokia; George Kurian, CEO, NetApp; Vanitha Narayanan, MD, IBM India; Neelam Dhawan, MD, HP India; Aruna Jayanthi, CEO, Capgemini India; Kirthiga Reddy, Head of Office, Facebook India; Kuma Srinivasan, President, Intel India; Sabeer Bhatia, co-founder of Hotmail; Padmasree Warrior, Chief Technology Officer, Cisco.

171. Tharoor, Shashi. 2012. Pax Indica: India and the World of the Twenty-First Century. New Delhi: Penguin Global.

Chapter 4

172. Three examples have been used in this book to illustrate the discriminatory

environment during the colonial period: (i) English professors' resentment of "working under an Indian director" Parameswaran, Uma. 2011. *C.V. Raman: A Biography*. New Delhi: Penguin Books India. P174; (ii) Lord Curzon's likening higher education for Indians to "presenting a naked man with a top-hat when what he wants is a pair of trousers" Dilks, David. 1969. *Curzon in India: Achievement*, Vol. 1. London: Rupert Hart-Davis, P244 and (iii) J.C. Bose being paid a salary "half of what the British teachers were paid" INSA. 2001. *Pursuit and Promotion of Science: The Indian Experience*. Indian National Science Academy. P23. Divisions and discrimination persist even today, in independent India. However, now the source is not a foreign occupying power but age-old divisions based on deep-set caste hierarchy.

173. Oxford historian Dr. Alan Chapman in his 1998 book, *The Victorian Amateur Astronomer*, calls them 'Grand Amateurs', rich and powerful individuals who built, or commissioned, and then used large telescopes to undertake independent astronomical research. They were merchants or engineers who became wealthy during the Industrial Revolution or married into rich families. William Lassell from Manchester discovered the largest Moon of Neptune in 1846 a few weeks after Neptune's discovery. Engineer James Naismith, originally from Scotland, built machines and tools to build large telescopes. He was one of the first to privately own a 20-inch (50.8 cm) diameter mirror telescopes in 1840. Although there were rich Indians (especially the Princes and Rajas) some of who financially supported institutions such as IISc and IACS, none of them apparently had the aptitude or the inclination to invest and conduct science personally.

174. Birchenough, Tom. 14 October 2014. Cosmonauts: How Russia Won the Space Race. *Theartsdesk.com.* Retrieved from http://www.theartsdesk.com/tv/cosmonauts-how-russia-won-space-race-spaceman-afghanistan-bbc-four

175. Chatterjee, Santimay and Enakshi Chatterjee. 1976. *Satyendra Nath Bose*. New Delhi: National Book Trust, India.

176. In one instance, the police had to step in to control the crowds in Manchester. "The aid of the police was required to make way for him to the manufactories and when he had entered, it was necessary to close and bolt the gate to keep out the mob." Collet, Sophia Dobson. 1914. *The Life and Letters of Raja Rammohun Roy*. Edited by Hem Chandra Sarakar. 2nd Edition. Calcutta: A. C. Sarkar.

177. As a result of the reform bill, the number of voters increased from 500,000 to 800,000. This was still only a small percentage of the British population of 14 million at the time. The passing of the bill in the third attempt was considered a victory against corruption and for liberal values. Roy had publicly declared "in the event of the Reform Bill being defeated I would renounce my connection with this country". This reflects Roy's position on the bill and spoke clearly of his liberal political values.

178. Collet, Sophia Dobson. 1914. *The Life and Letters of Raja Rammohun Roy*. Edited by Hem Chandra Sarakar. 2nd Edition. Calcutta: A. C. Sarkar.

179. Wali, Kameshwar C. 1992. *Chandra: A Biography of S. Chandrasekhar*. New edition. Chicago: University of Chicago Press. P258.

180. INSA. 2001. *Pursuit and Promotion of Science: The Indian Experience*. New Delhi: Indian National Science Academy. P23.

181. Mahanti, Subodh. 2002. *Acharya Jagadish Chandra Bose. A Pioneer of Modern Indian Science*. Retrieved from http://www.alternativaverde.itwww.alternativaverde.it/sttlcing/documenti/Bose/bos e.pdf

182. Singh, Rajinder. 2004. Nobel Laureate C.V. Raman's Work on Light Scattering: Historical Contributions to a Scientific Biography. Berlin: Logos Verlag. P9.

183. Belrose, John. 1995. Fessenden and Marconi: Their Differing Technologies and Transatlantic Experiments during the First Decade of This Century. *Proceedings of the 1995 International Conference on 100 Years of Radio*.

184. The following article asserts that the key element, the mercury autocoherer, used by Marconi was invented by Jagadish Chandra Bose. Aggarwal, Varun. 2006. *Jagadish Chandra Bose: The Real Inventor of Marconi's Wireless Detector*. Massachusetts Institute of Technology. P1. Retrieved from http://web.mit.edu/varun_ag/www/bose_2006.pdf

185. Bondyopadhyay, Probir K. 1998. Under the Glare of a Thousand Suns – The Pioneering Works of Sir J. C. Bose. *Proceedings of the IEEE* 86 (1): 218–85.

186. The paper is available online. In it, Bose describes in detail the use of different metals (sodium, potassium, iron aluminium, copper, silver, gold and others) as coherers. See http://web.mit.edu/varun_ag/www/bose_original.pdf

187. J.C. Bose. 17 May 1901. Personal Letter to Rabindranath Tagore. Archives of Rabindra Bhavan, Santiniketan. The patent that Bose talks about in his letter to Tagore is the patent he chose not to file but Marconi did two days before his landmark transatlantic experiment (G. Marconi, British Patent 18 105, applied for Sept. 10, 1901). Interestingly, Bose did succumb and file a patent for the Detector for Electrical Disturbances under his own name on 29 March 1904, but let it lapse. Retrieved from http://web.mit.edu/varun_ag/www/jcbosepatent.pdf

188. A copy of the original patent no. 755 840 dated 29 March 1904 is available here: http://web.mit.edu/varun_ag/www/jcbosepatent.pdf

189. Gardiol, Fred E. 2011. About the Beginnings of Wireless. *International Journal of Microwave and Wireless Technologies* 3 (04). doi:10.1017/S1759078711000444.

190. The reference below indicates that the time of day and the physical characteristics of the (predominately) antenna appear to indicate that the famous broadcast is more likely to have used shortwave rather than long wavelengths. Bondyopadhyay, P. K. 2001. Marconi's 1901 Transatlantic Wireless Communication Experiment. *31st European Microwave Conference*, London.

191. Banerjee, Biswanath. 2010. The Scientist and the Poet: Acharya Jagadish

Chandra Bose and Rabindranath Tagore. *Rupkatha Journal of Interdisciplinary Studies in Humanities* 2 (4). Retrieved from http://rupkatha.com/V2/n4/08JCBoseandTagore.pdf

192. At that time, smallpox and cholera epidemics claimed the lives of almost half of the children before they reached their first birthday. Kanigel, Robert. 1992. *The Man Who Knew Infinity: Life of the Genius Ramanujan*. New edition. London: Abacus. P12.

193. Rao, Srinivasan. 2000. *Life and Work of the Mathemagician Srinivasa Ramanujan*. Chennai: The Institute of Mathematical Sciences. P5. Retrieved from http://arxiv.org/pdf/math/0003184.pdf

194. The Hindu. 25 December 2011. Ramanujan Lost and Found: a 1905 letter *from Hindu*. Retrieved from http://www.thehindu.com/features/friday-review/history-and-culture/ramanujan-lost-and-found-a-1905-letter-from-the-hindu/article2745164.ece.

195. Rao, Srinivasan. 2000. *Life and Work of the Mathemagician Srinivasa Ramanujan*. Chennai: The Institute of Mathematical Sciences. P8 and P10. Retrieved from http://arxiv.org/pdf/math/0003184.pdf.

196. Kanigel, Robert. 1992. *The Man Who Knew Infinity: Life of the Genius Ramanujan*. New edition. London: Abacus. P175.

197. Berndt, Bruce C. and Robert Alexander Rankin. 2001. *Ramanujan: Essays and Surveys*. American Mathematical Society. P108.

198. Kanigel, Robert. 1992. *The Man Who Knew Infinity: Life of the Genius Ramanujan*. New edition. London: Abacus. P177.

199. In the letter, Walker acknowledged that Ramanujan had already been in contact with G.H. Hardy in Cambridge and encouraged Madras University to support him. G. Walker, Gilbert. 26 February 1913. Comparable to a mathematics fellow of Cambridge. Letter to the registrar of the University of Madras.

200. Ramanujan, S. 16 January 1913. Letter to G.H. Hardy. Cited in Rao, Srinivasan. 2000. *Life and Work of the Mathemagician Srinivasa Ramanujan*. Chennai: The Institute of Mathematical Sciences. P10. Retrieved from http://arxiv.org/pdf/math/0003184.pdf. A transcript of this letter is available in many sources.

201. Aiyangar, Srinivasa Ramanujan. 1995. *Ramanujan: Letters and Commentary*. American Mathematical Society. P118.

202. Rao, Srinivasan. 2000. *Life and Work of the Mathemagician Srinivasa Ramanujan*. Chennai: The Institute of Mathematical Sciences. P18. Retrieved from http://arxiv.org/pdf/math/0003184.pdf

203. Darling, David. 2004. The Universal Book of Mathematics: From Abracadabra to Zeno's Paradoxes. New Jersey: John Wiley & Sons. P266.

204. Aiyangar, Srinivasa Ramanujan. 1995. *Ramanujan: Letters and Commentary*.

American Mathematical Society. P3.

205. Wali, Kameshwar C. 1992. *Chandra: A Biography of S. Chandrasekhar*. New edition. Chicago: University of Chicago Press. P262.

206. Kanigel, Robert. 1992. *The Man Who Knew Infinity: Life of the Genius Ramanujan*. New edition. London: Abacus. P317.

207. Ramaseshan, S. 1990. Srinivasa Ramanujan. *Current Science* 59 (24): 1309–16, 1311.

208. A fascinating story of this discovery is recounted by Professor Andrews in the transcript of a video recorded in 2002. Retrieved from http://www.imsc.res.in/~rao/ramanujan/interviewindex.htm Andrews, George. 2002. Discovery of Ramanujan's Lost Notebook.

209. Some of the topics that Ramanujan was working on included the following: elementary mathematics, number theory, infinite series, integrals, asymptotic expansions and approximations, hypergeometric functions, q-series, continued fractions, theta functions and modular equations, elliptic functions to alternative bases, class invariants. The following paper provides an overview for readers with a mathematical insight. Berndt, Bruce C. 1998. *An Overview of Ramanujan's Notebooks*. Retrieved from http://www.math.uiuc.edu/~berndt/articles/aachen.pdf

210. Mark Zuckerberg, founder of Facebook names Ramanujan as one of his favourite mathematicians. See https://www.youtube.com/watch?v=HNA6TXyWUNE

211. For a long time, scientists believed that nothing could ever escape from black holes. Then, in 1974, Professor Stephen Hawking proposed a mechanism by which mass could escape from black holes (known as Hawking Radiation). Mock modular function is being used to help understand how this could work. Aron, Jacob. 2016. Mathematical Proof Reveals Magic of Ramanujan's Genius. *New Scientist*. Retrieved from https://www.newscientist.com/article/mg21628904-200-mathematical-proof-reveals-magic-of-ramanujans-genius/

212. She received a small pension from the University of Madras until her death in 1994 at the age of 94. During her life, she lived independently. At one stage, she trained and then worked as a tailor. In 1950, her close friend had died leaving her 7-year-old son for whom she became a foster mother. She raised and educated him. He eventually became an officer at the State Bank of India. In 1980s, he took early retirement to look after her in her final few years. Rao, Srinavasan. 2002. *Ramanujan's Wife: Janakiammal (Janaki)*. Chennai: Institute of Mathematical Sciences.

213. Murty, M. Ram and V. Kumar Murty. 2012. *The Mathematical Legacy of Srinivasa Ramanujan*. India: Springer Science & Business Media. P9.

214. Raman is sometimes reported as the first Indian to be elected as a Fellow of the Royal Society, but he was the second. The first Indian to become a Fellow of the Royal Society was Ardaseer Cursetjee. He was elected a Fellow on 27 May 1841 with the support of the East India Company for his services to naval architecture.

215. Kademani, B. S., V. L. Kalyane and A. B. Kademani. 1994. Scientometric Portrait of Nobel Laureate Dr. C.V. Raman. *Indian Journal of Information, Library and Society* 9 (2): 125–50. Retrieved from http://core.ac.uk/download/pdf/11877009.pdf

216. Documentary of Sir C.V. Raman. 20 June 2006. Retrieved from http://www.thehindu.com/todays-paper/tp-national/tp-tamilnadu/documentary-on-sir-cv-raman/article3122049.ece.

217. This paper was titled 'Unsymmetrical diffraction bands due to a rectangular aperture'. It was published in the *Philosophical Magazine* (London) in 1906.

218. Raman had rented a house on Scots Lane, which ran off Bowbazzar Street. The IACS was located on 210 Bowbazzar Street. It is interesting to speculate the direction in which the IACS, Raman and indeed Indian science would have developed had Raman not encountered the IACS. Many of the details in this section come from a special publication to mark the centenary of Raman's birth. Ramaseshan, S. and C. Ramachandra Rao. 1988. *C.V. Raman: A Pictorial Biography*. Bangalore: Indian Academy of Sciences.

219. Singh, Rajinder. 2010. C.V. Raman's Research in Astronomy. *Current Science* 99 (8): 1127–32, 1130.

220. *Ibid.*

221. A reciprocal arrangement existed between countries to subscribe to each other's scientific publications. For example, when Raman published a paper in the *Philosophical Magazine* in Britain, it would be available in Science Academies of the German, Swedish and Dutch Philosophical Societies. Singh, Rajinder. 2004. *Nobel Laureate C.V. Raman's Work on Light Scattering: Historical Contributions to a Scientific Biography*. Berlin: Logos Verlag. P11.

222. Raman, C. V. 1921. Colour of the Sea. *Nature* 108: 367. Retrieved from http://dspace.rri.res.in/bitstream/2289/2066/1/1921%20Nature%20V108%20p367.pdf

223. Raman's letters in the IISc archives.

224. Names of all those who nominated Raman are listed here: http://www.nobelprize.org/nomination/archive/show_people.php?id=7540

225. This is how S. Ramaseshan, Raman's nephew, recalls it in Ramaseshan, S. and C. Ramachandra Rao. 1988. *C.V. Raman: A Pictorial Biography*. Bangalore: Indian Academy of Sciences. P16. However, science historian Rajinder Singh (in personal communication with the author in January 2016) considers this more likely to be an exaggeration since the telegram in question has not been evidenced.

226. Formal Agreement Signed by Raman on 6 October 1932 for the Post of the Director of the IISc. IISC Archives.

227. A list of all the applicants is included in the appointment letter dated 19 May 1932 from William Henry Bragg, who chaired the committee responsible for selecting the director. Archives of the IISc.

228. Most scientists went to the US. By the late 20th century, two of their collective achievements arguably resulted in the US becoming the dominant nation on the planet. The Manhattan project built the first atomic bomb in 1945, and by 1972, the Apollo programme had taken two dozen men to the Moon and back. Most of the scientists and engineers behind these successes were brought together in the US from different parts of the world. Raman's cousin Chandrasekhar was asked to join the Manhattan Project in March 1944 by Hans Beth and John von Neumann. Chandrasekhar agreed initially, but by the time the clearance came through in September 1944, he changed his mind and did not go. Wali, Kameshwar C. 1992. *Chandra: A Biography of S. Chandrasekhar*. New edition. Chicago: University of Chicago Press. P196.

229. Parameswaran, Uma. 2011. *C.V. Raman: A Biography*. New Delhi: Penguin Books India. P173.

230. Raman's reference on 30 July 1966 to "shoot men into space and make them walk there" is probably a reference to Alexei Leonov's spacewalk in March 1965. One of the two cosmonauts on the Voskod spacecraft, Leonov was the first human to walk in space. NASA. 1968. Astronautics and Aeronautics, 1966. *Chronology on Science, Technology and Policy*. P253.

231. Thapar, Romila. 2015. *The Public Intellectual in India*. Aleph Book Company.

232. A total of 6 bosons and 12 fermions make up everything in the universe. Matter is made up of leptons or and quarks, collectively known as fermions. Although leptons can exist as free particles, quarks cannot. Multiple quarks stick together to form familiar particles, like neutrons and protons. There are 6 leptons and 6 quarks making up a total of 12 (or 24 if the antiparticles are included) particles. Each force, gravity, strong nuclear force, weak nuclear force and the electromagnetic force, has an associated boson. The electromagnetic force boson is the photon; the weak nuclear force has three variants, W+, W-, z; the strong nuclear force has the gluon and gravity has the graviton (yet to be discovered). The fundamental particles can be categorised in many ways. One is based on the concept of spin. If a particle has half-integral integer spin (that is, -0.5, -1.5, 0.5 or 1.5), it is known as a fermion, which includes the electron, neutrino and quarks. If a particle has an integral integer spin (that is, 0, 1 or 2), it is known as a boson. The Higgs particle is considered to be a boson because it has an integer spin; it is not associated with a force but endows the attribute of mass to other particles. Bosons can share the same quantum states, but fermions cannot (prohibited by the Pauli Exclusion Principle). A concise description of this Standard Model is available here: http://www.physik.uzh.ch/groups/serra/StandardModel.html

233. Chatterjee, Santimay, C.K. Majumdar and S.N. Bose National Centre for Basic Sciences. 1994. *S N Bose: The Man and His Work*. Calcutta: S. N. Bose National Centre for Basic Sciences. P55.

234. *Ibid*. P33.

235. Wali, Kameshwar C. 2009. *Satyendra Nath Bose – His Life and Times: Selected Works*. NJ: World Scientific Publishing Company. Pxx

236. Einstein published his Special Theory of Relativity in 1905, General Theory of Relativity in 1915, acquired experimental evidence for his Theory of General Relativity in 1919 and received the Nobel Prize in 1921.

237. The translation was published as Original Papers by A. Einstein and H. Minkowski, translated into English by M.N. Saha and S.N. Bose with a historical introduction by P.C. Mahalanobis. Translated into English for the first time, it includes the original papers with additions. The published version contained seven sections: (i) Historical Introduction (By Mr. P. C. Mahalanobis), (ii) On the Electrodynamics of Moving Bodies, (iii) Albert Einstein (A short biographical note by Dr. Meghnad Saha), (iv) Principle of Relativity (H. Minkowski's original paper on the restricted Principle of Relativity, first published in 1909. Translated from the German original by Dr. Meghnad Saha), (v) Appendix to the above by H. Minkowski (Translated by Dr. Meghnad Saha), (vi) The Foundation of the Generalized Theory of Relativity. (A. Einstein's second paper on the Generalised Principle, first published in 1916. Translated from the German original by Mr. Satyendra Nath Bose) and (vii) Notes. An interesting source of information on S. N. Bose is the S. N. Bose Biography Project, blog.snbose.org, maintained by Bose's grandson Falguni Sarkar.

238. Wali, Kameshwar C. 2009. *Satyendra Nath Bose – His Life and Times: Selected Works*. NJ: World Scientific Publishing Company. Pv.

239. Although the existence of dark matter is well established, what it actually is not known. Freitas, R. C. and Gonçalves S. V. B. 2013. Cosmological perturbations during the Bose-Einstein condensation of dark matter. Journal of Cosmology and Astroparticle Physics. Retrieved from https://www.researchgate.net/publication/233780309_Cosmological_perturbations _during_the_Bose-Einstein_condensation_of_dark_matter

240. A fascinating record of personal recollections has been gathered by Bose's grandson and Bose's students. http://newweb.bose.res.in/Prof.S.N.Bose-Archive/objects/0022.pdf

241. This recollection and other fascinating details have been gathered by Bose's grandson and Bose's students. A transcript of the interview is available online: https://sites.google.com/site/snbproject/purnimasinha

242. *Ibid.*

243. *Ibid.*

244. *Ibid.*

245. By all accounts, Bose was a quiet, mild-mannered individual, but two interesting references help shed some light on Bose's political views. One is in the book *S N Bose: The Man and His Work* (Chatterjee, Santimay, C. K. Majumdar and S. N. Bose National Centre for Basic Sciences. 1994. *S N Bose: The Man and His Work*. Calcutta: S. N. Bose National Centre for Basic Sciences. P107.) Another indicates that his visit to Europe also had a political motivation. Dr. Purnima Sinha had recorded three long interviews with Bose during the last eight months before he

died in 1974. In one of them, she asked Bose why he had gone to Paris. He replied "I was informed that my friend Abani Mukherjee (a terrorist nationalist leader who was absconding) was in trouble. I had taken some money for him from the country. After meeting Abani, I thought that I will stay in Paris for a while". Retrieved from https://sites.google.com/site/snbproject/purnimasinha

246. Wali, Kameshwar C. 2009. *Satyendra Nath Bose – His Life and Times: Selected Works*. NJ: World Scientific Publishing Company. Pxxxiii.

247. Deshmukh, Chintamani. 2010. *Homi Jahangir Bhabha*. New Delhi: National Book Trust. P6.

248. Singh, Rajinder. 2008. Indo-American Relations with Reference to Bernard Peters. *Indian Journal of History of Science* 43.3 (437–454). PP440.

249. Antiparticles had been implied in the work of Paul Dirac.

250. Pickering, William. 1998. *Carl David Anderson 1905-1991*. Washington, D. C.: National Academy of Science. P7. Retrieved from http://www.nasonline.org/publications/biographical-memoirs/memoir-pdfs/anderson-carl-d.pdf

251. Bhabha's paper describing this scattering was read out at the Royal Society on 20 October 1935 and published in the proceedings in the following year. Bhabha, H. J. 1936. The Scattering of Positrons by Electrons with Exchange on Dirac's Theory of the Positron. *Proceedings of the Royal Society of London. Series A, Mathematical and Physical Sciences* 154 (881): 195–206.

252. Anderson, Robert S. 2010. *Nucleus and Nation: Scientists, International Networks, and Power in India*. Chicago: University of Chicago Press. P100.

253. Deshmukh, Chintamani. 2010. *Homi Jahangir Bhabha*. New Delhi: National Book Trust. P9.

254. Bhabha was one of the four Indians who had been elected as a Fellow of the Royal Society but had not gone through the formal signing-in ceremony that was always conducted in London. Perhaps because Britain was in a state of war in 1941, the Royal Society for the first time conducted the admissions ceremony in India. Deshmukh, Chintamani. 2010. *Homi Jahangir Bhabha*. New Delhi: National Book Trust. P19.

255. Anderson, Robert S. 2010. *Nucleus and Nation: Scientists, International Networks, and Power in India*. Chicago: University of Chicago Press. P192.

256. In 1953, Blackett presented a paper in the Australian capital highlighting his concern on the increasing gap between the rich and poor countries. He asserted that "scientists are responsible to a considerable degree" and that "this vast difference will become the outstanding international problem of the next decades." Blackett, Patrick. 1953. Rich Countries Are Getting Richer. Text of Presentation - Canberra Australia. D-2004-00200-110 (1-6) 31/3/53. TIFR Archives.

257. Bhabha, Homi. 1948. Letter from Bhabha to Blackett Requesting a Meeting to Explore Potential Collaboration in Nuclear Energy Research. D-2004-00200-15

1 – 2. TIFR Archives.

258. Raja, Jay and Manoranjan Rao. 18 November 2002. When Thumba Took Off. *The Hindu.*

259. Shah, Amrita. 2007. *Vikram Sarabhai: A Life.* Illustrated edition. Viking (India). P122.

260. Hore, Peter. 2004. Patrick Blackett: Sailor, Scientist, Socialist. Routledge. P258.

261. This crash, like most, was the result of compound failures. One of the navigational instruments was not working from the start of the flight, and there was ambiguous air traffic communication, but primarily, the crew failed to correctly identify their location. Descending through the clouds to Geneva airport, the aircraft crashed into Mont Blanc with a loss of all 11 crew and 106 passengers. The harsh winter weather prevented all the wreckage from being recovered at the time of the crash. At the beginning of the 20th century, the Bosons Glacier stretched from the summit of Mont Blanc right down to the valley floor, but now, it has receded to about 1400 m above the valley. Almost half a century after the event, the slow-moving glacier revealed debris from a crash. On 21 August 2012, mountain rescue workers Arnaud Christmann and Jules Bergernear recovered a jute bag on the slopes of the French Ski resort, Camonix. The bag had unexpected markings. Sections of the text that were legible included 'Diplomatic mail', 'Ministry of External Affairs' and 'Indian Government Services'. It was quickly identified as debris from the crash of Boeing 707-437 Air India flight AI 101 on the morning of 24 January 1966. Retrieved from http://abcnews.go.com/blogs/headlines/2012/08/46-years-later-diplomatic-mail-found-in-alpine-wreckage/.

262. Wali, Kameshwar C. 1992. *Chandra: A Biography of S. Chandrasekhar.* New edition. Chicago: University of Chicago Press. P288. Sarabhai was not the first choice to succeed Bhabha as the Chairman of the AEC. In their haste to fill the hole left by the sudden death of Homi Bhabha, the Indian officials had offered the post to Punjab-born astrophysicist Subrahmanyan Chandrasekhar then based at the University of Chicago, only to realise later that Chandrasekhar, as an American citizen, was not eligible and should not have been offered it. Chandrasekhar declined the post, paving the way for Sarabhai.

Chapter 5

263. Jatia, D. N. 1980. *From the Diary of Stephen Smith.* New Delhi: The Congress. P57.

264 . One such mention was Edward Pendray's writing in March 1945 in Harper's Magazine. Retrieved from http://epizodsspace.no-ip.org/bibl/inostr-yazyki/harper/1945/pendray.pdf.
http://www.flightglobal.com/pdfarchive/view/1934/1934%20-

%201257.html?search=rocket%20mail. Pendray's publication in the *Journal of the American Rocket Society*, March 1936, page 19, also records this achievement. Flight International, too, records Smith's Ship to Shore rocket mail experiments although his name is not mentioned.

265. Imperial Airways was a commercial airways company that operated between 1924 and 1939. It was first merged into the British Overseas Airways Corporation (BOAC). In 1974, BOAC merged with the British European Airways Corporation to create British Airways.

266 . Hurren, B. J. 1958. *Flight Global.* P34. http://www.flightglobal.com/FlightPDFArchive/1958/1958-1-%20-%200022.PDF.

267 . Bertolotti, M. 2013. The Discovery: Victor F. Hess and the Balloon Ascents. In *Celestial Messengers: Cosmic Rays—The Story of a Scientific Adventure.* Berlin, Heidelberg: Springer–Verlag. PP33–44. http://www.physics.princeton.edu/~mcdonald/examples/EP/hess_au_33_13.pdf

268 . Smith, S. H and F. Billig. 1955. Billig's Specialized Catalogues: Rocket Mail Catalogue, Vol. 8. P62.

269 . *Ibid.* P68.

270 . Putnam, Christopher S. 10 August 2007. One Small Step for Mail. *Damn Interesting.* Retrieved from http://www.damninteresting.com/one-small-step-for-mail/

271. Zucker, J. and I. Baddiel. 2011. *Never in a Million Years.* Hachette UK.

272. Retrieved from http://www.spaceflownartifacts.com/flown_apollo15_covers.html.

273. Interview with Al Worden. 22 May 2011. Retrieved from http://astrotalkuk.org/2011/07/04/episode-45-4-july-2011-apollo-15-command-module-pilot-al-worden/

274. Retrieved from https://astrotalkuk.org/falling-to-earth-book-review/

275. An interesting aside. Max Kronstein had been actively involved in the development of rocket mail and knew of Smith's work. Smith named his rocket number 135 that fired on 1 February 1937 as 'Dr M Kronstein'. He named rocket number 169 that fired on 25 June 1938 as 'Marienne Kronstein', after Max Kronstein's daughter. Smith, S. H. and F. Billig. 1955. *Billig's Specialized Catalogues: Rocket Mail Catalogue*, Vol. 8. Fritz Billig. PP46, 52.

276. Oakland, Rod. 16 August 2012. Propaganda Leaflets of the Spanish Civil War, 1936–1939. *Psywar*.org. Retrieved from http://www.psywar.org/spanishcivilwar.php

277. Smith, S. H. and F. Billig. 1955. Billig's Specialized Catalogues: Rocket Mail Catalogue, Vol. 8. P55.

278. Ibid.

279. *Ibid.*

280. *Ibid.* P57.

281. Putnam, Christopher S. 10 August 2007. One Small Step for Mail. *Damn Interesting.* Retrieved from http://www.damninteresting.com/one-small-step-for-mail/

282. Jatia, D. N. 1980. *From the Diary of Stephen Smith.* New Delhi: The Congress. P22.

283. Retrieved from http://alphabetilately.org/airgraph.html.

284. E-mail correspondence with Smith's granddaughter who now lives in London. 5 July 2016.

285. James, S. Pais 2001. *Anglo-Indians: The Dilemma of Identity.* Retrieved from http://ehlt.flinders.edu.au/projects/counterpoints/PDF/A7.pdf.

286. Kitson, S. 1992. *Policeman Rocketeer.* Indian Stamp Dealers Association.

287. Interview with Melvyn Brown. 2 April 2014. Brown is a writer on Anglo-Indian affairs and near neighbour of the Smith family who met Hector in 1970s.

288. Correspondence with Smith's granddaughter who now lives in London. 5 July 2016.

289. Smith, S. H. 1931. Bulletin of the Indian Airmail Society V1(1): 369.

290. Smith, S. H. 1931. Bulletin of the Indian Airmail Society V5(4): 333.

291. Aerophilatelic Hall of Fame. Retrieved from http://www.americanairmailsociety.org/html/aerophilatelic_hall_of_fame.html.

Chapter 6

292. Despite still being male dominated, the DOS (where ISRO is located within the government) is a more equal employer compared to other parts of the government. Approximately 20% of the 15,656 employees in the 17 ISRO centres are women. Although there are not many women in senior positions within ISRO, their contribution is publicly recognised. For a breakdown of male/female employees, see Government of India, Department of Space. 2016. ISRO Annual Report 2015-2016. Table iv on P121. http://isro.gov.in/sites/default/files/flipping_book/Annual_Report_2016_1/files/assets/common/downloads/Annual%20Report%202016.pdf

293. Harvey, Brian, Henk H. F. Smid, and Theo Pirard. 2011. *Emerging Space Powers: The New Space Programs of Asia, the Middle East and South-America.* Springer Science & Business Media.

294. ISRO. 2016. *From Fishing Hamlet to Red Planet: India's Space Journey.* Noida, Uttar Pradesh: Harper Collins India. P58.

295. Shah, Amrita. 2007. *Vikram Sarabhai: A Life*. Illustrated edition. New Delhi: Viking (India). P14.

296. The German scientist Alfred Wegener had written about the concept of continental drift in his 1915 book *Die Entstehung der Kontinente und Ozeane*, but his ideas were not formally accepted until the 1950s.

297. Deshmukh, Chintamani. 2010. *Homi Jahangir Bhabha*. New Delhi: National Book Trust. P14.

298. RT News. First Lady of Space: Tereshkova's Flight TIMELINE. Retrieved from http://rt.com/news/tereshkova-first-space-timeline-726/

299. Osborne, Z. et al. 1975. *Apollo Light Flash Investigations*. Houston, Texas: NASA Johnson Space Center.

300. Harvey, Brian, Henk H. F. Smid and Theo Pirard. 2010. *Emerging Space Powers: The New Space Programs of Asia, the Middle East and South-America*. Berlin, New York: Springer. P143.

301. Shah, Amrita. 2007. *Vikram Sarabhai: A Life*. Illustrated edition. New Delhi: Viking (India). P52.

302. Sarabhai, Mrinalini. 2004. *The Voice of the Heart: An Autobiography*. New Delhi: Harper Collins. P112.

303. *Ibid*.

304. Harvey, Brian, Henk H. F. Smid and Theo Pirard. 2010. *Emerging Space Powers: The New Space Programs of Asia, the Middle East and South-America*. Berlin, New York: Springer. P142.

305. Gandhi, R. 2006. *Gandhi: The Man, His People, and the Empire*. Berkeley: University of California Press. P183.

306. Harvey, Brian, Henk H. F. Smid and Theo Pirard. 2010. *Emerging Space Powers: The New Space Programs of Asia, the Middle East and South-America*. Berlin, New York: Springer. P142.

307. Lennon, Joseph. 2007. Fasting for the Public: Irish and Indian Sources of Marion Wallace Dunlop's 1909 Hunger Strike. In Eóin Flannery and Angus Mitchell (Eds) *Enemies of Empire: New Perspectives on Imperialism, Literature and Historiography*. Dublin: Four Courts Press. Retrieved from http://www.academia.edu/6073253/Fasting_for_the_public_Irish_and_Indian_sources_of_Marion_Wallace_Dunlops_1909_hunger_strike

308. Some of Harry Turner's writings, paintings and photographs are available online at http://www.htspweb.co.uk/fandf/romart/het/history.htm

309. Sarabhai, Mrinalini. 2004. *The Voice of the Heart: An Autobiography*. New Delhi: Harper Collins. P70.

310. *Ibid*. P162.

311. Shah, Amrita. 2007. *Vikram Sarabhai: A Life*. Illustrated edition. New Delhi:

Viking (India). P110.

312. *Ibid.* P111.

313. Sarabhai, Mrinalini. 2004. *The Voice of the Heart: An Autobiography.* New Delhi: Harper Collins. P175.

314. Kalam, A. P. J. Abdul and A. Tiwari. 1999. *Wings of Fire: An Autobiography.* Hyderabad: Universities Press. P31.

315. Officially, Israel's policy is of nuclear ambiguity, not admitting or denying its position on nuclear weapons. Mordechai Vanunu, an Israeli whistle blower, maintains a website recording his experience of exposing Israel's nuclear arsenal and his efforts in achieving a nuclear free world. See http://www.vanunu.com/nukes/20011111globe.html

316. Sarabhai came from a wealthy family, which had grown through his contribution. When he took on the roles at the DEA and AEC, he took a salary of 1 rupee. This is according to a small piece "Dr. Vikram Sarabhai – Some Reminiscences" written by P. Radhakrishnan. In 1966, he was interviewed by Sarabhai for a post in ISRO. Author's communication with P. Radhakrishnan. 13 September 2013.

317. Peaceful Nuclear Explosions were proposed for building harbours, reservoirs and the Panama Canal. See http://www.chymist.com/The%20Plan%20to%20Nuke%20Panama.pdf.

318. Wittner, Lawrence S. 1993. *The Struggle Against the Bomb.* California: Stanford University Press. P150.

319. Shah, Amrita. 2007. *Vikram Sarabhai: A Life.* Illustrated edition. New Delhi: Viking (India). P159. Also, see http://www.nuclearfiles.org/menu/key-issues/nuclear-weapons/issues/proliferation/india/index.htm

320. *Ibid.* P189.

321. Anderson, R. S. 2010. *Nucleus and Nation: Scientists, International Networks, and Power in India.* Chicago, London: University of Chicago Press. P441.

322. ISRO. 2015. *From Fishing Hamlet to Red Planet: India's Space Journey.* Noida, Uttar Pradesh: Harper Collins India. P77.

323. Kalam, A. P. J. Abdul and A. Tiwari. 1999. *Wings of Fire: An Autobiography.* Hyderabad: Universities Press. P26.

324. *Ibid.* P45.

325. Sarabhai, Mrinalini. 2004. *The Voice of the Heart: An Autobiography.* New Delhi: Harper Collins. P235.

326. Author's conversation with Jacques Blamont. 18 April 2013.

327. Shah, Amrita. 2007. *Vikram Sarabhai: A Life.* Illustrated edition. New Delhi: Viking (India). P208.

328. Sarabhai, Vikram. December 1963. Significance of Sounding Rocket Range in

Kerala. In ISRO. 2015. *From Fishing Hamlet to Red Planet: India's Space Journey*. Noida, Uttar Pradesh: Harper Collins India. Reproduced from *Nuclear India* 2 (4), December 1963.

329. There are several statues commemorating the work of Sarabhai. ISRO's headquarters has a bust of Sarabhai in the garden with his words "we must be second to none in the application of advanced technologies to the real problems of man and society."

330. Shah, Amrita. 2007. *Vikram Sarabhai: A Life*. Illustrated edition. New Delhi: Viking (India). P179.

331. Rao, U. R. December 2001. Vikram Sarabhai: The Scientist. *Resonance*. Retrieved from http://www.ias.ac.in/article/fulltext/reso/006/12/0010-0018

Chapter 7

332. Scott, D.,Leonov, A and C. Toomey. 2011. Two Sides of the Moon: Our Story of *the Cold War Space Race* (Foreword by Neil Armstrong). London: Simon & Schuster.

333. Doraisamy, Ashok Maharaj. 2011. Space for Development: US-Indian Space Relations 1955–1976. P30. Retrieved from https://smartech.gatech.edu/bitstream/handle/1853/45973/Maharaj_Ashok_201112_Phd.pdf

334. Harvey, Brian, Henk H. F. Smid and Theo Pirard. 2011. Emerging Space Powers: The New Space Programs of Asia, the Middle East and South-America. Springer Science & Business Media. A detailed piece written by M. Nicolet is available here: https://web.archive.org/web/20130728091015/http://www.wmo.int/pages/mediacentre/documents/Int.GeophysicalYear.pdf

335. Bronk, Detlev. 1954. Letter from the National Academy of Sciences. National Academy of Sciences. https://www.eisenhower.archives.gov/research/online_documents/igy/1954_4_21.pdf

336. National Science Foundation Proposal. 1954. http://www.eisenhower.archives.gov/research/online_documents/IGY/1954_6_2.pdf.

337. In 1996, the name of the International Council of Scientific Unions was simplified to International Council for Science (ICS).

338. The Sun goes through a period of increased activity every 11 years. The activity is manifested as variations in its magnetic field, increase in sunspots and solar storms that give rise to high energy solar particles arriving at the Earth. In addition to generating a spectacular display of aurora, it can also increase radiation risks for satellites and astronauts in Earth orbit. Occasionally, as in Canada in 1989, solar magnetic storms (Coronal Mass Ejections) can result in large-scale damage to

electricity power supplies on Earth.

339. Dickson, Paul. 2001. Sputnik: The Shock of the Century. First Printing. New York: Walker Publishing Company. P77.

340. Sarabhai, Vikram. Significance of Sounding Rocket Range in Kerala. Reproduction of Nuclear India. Mumbai: Department of Atomic Energy (DAE).

341. Zhou, Yong. 2013. The Histories of the International Polar Years and the Inception and Development of the International Geophysical Year: Annals of The International Geophysical Year. Elsevier. P403

342. Whitehouse Statements. First Announcement of US's Plans for Satellite Launch. Retrieved from http://www.eisenhower.archives.gov/research/online_documents/NASA/Binder1.pdf.

343. Satellites get their names close to their launch date. Sputnik in the USSR was initially known as PS, Preliminary Satellite and later Artificial Earth Satellite but the name Sputnik stuck, even in the USSR, after it was popularised in the West. The name Explorer for the first US satellite was selected by Eisenhower after launch. Aryabhata in India initially was 'the Indian science satellite.' Similarly, the name 'Explorer-1' was President Eisenhower's idea to make it sound scientific and civilian. For greater details on naming satellite conventions used by the USSR Siddiqi, Asif A. 2010. *The Red Rockets' Glare: Spaceflight and the Soviet Imagination, 1857-1957*. First edition. Cambridge: Cambridge University Press. P334.

344. Dickson, Paul. 2001. Sputnik: The Shock of the Century. First Printing. New York: Walker & Company. P11.

345. Siddiqi, Asif A. 2010. The Red Rockets' Glare: Spaceflight and the Soviet Imagination, 1857-1957. First edition. Cambridge: Cambridge University Press. P342.

346. Ibid. P358.

347. Professor Praful Bhavsar. 16 January 2013. Communication with the author.

348. Telephone interview with Professor Praful Bhavsar. 18 December 2012.

349. Rao, U. R. 2001. Vikram Sarabhai, The Scientist. Resonance—Journal of Science Education 6(12). Retrieved from http://www.ias.ac.in/article/fulltext/reso/006/12/0010-0018.

350. This figure for 2015–16 is from an official source and refers to the DOS under which ISRO operates. The figures also provide a breakdown (legally required in India) of the number of employees belonging to Scheduled Castes and Scheduled Tribes. Retrieved from http://isro.gov.in/sites/default/files/flipping_book/Annual_Report_2016_1/files/assets/common/downloads/Annual%20Report%202016.pdf P115.

351. This high-speed and high-altitude (between 7 and 16 km) ribbon of wind, known as the jet stream, that snakes around the Earth had been speculated since the 1883 volcanic eruption of Krakatoa. It was eventually detected in the middle of the 20th century. The northern and southern hemispheres have polar and

subtropical jet streams (commonly referred to in the singular) that vary in precise location and intensity but are typically a narrow (1 km to 5 km wide) band that girdles the Earth at up to 400 km/h from west to east. Flight plans routes used by commercial and military pilots routinely exploited the jet stream as tail wind and avoid it as a head wind.

352. Raja, Jay and Manoranjan Rao. 18 November 2002. When Thumba took off. The Hindu. Retrieved from http://www.hindu.com/thehindu/mp/2002/11/18/stories/2002111801050200.htm.

353. Most sources record decisions of the key Indian stakeholders, like Sarabhai and Bhabha, but the following work indicates the influence of the scientists from the US. Doraisamy, Ashok Maharaj. 2011. Space for Development: US-Indian Space Relations 1955 -1976. P63. Retrieved from https://smartech.gatech.edu/bitstream/handle/1853/45973/Maharaj_Ashok_201112_Phd.pdf.

354. Sarabhai, Vikram. 1963. Significance of Sounding Rocket Range in Kerala. Reproduction of Nuclear India. Mumbai: Department of Atomic Energy (DAE).

355. Thumba's precise location is 8°32'28" N and 76°51'38" E, and a magnetic latitude of 0°24'S, that is, not on the equator but half a degree north from it. The Earth's magnetic field is not fixed. The value of 0°24'S was accurate in 1964.

356. Raj, Gopal. 2003. Reach for the Stars: The Evolution of India's Rocket Programme. New Delhi: Penguin Books India. P11.

357. Rao, P. V. Manoranjan and P. Radhakrishnan. 2012. A Brief History of Rocketry in ISRO. Hyderabad: Universities Press (India) Private Limited.

358. Vikram Sarabhai's speech on the dedication of TERLS to the UN on 2 February 1968.

359. Raj, Gopal. 2003. Reach for the Stars: The Evolution of India's Rocket Programme. New Delhi: Penguin Books India. P26.

360. Ibid. P15.

361. Rao, P. V. Manoranjan and P. Radhakrishnan. 2012. A Brief History of Rocketry in ISRO. Hyderabad: Universities Press (India) Private Limited.

362. Kalam, A. P. J. Abdul and Arun Tiwari. 1999. Wings of Fire: An Autobiography. Hyderabad: Universities Press. P22.

363. ISRO. 2016. From Fishing Hamlet to Red Planet: India's Space Journey. Noida, Uttar Pradesh: Harper Collins India. P83.

364 Prime Minister Jawaharlal Nehru told Indian Parliament on 8 Aug 1961 that India had agreed to have a rocket launching station on her territory under TJ.N. auspices for international use. https://history.nasa.gov/AAchronologies/1962.pdf

365. Massey, Sir Harrie Stewart Wilson and Malcolm Owen Robins. 1986. History of British Space Science. Cambridge: Cambridge University Press. P175.

366. Treaty Series. 1971. Treaties and International Agreements Registered or Filed and Recorded with the Secretariat of the United Nations, Volume 778. New York:

United Nations. P215.

367. From 1977 to 1981, Soviet and Indian specialists made joint experiments in the area of extra-atmospheric astronomy with Soviet Y-telescopes installed on Indian high-altitude balloons. Nikitin, S. A. 1985. The Space Flight of the Soviet-Indian Crew. NASA TM-77615. Retrieved from http://ntrs.nasa.gov/archive/nasa/casi.ntrs.nasa.gov/19850012916.pdf.

368. These figures come from the presentation made by the ISRO Chairman Kiran Kumar on 18 August 2015 at the Inter-University Centre for Astronomy and Astrophysics. Retrieved from https://www.youtube.com/watch?v=4gfAnfOsjHI

369. Pakistan Space and Upper Atmosphere Research Commission (SUPARCO). Retrieved from http://www.suparco.gov.pk/pages/history.asp.

370. Wallace, Harold D., Jr. 1997. Wallops Station and the Creation of an American Space Program. NASA History Series. P91. Retrieved from http://ntrs.nasa.gov/archive/nasa/casi.ntrs.nasa.gov/19970037643.pdf

371. Interview with Professor U.R. Rao. 16 August 2013.

372. Ibid.

373. Massey, Sir Harrie Stewart Wilson and Malcolm Owen Robins. 1986. History of British Space Science. Cambridge: Cambridge University Press. P168.

374. The Indian engineer was R. Aravamudan. He concludes that "without Sarabhai, India's space program might have ended in the same doldrums as Pakistan's". Aravamudan, R.; Aravamudan, Gita. ISRO: A Personal History (Kindle Location 248). HarperCollins Publishers India.

375. The complete memorandum of understanding is relocated in Annex 13. Massey, Sir Harrie Stewart Wilson and Malcolm Owen Robins. 1986. History of British Space Science. Cambridge: Cambridge University Press. P1488.

376. Massey, Sir Harrie Stewart Wilson and Malcolm Owen Robins. 1986. History of British Space Science. Cambridge University Press. P168

377. Crilly, Rob. 2012. Why Abdus Salam, Pakistan's Great Physicist, Has Been Written Out of History by His Own Country. Retrieved from http://timesofahmad.blogspot.co.uk/2012/08/why-abdus-salam-pakistans-great.html/

378. Interview with Professor U.R. Rao. 16 August 2013.

379. 2004, Dyson, Freeman. A special publication to mark the 40th anniversary of the "Abdus Salam international centre for theoretical research. Retrieved form http://users.ictp.it/~pub_off/books/100_reasons.pdf P81

380. This article also discusses the role of ISRO's IRNSS and other satellites for national defence. Zahid, Ahsan Ali. 27 May 2016. Indian Space Odyssey. Modern Diplomacy. Retrieved from http://moderndiplomacy.eu/index.php?option=com_k2&view=item&id=1451:indian-space-odyssey&Itemid=723.

381 Vladimir points out in his chapter "Evolution of India-Russia Partnership Vladimir" that the UK and Canada became space nations in 1962, Italy in 1964,

France in 1965, Australia in 1967 and West Germany in 1969. Rajagopalan, Rajeswari Pillai and Narayan Prasad. 2017. Space India 2.0: Commerce, Policy, Security and Governance Perspectives. Observer Research Foundation. P250

382. Although India became independent in 1947 when the British left, the city of Pondicherry and several other regions on the South Coast that had been French remained under French control until the treaty of cession was signed by the two countries in May 1956 and subsequently ratified in 1963. The city of Pondicherry and the other enclaves of Karika (Karaikal), Mahé and Yanam (Yanaon) came to be administered as the Indian Union Territory of Puducherry from 1 July 1963.

383. Telephone conversation with Jacques Blamont. 19 April 2013. Interestingly, he also said that he was not present for the launch on 21 November 1961 and arrived in India for the first time only in January 1964. Other sources indicate that he was present. For example, Doraisamy, Ashok Maharaj. 2011. Space for Development: US-Indian Space Relations 1955 -1976. Retrieved from https://smartech.gatech.edu/bitstream/handle/1853/45973/Maharaj_Ashok_201 112_Phd.pdf. P73

384. Apparently, the mobile radar unit used to measure altitude, half a century ago, was still in operational service in VSSC. Author's conversation with U. R. Rao. 19 April 2013.

385. Baskaran, A. 2000. Technology Accumulation in India's Space Programme Ground Systems: The Contributions of Foreign and Indigenous Input. Discussion Paper Series: Economics. Middlesex University Business School. Retrieved from http://www.researchgate.net/profile/Angathevar_Baskaran/publication/22338978 3_Technology_accumulation_in_the_ground_systems_of_India%27s_space_progr am_the_contribution_of_foreign_and_indigenous_inputs/links/00463524d27640a 4eb000000.pdf

386. These statistics come from the 14th January entry in NASA's Astronautics and Aeronautics. 1963. Chronology on Science, Technology and Policy. Washington, DC: NASA

387. Wallace, Harold D., Jr. 1997. Wallops Station and the Creation of an American Space Program. NASA History Series. P105. Retrieved from http://ntrs.nasa.gov/archive/nasa/casi.ntrs.nasa.gov/19970037643.pdf. Starting 1961, Wallops hosted training courses for engineers and researchers for many years in many countries including Japan, Sweden and Pakistan.

388. ISRO. 2016. From Fishing Hamlet to Red Planet: India's Space Journey. Noida, Uttar Pradesh: Harper Collins India. P74.

389. Kalam, A. P. J. Abdul and Arun Tiwari. 1999. Wings of Fire: An Autobiography. Hyderabad: Universities Press. P25.

390. Numerous Nike Ajax surface to air missiles were recommissioned as the first of two stage sounding rockets. Details here come from Crough, R. A. and H. L. Galloway. 1966. Nike Apache Performance Handbook. NASA. Retrieved from https://ntrs.nasa.gov/search.jsp?R=19670015760.

391. Crough, R. A. and H. L. Galloway. 1966. Nike Apache Performance Handbook. NASA. Retrieved from http://ntrs.nasa.gov/search.jsp?R=19670015760. One of the authors, Howard

Galloway, later became NASA's representative in India for the duration of the SITE programme.

392. Astronautics and Aeronautics. 1963. Chronology on Science, Technology and Policy. Washington, DC: NASA. P336.

393. ISRO. 2015. From Fishing Hamlet to Red Planet: India's Space Journey. Noida, Uttar Pradesh: Harper Collins India. P36.

394. Bhavsar, P. 2013. The First Rocket Launch from TERLs. Personal communication between Professor Bhavsar and the author.

395. Kalam, A. P. J. Abdul and Arun Tiwari. 1999. Wings of Fire: An Autobiography. Hyderabad: Universities Press. P26.

396. Raj, Gopal. 2003. Reach for the Stars: The Evolution of India's Rocket Programme. New Delhi: Penguin Books India. P17.

397. Jenkins, Reed B. 1963. Nike Apache Performance Handbook. Washington, DC: NASA.

398. A recollection by K. Narayana Kurup on the 40th anniversary of the event on 21 November 2003. India's Space Odyssey—Church to Chandrayaan. 22 October 2008. Retrieved from http://indiatoday.intoday.in/story/India's+space+odyssey+-+church+to+Chandrayaan/1/18262.html

399. Sarabhai, Vikram. 1963. Significance of Sounding Rocket Range in Kerala. Reproduction of Nuclear India. Mumbai: Department of Atomic Energy (DAE).

400. Raj, Gopal. 2003. Reach for the Stars: The Evolution of India's Rocket Programme. New Delhi: Penguin Books. P17.

401. ISRO. 2016. From Fishing Hamlet to Red Planet: India's Space Journey. Noida, Uttar Pradesh: Harper Collins India. P75.

402. The term non-alignment was coined by Indian diplomat V. K. Menon at the United Nations in 1953.

403. Personal communication between Jacques Blamont and the author. 19 April 2013.

404. ISRO. 2016. From Fishing Hamlet to Red Planet: India's Space Journey. Noida, Uttar Pradesh: Harper Collins India. P50.

405. Sarabhai, Mrinalini. 2004. The Voice of the Heart: An Autobiography. New Delhi: Harper Collins. P202.

406. Clarke, Arthur C. 1993. How the World Was One: Turbulent History of Global Communications. New edition. London: Gollancz. P216.

407. Clarke is credited with the idea of geostationary orbits for use by communication satellites. He acknowledged the pioneering work by Konstantin Tsiolkovskii and Hermann Oberth, but credited Hermann Noordung with the original idea. Noordung, Hermann and Ernst Stuhlinger. 2011. The Problem of Space Travel: The Rocket Motor. NASA History Series. Pxxi. Retrieved from https://wordery.com/the-problem-of-space-travel-hermann-noordung-9781780392745

408. A personal recollection by E. V. Chitnis in ISRO. 2016. From Fishing Hamlet to Red Planet: India's Space Journey. Noida, Uttar Pradesh: HarperCollins India. P53.

409. Clarke, Arthur C. 1993. How the World Was One: Turbulent History of Global Communications. New edition. London: Gollancz. P218.

410. Ibid.

411. Wales, O. Robert and NASA Scientific and Technical Information Branch. 1981. ATS-6 Final Engineering Performance Report. Washington, DC: NASA.

412. ISRO. 2016. From Fishing Hamlet to Red Planet: India's Space Journey. Noida, Uttar Pradesh: HarperCollins India. P53.

413. Harvey, Brian, Henk H. F. Smid, and Theo Pirard. 2011. Emerging Space Powers: The New Space Programs of Asia, the Middle East and South-America. Springer Science & Business Media. P149.

414. Wales, O. Robert and NASA Scientific and Technical Information Branch. 1981. ATS-6 Final Engineering Performance Report. Washington, DC: NASA. P41.

415. Dick, Steven J. 2009. NASA's First 50 Years: Historical Perspectives. Washington, DC: NASA. P134.

416. Arnold Frutkin, the Deputy Director at NASA for International Affairs, had determined that India would be best placed to participate in what became the SITE programme. The US State Department prevented Frutkin from making the offer, lest the US was once more snubbed by India as it had been when requesting that Voice of America transmitters be set up in India. Frutkin had already been in contact with Sarabhai and knew about India's positive desire to engage. Frutkin's solution was to get India to make the initial request. This he did but not without a slight misunderstanding on Sarabhai's part as recalled by Frutkin in this lovely exchange "So I said, 'Write me a letter proposing Indian participation in the SITE programme which is being prepared.' He said, 'Sure.' So, he wrote me a letter and said, 'In accordance with your suggestion, we would like.' I called him up and said, 'No, Vikram. Write another letter that doesn't refer to my phone call.' So, he wrote another letter." ISRO. 2016. From Fishing Hamlet to Red Planet: India's Space Journey. Noida, Uttar Pradesh: Harper Collins India. P354.

417. NASA. 1970. Astronautics and Aeronautics, 1970. Washington, DC: NASA. P245. Retrieved from http://history.nasa.gov/AAchronologies/1970.pdf

418. Shah, Amrita. 2007. Vikram Sarabhai: A Life. Illustrated edition. New Delhi: Viking (India). P178.

419. Raman, Srinivasan. No Free Launch: Designing the Indian National Satellite. P6. In Beyond the Ionosphere: The Development of Satellite Communications. NASA History Series. Retrieved from http://history.nasa.gov/SP-4217/ch16.htm

420. Wales, O. Robert and NASA Scientific and Technical Information Branch. 1981. ATS-6 Final Engineering Performance Report. Washington, DC: NASA. P46. A slightly more detailed reference is included in Astronautics and Aeronautics. 1975. Chronology on Science, Technology and Policy.

421. NASA. 1975. Apollo Soyuz Test Project Press Kit. Retrieved from http://history.nasa.gov/astp/documents/astp%20press%20kit%20(us).pdf.

422. Wales, O. Robert and NASA Scientific and Technical Information Branch. 1981. ATS-6 Final Engineering Performance Report. Washington, DC: NASA. P18. Retrieved from http://ntrs.nasa.gov/archive/nasa/casi.ntrs.nasa.gov/19820008274.pdf

423. Galloway, a rocket engineer, worked extensively on the Nike-Apache rocket and was probably involved with India in 1963 when the first Nike-Apache rockets were delivered from NASA to India. He died suddenly in India after he had been working in different Indian locations for 15 months monitoring the SITE programme. In the introduction to one of Galloway's published reports, Dr. J. F. Clark, Director of the Goddard Space Flight Centre, pays a touching tribute to Galloway's contribution. Galloway, H. L. 1976. Satellite Instructional Television Experiment (SITE) Reports from NASA Representative. Retrieved from http://ntrs.nasa.gov/archive/nasa/casi.ntrs.nasa.gov/19760024290_1976024290.pdf P ii

424. Planning Commission (PEO). 1981. Evaluation Report on Satellite Instructional Television Experiment (SITE). Retrieved from http://planningcommission.nic.in/reports/peoreport/cmpdmpeo/volume2/erosi.pdf.

425. Galloway, H. L. 1976. SITE Reports from the NASA Resident Representative in India. Retrieved from http://www.researchgate.net/publication/234310489_Satellite_Instructional_Television_Experiment_%28SITE%29_Reports_from_the_NASA_resident_representative_in_India

426. Rao, U.R. 2001. Space Technology for Sustainable Development. New Delhi: McGraw-Hill Education. P235.

427. Galloway, H. L. 1976. SITE Reports from the NASA Resident Representative in India. P68. Retrieved from http://www.researchgate.net/publication/234310489_Satellite_Instructional_Television_Experiment_%28SITE%29_Reports_from_the_NASA_resident_representative_in_India.

428. This paper discusses the impact of space communication experiments on social development in India. SITE is just one of the many experiments the author examines. Chitnis, E. V. 1983. The Role of Space Communication in Promoting National Development with Specific Reference to Experiments Conducted in India. Advances in Space Research 3 (7): 125–32. doi:10.1016/0273-1177(83)90156-4.

429. The Prime Minister of Sri Lanka insisted on ISRO providing him with a satellite system as a prerequisite for allowing one to be presented to Arthur C. Clarke. Interview with Professor U.R. Rao. August 2013.

430. Galloway, H. L. 1976. SITE Reports from the NASA Resident Representative in India. P3. Retrieved from http://www.researchgate.net/publication/234310489_Satellite_Instructional_Television_Experiment_%28SITE%29_Reports_from_the_NASA_resident_representat

ive_in_India.

Chapter 8

431. ISRO's infrastructure is continuously evolving to meet the needs of the dynamic environment in which it operates. This chapter does not attempt to comprehensively cover all the activities undertaken by all ISRO centres across India. Instead, key activities of only a sample of the primary ISRO centres are covered. Other countries with an operational space programme include the US, Russia, Japan, China, North Korea, South Korea, Israel, Europe and Iran. Countries that are actively investing to develop a space industry include Brazil, Indonesia, Argentina and South Africa. Many other countries (around 70) are developing or already have some level of investment in space assets.

432. The interim Chairman M. G. K. Menon recalls during an interview on 27 February that the new ISRO structure (segregation from the AEC and Satish Dhawan being in-charge) was concluded during a meeting with the Indian Prime Minister a few days after Sarabhai had died. ISRO. 2016. *From Fishing Hamlet to Red Planet: India's Space Journey*. Noida, Uttar Pradesh: HarperCollins India. P45.

433. It was originally called the Indian Committee for Space Research (INCOSPAR) when founded in 1962 by Homi Bhabha. He immediately asked Vikram Sarabhai to lead INCOSPAR as Chairman. In 1966, Bhabha died suddenly. Sarabhai replaced him as the Chairman of AEC and also continued with his existing role as the Chairman of INCOSPAR.

434. Rao, U. R. 2013. *India's Rise as a Space Power*. Delhi: Cambridge University Press India Pvt Ltd. P195.

435. *Ibid*. P86.

436. *Ibid*. P107.

437. Professor Roddam Narasimha, the longest serving member of the Space Commission (until 2012). In conversation with the author. 1 February 2015.

438. One of the early announcements from ISRO on its desire to outsource was made in 2009. Abrar, Peerzada. 3 November 2009. ISRO to Outsource Rocket-work to Private Companies. The Economic Times. Retrieved from http://articles.economictimes.indiatimes.com/2009-11-03/news/27655023_1_unmanned-lunar-mission-pslv-earth-observation-satellites

439. A series of ISRO chairmen have made announcements to the effect that greater private sector involvement is essential for ISRO to grow. Now, ISRO is considering outsourcing the production of PSLV in its entirety. In practice, ISRO's engagement with the private sector remains lethargic. Saini, J. 30 June 2016. ISRO Throws Satellite Making Open to Private Sector. *The Talking Indian*. Retrieved from http://thetalkingindian.com/2016/06/isro-throws-satellite-making-open-to-private-sector/

440. Harvey, Brian, Henk H. F. Smid and Theo Pirard. 2010. Emerging Space Powers: The New Space Programs of Asia, the Middle East and South America. New York: Springer.

441. There are actually two tunnels a 1 m wind tunnel up to Mach 6 and a Shock tunnel up to Mach 8. http://www.vssc.gov.in/VSSC_V4/index.php/retired-employee-portal/58-infrastructure/1179-hypersonic-wind-tunnel-facility-2

442. Prior to his visit to Vienna, Sarabhai considered meteorology, communication and navigation as the primary satellite-based applications from which India could benefit the most. Joseph, George. 2016. *India's Journey Towards Excellence in Building Earth Observation Cameras*. First edition. Chennai: Notion Press. P4.

443. Rocchio, Laura. 19 April 2006. Landsat Island. *NASA Landsat Science*. Retrieved from http://landsat.gsfc.nasa.gov/?p=258

444. Joseph, George. 2016. India's Journey Towards Excellence in Building Earth Observation Cameras. First edition. Chennai: Notion Press. P21.

445. One such experiment, Balloon Resources Inventory Development Experiment, nicknamed 'BRIDE', imaged the Earth from 27 km using a payload of 70 kg for 3 hours on 2 February 1975. Joseph, George. 2016. *India's Journey Towards Excellence in Building Earth Observation Cameras*. First edition. Chennai: Notion Press. P23.

446. SAMIR provided information on the sea state and atmospheric water vapour content for use in meteorological studies. The charge-coupled device (CCD) camera, Side-Looking Radar and Colour Infrared (CIR) were developed by ISRO at SAC. When launched in 1995, IRS-1C provided the highest spatial resolution among civilian satellites. Joseph, George. 2016. *India's Journey Towards Excellence in Building Earth Observation Cameras*. First edition. Chennai: Notion Press. P110.

447. Some of these projects are already under way others have yet to start. Summary of all the 170 projects is available here. http://www.livemint.com/Politics/lP1TRsmiAhFaecOouZdi2M/Govt-partners-Isro-on-170-projects-to-use-space-technology-f.html

448. Bagla, Pallava (PTI). 10 April 2016. India, China Set Eyes on Joint Satellite Constellation for BRICS. *The Economic Times*. Retrieved from http://economictimes.indiatimes.com//articleshow/51763274.cms.

449. An alternate power source used by spacecraft that must operate in environments devoid of sufficient sunlight is Radio Thermal Generators (RTG). RTGs are expensive and complex. Their deployment is subject to political and environmental controversy.

450. Typically, the mass the mass of the spacecraft (dry) is about 1% of the total (including the propellant) at launch. As an illustration (Astrosat is in Earth orbit – not as far to travel as Mars)

Spacecraft	Total Mass(kg)	Fuel(kg)	Ratio
Maven	2454	809	0.33

MOM	1340	488	0.36
Astrosat	1513	1470	0.98

451. This is a very delicate process. In addition to the altitude control thrusters, of which there can be about a dozen, a satellite uses magnetic torquers and reaction wheels in the manoeuvre.

452. A PSLV uses identical solid propellant in its first stage (150 tonnes including the strap-ons) and the third stage (8 tonnes). The second stage typically uses 40 tonnes of UDMH and nitrogen tetroxide and the fourth stage monomethylhydrazine and mixed oxides of nitrogen (MON-3).

453. For example, the Isp for the PSLV first stage (solid) is 237, PSLV second stage (liquid) is 293, and for a typical semi-cryogenic engine, it is 320. The most efficient category of rocket engine uses liquid oxygen and liquid hydrogen cryogenic propellants, for example, GSLV-Mk3 upper stage with an Isp of 443.

454. DOS. ISRO Annual Report 2014–2015. P15. Retrieved from http://www.isro.gov.in/sites/default/files/pdf/AR2014-15.pdf. Also, there is a proposal (in very early stages) for a ground station in Vietnam. Kumari, Chetan. 25 May 2015. ISRO Moots Ground Station in Vietnam. The Times of India. Retrieved from http://timesofindia.indiatimes.com/india/Isro-moots-ground-station-in-Vietnam/articleshow/47410113.cms

455. The other two sites were near Kanakpura road and Solur. Byalalu was selected for its minimal aviation traffic overhead, low industrial installation, good road access and a natural bowl-shaped valley into which the dish could be situated providing further radio shielding. The most detailed account of the building of Byalalu is provided by Hathwar, G. R. 2016. *Indian Deep Space Network: Mars Not Too Far*. Chennai: Notion Press. Incidentally, if you ever visit Byalalu, make time for the local tourist attraction known as the Bib Banyan Tree.

456. ISRO has established TT&C at higher latitudes to support its remote-sensing satellites. Provision for TT&C and data services for Indian remote-sensing satellites from a high-latitude TT&C network maintained by Kongsberg Satellite Services for a medium term has been a notable activity. Strategic dialogues have been progressing well with Angkasa (the Malaysian Space Agency), Swedish Space Corporation, United S-band Network (USN) of the US, INPE of Brazil and JPL for future ISRO missions. ISRO. 2016. *From Fishing Hamlet to Red Planet: India's Space Journey*. Noida, Uttar Pradesh: Harper Collins India. P301.

457. International Telecommunications Union mandated a minimum distance of 500 km between an uplinking and a ground station. INTELSAT had a station in Jaffna in Ceylon, which was only 450 km from Bangalore, so an alternative location had to be found. Rao, U. R. 2013. India's Rise as a Space Power. Delhi: Cambridge University Press India Pvt Ltd. P98.

458. Interview with S.V. Kumar. Available here https://astrotalkuk.org/episode-70-indias-deep-space-network-and-isro-satellite-centre/.

459. Rangarajan, S. 1995. INSAT Master Control Facility: An Insight. Current Science 69 (5): 410–415. Retrieved from http://www.currentscience.ac.in/Downloads/article_id_069_05_0410_0415_0.pdf

460. Stoney, William. 15 April 2003. Civil Land Observation Satellites. Encyclopaedia of Space Science and Technology. Retrieved from ftp://landsat-legacy.gsfc.nasa.gov/outgoing/documents/LL-25340915497.pdf.

461. Rao, U. R. 2013. India's Rise as a Space Power. Delhi: Cambridge University Press India Pvt Ltd. P134.

462. Ibid.

463. TNN. 9 February 2011. Antrix Responsible for Marketing ISRO Tech. The Times of India. Retrieved from http://timesofindia.indiatimes.com/india/Antrix-responsible-for-marketing-ISRO-tech/articleshow/7457366.cms.

464. http://www.oecd-ilibrary.org/economics/the-space-economy-at-a-glance-2014_9789264217294-en.

465. Antrix Corporation Limited Annual Report 2009-2010. P1 Retrieved from http://www.antrix.gov.in/sites/default/files/financials/ANNUAL_REPORT_2009_2010_English.pdf

466. DOS. 17 July 2014. Income Through Launching of Satellite. Retrieved from http://pib.nic.in/newsite/PrintRelease.aspx?relid=106824.

467. The Tauri Group. 2014. State of the Satellite Industry Report. Satellite Industry Association. Retrieved from http://www.sia.org/wp-content/uploads/2014/05/SIA_2014_SSIR.pdf.

468. Patil, P. A. 2014. Antrix Corporation: Can It Provide Impetus to Indian Space Market. In Focus. Centre for Air Power Studies. Retrieved from http://capsindia.org/files/documents/CAPS_Infocus_PP_2.pdf

469. Gopinath, C. and L. Surendra. 2008. Antrix Corporation Limited: A Strategy for the Global Market. Case Research Journal 28 (3-4 Summer/Fall). Retrieved from http://cgopinath.brinkster.net/media/f23d337e3c458d7ffff8314a00140f.pdf

Chapter 9

470. These figures are from the DOS http://pib.nic.in/newsite/PrintRelease.aspx?relid=111956

471. Department of Atomic Energy. 1970. Annual Report of the Department of Atomic Energy, Government of India 1969–70. P70. Retrieved from http://www.iaea.org/inis/collection/NCLCollectionStore/_Public/01/001/1001868.pdf

472. ISRO. 2016. *From Fishing Hamlet to Red Planet: India's Space Journey.* Noida, Uttar Pradesh: HarperCollins India. P52.

473. In the hot Indian summer, the friction generated by the wheels of the jeep ignited the leaves occasionally, setting the wheels on fire. See Raj, Gopal. 2003. *Reach for the Stars: The Evolution of India's Rocket Programme.* New Delhi: Penguin Books India. P48.

474. In addition to Sriharikota, India has used Thumba and Balasore for launching sounding rockets. Only Sriharikota, however, has the required facilities for payloads destined for Earth orbit.

475. Indian/Soviet Satellite Launch Range. 21 Feb 1974 https://www.cia.gov/library/readingroom/docs/CIA-RDP79B01709A001500010029-5.pdf

476. Rajaraman, V. 2012. History of Computing in India 1955 – 2010. P26. Retrieved from http://www.cbi.umn.edu/hostedpublications/pdf/Rajaraman_HistComputingIndia.pdf

477. Baskaran, A. 2000. Technology Accumulation in India's Space Programme Ground Systems: The Contribution of Foreign and Indigenous Input. *Discussion Paper Series.* London: Middlesex University Business School. P8. Retrieved from http://www.researchgate.net/profile/Angathevar_Baskaran/publication/223389783 _Technology_accumulation_in_the_ground_systems_of_India%27s_space_ programme_the_contribution_of_foreign_and_indigenous_inputs/links/00463524 d27640a4eb000000.pdf

478. ISRO demonstrates its equal opportunity employer status by providing a breakdown of the employees at Sriharikota. In summary, this figure of 1958 includes 360 women, three deaf and dumb, one blind, 47 orthopedically handicapped, 350 from scheduled castes and 89 from scheduled tribes. Department of Space. 2016. Annual Report 2015-2016. P118. Retrieved from http://www.isro.gov.in/sites/default/files/article-files/right-to-information/annual_report-15-16.pdf

479. Once the designated transfer orbit is attained about 10 minutes after launch, the control and management of the spacecraft is passed on to the host nation for a foreign spacecraft or to ISRO's MCF in Hassan 550 km west of Sriharikota. MCF is responsible for delivering it to its final orbit and for subsequent station-keeping, management and operations.

480. The final constitutional link of Australia with Britain ended with the passing of the Australia Act 1986.

481. ISRO. 2016. *From Fishing Hamlet to Red Planet: India's Space Journey.* Noida, Uttar Pradesh: HarperCollins India. P77.

482. *Millennium Post.* 14 May 2014. DRDO Chief Opens Test Facility for Rocket Sled Track in Chandigarh. Retrieved from http://www.millenniumpost.in/NewsContent.aspx?NID=58074.

483. The MCC was formally dedicated to the nation by the President of India Mrs Pratibha Devisingh Patil on 2 January 2012. Retrieved from

http://www.isro.gov.in/update/02-jan-2012/president-of-india-dedicates-new-mission-control-centre-and-launch-control-centre

484. ISRO. 2016. *From Fishing Hamlet to Red Planet: India's Space Journey*. Noida, Uttar Pradesh: Harper Collins India. P159.

485. Harvey, Brian, Henk H. F. Smid, and Theo Pirard. 2011. *Emerging Space Powers: The New Space programmes of Asia, the Middle East and South-America*. Springer Science & Business Media. P245–253.

486. Commercial space operations by the private sector are firmly in place, particularly in the US. So far only spacecraft and not human crew have been launched into space. Some of the "operational shortcuts", for example, prelaunch testing of a fully fuelled launch vehicle with the payload in-situ and plans for fuelling a launch vehicle with human crew onboard have been challenged by NASA. In all previous human spaceflight programmes, the crew entered the launch vehicle after fuelling was complete. This is an interesting letter from a former Apollo Astronaut to SpaceX as SpaceX finalise their plans for human spaceflight: https://www.hq.nasa.gov/office/pao/FOIA/17-HQ-F-00079-ID.pdf

487. This was the period soon after the demise of the USSR, and the US was limiting India's imports using the Missile Technology Control Regime.

488. Rao, U. R. 2013. *India's Rise as a Space Power*. Delhi: Cambridge University Press India Pvt Ltd. P171.

489. Doraisamy, Ashok Maharaj. 2011. Space for Development: US-Indian Space Relations 1955-1976. A Thesis Presented to The Academic Faculty, Georgia Institute of Technology, Atlanta, GA. Retrieved from https://smartech.gatech.edu/bitstream/handle/1853/45973/Maharaj_Ashok_201112_Phd.pdf

490. Baskaran, A. 2011. Competence Building in Complex Systems: The Case of Satellite Launch Vehicles in the India's Space programme. *Technovation* 21 (2): 109–21.

491. Baskaran, Angathevar. 'Competence Building in Complex Systems: The Case of Satellite Launch Vehicles in the India's Space Program'. The 9th Globelics International Conference: Creativity, Innovation and Economic Development, 15 November 2011.

492. Kalam, A. P. J. Abdul and Arun Tiwari. 1999. *Wings of Fire: An Autobiography*. Hyderabad: Universities Press. P29.

493. The Hindu: 6 Killed in Explosion at Sriharikota Space Centre. http://www.thehindu.com/2004/02/24/stories/2004022406180100.htm

494. ISRO. 29 August 2013. Restoration of GSLV-D5 Mission. ISRO Press Release. Retrieved from http://isro.gov.in/update/29-aug-2013/restoration-of-gslv-d5-mission

495. I was at the Sriharikota media centre at the time hoping to experience my first live rocket launch in India. The media centre was a hive of activity with TV and

press media. There were TV monitors showing live images of the rocket on the launch pad going through the final stages of preparations. Some of these images were broadcast live. When the fuel tanks of the second stage were pressurised in readiness for launch, the tank leaked fuel. A white mist developed and began to surround the launch vehicle. The live images to the monitors were suddenly stopped. The launch was aborted, rescheduled and completed in January 2014. On 24 October 1960, a launch pad accident that came to be known as the Nedelin disaster occurred at the Baikonur test range in the USSR. The incident remained a Soviet Union secret on Khrushchev's orders for many years. The precise number that perished is unknown. The general consensus has it that it was over 100.

496. Kanavi, S. 16 February 2014. ISRO: Cryogenic Success. *Reflections*. Retrieved from http://reflections-shivanand.blogspot.co.uk/2014/02/isro-cryogenic-success.html.

497. Electric propulsion is arguably more useful for ISRO since its heavy-lift launch vehicle has a lower specification than most. This example illustrates the level of efficiency 'The Prius of Space'. NASA/JPL. Accessed 14 August 2017. http://www.jpl.nasa.gov/news/features.cfm?feature=1468.

498. The 80 kg figure was not included in the official GSAT-9 brochure but is available here. http://spaceflight101.com/gslv-gsat-9/gsat-9/

499. Vikram Sarabhai was in Thumba to attend the opening of the new Thumba railway station when he suddenly died on 30 December 1971.

500. This is simply a result of the Earth's daily rotation at the equator, that is, the Earth's circumference at the equator divided by the time of one Earth rotation (40,075 km/23.93 hours = 1,675 km/h or 465 m/s). This amounts to about 6% of the required speed to reach Earth orbit.

501. D. S., Madhumathi. 13 March 2015. ISRO's Second Launcher Assembly Unit Gets a Budget Boost. *The Hindu*. Retrieved from http://www.thehindu.com/todays-paper/tp-national/isros-2nd-launcher-assembly-unit-gets-a-budget-boost/article6988134.ece

502. D. S., Madhumathi. 6 January 2016. Sriharikota Space Port Scores 50. *The Hindu*. Retrieved from http://www.thehindu.com/todays-paper/tp-national/sriharikota-space-port-scores-50/article8070373.ece

503. ISRO. 2016. *From Fishing Hamlet to Red Planet: India's Space Journey*. Noida, Uttar Pradesh: Harper Collins India. P131.

504. Karthikeyan K. 13 November 2013. Kulasekarapattinam's Ideal Launch Pad. *Deccan Chronicle*. Retrieved from http://www.deccanchronicle.com/131113/news-current-affairs/article/kulasekarapattinam%E2%80%99s-ideal-launch-pad

505. Kulasekarapattinam in Tamil Nadu is closer to the equator (8° instead of 13 for Sriharikota) and thus capable of carrying an additional 350kg of payload to GEO. This could be additional transponder capacity of additional fuel to prolong operating lifetime. TN Ideal Spot for 3rd Rocket Launch Pad: Scientists, dtNext.in, 18 July 2016. http://www.dtnext.in/News/City/2016/07/18014637/TN-ideal-

spot-for-3rd-rocket-launch-pad-Scientists.vpf

Chapter 10

506. Named after the American-Hungarian engineer Theodore von Karman, who calculated that at 100 km the air is too thin to support conventional flight. It is also thin enough to minimise friction for a vehicle to attain Earth's orbit. Like the Earth's equator, nothing physical marks the Karman line. In fact, the ionosphere extends beyond this boundary. Although tenuous, the atmosphere is present and will cause any satellite in orbit at this altitude to re-enter within days. Retrieved from https://en.wikipedia.org/wiki/K%C3%A1rm%C3%A1n_line

507. Günther Seibert's 2006 paper captures the development of sounding rockets during this early phase of rocketry. Günther Seibert. 2006. The History of Sounding Rockets and Their Contribution to European Space Research. ESA. Retrieved from http://www.esa.int/esapub/hsr/HSR_38.pdf

508. Raj, Gopal. 2003. *Reach for the Stars: The Evolution of India's Rocket Programme*. New Delhi; New York, NY: Penguin Books India. P33.

509. The total number of variants that can exist is 15, although only three are still in regular use today. The single stage RH-75 could lift 1 kg payload to an altitude of 10 km. The last in the series was the two-stage RH-560/300 Mk2, which could deliver 100 kg to an altitude of 500 km. Rao, P. V. Manoranjan and P. Radhakrishnan. 2012. *A Brief History of Rocketry in ISRO*. Hyderabad: Universities Press (India) Private Limited.

510. ISRO. 21 July 2015. 100th Consecutively Successful Launch of Sounding Rocket RH-200 from TERLS. *Spaceref.* Retrieved from http://spaceref.com/news/viewsr.html?pid=47521.

511. ISRO. 2016. *From Fishing Hamlet to Red Planet: India's Space Journey*. Noida, Uttar Pradesh: Harper Collins India. P90. A more detailed account of the event here http://www.spansen.com/2017/02/when-indian-rocket-shot-up-from.html

512. Sarabhai, Vikram. 1970. Atomic Energy and Space Research: A Profile for the Decade 1970-80. Mumbai: Atomic Energy Commission, Government of India. P27. Retrieved from http://www.iaea.org/inis/collection/NCLCollectionStore/_Public/02/006/2006423.pdf

513. This example comes from Raj, Gopal. 2003. *Reach for the Stars: The Evolution of India's Rocket Programme*. New Delhi; New York, NY: Penguin Books India. P106.

514. Where solid propellants are used, it is a rocket motor; otherwise, it is a rocket engine. This distinction of terms is almost universally accepted.

515. Just as the efficiency of petrol or diesel is measured in terms of miles/gallon or

km/h, the efficiency of a rocket engine is measured in terms of Isp, the ratio of the thrust produced by the engine to the rate of fuel consumption. It is measured in units of time in seconds. This is the length of time that unit weight of propellant would last if used to produce one unit of thrust continuously.

516. By the end June 2017, this amounts to a total of 60 flights from Sriharikota. In all there have been 64 flights ATV-D01 03/03/201, LVM-3 18/12/2014, RLV 23/05/2016 and the ATV-D02 28/08/2016 are not listed.

517. How ISRO addressed this particular hurdle of developing the inertial guidance system is described in detail in Chapter 7 of Raj, Gopal. 2003. *Reach for the Stars: The Evolution of India's Rocket Programme*. New Delhi; New York, NY: Penguin Books India. P148.

518. Here 'small' is a relative term. Strontium perchlorate is used for directional control for solid-fuel stages; it is a liquid. In the case of the first stage on the PSLV, up to 12 L/s is required on demand. ISRO. 2016. *From Fishing Hamlet to Red Planet: India's Space Journey*. Noida, Uttar Pradesh: Harper Collins India. P142.

519. The flex nozzle used on the 22m long S200 boosters for the LVM3 can move up to + or − 7.8° using electro-hydraulic actuators https://web.archive.org/web/20101125141844/http://isro.org/pressrelease/contents /2010/pdf/S200_STATIC_TEST-01.pdf

520. Another potential reason for the choice of solid propellant was the strong connection with the space programme of Japan that Sarabhai had cultivated through Professor Hideo Itokawa (1912–1999) from the Institute of Space and Aeronautical Science in Japan. Just as SLV-3 was getting underway, Japan succeeded in launching its first satellite (23 kg Osumi) using the Lambda-4S launcher, an all-solid-fuel launcher.

521. At this time, Sarabhai was the Chairman of AEC, where organisationally ISRO sat. The DOS had not yet come into existence. This paper is in two sections: the first is devoted to Atomic Energy and the second to Space Research. Sarabhai, Vikram. 1970. Atomic Energy and Space Research: A Profile for the Decade 1970-80. Mumbai: Atomic Energy Commission, Government of India. http://www.iaea.org/inis/collection/NCLCollectionStore/_Public/02/006/2006423 .pdf

522. By December, Sarabhai negotiated a deal whereby ISRO would build its own satellite that USSR would launch. He had delegated the responsibility of designing and building the satellite to Professor U. R. Rao, who completed this task with the launch of India's first satellite, Aryabhata, by the USSR in 1975. At 360 kg, it was 10 times the mass that the SLV-3 was capable of.

523. The SLV-3 Project Manager Abdul Kalam records this as 0.65 m diameter on page 113 of ISRO. 2016. *From Fishing Hamlet to Red Planet: India's Space Journey*. Noida, Uttar Pradesh: Harper Collins India. P113.

524. One of the most detailed descriptions of the development of ISRO's launchers can be found in Raj, Gopal. 2003. *Reach for the Stars: The Evolution of India's*

Rocket Programme. New Delhi; New York, NY: Penguin Books India. P54.

525. Dick, Steven J. ed. 2013. Remembering the Space Age: Proceedings of the 50th Anniversary Conference. Washington, DC: NASA. P45.

526. The terminology used by rocket scientists globally is under constant evolution. A rocket motor is synonymous with a rocket engine. Initially, it was a term used when referring to solid rocket motors, but today, it has a more ambiguous use.

527. During these early days of liquid propellants, ISRO acquired the red fuming nitric acid from the IAF. The quality was so poor that one ISRO scientist recalled that it was "neither red nor fuming". Raj, Gopal. 2003. *Reach for the Stars: The Evolution of India's Rocket Programme*. New Delhi; New York, NY: Penguin Books India. P107.

528. ISRO. 2016. *From Fishing Hamlet to Red Planet: India's Space Journey*. Noida, Uttar Pradesh: Harper Collins India. P118.

529. *Ibid*. P120.

530. Zykofsky, Paul. February 1982. Program Contributes to Nation-Building. *Executive Intelligence Review*. Volume 9(5). P38. Retrieved from http://www.larouchepub.com/eiw/public/1982/eirv09n05-19820209/index.html

531. Menon, A. K. 15 May 1983. SLV-3: A Space Odyssey. *India Today*. Retrieved from http://indiatoday.intoday.in/story/india-successfully-launches-slv-3-rocket-with-rohini-satellite-payload/1/371567.html.

532. Following this success, Kalam became the centre of media attention. This generated envy among his senior colleagues, who too had contributed to the success of SLV-3 but had not received appropriate recognition. Kalam talks about this bitterness in his autobiography. Kalam, A. P. J. Abdul and Arun Tiwari. 1999. *Wings of Fire: An Autobiography*. Hyderabad: Universities Press. P55.

533. The original First Launch Pad used by SLV-3 and ASLV was decommissioned in 1994.

534. December 1985, Countdown number 68. The House Journal of Vikram Sarabhai Space Centre. A video of the launch is available here: https://www.youtube.com/watch?v=-BLNhTU-F-o also see https://www.reddit.com/r/ISRO/wiki/resources/suborbital_flights/so-300-200

535. The booster nozzles were canted by 9° to the vertical to prevent damage from the instability that may be caused by the expansion of the exhaust gas at the core stage.

536. This technique is known as the Secondary Injection Thrust Vector Control for yaw and pitch control using strontium perchlorate and a separate RCS for roll.

537. Rao, U. R. 2013. *India's Rise as a Space Power*. Delhi: Cambridge University Press India Pvt Ltd. P114.

538. Although HEF-20 has its origins in the US, during the late 1960s, it was not easily available. ISRO tasked the Propellant Engineering Division to indigenise the

production of HEF-20 and other propellants with the aim of becoming self-reliant. Kurup, M. R., V. N. Krishnamoorthy and M. C. Uttam. 1 March 1988. Development of Solid Propellant Technology in India. *Sadhana* 12(3): 229–34. P230. doi:10.1007/BF02812029.

539. The numbers represented here come from a variety of published sources including A Brief History of Rocketry in ISRO and ISRO's website. There are discrepancies typically around 10%, especially with SLV-3).

540. Some accounts record that the interval between the first-stage shutting down and the second-stage starting was 0.5 second. The figure of 1.5 seconds I have used comes from Raj, Gopal. 2003. *Reach for the Stars: The Evolution of India's Rocket Programme*. New Delhi; New York, NY: Penguin Books India. P135.

541. The following source indicates a velocity of 6 km/s at the time of stage 1 burnout, which is too high to be accurate. ISRO. 2016. *From Fishing Hamlet to Red Planet: India's Space Journey*. Noida, Uttar Pradesh: Harper Collins India. P126.

542. This was a collection of spacecraft with gamma-ray detectors. Gamma rays are an extremely powerful form of electromagnetic radiation emanating from the most powerful events (black holes and supernovae) in the universe. The Earth's magnetic fields prevent gamma rays from reaching the Earth, so they are best investigated from space. Retrieved from http://www.ssl.berkeley.edu/ipn3/

543. The Inertial Measurement Unit used in the SLV-3 Open Loop Guidance System was provided by SAGEM of France. "ISRO had approached US companies, but they refused to provide even catalogues about their gyros". Raj, Gopal. 2003. *Reach for the Stars: The Evolution of India's Rocket Programme*. New Delhi; New York, NY: Penguin Books India. P147.

544. Gupta, S. C. 1995. Growth of Capabilities of India's Launch Vehicles. *Current Science* 68 (7): 687–91. P690.

545. Sarabhai, Vikram. 1970. Atomic Energy and Space Research: A Profile for the Decade 1970-80. Mumbai: Atomic Energy Commission, Government of India. P28. Retrieved from http://www.iaea.org/inis/collection/NCLCollectionStore/_Public/02/006/2006423.pdf

546. Sarabhai clearly states, 'synchronous orbit' rather than polar orbit, but that capability came later with the GSLV. By the late 1980s, remote sensing satellites, such as Aryabhata and SROSS, were considered to provide data of higher domestic value. A remote-sensing satellite, typically about 1,500 kg, requires a lower orbit of around 900 km. Communication satellites are heavier, around 2,500 kg, and require a higher orbit of 36,000 km. The Indian remote satellite (IRS) programme was initiated in the early 1980s. By then, ISRO could build and launch them without foreign partners. Remote-sensing satellites required a polar orbit, and the launcher became the Polar Satellite Launch Vehicle (PSLV). The GSLV would become the next step in the evolution of ISRO's launchers. ISRO developed the expertise to build communication satellites around the same time as IRS but chose to incorporate more functionality, television broadcast, weather forecasting, disaster

warning and search and rescue, and not just communication. This increased their weight beyond the PSLV limit. ISRO turned to the US and France to put them in orbit. Until GSLV-Mk3 becomes operational, ISRO will continue to engage ESA's Ariane 5 to launch satellites heavier than 2,000 kg.

547. National Photographic Interpretation Centre, National Security Information. 24 February 1984. https://www.cia.gov/library/readingroom/docs/CIA-RDP84T00491R000100410001-1.pdf. A reference to PSLV in 1981 in a New scientist report with an early configuration. https://books.google.co.in/books?id=IbbMj56ht8sC&lpg=PP1&pg=PA215#v=one page&q&f=false

548. Baskaran, Angathevar. 2011. Competence Building in Complex Systems: The Case of Satellite Launch Vehicles in India's Space Programme. The 9th Globelics International Conference: Creativity, Innovation and Economic Development.

549. For 16 months, about 40 Indian engineers were trained in Russia in cryogenic engine technology. Interview with Professor U. R. Rao. 2013. See http://astrotalkuk.org/2013/12/05/isro-the-early-years/

550. Raj, Gopal. 2003. *Reach for the Stars: The Evolution of India's Rocket Programme*. New Delhi; New York, NY: Penguin Books India. P120.

551. Gopal Raj offers a persuasive case about why it was in France's interest to engage in this arrangement with India. Raj, Gopal. 2003. *Reach for the Stars: The Evolution of India's Rocket Programme*. New Delhi; New York, NY: Penguin Books India. P123.

552. Baskaran, Angathevar. 2011. Competence Building in Complex Systems: The Case of Satellite Launch Vehicles in India's Space Programme. The 9th Globelics International Conference: Creativity, Innovation and Economic Development.

553. N. Narayanamoorthy, who joined VSSC in 1971 and had been associated with the development of PSLV right from its formative years, states that the actual deliverables resulting from this agreement were as follows: (1) ISRO would supply 10,000 space-qualified pressure transducers to CNES, and (2) in return, France would transfer the know-how for their liquid engine, Viking, to ISRO. ISRO. 2016. *From Fishing Hamlet to Red Planet: India's Space Journey*. Noida, Uttar Pradesh: Harper Collins India. P131. Another account states that it was not 10,000 but 7,000. Raj, Gopal. 2003. *Reach for the Stars: The Evolution of India's Rocket Programme. New Delhi; New York, NY: Penguin Books India. P123*. The following source states that it was 7,000, not 10,000 transducers. Nagappa, Rajaram. 2016. Development of Space Launch Vehicles in India. *Astropolitics* 14(2-3), 158-176. P165. DOI: 10.1080/14777622.2016.1244877

554. Rao, P. V. Manoranjan and P. Radhakrishnan. 2012. *A Brief History of Rocketry in ISRO*. Hyderabad: Universities Press (India) Private Limited.

555. ISRO. 2016. *From Fishing Hamlet to Red Planet: India's Space Journey*. Noida, Uttar Pradesh: Harper Collins India. P134.

556. Rao, P. V. Manoranjan and P. Radhakrishnan. 2012. *A Brief History of

Rocketry in ISRO. Hyderabad: Universities Press (India) Private Limited. P136.

557. The US secret service recorded one of the earliest references to the PSLV (they referred to as the Polar Space Launch Vehicle) in a now declassified as SECRET a report dated 24 February 1984. The evidence came from satellite imagery of what looked like a static display of PSLV mock-up components. National Photographic Interpretation Centre, National Security Information. 24 February 1984.

558. A derivative from the more commonly used UDMH and N2O2. Monomethylhydrazine fuel and mixed oxides of nitrogen.

559. For roll control, the PSLV first stage used an RCS. A redesigned and enhanced version of the RCS was developed as the PSLV fourth stage.

560. ISRO. 2016. *From Fishing Hamlet to Red Planet: India's Space Journey*. Noida, Uttar Pradesh: Harper Collins India. P145.

561. *Ibid*. P153.

562. Rao, U. R. 2013. *India's Rise as a Space Power*. Delhi: Cambridge University Press India Pvt Ltd. P177.

563. Raj, Gopal. 2003. *Reach for the Stars: The Evolution of India's Rocket Programme*. New Delhi; New York, NY: Penguin Books India. P237.

564. Interview with SDSC-SHAR Director Dr. M. Y. S. Prasad. 12 January 2015. See Episode 72 on www.astrotalkuk.org

565. Raj, Gopal N. 23 May 2016. After Successful Test of Reusable Vehicle, ISRO Has Further Plans for Slashing Launch Costs. *The Wire*. Retrieved from http://thewire.in/2016/05/23/not-just-the-rlv-td-isro-has-more-plans-for-slashing-launch-costs-37943/

566. In 2011, NASA stopped operating the Space Shuttle, and since then, the only way astronauts and cosmonauts can get to the ISS is via the Russian Soyuz launcher. Soyuz is an enhanced version of the Vostok launcher used by Yuri Gagarin. The launch site used for modern launches is the same one that Gagarin used.

567. ISRO does not have a formal process or platform for announcing its strategic choices. Much of the information comes from lectures and presentations delivered by senior ISRO personnel. Like this one available on YouTube by Dr V Narayanan from the LPSC spoking about "Cryogenic Propulsion Systems for ISRO Launch Vehicles" in February 2017 https://www.youtube.com/watch?v=7Jy-fXBTlIE

568. http://www.kmu.gov.ua/control/uk/publish/article?art_id=246186471. This short report is in Russian. Google translate does an adequate job.

569. ISRO to Bank on Semi-Cryogenic Engine for Heavy Lift Rockets - The Hindu Retrieved from http://www.thehindu.com/sci-tech/science/isro-to-bank-on-semi-cryogenic-engine-for-heavy-lift-rockets/article19095367.ece. For more technical details of ISRO's Semi Cryogenic engine design see https://www.ijirset.com/upload/2013/special/energy/16_DESIGN.pdf

570. S, Madhumathi D. 'Russian Tie-up to Boost ISRO's Semi Cryogenic Launcher Plan'. The Hindu. 14 August 2015, sec. Science. http://www.thehindu.com/sci-tech/science/russian-tieup-to-boost-isros-semicryogenic-launcher-plan/article7536263.ece.

571. This table, along with some of the other technical specifications in this chapter, comes from Norbert Brugge. Based in Germany, he hosts a website providing details on the development of space launch vehicles in countries around the world. http://www.b14643.de/Spacerockets_1/index.htm

572. These figures come from the CEO of SpaceX Elon Musk. SpaceX has successfully demonstrated the safe return of the first stage of its Falcon rocket to land on Earth and a floating platform at sea. AFP. 9 April 2016. SpaceX Lands Rocket on Ocean Platform for the First Time. *The Express Tribune.* Retrieved from http://tribune.com.pk/story/1081654/spacex-lands-rocket-on-ocean-platform-for-first-time/

573. This table, along with some of the other technical specifications in this chapter, comes from Norbert Brugge. Based in Germany, he hosts a website providing details on the development of space launch vehicles in countries around the world. See http://www.b14643.de/Spacerockets_1/index.htm

574. RESPOND is ISRO's research programme under which it uses part of its funding to sponsor external research. VSSC. June 2013. Titles for Respond Project/Research Area Document. P7. Retrieved from http://www.vssc.gov.in/VSSC_V4/images/pdf_files/respond/Research_Area.pdf

575. DOS. ISRO Annual Report 2014–2015. P75. Retrieved from http://www.isro.gov.in/sites/default/files/pdf/AR2014-15.pdf

576. ISRO maintained a low profile during the lead-up to the RLV-TD flight. Unlike other launches, it was not broadcast live. Nothing appeared on ISRO's Facebook, Twitter or website until after the flight. A video of the launch looking down at the booster was published on ISRO's Facebook page: https://www.facebook.com/ISRO/videos/1733895800167108/

577. ISRO formally acknowledged this support via its only press release issued after the mission had been completed. ISRO. 23 May 2016. India's Reusable Launch-Vehicle Technology Demonstrator (RLV-TD), Successfully Flight Tested. Retrieved from http://www.isro.gov.in/update/23-may-2016/india%E2%80%99s-reusable-launch-vehicle-technology-demonstrator-rlv-td-successfully

578. Ramachandran, R. and T. S. Subramanian. 24 June 2016. "Design Process Has Been Validated". *Frontline.* Retrieved from http://www.frontline.in/science-and-technology/design-process-has-been-validated/article8704727.ece

579. VSSC Director K. Sivan stated that "As per data the RLV-TD landed softly in the Bay of Bengal. As per our calculations, it would have disintegrated at the speed at which it touched the sea". ISRO to Test Next Reusable Launch Vehicle After Studying Data of May 23 Flight. 23 May 2016. Retrieved from http://zeenews.india.com/news/space/isro-to-test-next-reusable-launch-vehicle-

after-studying-data-of-may-23-flight_1888916.html. See also http://www.msn.com/en-us/news/technology/indias-budget-mini-space-shuttle-blasts-off/ar-BBtm0dn?ocid=ansmsnnews11. Reports on what survived of the RLV-TD were inconsistent. Within a month of the RLV-TD launch, the ISRO Chairman stated that the vehicle "cannot survive". Ramachandran, R. and T. S. Subramanian. 24 June 2016. "Design Process Has Been Validated". *Frontline*. Retrieved from http://www.frontline.in/science-and-technology/design-process-has-been-validated/article8704727.ece

580. A coast guard helicopter, which was employed by ISRO, located the RLV-TD floating in the sea around 20 minutes after the splashdown. The coast guard reportedly took pictures, and sources said the vehicle didn't suffer major damage as against the popular view that the winged body would disintegrate at the point of touchdown. For full story, see http://www.newindianexpress.com/cities/chennai/ISROs-new-desi-reusable-wings-taste-success-in-maiden-flight/2016/05/24/article3448063.ece

581. This press release was issued on the day of the mission shortly after it was successfully completed. ISRO. 23 May 2016. India's Reusable Launch Vehicle-Technology Demonstrator (RLV-TD), Successfully Flight-Tested. Retrieved from http://www.isro.gov.in/update/23-may-2016/india%E2%80%99s-reusable-launch-vehicle-technology-demonstrator-rlv-td-successfully

582. Express News Service. 30 May 2015. Project Report Ready for Airstrip at Sriharikota. *The New Indian Express*. http://www.newindianexpress.com/cities/thiruvananthapuram/2015/may/30/Project-Report-Ready-for-Airstrip-at-Sriharikota-767516.html

583. The UK-based company Reaction Engines is working on a Single-Stage-To-Orbit technology where a traditional runway is used for take-off, rather than a launch pad. Once payload has been delivered to space, the launch vehicle lands at the same runway in its entirety. Following refurbishment, it can be reused within a few weeks. Parkinson, R. C. 21 February 2011. Space: The Development of Single-Stage Space Flight. *The Global Herald*. Retrieved from http://theglobalherald.com/space-the-development-of-single-stage-space-flight/11656/

584. ISRO published the following press release once the flight was completed. ISRO. 28 August 2016. Successful Flight Testing of ISRO's Scramjet Engine Technology Demonstrator. Retrieved from http://www.isro.gov.in/update/28-aug-2016/successful-flight-testing-of-isros-scramjet-engine-technology-demonstrator. Find more details here: http://www.isro.gov.in/isro%E2%80%99s-scramjet-engine-technology-demonstrator-successfully-flight-tested

585. The Russian military recently announced the development of a strategic bomber using hypersonic engine technology. The high altitude and speed will allow these jets to deliver nuclear weapons anywhere on Earth within two hours of launch from the Russian territory. The announcement states that the nuclear-capable bomber will be developed by 2020. Russian Aviation. 14 July 2016. New Russian

Bomber to Be Able to Launch Nuclear Attacks from Outer Space. *Russian Aviation*. Retrieved from http://www.ruaviation.com/news/2016/7/14/5989/?h.

586. Speaking a few weeks after the success of the RLV-TD, the ISRO Chairman stated, "Today one of the arguments is whether it is really going to happen or not". Ramachandran, R. and T. S. Subramanian. 24 June 2016. "Design Process Has Been Validated". *Frontline*. Retrieved from http://www.frontline.in/science-and-technology/design-process-has-been-validated/article8704727.ece

Chapter 11

587. Maitra, R. and S. Maitra. 20 August 1993. Another Indian Multi-Purpose Satellite. *Executive Intelligence Review* 20 (32):8–9. Retrieved from http://www.larouchepub.com/eiw/public/1993/eirv20n32-19930820/eirv20n32-19930820_008-another_indian_multi_purpose_sat.pdf

588. Space Foundation. 22 June 2016. Space Foundation Report Reveals Global Space Economy at \$323 Billion in 2015. *Space Foundation*. Retrieved from https://www.spacefoundation.org/media/press-releases/space-foundation-report-reveals-global-space-economy-323-billion-2015

589. The Ariane 5 launch capacity to GTO mentioned is that of Ariane 5 G variant and is an approximate figure. Other Ariane 5 variants can deliver larger loads to GTO. In addition to an increased capacity to GTO, Ariane 6 is designed to cut the cost of such launches by about 50%. For Ariane 5 specifications and launch history, see http://www.spacelaunchreport.com/ariane5.html

590. This is a qualitative rather than a quantitative distinction, for there are many variables. For example, Vehicle: Total: Payload (weight in tonnes) ratios, Saturn 5, 3038:118:0.03; Ariane 5, 746:17:0.3; PSLV XL, 300:3.2:0.01.

591. Rao, U. R. 2013. *India's Rise as a Space Power*. Delhi: Cambridge University Press India Pvt Ltd. P178.

592. Rao, P. V. Manoranjan and P. Radhakrishnan. 2012. *A Brief History of Rocketry in ISRO*. Hyderabad: Universities Press (India) Private Limited. P172.

593. A member of that team recalls that the LOX containers were of so poor quality that only 75% of the LOX would escape from the containers over a period of 24 hours. Raj, Gopal. 2003. *Reach for the Stars: The Evolution of India's Rocket Programme*. New Delhi; New York, NY: Penguin Books India. P236.

594. The preliminary design for a 75-kN LOX-LH2 engine was completed. The shower head injector elements of a 30-kN LOX-kerosene engine was also tested at the then National Aeronautical Laboratory. ISRO. 2016. *From Fishing Hamlet to Red Planet: India's Space Journey*. Noida, Uttar Pradesh: Harper Collins India. P156.

595. Raj, Gopal. 2003. *Reach for the Stars: The Evolution of India's Rocket*

Programme. New Delhi; New York, NY: Penguin Books India.

596. In Chapter 9 of his book, Gopal Raj provides one of the most detailed, independently researched analysis of ISRO's pursuit of cryogenic engine technology. Raj, Gopal. 2003. *Reach for the Stars: The Evolution of India's Rocket Programme*. New Delhi; New York, NY: Penguin Books India.

597. There was a dramatic decline in the number of launches in the final decade of the USSR. From a peak of 102 launches in 1982, it dropped to 23 in 1996. Harvey, Brian. 2007. *The Rebirth of the Russian Space Programme: 50 Years After Sputnik, New Frontiers*. Springer Science & Business Media. P312.

598. Much of the material in this section comes from interviews with Professor U. R. Rao by the author and his 2013 book *India's Rise as a Space Power*. As the Chairman of ISRO at that time, Rao was at the heart of this controversial and politically charged period. Having worked closely with the USSR during the building and launching of ISRO's first satellite, Aryabhata, Rao had an intimate knowledge of and close connections with many of the key people in the space programme of the USSR. Although authoritative, Rao's views must have been informed by his position as a key stakeholder. His book is a key source of important information. Rao, U. R. 2013. *India's Rise as a Space Power*. Delhi: Cambridge University Press India Pvt Ltd. Another well-researched source providing a valuable independent perspective is Raj, Gopal. 2003. *Reach for the Stars: The Evolution of India's Rocket Programme*. New Delhi; New York, NY: Penguin Books India.

599. The name of the USSR/Russian space agency has been changed three times, from the state-owned Glavkosmos (sometimes also spelt as Glavkosmos) founded in 1985 to the Russian Federal Space Agency, a private concern, in 1992 and once more state-owned following the presidential decree of 2015 No. 666 as Roscosmos State Corporation for Space Activities (normally abbreviated to Roscosmos).

600. Rao, U. R. 2013. *India's Rise as a Space Power*. Delhi: Cambridge University Press India Pvt Ltd. P180.

601. ISRO. 2016. *From Fishing Hamlet to Red Planet: India's Space Journey*. Noida, Uttar Pradesh: Harper Collins India. P157.

602. Rao, U. R. 2013. *India's Rise as a Space Power*. Delhi: Cambridge University Press India Pvt Ltd. P186 and 1987.

603. The following reference states it as four years. During the entire test programme in Russia, which extended over a period of four years, mostly in severe winter, the performance of ISRO electronics was highly satisfactory. ISRO. 2016. *From Fishing Hamlet to Red Planet: India's Space Journey*. Noida, Uttar Pradesh: Harper Collins India. P157.

604. Raj, Gopal. 2003. *Reach for the Stars: The Evolution of India's Rocket Programme*. New Delhi; New York, NY: Penguin Books India. P249.

605. *Ibid*. P256.

606. For example, Delta IV Heavy has a capacity to deliver over 14 tonnes to GTO.

Retrieved from
https://en.wikipedia.org/wiki/Comparison_of_orbital_launch_systems

607. The very first instance that the US denied rocket technology to India was in 1961. Because of the military connotations of many items of rocket equipment and technology, restrictions had been imposed on the export of many classes of items. Krige, John, Angelina Long Callahan and Ashok Maharaj. 2013. *NASA in the World: Fifty Years of International Collaboration in Space.* Palgrave Macmillan. P225. This recent publication contains extensive information regarding Indo-US collaboration. Also, a declassified report from the US State Department indicates that the US was aware of India's missile capability from an interview with the ISRO Chairman Satish Dhawan in June 1974. The report filed by the US Ambassador to India Patrick Moynihan refers to the transcript of the interview where Dhawan states clearly that India has the capacity to produce the solid fuel necessary for an intermediate range ballistic missile. The full transcript of the interview is available here: https://aad.archives.gov/aad/createpdf?rid=157717&dt=2474&dl=1345

608. Chengappa, Raj. 2000. Weapons of Peace: The Secret Story of India's Quest to Be a Nuclear Power. Harper Collins Publishers, India. P167

609. Kalam, A. P. J. Abdul and Arun Tiwari. 1999. *Wings of Fire: An Autobiography.* Hyderabad: Universities Press. P35.

610. Milhollin, Gary. 25 June 1998. Testimony of Gary Milhollin. *GlobalSecurity.org.* Retrieved from http://www.globalsecurity.org/space/library/congress/1998_h/980625-milhollin.htm

611. Harvey, Brian, Henk H. F. Smid and Theo Pirard. 2011. *Emerging Space Powers: The New Space Programmes of Asia, the Middle East and South-America.* Springer Science & Business Media.

612. Rao, U. R. 2013. *India's Rise as a Space Power.* Delhi: Cambridge University Press India Pvt Ltd. P136.

613. Raj, Gopal. 2003. *Reach for the Stars: The Evolution of India's Rocket Programme.* New Delhi; New York, NY: Penguin Books India. P245.

614. Following five tests in 1998, known as Pokhran-2, further sanctions followed. The UN resolution 1172 encouraged "all states to prevent the export of equipment, materials or technology" to India. This diminished India's ability to grow its space programme.

615. Although no names were mentioned in this exchange, high on Gore's list would have been Abdul Kalam. Kalam, who started his career at Thumba as a Project Manager, developed SLV-3, India's first launcher, and placed a satellite in Earth's orbit in 1980. Two years later, he left ISRO for DRDO and helped develop missiles for the Department of Defence.

616. Rao, U. R. 2013. *India's Rise as a Space Power.* Delhi: Cambridge University Press India Pvt Ltd. P1137.

617. The precursor to ESA was the ELDO established in 1961 consisting of Britain, France, West Germany, Belgium, Italy, the Netherlands and Australia as associate members. Following delays and cost overruns and UK's decision to leave, ELDO was merged with the European Space Research Organisation to create ESA in 1975. The relationship between the US and France deteriorated during the mid-1960s. The US imposed restrictions on technology transfer through a National Security Action Memorandum 294 in April 1964 to ensure that the French military nuclear capability did not benefit from US technology. In 1966, France temporarily became inactive in NATO further isolating it from the US.

618. Lele, Ajey. 2016. Power Dynamics of India's Space Program. *Astropolitics* 14 (2–3): 120–34. P127. doi:10.1080/14777622.2016.1237212.

619. In 1969, the US signed an agreement with Japan for cooperation in space activities. One of the consequences of this agreement was that "Japan got the best of both worlds, maintenance of its own solid-rocket development programme and acquisition of advance US liquid technology that kick-started a new generation of highly sophisticated technologies leading to the H-IIA's advance cryogenic engines". Pekkanen, Saadia and Paul Kallender-Umezu. 2010. *In Defence of Japan: From the Market to the Military in Space Policy.* Stanford, CA: Stanford University Press. P114.

620. Section 539 of the 112th Congress Public Law 55 states (a) None of the funds made available by this Act may be used for the National Aeronautics and Space Administration (NASA) or the Office of Science and Technology Policy (OSTP) to develop, design, plan, promulgate, implement, or execute a bilateral policy, program, order, or contract of any kind to participate, collaborate, or coordinate bilaterally in any way with China or any Chinese-owned company unless such activities are specifically authorized by a law enacted after the date of enactment of this Act. Retrieved from https://www.gpo.gov/fdsys/pkg/PLAW-112publ55/html/PLAW-112publ55.htm

621. India: Govt. Silent over F.B.I. Charges on Two Key Units. Released on 9 September 2013. *WikiLeaks.* Retrieved from https://wikileaks.org/gifiles/docs/32/321952_-os-india-govt-silent-over-f-b-i-charges-on-two-key-units-.html

622. Publicly, the US sanctions were targeting India's military potential. In practice, the target was the space programme of India. Response to Germany's Request for Guidance on Export with MTCR "No Undercut" Policy Implications (C). 16 November 2009. *WikiLeaks.* Retrieved from https://wikileaks.org/plusd/cables/09STATE117929_a.html

623. In this instance, the sanctions were breached but by a US company. Ukraine: Appeal for USG Forbearance on India Space Program Cooperation. 7 September 2007. *WikiLeaks.* Retrieved from https://wikileaks.org/plusd/cables/07KYIV2245_a.html

624. Raj, Yashwant. 25 January 2011. US Ends Export Controls for India, Lifts Ban on ISRO, DRDO. *Hindustan Times.* Retrieved from

http://www.hindustantimes.com/world-news/us-ends-export-controls-for-india-lifts-ban-on-isro-drdo/article1-654333.aspx

625. Space Foundation. 22 June 2016. Space Foundation Report Reveals Global Space Economy at $323 Billion in 2015. *Space Foundation*. Retrieved from https://www.spacefoundation.org/media/press-releases/space-foundation-report-reveals-global-space-economy-323-billion-2015

626. The nine international customers were Austria, Canada, Denmark, France, Germany, Indonesia, Singapore, the UK and US. By 2015, the total number of foreign satellites launched by ISRO was 51. The list of these 51 satellites and their dates of launch are listed here: http://www.isro.gov.in/isro-crosses-50-international-customer-satellite-launch-mark

627. Despite the sanctions, the US was willing to assist India if it benefited commercially. India Rolls Out the Red Carpet for NASA Administrator. 11 May 2006. *WikiLeaks*. Retrieved from https://wikileaks.org/plusd/cables/06NEWDELHI3272_a.html

628. Rao, U. R. 2013. *India's Rise as a Space Power*. Delhi: Cambridge University Press India Pvt Ltd. P197.

629. These dates refer to the first use of cryogenic engines in an actual launch or the engine's first successful test on the ground. The KVD-1 engine, for example, was built by the USSR but never used in space. This was the cryogenic engine that India purchased from the USSR. Brian Harvey refers to the date for the first successful test of the KVD-1, June 1967. Harvey, Brian. 2007. *The Rebirth of the Russian Space Programme: 50 Years After Sputnik, New Frontiers*. Springer Science & Business Media. P310.

630. Kumar, Kiran. 2016. Indigenous Development of Materials for Space Programme. https://youtu.be/T7FasuzmLH0?list=PLHQBBITigBvrfSspQvLP6L_dZ7dIbufBq 22:20 minutes in

631. Kumar, Kiran. 2016. Indigenous Development of Materials for Space Programme. https://youtu.be/T7FasuzmLH0?list=PLHQBBITigBvrfSspQvLP6L_dZ7dIbufBq 48:50 minutes in

632. The DMC-3 constellation offers images with an optical resolution of 1 m. The three satellites at 120° on 651 km Sun-synchronous Low Earth Orbit were placed in orbit by ISRO in July 2015. SSTL. 16 September 2015. SSTL's DMC3 Constellation Demonstrates 1-metre Capability. *Surrey Satellite Technology Limited*. Retrieved from http://www.sstl.co.uk/Press-en/2015-News-Archive/SSTL-s-DMC3-Constellation-demonstrates-1-metre-cap

633. The MTCR is one of the four global export control instruments that are of interest to India. The others are Nuclear Suppliers Group (for nuclear and related items), the Australia Group (for chemical and biological items), and the Wassenaar Arrangement (for conventional arms and dual-use goods, including information

security). Following an unsuccessful attempt in 2015 that was blocked by Italy, India became an MTCR member nation in 2016. MTCR. 27 June 2016. Chairs' Statement on the Accession of the Republic of India to the Missile Technology Control Regime (MTCR). *MTCR.* Retrieved from http://mtcr.info/wordpress/wp-content/uploads/2016/06/160627-India.pdf

Chapter 12

634. Bill Anders was one of the crew of three men in Apollo 8, who left Earth to visit the Moon in December 1968. No human being had ever left Earth prior to this mission. Reynolds, David West. 2013. Apollo: The Epic Journey to the Moon, 1963-1972. MBI Publishing Company.

635. ISRO's full vision statement is "Our vision is to harness space technology for national development while pursuing space science research and planetary exploration". It can be seen on its homepage at www.isro.gov.in

636. The title of this chapter is taken from Arthur C. Clarke's 1971 article. Clarke is credited with conceiving the concept of satellite communication although he was not the first. He happened to be in Sri Lanka (then known as Ceylon) during the first large-scale satellite TV experiment and participated in it. Clarke, Arthur C. 1977. *The View from Serendip.* Random House.

637. In this paper, Sarabhai articulates his ideas for India to develop satellites to serve a variety of functions, including meteorology, communication and broadcast. Two years earlier, he had assigned Professor U. R. Rao with the task of leading the Indian satellite programme. ISRO. 2016. *From Fishing Hamlet to Red Planet: India's Space Journey.* Noida, Uttar Pradesh: HarperCollins India. P211.

638. India had already announced plans for the launch of an Indian-built satellite prior to the Soviet offer. The plan was for a modest 30-kg spherical, almost Sputnik-like, spacecraft launched using its own four-stage launcher within the next three to four years. Vikram Sarabhai said that the satellite would weigh 29.9 kg (66 Ibs), be football size and be launched by a four-stage booster. *Astronautics and Aeronautics, 1970: Chronology of Science, Technology, and* Policy. (NASA SP–4015, 1972). https://history.nasa.gov/AAchronologies/1970.pdf P239.

639. Rao, U. R. 2013. *India's Rise as a Space Power.* Delhi: Cambridge University Press India Pvt Ltd. P27.

640. Rao sought instructions from Sarabhai on how the negotiations with the USSR should proceed, but Sarabhai gave him a free hand. Rao, U. R. 2013. *India's Rise as a Space Power.* Delhi: Cambridge University Press India Pvt Ltd. P28. However, writing elsewhere, Rao suggests that India had to persuade the USSR "We had a four-day meeting at the Academy in which the USSR finally agreed, after considerable discussion, to provide a free launch for our first satellite." ISRO. 2016. *From Fishing Hamlet to Red Planet: India's Space Journey.* Noida, Uttar Pradesh: HarperCollins India. P212.

641. As a result of his death, quite the opposite happened with the project on the semi-cryogenic engine. Upon Sarabhai's death, it ceased. ISRO. 2016. *From Fishing Hamlet to Red Planet: India's Space Journey.* Noida, Uttar Pradesh: HarperCollins India. P156.

642. Perhaps, Rao's memories of the loss of another visionary leader provided a stronger emotional commitment. Rao had started out as an Assistant Professor at the University of Dallas on the very day that Kennedy was assassinated. Kennedy was on his way to the University where Rao and others were waiting to meet him.

643. Joseph, George. 2016. India's Journey towards Excellence in Building Earth Observation Cameras. First edition. Notion Press. P45

644. ISRO. 2016. *From Fishing Hamlet to Red Planet: India's Space Journey.* Noida, Uttar Pradesh: HarperCollins India. P212.

645. Jayaraman, V. 10 June 2014. Living Legends in Indian Science: Udupi Ramachandra Rao. *Current Science* 106 (11): 1581–1591. Retrieved from http://www.currentscience.ac.in/Volumes/106/11/1581.pdf. Explorer 34 and 41 were launched after U. R. Rao had returned to India in 1966. Rao is one of ISRO's most industrious staff members having been in post for half a century. He had had some remarkable experiences working with many US and Russian scientists and engineers throughout his eventful career, which continues to this day. I have had the privilege of interviewing him in his office at the ISRO HQ several times. One such interview is available online here: https://astrotalkuk.org/isro-the-early-years

646. In M. G. K. Menon's words "This (satellite) needs a different type of programme because of electronics support". ISRO. 2016. *From Fishing Hamlet to Red Planet: India's Space Journey.* Noida, Uttar Pradesh: HarperCollins India. P47.

647. Rao, U. R. 2013. *India's Rise as a Space Power.* Delhi: Cambridge University Press India Pvt Ltd. P37. This purchase mechanism was subsequently copied by other teams in ISRO.

648. R. Aravamudan, one of the engineers that Sarabhai handpicked in 1962 for India's space programme, recalls how SDSC-SHAR acquired its first Satellite Telemetry Ground Station that was used to receive signals from Aryabhata. Under Sarabhai's guidance, he bought a fully functioning British Satellite Tracking Station from Australia following the termination of the ELDO programme. ISRO. 2016. *From Fishing Hamlet to Red Planet: India's Space Journey.* Noida, Uttar Pradesh: HarperCollins India. P77.

649. In an interview with Professor U. R. Rao on 16 August 2013, he stated "I knew the prime minister was most likely to pick Aryabhata because Mitra was a little obscure and she would not have selected Jawahar because of the resemblance to her father's name." Jawaharlal Nehru was India's first post-independence Prime Minister.

650. It was agreed that India would issue postal stamps on the day of the launch to commemorate the successful launch of Aryabhata. As with all rocket launches, there was a potential for delay, and therefore, stamps with two launch dates 19 and 20

April were printed. Following the successful launch on the 19 April, stamps bearing the launch date of 20 April were destroyed. ISRO. 2016. *From Fishing Hamlet to Red Planet: India's Space Journey*. Noida, Uttar Pradesh: HarperCollins India. P214.

651. Rao, U. R. September 1978. An Overview of the 'Aryabhata' Project. *Proceedings of the Indian Academy of Sciences* (Engineering Sciences) C 1(2): 117–33. P127. doi:10.1007/BF02843538.

652. In his book *Emerging Space Powers*, Brian Harvey records details of the technical problems related to the TV camera this mission suffered "Although the radiometers worked properly, the television cameras were not turned on for nearly a year because of gas trapped in the camera system. They were used only during passes over India and were switched off otherwise to save power. About ten pictures were received every day. The cameras provided information on snow melting in the Himalayas, river flooding in northern India, desertification in Rajasthan, rainfall off the coast of India and mineral resources in Gujarat". Harvey, Brian, Henk H. F. Smid and Theo Pirard. 2011. *Emerging Space Powers: The New Space Programmes of Asia, the Middle East and South-America*. Springer Science & Business Media.

653. For example, the following paper published in 1980 was still reporting erroneously that Aryabhata had ended after the first five days. Velupillai, David. 28 June 1980. ISRO: India's Ambitious Space Agency. *FLIGHT International*. P1466.

654. Perhaps, it was in response to the misreporting that Rao and Kasturirangan penned the following paper with the title carrying their central message. Rao, U. R. and K. Kasturirangan. 1 December 1975. Aryabhata, Eight Months of Life. *Bulletin of the Astronomical Society of India* Volume 3: 75.

655. Rao, U. R. 2013. *India's Rise as a Space Power*. Delhi: Cambridge University Press India Pvt Ltd. P39.

656. Soviet Launch of Aryabhata. 30 April 1975. *WikiLeaks*. Retrieved from https://wikileaks.org/plusd/cables/1975NEWDE05789_b.html

657. Harvey, Brian, Henk H. F. Smid, and Theo Pirard. 2011. *Emerging Space Powers: The New Space Programs of Asia, the Middle East and South-America*. Springer Science and Business Media. P149.

658. The United Nations defines remote sensing of the Earth from space as "the sensing of the Earth's surface from space by making use of the properties of electromagnetic waves emitted, reflected or diffracted by the sensed objects, for the purpose of improving natural resources management, land use and the protection of the environment". Retrieved from http://www.un.org/documents/ga/res/41/a41r065.htm

659. In the 1970s, the recently established National Remote Sensing Agency used several aircraft, including Canberra, DC3 and a HS748, to conduct aerial remote sensing from aircraft. ISRO. 2016. *From Fishing Hamlet to Red Planet: India's Space Journey*. Noida, Uttar Pradesh: HarperCollins India. P309.

660. This paper has been published multiple times, first in 1971. NASA saw ATS-6 as a realisation of Clarke's vision, originally published in 1945. It was also added as

the foreword to the ATS-6 final engineering report published by NASA. Wales, Robert O. (Ed.) November 1981. ATS-6 Final Engineering Performance Report. NASA Reference Publication 1080. Pxviii. Retrieved from http://ntrs.nasa.gov/archive/nasa/casi.ntrs.nasa.gov/19820008274.pdf

661. P. R. Pisharoty recalls a meeting with Prime Minister Indira Gandhi organised by Vikram Sarabhai on the topic of remote sensing. The meeting was at her private residence, late in the evening with a small group. One of those in attendance was the Cabinet Secretary M. S. Swaminathan. At the end of the presentation, Swaminathan asked Pisharoty if remote sensing could assist with early detection of coconut wilt disease in Kerala. It was this meeting that initiated the use of remote-sensing technology in the detection of coconut wilt disease. ISRO. 2016. *From Fishing Hamlet to Red Planet: India's Space Journey*. Noida, Uttar Pradesh: HarperCollins India. P320. A copy of the letter from Vikram Sarabhai dated 22 February 1969 inviting the Prime Minister to this meeting is included on page 123 in Joseph, George. 2016. *India's Journey towards Excellence in Building Earth Observation Cameras*. First edition. Chennai, Tamil Nadu: Notion Press.

662. Krige, John, Angelina Long Callahan and Ashok Maharaj. 2013. *NASA in the World: Fifty Years of International Collaboration in Space*. NY: Palgrave Macmillan. P224.

663. Since its first EO satellite IRS-1A launched in 1988, India has developed expertise and associated ground infrastructure. In early 2015, ISRO had 11 operational EO satellites in orbit, Resourcesat-1 and 2, Cartosat-1, 2, 2A, 2B, RISAT-1 and 2, OCEANSAT-2, Megha-Tropiques and SARAL.

664. The UNGA resolution 2600 (XXIV) of the International Co-operation in the Peaceful Uses of Outer Space, 16 December 1969, contains the following "Invites Member States with experience in the field of remote earth resources surveying to make such experience available to other Member States which do not have such experience... taking particular interest of the needs of developing countries". A few months earlier, the US President had told the UNGA that America would proceed with its Earth Resources programme so as to share the benefits of its work in this field with other nations "as this program proceeds and fulfils its promise." Chapter 11 in the following book provides one of the clearest and detailed accounts of the collaboration NASA provided to the early Indian space programme. Krige, John, Angelina Long Callahan and Ashok Maharaj. 2013. *NASA in the World: Fifty Years of International Collaboration in Space*. NY: Palgrave Macmillan.

665. The historic Apollo 11 mission was launched on 16 July and returned on 21 July 1969. Apollo 12 was launched on 14 November, and it returned on 24 November 1969.

666. Although the term non-aligned was coined prior to the 1955 Bandung conference, at the end of this conference, India emerged as the leader of the group of countries that considered themselves to be non-aligned. Shimazu, Naoko. 2011. 'Diplomacy as Theatre': Recasting the Bandung Conference of 1955 as Cultural History. Singapore: Asia Research Institute, National University of Singapore. P5.

667. Joseph, George. 2016. India's Journey towards Excellence in Building Earth Observation Cameras. First edition. Notion Press. P46

668. *Ibid.* P49.

669. *Ibid.*

670. George Joseph describes the moment leading up to the switching on of the camera on Bhaskara-1 almost a year after the launch. Among those presents were Mr Goel, Head of the Controls Division, and the ISRO Chairman Satish Dhawan. The following conversation resulted in a bet.

Mr Goel was sitting next to Professor Dhawan and having a conversation.

Mr Goel: Sir, I am sure TV camera will work.

Prof. Dhawan: Are you very confident?

Mr Goel: Yes Sir.

Prof. Dhawan: Let us have a bet.

Mr Goel: How much?

Prof. Dhawan: All right, if the camera works I will give you half of my one-year salary I get from the Department of Space. At the time, Dhawan was receiving a token salary of 1 rupee as explained in Joseph, George. 2016. India's Journey towards Excellence in Building Earth Observation Cameras. First edition. Chennai, Tamil Nadu: Notion Press. P56.

671. Kramer, Herbert J. 2002. Observation of the Earth and Its Environment: Survey of Missions and Sensors. Springer Science & Business Media. P348.

672. For those interested, the specifications were: Bhaskara-1: TV camera (two channels visible and infrared) 0.54-0.66 micron, 0.75-0.85 micron, SAMIR satellite microwave radiometer 19 + 23 GHz; Bhaskara-2: TV cameras (two channels) 0.54-0.66 micron, 0.75-0.85 micron, SAMIR microwave radiometers (three channels) 19.24, 22.235 and 31.4 GHz

673. Joseph, George and Praful Bhavsar. 1984. Activities at Indian Space Research Organisation (ISRO) on Development of Space-Borne Remote Sensing Sensors. *ISPRS Archives* Volume XXV Part A1: 151-161. Rio de Janeiro, Brazil: ISPRS. P154. Retrieved from http://www.isprs.org/proceedings/XXV/congress/part1/151_XXV-part1.pdf.

674. Professor S. Kalyana Raman, the Deputy Project Director of IRS-1A, recalls "Major items were satellite components including CCDs (charge-coupled devices) required for payload sensor, payload optics, Travelling Wave Tube Amplifier (TWTA), solar cells, Ni-Cd batteries on the satellite side and high-density digital tape recorders for payload data reception station at Hyderabad". ISRO. 2016. *From Fishing Hamlet to Red Planet: India's Space Journey.* Noida, Uttar Pradesh: HarperCollins India. P249.

675. In the state of Odisha, only 44 deaths were reported by 18 October 2013. The cyclone arrived on the 4 October. e

http://www.thehindu.com/news/national/other-states/odishas-death-toll-after-cyclone-floods-climbs-to-44/article5247992.ece

676. Hegde, V. S. 2014. *Indian Earth Observations Satellites and Applications: Reaping Social Benefits.* Bangalore, Karnataka: Antrix Corporation Limited. P7. Retrieved from http://fgks.in/images/pdf/conf/Hegde-ISRO_Case_Study.pdf

677. In a similar event in 1999, a cyclone killed more than 10,000 people in Orissa. By 2013, the space-based assets and the experience of using them had matured so much that the impact could be dramatically reduced. Francis, S., Prasad V. S. K. Gunturi and Munish Arora 2001. Performance of Built Environment in the October 1999 Orissa Super Cyclone. In Conference on Disaster Management. Pilani, Rajasthan: Birla Institute of Technology and Sciences.

678. Hegde, V. S. 2014. *Indian Earth Observations Satellites and Applications: Reaping Social Benefits.* Bangalore, Karnataka: Antrix Corporation Limited. P8. Retrieved from http://fgks.in/images/pdf/conf/Hegde-ISRO_Case_Study.pdf

679. *Ibid.* P9.

680. Press Information Bureau, Government of India and Dr Jitendra Singh in written reply to a question in the Lok Sabha. 16 March 2016. Forecasting of Natural Disasters. Retrieved from http://pib.gov.in/newsite/PrintRelease.aspx?relid=138072

681. Govt Unveils Satellite Surveillance to Curb Illegal Mining. 15 October 2016. *The Hindu.* Retrieved from http://m.thehindu.com/news/national/piyush-goyal-unveils-satellite-surveillance-to-curb-illegal-mining/article9224543.ece

682. GAGAN to save lives at unmanned railway crossings - http://icast.org.in/news/2015/feb15/feb27BSa.pdf

683. Interestingly, the PFZs are communicated to fishermen in a variety of ways. They are prepared in a number of languages (English, Hindi, Gujarati, Marathi, Kannada, Malayalam, Tamil, Telugu, Oriya and Bengali) and (i) displayed through electronic display boards installed at major fishing harbours, (ii) broadcast through electronic and print media and (iii) made available online and via SMS. The PFZs have increased the chances of an improved catch from around 30% to 90% generating around $700 million. ISRO. 2016. *From Fishing Hamlet to Red Planet: India's Space Journey.* Noida, Uttar Pradesh: HarperCollins India. P345.

684. Rao, U. R. 2013. *India's Rise as a Space Power.* Delhi: Cambridge University Press India Pvt Ltd. P60

685. Latitude in degrees for some of the most common launch sites: Kourou 5.2, SDSC-SHAR 13.7, Kennedy Space Centre 28.5, Baikonur 45.9 and Kapustin Yar 48.4.

686. In telling the INSAT-2 story, P. Ramachandran explains that launching from near the equator can add two years of additional operational lifetime. "INSAT-2 satellites should have a design life of at least 7 years, the actual operating life being determined by choice of launcher. Launches from Cape Kennedy on STS or Delta

would give a life of over 7 years whereas a launch from Kourou on Ariane or from SDSC-SHAR on GSLV-Mk I would give a life of over 9 years." ISRO. 2016. *From Fishing Hamlet to Red Planet: India's Space Journey*. Noida, Uttar Pradesh: HarperCollins India. P236.

687. The launch site was established as a French facility in 1968, a function that it still fulfils, but it is now the premier launch site for ESA. Guiana is a former French colony located on the north-east coast of South America just north of Brazil. It is also known as Guiana Space Centre. It has its headquarters in the nearby town of Kourou and is sometimes referred to as the Kourou Space Centre. Following a Russian/ESA agreement, an additional launch facility has been developed for use by Russia to launch the Soyuz launcher, the first one of which was launched from there in 2011.

688. Ariane Payload Gets the Shakes. 20 January 1979. *FLIGHT International*. Retrieved from https://www.flightglobal.com/pdfarchive/view/1979/1979%20-%200215.html

689. This is how Dr R. M. Vasagam, the APPLE Project Director, recollected this iconic bullock-cart incident. ISRO. 2016. *From Fishing Hamlet to Red Planet: India's Space Journey*. Noida, Uttar Pradesh: HarperCollins India. P220.

690. Velupillai, David. 23 June 1980. ISRO: India's Ambitious Space Agency. *FLIGHT International*. P1469. Retrieved from https://www.flightglobal.com/pdfarchive/view/1980/1980%20-%201537.html.
The momentum wheel produced at VSSC fulfilled the role as a back-up. However, after the first year of APPLE operation, it was assigned as primary to assess its performance and data collection.

691. Professor Rao describes this as a passive thermal control system but goes on to add "along with the heaters to maintain the temperature of the satellite within the required range". Rao, U. R. 2013. *India's Rise as a Space Power*. First edition. Delhi: Cambridge University Press India Private Limited. P67. It seems more a hybrid rather than a passive thermal control system. However, the following reference indicates that the thermal covering was provided by the USSR. Nikitin, S. A. 1985. The Space Flight of the Soviet-Indian Crew. NASA TM-77615. P4. http://ntrs.nasa.gov/archive/nasa/casi.ntrs.nasa.gov/19850012916.pdf

692. Raj, Gopal. 2003. *Reach for the Stars: The Evolution of India's Rocket Programme*. New Delhi; New York, NY: Penguin Books India. P119.

693. In Chapter 9, Professor U. R. Rao provides a detailed record of the events surrounding the design, building, launching and operationalising of APPLE. Rao, U. R. 2013. *India's Rise as a Space Power*. First edition. Delhi: Cambridge University Press India Private Limited. P65.

694. Metosat-2 was one of ESA's earliest weather satellites operating from GEO. CAT-3 was one of a series of spacecraft designed to help ESA gather characteristics about the atmosphere that the launch vehicle Ariane had to support. It contained sensors for acceleration, noise, pollution and vibrations. Metosat-2, like APPLE, got to GEO with their internal rockets, but CAT-3 was designed to remain in GTO. It

provided telemetry data using onboard batteries, which limited the mission duration to 60 days.

695. Harvey, Brian, Henk H. F. Smid and Theo Pirard. 2011. *Emerging Space Powers: The New Space Programs of Asia, the Middle East and South-America.* Springer Science & Business Media.

696. Specifically, APPLE was put through the following "A pitch rotation manoeuvre was designed to rotate the spacecraft like rotisserie manoeuvre for 4 hours every day in the night time and to re-acquire Earth lock after 4 hours and continue normal operations for the rest of the day". ISRO. 2016. *From Fishing Hamlet to Red Planet: India's Space Journey.* Noida, Uttar Pradesh: HarperCollins India. P219.

697. Rao, U. R. 2013. *India's Rise as a Space Power.* First edition. Delhi: Cambridge University Press India Private Limited, 2014. P70.

698. *Ibid.* P64.

699. A communication satellite has a unique orbit located 36,000 km, directly above the equator, a Geostationary Orbit (GEO). Any satellite at this distance takes 24 hours to orbit the Earth and remains in the same place in the sky at all times. An ideal launch site on the Equator provides a free 1.5 km acceleration of the required 4 km/s (GEO) or 7 km/s velocity (for LEO) required for orbit.

700. Sarabhai, Vikram. 1970. Atomic Energy and Space Research. A Profile for the Decade 1970-80. Mumbai: Atomic Energy Commission, Government of India. P29. Retrieved from http://www.iaea.org/inis/collection/NCLCollectionStore/_Public/02/006/2006423.pdf

701. Agrawal, Binod C. and Arbind K. Sinha (Eds). *SITE to INSAT: Challenges of Production and Research for Women and Children.* New Delhi: Concept Publishing Co. P13.

702. Raman, Srinivasan. 1997. No Free Launch: Designing the Indian National Satellite. In Butrica, Andrew J. (Ed.) *Beyond the Ionosphere: Fifty Years of Satellite Communication.* NASA. P12. Retrieved from http://history.nasa.gov/SP-4217/ch16.htm. Pramod Kale has interesting details in his chapter entitled Origins of INSAT-1 in ISRO. 2016. *From Fishing Hamlet to Red Planet: India's Space Journey.* Noida, Uttar Pradesh, India: HarperCollins India. P224

703. Rao, U. R. 2013. *India's Rise as a Space Power.* Delhi: Cambridge University Press India Pvt Ltd. P92.

704. Writing for NASA in 1997, S. Raman captures the complex consideration that must have dominated the meetings during the design phase. Raman, Srinivasan. 1997. No Free Launch: Designing the Indian National Satellite. In Butrica, Andrew J. (Ed.) *Beyond The Ionosphere: Fifty Years of Satellite Communication.* NASA. Retrieved from http://history.nasa.gov/SP-4217/ch16.htm.

705. This case was championed by P. R. Pisharoty. ISRO. 2016. *From Fishing*

Hamlet to Red Planet: India's Space Journey. Noida, Uttar Pradesh, India: Harper Collins India. P228.

706. This payload description comes from the INSAT-1A Project Director Pramod Kale, who had served as the Project Manager for SITE. ISRO. 2016. *From Fishing Hamlet to Red Planet: India's Space Journey*. Noida, Uttar Pradesh: HarperCollins India. P229.

707. Rao, U. R. 2013. *India's Rise as a Space Power*. Delhi: Cambridge University Press India Pvt Ltd. P95.

708. *Ibid*, P98.

709. This source indicates that ISRO recovered $70 million. Evans, Ben. 2012. *Tragedy and Triumph in Orbit: The Eighties and Early Nineties*. Springer Science & Business Media. P116.

710. Lunney, Glynn S. 2000. NASA Johnson Space Center Oral History Project. Edited Oral History Transcript. Retrieved from http://www.jsc.nasa.gov/history/oral_histories/LunneyGS/LunneyGS_1-28-99.htm. Incidentally, INSAT-1B was deployed from the Space Shuttle cargo bay by mission specialists Dale A. Gardner and Dr Guion S. Bluford, Jr. Bluford was NASA's first African-American astronaut in space.

711. This is a digest of some of the services initiated by INSAT-1B as listed by RAO and other sources. Rao, U. R. 2013. *India's Rise as a Space Power*. Delhi: Cambridge University Press India Pvt Ltd. P146.

712. The declassified SECRET report states that India's 160 ground station using S-band (2.5 GHz) would need to be modified to either S-band (6/4 GHz) or Ku-band (14/11 GHz) at a cost of at least $50 million. CIA Report "India: Space Satellite Options), 23 July 1986, https://www.cia.gov/library/readingroom/docs/CIA-RDP86T01017R000302820001-5.pdf P4

713. Email exchange with the author in July 2016. Dr David Baker, who worked as Managing Director for Space Consultants International during the 1980s, came to this conclusion on the basis of the information made available to him during his consultancy.

714. All five of the INSAT-2 satellites were equipped with a Satellite Assisted Search and Rescue (SASAR), which is part of the international satellite-aided search and rescue programme. India has signed an agreement with the International COSPAS–SARSAT Council for the use and operation of Local User Terminals (LUTs) and an INMCC in Bangalore and Lucknow. The LUTs provide substantial coverage of the Indian Ocean region, as well as other countries: Bangladesh, Bhutan, Maldives, Nepal, Sri Lanka, Seychelles and Tanzania. Retrieved from https://directory.eoportal.org/web/eoportal/satellite-missions/i/insat-2

715. INSAT-2D was launched in July 1997 and was terminated in October 1997 following a problem with an electrical power system. To plug the gap, ISRO procured Arabsat-1C (that had been in orbit since 1992) in November 1997.

716. *Ibid.* P247 and 257.

717. These are some of the innovations described by K. Kasturirangan, who at the time of writing this paper in 1988 was the ISRO Chairman. Kasturirangan, K. and Sridharamurthy K. March 1988. ISRO Spacecraft Technology Evolution. *Sadhana* 12 (3): 251–88. See also ISRO. 2016. *From Fishing Hamlet to Red Planet: India's Space Journey.* Noida, Uttar Pradesh: HarperCollins India. P244.

718. ISRO. 2016. *From Fishing Hamlet to Red Planet: India's Space Journey.* Noida, Uttar Pradesh: HarperCollins India. P365.

719. The satellite capacity to deliver tele-education remains low. ISRO. 2016. *From Fishing Hamlet to Red Planet: India's Space Journey.* Noida, Uttar Pradesh: HarperCollins India. P367.

720. Dr Prem Shanker Goel, writing about Operational Satellites of ISRO, relates the spirit of ISRO engineers building TES to that of the European scientists during World War II. ISRO. 2016. *From Fishing Hamlet to Red Planet: India's Space Journey.* Noida, Uttar Pradesh: HarperCollins India. P263. Also, see P258. "It was a near-miracle to build and launch TES in 2 years, incorporating 11 new technologies like phased-array antenna, step and stare mode for apparent velocity reduction by a factor of 5.8".

721. The six GEO slots referred to in the reference below have now increased with the advent of the IRNSS programme. By mid-2016, two slots, in particular, were becoming congested, 74°E had five operational satellites and 55°E had 4. These numbers will change over time. Kasturirangan, K. and Sridharamurthy K. March 1988. ISRO Spacecraft Technology Evolution. *Sadhana* 12 (3): 251–88. See also ISRO. 2016. *From Fishing Hamlet to Red Planet: India's Space Journey.* Noida, Uttar Pradesh: HarperCollins India. P306.

722. The beacon transmits its unique identification, along with its coordinates, which can be used during the search and rescue operation. These are three types of beacons: (i) EPIRB, Emergency Position Indicating Radio Beacon used in maritime applications. It emits a 5-watt burst about every 52 seconds at 406 MHz; (ii) ELT, Emergency Locator Transmitter for aviation use. Older versions transmit only on 121.5 MHz, but latest models use both 121.5 and 406 MHz; (iii) PLB, Personal Locator Beacon. These are small handheld devices for personal use and not associated with an aircraft or a ship. In October 2000, the International Council of COSPAS–SARSAT ceased processing the 121.5 MHz signal via satellite due to the extremely high level of false alarms.

723. ISRO. 2016. *From Fishing Hamlet to Red Planet: India's Space Journey.* Noida, Uttar Pradesh: HarperCollins India. P302.

724. As reported by the end of 2015. COSPAS–SARSAT Information Bulletin. Issue 26. 2015. Retrieved from https://www.cospas-sarsat.int/images/stories/SystemDocs/Current/Bul26-FINAL-v-091015.pdf

725. India had been participating since 1983 with a ground-based Mission Control Centre, along with Australia, Norway and the UK. Barnes, Richard J. H. and

Jennifer Clapp. November 1995. COSPAS–SARSAT: A Quiet Success Story. *Space Policy* 11 (4): 261–68: 263. The Indian Department of Space makes a passing reference on page 58 in its Annual Report 2014-2015. Government of India, Department of Space. Retrieved from http://www.isro.gov.in/sites/default/files/article-files/right-to-information/AR2014-15.pdf

726. Although India is responsible for registering user details for each COSPAS–SARSAT beacon, it can detect signals from any compatible transmitter. The frequency of 121.5 MHz used by all seafarers and aviators as the emergency frequency was not a frequency best suited for a space-based system, which required the signal to leave and then return through the Earth's atmosphere, a round-trip potentially of almost 80,000 km.

727. The global population of the 406 MHz distress beacon was estimated at 600,000 as of December 2007. Retrieved from https://inmcc.istrac.org/brochurehtml/page1.htm

728. ISRO. 18 March 2002. United Nations, India Workshop on Satellite Aided Search and Rescue (SASAR) Inaugurated. Retrieved from http://www.isro.gov.in/update/18-mar-2002/united-nations-india-workshop-satellite-aided-search-and-rescue-sasar-inaugurated.

Chapter 13

729. Pace, Scott, Gerald Frost, Irving Lachow et al. 1995. *The Global Positioning System: Assessing National Policies*. Santa Monica, CA: RAND. P238.

730. See http://nssdc.gsfc.nasa.gov/nmc/spacecraftDisplay.do?id=1960-003B

731. In September 1993, the US announced at the Annual Conference of the International Civil Aviation Organization that the Global Positioning System established by the US for use by its military would be available for civilian use for free worldwide. See Pace, Scott, Gerald Frost, Irving Lachow et al. 1995. *The Global Positioning System: Assessing National Policies*. Santa Monica, CA: RAND. P248. This was in part triggered by the USSR shooting down a civilian airline KAL007 when it entered the USSR airspace as a result of a navigational error. Degani, A. 2001. Korean Airlines Flight 007: Lessons from the Past and Insights for the Future. Retrieved from ntrs.nasa.gov/archive/nasa/casi.ntrs.nasa.gov/20020043310.pdf

732. ISRO. 28 April 2016. PSLV-C33 Successfully Launches India's Seventh Navigation Satellite IRNSS-1G. Retrieved from http://www.isro.gov.in/update/28-apr-2016/pslv-c33-successfully-launches-indias-seventh-navigation-satellite-irnss-1g

733. ISRO Chairman speaking on January 2014. Jagannathan, Venkatachari. 4 April 2014. ISRO Examining Business Model for Industries in Satellite and Rocket Production. *Two Circles*. Retrieved from

http://twocircles.net/2014apr04/isro_examining_business_model_industries_satelli te_rocket_production.html#.VPOv4FNIKBY.

734. A comprehensive description of the IRNSS program on Earth Observation Portal. https://eoportal.org/web/eoportal/satellite-missions/i/irnss

735. Sometimes, the 35,786-km distance to GSO is reported as 42,164 km. The difference of 6,378 km is equal to the radius of the Earth. Orbital dynamics assume that the Earth is a single point with the mass of the Earth concentrated at that point. The distance of 42,164 km is the distance between the centre of the Earth and the satellite's orbit, and 35,678 km is the distance from the surface of the Earth to the orbit. In addition, all seven of the IRNSS satellites are located at GSO/GEO from where they can remain over India at all times. GPS systems for China, Europe, Russia and the US all use medium Earth orbit, GPS 20,180 km, GLONASS 19,130 km, Beidou 21,150 km and Galileo 23,222 km.

736. Saikiran, Byroju and Vippula Vikram. 2013. IRNSS Architecture and Applications. *KIET International Journal of Communications and Electronics* 1 (3): 25.

737. The report entitled "Economic impact to the UK of a disruption to GNSS", was published in April 2017 in London. Retrieved from http://www.insidegnss.com/node/5520

738. Founders of the space programme of India insisted that its primary objective was societal development. These are just three examples of how the IRNSS service will be exploited: Satellite Mapping to Boost Diary Farming (http://www.business-standard.com/article/current-affairs/satellite-mapping-to-boost-dairy-farming-116033000465_1.html), ISRO Conducts 1st Satellite-based Warning System Trial for Railways (http://indianexpress.com/article/cities/ahmedabad/isro-conducts-1st-satellite-based-warning-system-trial-for-railways/) and ISRO to Map and Provide Management Plans for Heritage Sites and Monuments of National Importance (http://www.isro.gov.in/isro-to-map-and-provide-management-plans-heritage-sites-and-monuments-of-national-importance).

739. The scope of the IRNSS appears to be fluid. The number of spare satellites appears to have increased from two to four, and the expected lifetime of the satellites is somewhere between 5 and 12 years. ISRO Works on 4 Back-up Satellites for IRNSS. 8 January 2016. *The Indian Express.* Retrieved from http://indianexpress.com/article/india/india-news-india/isro-works-on-4-back-up-satellites-for-irnss/. Initially, ISRO had published a lifetime period of 10 years, but the 12-year figure has now become more established. By the time, the system became operational, one of the seven satellites would have been in orbit for three years and another two for at least a year. http://ilrs.gsfc.nasa.gov/docs/2015/ilrsmsr_1106_org_Version1withsignature_for_I RNSS-1D-final.pdf

740 The clock failures are attributed to faulty components. Rubidium clocks in future NavIC satellites will be checked for quality assurance prior to launch. Retrieved from http://www.india.com/news/agencies/navigation-satellite-clocks-

ticking-system-to-be-expanded-isro-2221095/. The Galileo satellites have 4 (2 rubidium and 2 hydrogen maser) clocks onboard each satellite. See https://phys.org/news/2017-07-europe-galileo-satnav-problems-clocks.html

741. ISRO. *ISRO Annual Report 2014-2015*. Government of India, Department of Space. P29. Retrieved from http://www.isro.gov.in/sites/default/files/pdf/AR2014-15.pdf

742. Products are now commercially available where a single device can interface simultaneously with IRNSS, GPS, GLONASS and GAGAN. See http://www.accord-soft.com/IRNSS_systems.html

743. Seventh BRICS Summit. 9 July 2015. Ufa Declaration. Ufa, the Russian Federation. P16. http://brics2016.gov.in/upload/files/document/5763c20a72f2d7thDeclarationeng.pdf

744. A time source designated Stratum 0 generates the most accurate time signal possible. A stratum zero time source is not synchronised with any other time source. A stratum-1 time source is synchronised with a stratum-0 time source. Stratum 2 is synchronised with stratum-1 and so on. This hierarchy is the backbone of the Network Time Protocol and drives the time value around the world. For more, see http://www.ntp.org/ntpfaq/NTP-s-algo.htm#Q-ACCURATE-CLOCK

745. To compensate, the IRNSS onboard clocks are not tuned to the nominal frequency of 10.23 MHz but reduced by the so-called 'factory offset' of 10.229999994484488852 MHz. For more details, see Babu, R., T. Rethika and S. C. Rathnakara. 2012. Onboard Atomic Clock Frequency Offset for Indian Regional Navigation Satellite System. *International Journal of Applied Physics and Mathematics* 2 (4): 270–272. Retrieved from http://www.ijapm.org/papers/109-P20007.pdf

746. Corner cubes were delivered to the Moon by Apollo 11, 14 and 15, as well as USSR's Lunakhood 2 spacecraft. These passive devices continue to operate decades after they landed. The distance between the Earth and the Moon is frequently measured with an accuracy of 3 cm. See http://www.lpi.usra.edu/lunar/missions/apollo/apollo_11/experiments/lrr/

747. Retroreflector Array (RRA) Characteristics. *ILRS*. Retrieved from http://ilrs.gsfc.nasa.gov/missions/satellite_missions/future_missions/irns_reflector.html.

748. Ganeshan, A.S. et al. First Position Fix with IRNSS. July/August 2015. *Inside GNSS*. Retrieved from http://www.insidegnss.com/auto/julyaug15-IRNSS.pdf. P50

749. Laser ranging is a critical element of the IRNSS infrastructure, India is not new to laser ranging. https://cddis.nasa.gov/lw20/docs/2016/posters/P62-Elango_poster.pdf.

750. Interference is the inability of a satnav receiver to receive a usable signal continuously. This can be the (deliberate or unintentional) consequence of a nearby transmitter. Jamming is interference introduced by a device designed to interfere

and prevent reception of the intended signal. Spoofing is also a deliberate action but intended to mislead. It does not prevent the reception of a signal, but a local transmitter generates a false signal posing as the satnav signal. See https://www.newscientist.com/article/dn20202-gps-chaos-how-a-30-box-can-jam-your-life/

751. In early 2017, ISRO reported faults with all three of the rubidium atomic clocks on IRNSS-1A. 3 Atomic Clocks Fail on 1 Indian Satellite, Replacement Prepped. 30 January 2017. GPS World. Retrieved from http://gpsworld.com/3-atomic-clocks-fail-on-1-indian-satellite-replacement-prepped/ ESA's Galileo programme also reported similar problems (ten failures on five satellites), but the built-in redundancy has ensured that all satellites in the Galileo constellation continue to operate.

752. The Airports Authority of India quantifies this reliability as "the position errors with probability greater or equal to 0.9999999 in one hour for en-route through Non-Precision Approach operations and for Precision Approach in 150 seconds". See http://www.aai.aero/public_notices/aaisite_test/faq_gagan.jsp.

753. GAGAN provides a Non-Precision Approach. The standard deviation off latitude, longitude and altitude were found to be less than 4 m, which indicates that the position accuracies of GAGAN are well within the 7.6 m requirement. Ganeshan, A.S. et al. February 2016. GAGAN, India's SSBAS: Redefining Navigation over the Indian Region. *Inside GNSS*. Retrieved from http://www.insidegnss.com/node/4788

754. Picture presented by C. L. Indi, Jt. GM (GAGAN) Surendra Sunda, Manager (GAGAN). Airports Authorities of India. See http://slideplayer.com/slide/752290/

755. This blog post details the various SBAS standards. Prasad, Vasuki, C. R. Sudhir. 28 July 2014. GAGAN: India's First Step to a Future Air Navigation System (FANS). *The Flying Engineer*. Retrieved from http://theflyingengineer.com/flightdeck/gagan-indias-first-step-to-a-future-air-navigation-system-fans/

756. Figure 2 in the following paper provides a map of the 18 TEC stations. Chakravarty, S. C. February 2014. A Novel Approach to Study Regional Ionospheric Variations Using a Real-Time TEC Model. *Positioning* 5 (1): 1–11. doi:10.4236/pos.2014.51001.

757. The mysterious loss that befell the Malaysian Airlines Flight MH370 in March 2014 could not happen to an aircraft using a GAGAN system, which continuously emits location data.

758. Vaid, Rohit. 8 August 2014. New Indian Navigation Technology 'Gagan' to Be Offered to Partner Countries. *Indo-Canadian*s. Retrieved from http://www.indocanadians.ca/featured-stories/new-indian-navigation-technology-gagan-offered-partner-countries/.

759. The number of satellites that are active at any one point in time is subject to change as satellites are decommissioned either because their lifespan has come to an

end or because they have malfunctioned. Additional satellites are launched over time. Thus, the numbers quoted in the table and elsewhere are subject to change. Also, as systems develop towards completion, the precision of data available varies. For example, the number of BeiDou satellites in GSO/GEO orbit is either 5, 8 or 10 depending on the source: (a) http://www.navipedia.net/index.php/BeiDou_Space_Segment, (b) Slide 7 in Update of BeiDou Navigation Satellite System. China Satellite Navigation Office, 9 February 2015 or (c) https://directory.eoportal.org/web/eoportal/satellite-missions/c-missions/cnss

760. This animation provides a helpful description of the unusual orbits: https://www.youtube.com/watch?v=7ZRH4-SPtOU

761. UN Office for Outer Space Affairs. January 2016. International Committee on Global Navigation Satellite Systems: The Way Forward. United Nations. Retrieved from http://www.unoosa.org/res/oosadoc/data/documents/2016/stspace/stspace67_0_ht ml/st_space_67E.pdf. p17

762. The IRNSS Interface Control document is available here: http://irnss.isro.gov.in/register.aspx. There is no charge, but online registration is required.

Chapter 14

763. Press release from ISRO http://www.isro.gov.in/update/18-aug-2017/irnss-signal-space-interface-control-document-icd-ver-11-released

764. To quote Boris Chertok "I contend that if Gagarin's flight on April 12th, 1961 had ended in failure, U.S. astronaut Neil A. Armstrong would not have landed on the Moon on July 20th, 1969." Chertok, Boris and Asif A. Siddiqi. 2009. *Rockets and People Volume III: Hot Days of the Cold War*. NASA. Retrieved from http://history.nasa.gov/SP-4110/vol3.pdf. P79

765. The terms astronaut and cosmonaut are currently used interchangeably depending on where and who is using them. The US uses the term astronaut, the Russians use cosmonaut, the Chinese taikonaut, and it may turn out that India may use the term Vyomanout. In China, the term hangtianyuan is also gaining popularity. See http://archive.defense.gov/pubs/20030730chinaex.pdf. I use the term astronaut only for convenience and consistency.

766. Quoted from Sarabhai's speech on 2 February 1968 during the ceremony for formally dedicating the Thumba launch site to the United Nations. Prime Minister Indira Gandhi was present for the ceremony.

767. Rao, U. R. 2014. *India's Rise as a Space Power*. First edition. Delhi: Cambridge University Press India Private Limited. P86

768. Interview with Rakesh Sharma in August 2013. See

http://astrotalkuk.org/2013/11/03/rakesh-sharma/. For fascinating details on the USSR's secret plan under which the first six cosmonauts, including Gagarin, were selected, see Burgess, Colin and Rex Hall. 2009. *The First Soviet Cosmonaut Team: Their Lives and Legacies.* Springer Science & Business Media. P18

769. Me too in 1958.

770. Interview with Ravish Malhotra. 11 October 2013.

771. Nikitin, S. A. 1985. The Space Flight of the Soviet-Indian Crew. NASA TM-77615.   P6.   Retrieved   from http://ntrs.nasa.gov/archive/nasa/casi.ntrs.nasa.gov/19850012916.pdf. The original report was produced by the Interkosmos Council, USSR Academy of Sciences, in July 1984. The English version was made available by NASA at the link given.

772. Gennady Strekalov, the flight engineer aboard Sharma's flight, had narrowly escaped a launch pad fire and explosion on 26 September 1983, thanks to the Soyuz abort system.

773. On 18 March 1965, two cosmonauts in Voskhod-2 landed 400 km away from their designated target and had to spend two nights in their capsule with night-time temperature of -25°C. Survival training and their familiarity with the region was crucial in the eventual safe recovery.

774. Interview with Rakesh Sharma in August 2013. See http://astrotalkuk.org/2013/11/03/rakesh-sharmaa

775. Email exchange with Rakesh Sharma. April 2016

776. See https://en.wikipedia.org/wiki/Salyut_7 for a list of all the Salyut 7 missions that took up 27 cosmonauts between 1982 and 1986.

777. Despite the extensive speculation to the contrary in the media, these words were not rehearsed. Interview with Rakesh Sharma in August 2013. See http://astrotalkuk.org/2013/11/03/rakesh-sharma/

778. Nikitin, S. A. 1985. The Space Flight of the Soviet-Indian Crew. NASA TM-77615.   P9.   Retrieved   from http://ntrs.nasa.gov/archive/nasa/casi.ntrs.nasa.gov/19850012916.pdf.

779. Ibid

780. Interview with Rakesh Sharma in August 2013. See http://astrotalkuk.org/2013/11/03/rakesh-sharma

781. Ibid

782. Personal communication with Rakesh Sharma dated 9/10/2013.

783. This was the conclusion of Professor U.R. Rao, who became the ISRO Chairman in 1984, the same year as Sharma's flight. Rao, U. R. 2014. *India's Rise as a Space Power.* First edition. Delhi: Cambridge University Press India Private Limited. P90

784. The astronaut flight would cost an additional Rs. 2 crore (about $300,000)

according to an undated India Today clipping provided by N.C. Bhat.

785. This information has been sourced from NASA's Flight Assignment Baseline (FAB) document. It shows that the booking date for the launch was 13 November 1982, though it was common for the FAB to be amended multiple times to reflect operational requirements. Dr David Baker, who worked on flight manifests at NASA headquarters during the early 1980s, supplied this information from his personal archives. Following the loss of Challenger on 28 January 1986, the mission STS-61-H was reassigned to another crew and payload but was eventually cancelled altogether. See https://en.wikipedia.org/wiki/STS-61-H

786. Email correspondence with P. Radhakrishnan. 13 September 2013

787. These details were included in an article published in an ISRO in-house magazine called *Upagrah* made available by N. C. Bhat.

788 In February 1985, less than a year after Sharma's flight, Salyut 7 was almost lost. Whilts unmanned, it lost communication with Earth and a special rescue mssion was launched to rescue it. In the meantime, a fantastical and false story emerged in which the American Space Shuttle was to be used to kidnap Salyut 7 from orbit to steal sensitive military secrets. This was further promoted by a Russian documentory by Roscomos. Bart Hendrickx wrote a factual account available here http://www.thespacereview.com/article/2554/1

789. Ibid

790. This quote comes from an article published in an Indian newspaper, *The Daily Telegraph*, on 9 March 1986.

791. A classified US report indicated that a "New Delhi is likely to keep its payload specialist on standby for a future shuttle flight - perhaps to launch the INSAT-1D satellite scheduled to be ready in 1989". CIA Report "India: Space Satellite Options), 23 July 1986, https://www.cia.gov/library/readingroom/docs/CIA-RDP86T01017R000302820001-5.pdf P4

792. Email communication with P. Radhakrishnan. 13 September 2013. The full quote is as follows "Bhat and I were in Ford Aerospace, USA, for familiarisation with INSAT spacecraft, when we watched on the TV the Challenger flight on January 28, 1986. A little over a minute into the flight, the Shuttle blew into a ball of fire shattering my lifetime dream. Just as steeply as my hopes had soared high into space barely a year earlier, it now made a nosedive. I clearly remember the first question to me from the Selectionboard in 1985 "Why are you in this?" My answer was, "for thrill, excitement and adventure, and also to have something to tell my grandchildren." All that went up in smoke in that moment.

793. National Security Decision Directive Number 254 27/12/1986. 1986. National Security Council. Retrieved from https://reaganlibrary.archives.gov/archives/reference/Scanned%20NSDDS/NSDD254.pdf.

794. Fought, E. Bonnie. January 1988. Legal Aspects of the Commercialization of Space Transportation Systems. *Berkley Technology Law Journal* 3 (1). Retrieved

from
http://scholarship.law.berkeley.edu/cgi/viewcontent.cgi?article=1068&context=btlj.
P103

795. The following resource describes each Space Shuttle mission in thorough detail. https://spaceflight.nasa.gov/outreach/SignificantIncidents/assets/space-shuttle-missions-summary.pdf

796. From an account written in 2009. The full text is available in MS Word format here (with the consent of the author): https://gurbir.co.uk/wp-content/uploads/My_Flirtation_With_Space.doc

797. For ISRO's press release on this meeting, see http://www.isro.gov.in/update/07-nov-2006/scientists-discuss-indian-manned-space-mission

798. Post-independence, Indian Prime Minister Nehru chose to centralise India's national economy using the concept of a five-year plan that was common particularly among countries governed by socialist governments. There is not one but a series of five-year plans; each government department has one. The first plan covered 1951–56. The current 12th plan was published by the Planning Commission in 2011 and covers the period 2012–2017. With minor breaks, India has consistently pursued the five-year plan approach since independence. Previous plans are available here: http://www.planningcommission.nic.in/plans/planrel/fiveyr/welcome.html

799. Priyadarshini, Subhra. 3 January 2009. ISRO Unveils Manned Mission Design. *Nature India.* Retrieved from http://www.natureasia.com/en/nindia/article/10.1038/nindia.2009.2

800. India Plans to Hoist Tricolour on Moon by 2020. 4 January 2009. *Hindustan Times.* Retrieved from http://www.hindustantimes.com/india/india-plans-to-hoist-tricolour-on-moon-by-2020/story-kBha7UjIAPsXlPqTAlOK7H.html

801. During the 1960s, the US and USSR publicly denied there was a race to Moon at the time. Subsequent declassified documents leave no doubt that there was, in fact, a race. Similarly, today India is in a race with China. Even though India got to Mars before China, in pretty much every other respect, China's space programme is well ahead of India's. The report below reveals how India chose to go to Mars only after China's attempt had failed: http://zeenews.india.com/news/sci-tech/a-book-that-reveals-indias-journey-to-mars-and-beyond_1500017.html

802. USA International Business Publications (Ed). 2011. *India Space Programs and Exploration Handbook.* Int'l Business Publications. P125. See also media reports from the time, for example, *The Economic Times* article ISRO, Russian Space Agency Join Hands for Indian Man Mission (http://articles.economictimes.indiatimes.com/2008-12-13/news/27728784_1_moon-mission-human-space-flight-isro).

803. The following piece suggests that India probably could not agree to the price Russia put on the technology transfer. During the "USSR days", India benefited

from many free and very favourable deals. Post USSR days and presence of financially savvy Russian protagonists, favourable deals were no longer on the table for India. See http://www.thehindu.com/sci-tech/science/india-has-not-made-offer-to-russia-to-buy-soyuztma-isro/article369235.ece

804. PTI. 28 December 2012. IAF Developing Parameters for India's Manned Space Mission. *The Economic Times.* Retrieved from http://economictimes.indiatimes.com/news/politics-and-nation/iaf-developing-parameters-for-indias-manned-space-mission/articleshow/17798420.cms?intenttarget=no

805. Also, ISRO had concluded that India (like other nations) would select future astronauts with test pilot experience, so the IAF would be the primary source.

806. For ISRO's press release, see http://www.isro.gov.in/update/31-deCE-2013/media-reports-manned-mission-to-moon

807. The short but full statement is available here: http://pib.nic.in/newsite/erelease.aspx?relid=81367

808. The project director of India's HSF is S. Unnikrishnan. He was also the payload director for the 2014 Crew Module Atmospheric Re-entry Experiment. He describes the state of the HSF programme in section 8.10 entitled Initiatives on Indian Human Space Flight in the following publication: ISRO. 2015. *From Fishing Hamlet to Red Planet: India's Space Journey*. Noida, Uttar Pradesh, India: Harper Collins India.

809. An early design of a space suit that ISRO may use is based on the original Sokol space suits designed by the USSR in 1970. The current proposal for the Indian spacesuit is a modern version from Russia. For more information on this early version of the spacesuit, see http://danielmarin.naukas.com/2013/04/30/el-nuevo-traje-espacial-indio-que-en-realidad-es-ruso/. For the space food menu, ISRO has engaged the same company that supplied the food for Sharma's spaceflight in 1984.

810. This free return option relies on air drag slowing down the spacecraft in Low Earth orbit. After about a week, the speed loss would result in an automatic re-entry and a splash-down provided the crew onboard has sufficient supplies to survive that long. This precautionary approach was designed for Yuri Gagarin's historic 12 April 1961 spaceflight. However, Gagarin was launched into a higher than planned orbit. Re-entry would have occurred long after all the supplies were depleted.

811. Most rockets generate thrust at the bottom and "push" the payload. A Capsule Abort System "pulls" the crew capsule from the top. In the event of a fuel leak at the launch pad or an imminent explosion during the early phase of launch, the CAS, like an ejector seat, removes the crew capsule from the primary launch vehicle at high speed. This YouTube clip shows the CAS saving the crew of Soyuz T-10-1 in 1983: https://www.youtube.com/watch?v=UyFF4cpMVag

812. Even before the Space Age, a variety of insects and animals had made the journey to test rockets, and some even made it to space. In 1935, Stephen Smith in

India transported a small cock and a hen about 800 m across a river by a rocket. Immediately after the World War II, the US tested V2 rockets brought from Germany to the US using mice, monkeys and dogs. A monkey called Yorick and 11 mice were the first to survive a trip to space in September 1951. The first dogs to enter space were launched from Kapustin Yar in 1951(see 'Leading the Way: Soviet Space Dogs'. Accessed 8 August 2017. http://www.sciencemuseum.org.uk/visitmuseum/Plan_your_visit/exhibitions/cosm onauts/race-to-space/soviet-space-dogs). Many human lives have been lost in the pursuit of HSF since 1961, but not as many, had animals not been used. The first life forms to circle the Moon were turtles, wine flies, mealworms, plants, bacteria and other living matter in the USSR's Zond 5 spacecraft. It was launched on a free return orbit to the Moon in September 1968. For an interesting summary of animals used in space launch systems, see http://history.nasa.gov/animals.html. For the guinea pigs used in the Chinese space programme, see http://edition.cnn.com/2003/TECH/space/10/03/china.space.timeline/

813. A DOS audit report highlights the delays and cost overruns arising from the SRE-2:
http://www.cag.gov.in/sites/default/files/audit_report_files/Union_Compliance_Co mptroller_Auditor_General_27_2014_chap_4.pdf

814. ISRO. 2015. *From Fishing Hamlet to Red Planet: India's Space Journey*. Noida, Uttar Pradesh, India: Harper Collins India. P496

Chapter 15

815. The initial announcement indicated that the Rakesh Sharma will be played by one of Bollywood's leading actors, Amir Khan. Since then who will play the central role has become a little unclear. http://www.deccanchronicle.com/entertainment/bollywood/300617/confirmed-aamir-khan-to-star-in-astronaut-rakesh-sharmas-biopic.html

816. A public statement from her husband available here https://www.quora.com/Is-a-film-on-Kalpana-Chawla-being-made

817. Vikram Sarabhai. 1996. Sources of Man's Knowledge. Part of the National Programme of Talks. Series: Exploration in Space. 1966. Reproduced in *Resonance* December 2001. Volume 6(12):89–92. Retrieved from http://www.ias.ac.in/resonance/Dec2001/pdf/Dec2001Reflections.pdf.

818. In one of the more outspoken comments, a day after the launch of Chandrayaan-1Indian astrophysicist N. Sri Raghunandan Kumar declared "once Chandrayaan-1 relays its data on helium-3 stocks to ISRO's master control room, India will have a larger claim on natural lunar resources, when man begins to colonize it in the future." Maitra, Ramtanu. 31 October 2008. India Begins Its Journey to the Moon. *Executive Intelligence Review*. Volume 35(43): 64–66. P64

819. Kasturirangan, K. 20 July 2006. India's Space Enterprise—A Case Study in

Strategic Thinking and Planning. Dr K. Narayanan Oration. Australian National University. P9

820. Ibid

821. 2003 was a good year for the Chinese space programme. Not only did it get the approval for its first Moon mission, but that Moon mission was also part of a three-phase Chinese Lunar Exploration Program. Orbiter missions would be followed by landers and rovers and eventually sample return missions. It was also the same year that China successfully completed its first human spaceflight mission. See https://directory.eoportal.org/web/eoportal/satellite-missions/c-missions/chang-e-1

822. Just as the USSR denied that it was in a race to the Moon with the US during the 1960s, India too publicly denied participating in a race with China. For example, see http://www.washingtontimes.com/news/2005/sep/16/20050916-102917-9131r/. China did get to the Moon before India. However, India got to Mars in 2014 before China. ISRO declared that the remarkably successful MOM was a swift response to the failure of China's Yinghuo-1 mission to Mars. A Book that Reveals India's Journey to 'Mars and Beyond'. 17 November 2014. *Zee News*. Retrieved from http://zeenews.india.com/news/sci-tech/a-book-that-reveals-indias-journey-to-mars-and-beyond_1500017.html. China, in turn, was perhaps motivated to go to the Moon following Japan's success in 2003.

823. ISRO published a great deal of material on Chandrayaan-1, unlike other ISRO missions. Many of the details in this chapter come from Datta, J. and Chakravarty S. C. 2004. *Chandrayaan-1: India's First Mission to the Moon*. ISRO. P5. https://web.archive.org/web/20091012220210/http://www.isro.org/publications/pdf/Chandrayaan-1-booklet.pdf.

824. In 2009, the American Institute of Aeronautics and Astronautics (AIAA) selected ISRO's Chandrayaan-1mission as one of the recipients of its annual AIAA awards. See http://www.liquisearch.com/chandrayaan-1/awards_for_chandrayaan-1

825. ISRO. 2015. *From Fishing Hamlet to Red Planet: India's Space Journey*. Noida, Uttar Pradesh, India: Harper Collins India. P417

826. The Indian Space Science Data Centre (ISSDC) was built within the Byalalu site perimeter. The ISRO/NASA MoU, under which the NASA M3 instrument was delivered to lunar orbit, also included a replica of NASA's M3 Data Processing Subsystem a Payload Operating Centre (POC) workstation to ISSDC. NASA and ISRO. 2006. Memorandum of Understanding between the United States National Aeronautics and Space Administration (NASA) and the Indian Space Research Organisation (ISRO) on Cooperation Concerning NASA's Moon Mineralogy Mapper (M3) Instrument on ISRO's Chandrayaan-1 Mission. 09/05/2006.

827. Interview with the author. Director of the ISRO Satellite Centre SK Shivakumar 26 March 2014. See https://astrotalkuk.org/episode-70-indias-deep-space-network-and-isro-satellite-centre/

828. Ibid

829. Establishing a total cost of space missions can be far from clear. Although Byalalu is critical to other missions, such as MOM, its cost is accounted only once against Chandrayaan-1. Since then, an 11-m antenna has also been established at Byalalu, but it is unclear against which mission its costs are recorded. 32-M Antenna Set up to Track Chandrayaan I. 16 December 2007. *The Hindu.* Retrieved from http://www.thehindu.com/todays-paper/32m-antenna-set-up-to-track-chandrayaan-i/article1968628.ece.

830. With the Upcoming Launch of "Chandrayaan I" Moon Mission, India Aims to Be a Global Player in Space. 3. October 2008. *WikiLeaks.* Retrieved from https://wikileaks.org/plusd/cables/08NEWDELHI2641_a.html

831. This number of 126, as with all statistics, is open to interpretation. Not all these missions were successful; some barely managed to leave the launch pad. Some were fly-bys, and the Moon just happened to be on the way. Chandrayaan-1 for example, is listed twice, orbiter and the Moon Impact Probe.

832. Spurman, J. 2010. Lunar Transfer Trajectories. Retrieved from http://www.book.xlibx.info/bo-other/3569352-1-lunar-transfer-trajectories-spurmann-doc-10-02-version-date-feb.php.

833. This is the result of mass concentrations (or Mascons) that appear randomly below the visible surface of the Moon. Some parts of the lunar interior are denser than others. As a result, the gravitational force of the Moon is not uniform. Bhandari, Narendra. 2004. Scientific Challenges of Chandrayaan-1. The Indian Lunar Polar Orbiter Mission. *Current Science* 86 (11): 1489–98. P1496. Neil Armstrong was also concerned by the potential impact of Mascons on Apollo 11's ability to navigate to its precise landing point. Hansen, James R. 2006. *First Man: The Life of Neil A. Armstrong.* Simon and Schuster. P382. Astronaut Al Worden, who spent a week orbiting the Moon in Apollo 15 command module, was able to detect and map these Mascons by measuring the changes in the spacecraft during his 74 lunar orbits in 1975.

834. Vighnesam N. V. et al. 2009. Precise Orbit Computation of India's First Lunar Mission Chandrayaan-1. Using Accelerometer and Tracking Data during Early Phase. Retrieved from http://issfd.org/ISSFD_2009/OrbitDeterminationII/Vighnesam1.pdf.

835. PROCAD showed a maximum position difference of 0.32% while the minimum position difference was as little as 0.0002 %. Vighnesam N. V. et al. 2009. Precise Orbit Computation of India's First Lunar Mission Chandrayaan-1. Using Accelerometer and Tracking Data during Early Phase. Retrieved from http://issfd.org/ISSFD_2009/OrbitDeterminationII/Vighnesam1.pdf. P12

836. ISRO. 2015. *From Fishing Hamlet to Red Planet: India's Space Journey.* Noida, Uttar Pradesh, India: Harper Collins India. P430

837. India Aims to Map Moon in 2007 Voyage. 16 September 2005. *The Washington Times.* Retrieved from

http://www.washingtontimes.com/news/2005/sep/16/20050916-102917-9131r/

838. ISRO. 2015. *From Fishing Hamlet to Red Planet: India's Space Journey*. Noida, Uttar Pradesh, India: Harper Collins India. P187

839. The target mode used a compact three-mirror f/2.7 optical telescope with a field of view of 12°, which presents strips of 40 km width from 100 km. For details, see http://pds-imaging.jpl.nasa.gov/data/m3/CH1M3_0003/CATALOG/INST.CAT.

840. NASA and ISRO. 2006. Memorandum of Understanding between the United States National Aeronautics and Space Administration (NASA) and the Indian Space Research Organisation (ISRO) on Cooperation Concerning NASA's Moon Mineralogy Mapper (M3) Instrument on ISRO's Chandrayaan-1Mission. P4

841. Lundeen, S. et al. 22 November 2011. Data Product Software Interface Specification. Version 9.10. Brown University. P18

842. Ibid

843. Isaacson, Peter, Sebastien Besse, Noah Petrol, Jeff Nettles and the M3 Team. November 2011. M3 Overview and Working with M3 Data. NASA/ISRO. Retrieved from http://pds-imaging.jpl.nasa.gov/documentation/Isaacson_M3_Workshop_Final.pdf.

844. M3 collected data during five observational sessions: 18/11/08 – 24/01/09 119 images, 25/01/09 – 24/02/09 247 images, 15/04/09 – 27/04/09 197 images, 13/05/09 – 16/05/09 20 images and 20/5/09 – 16/08/09 375 images but from an orbit of 200 km. Isaacson, Peter, Sebastien Besse, Noah Petro, Jeff Nettles and the M3 Team. November 2011. M3 Overview and Working with M3 Data. NASA/ISRO. P13. Retrieved from http://pds-imaging.jpl.nasa.gov/documentation/Isaacson_M3_Workshop_Final.pdf.

845. The TMC captured images of northern Australia while 7,000 km from the Earth. See http://www.isro.gov.in/update/31-oct-2008/chandrayaan-1-camera-tested

846. Bhandari, Narendra. 2004. Scientific Challenges of Chandrayaan-1. The Indian Lunar Polar Orbiter Mission. *Current Science* 86 (11): 1489–98. P1492

847. Goswami, J. N. 2010. An Overview of The Chandrayaan-1 Mission. In *41st Lunar and Planetary Science Conference*, March 1–5, 2010. Retrieved from http://www.lpi.usra.edu/meetings/lpsc2010/pdf/1591.pdf.

848. ISRO. 2015. *From Fishing Hamlet to Red Planet: India's Space Journey*. Noida, Uttar Pradesh, India: Harper Collins India. P437

849. Athiray, P. S. January 2016. The First Detection of Sodium on Lunar Surface from the Chandrayaan-1 X-Ray Spectrometer (C1XS). *PLANEX Newsletter*. Volume 6(1). Retrieved from https://www.prl.res.in/~rajiv/planexnews/newarticles/Volume%20-6,%20Issue-1.7-10.pdf. P6

850 These 11-year cycles have been studied for over three centuries. Instances, where the cycle has not performed as expected, are not unusual. Cycle 24, which started in 2008, has been the weakest in 100 years. See http://www.skyandtelescope.com/astronomy-news/the-weakest-solar-cycle-in-100-years/

851. ESA. 26 January 2009. European Lunar X-ray Camera More Sensitive Than Expected. *Spaceref.* Retrieved from http://www.spaceref.com/news/viewpr.html?pid=27434

852. ESA. 19 October 2009. How the Moon Produces Its Own Water. *Science Daily.* Retrieved from http://www.sciencedaily.com/releases/2009/10/091015091605.htm

853. A signal would be transmitted from Mini-RF on Chandrayaan-1 and received by Mini-RF on LRO. Analysis of the returned backscatter signal as a function of the phase angle of the same area on the Moon would provide potentially the most definitive remote technique for discriminating between ice and rock units. Bussey, D. et al. 2008. Mini-RF Imaging Radars for Exploring the Lunar Poles. *Lunar and Planetary Science* XXXIX. Also, see https://directory.eoportal.org/web/eoportal/satellite-missions/l/lro

854. The following paper provides a detailed analysis of all the data collected by RADOM. Dachev, T., B.T. Tomov, Yu. N. Matviichuk, P. S. Dimitrov, S.V. Vadawale, J. N. Goswami, G. De Angelis and V. Girish. September 2011. An Overview of RADOM Results for Earth and Moon Radiation Environment on Chandrayaan-1 Satellite. *Advances in Space Research* 48 (5): 779–91. doi: 10.1016/j.asr.2011.05.009. P11

855. Goswami, J. N. 2010. An Overview of the Chandrayaan-1 Mission. In *41st Lunar and Planetary Science Conference,* March 1–5, 2010. Retrieved from http://www.lpi.usra.edu/meetings/lpsc2010/pdf/1591.pdf.

856. This orbit-raising manoeuvre was initially presented as an action to save fuel and get a better view of the lunar surface, but in reality, it was to help the spacecraft manage its temperature. Moon's Heat Hastened Indian Probe's Demise. 9 September 2009. *New Scientist.* Retrieved from https://www.newscientist.com/article/mg20327253.500-moons-heat-hastened-indian-probes-demise/

857. Hybrid DC-DC Converters. 2005. Modular Devices Inc. P153 http://www.mdipower.com/pdfcatalog/MDICat_06.pdf

858. Moon's Heat Hastened Indian Probe's Demise. 9 September 2009. *New Scientist.* Retrieved from https://www.newscientist.com/article/mg20327253.500-moons-heat-hastened-indian-probes-demise/

859. Chandrayaan-1Mission Terminated. 31 August 2009. *The Hindu.* Retrieved from http://www.thehindu.com/todays-paper/article216881.ece.

860 Using Earth based instruments NASA located Chandrayaan-1 in lunar orbit in mid-2017. 'New NASA Radar Technique Finds Lost Lunar Spacecraft'.

NASA/JPL, https://www.jpl.nasa.gov/news/news.php?feature=6769.

861. ISRO. 14 November 2007. India and Russia Sign an Agreement on Chandrayaan 2. Retrieved from http://www.isro.gov.in/update/14-nov-2007/india-and-russia-sign-agreement-chandrayaan-2

862. Laxman, Srinivas. 6 February 2012. India's Chandrayaan-2 Moon Mission Likely Delayed After Russian Probe Failure. *Asian Scientist.* Retrieved from http://www.asianscientist.com/2012/02/topnews/india-chandrayaan-2-moon-mission-delayed-after-russian-probe-failure-lev-zelyony-2012/

863. A spacecraft is powered by a battery charged by solar panels. Where mission profile prevents access to the Sun, for example, the lunar surface where the night lasts for 14 days or missions that venture away from the Sun to outer planets where the sunlight is weaker, an alternative power supply is required. Radioisotope Thermal Generators (a radioactive material that produces heat through natural radioactive decay) have been used by NASA and the former USSR in the past. India currently does not have a RTG provision for spacecraft. NASA's rover Curiosity operating on the surface of Mars since 6 August 2012 is fitted with a 5-kg supply of Plutonium 238 Dioxide. It can produce 100W of electricity with an expected lifetime of about 14 years. Sharma, R. 5 January 2016. ISRO Puts Off Nuclear Powered Space Mission. *The New Indian Express.* Retrieved from http://www.newindianexpress.com/nation/ISRO-Puts-Off-Nuclear-Powered-Space-Mission/2016/01/05/article3213005.

864. Successful Commercial Launches Boost ISRO's Reputation in 2015. 29 December 2015. *Business Standard.* Retrieved from http://www.business-standard.com/article/current-affairs/successful-commercial-launches-boost-isro-s-reputation-in-2015-115122900399_1.html

865. ISRO All Ready for Chandrayaan II, says Its Chairman Kiran Kumar. 29 February 2016. *The Times of India.* Retrieved from http://timesofindia.indiatimes.com/india/Isro-all-ready-for-Chandrayaan-II-says-its-chairman-Kiran-Kumar/articleshow/51189668.cms

866. Narasimhan, T. E. 8 February 2017. Chandrayaan-2 Launch Planned by First Quarter of 2018. *Business Standard.* Retrieved from http://www.business-standard.com/article/current-affairs/chandrayaan-2-launch-planned-by-first-quarter-of-2018-117020800902_1.html

867. The ISRO Chairman describes the mission profile as it was in August 2015. https://youtu.be/1TzL1UTELgc?list=PLsh2R7wQHlMVPIv2Et0Fmc-yV_uAhlhKj

868. Rekha, A. R., A. Abdul Shukkoor and P. P. Mohanlal. 2011. Challenges in Navigation System Design for Lunar Soft Landing. Retrieved from http://mohanlalpp.in/mysite/uploads/publish023.pdf. P1

869. Interestingly, this phenomenon does occur on the Earth but high in the Ionosphere. Communication between ground stations and satellites can be effected, so the Total Electron Count is continuously monitored on Earth. The TEC

measurements provide the correction for the Satellite-Based Augmentation System (SBAS) used in aviation to improve the accuracy of navigation satellite position information. See http://www.swpc.noaa.gov/phenomena/total-electron-content

870. ISRO Chairman makes the announcement that there is no race. We Are Not Engaged in Space Race with China: ISRO. 27 October 2013. *The Economic Times.* Retrieved from http://articles.economictimes.indiatimes.com/2013-10-27/news/43432744_1_mars-orbiter-mission-gslv-mom-spacecraft. That India's mission to Mars was motivated by the failure of Chinese Mars mission of 2011 was declared publicly in a book published in 2014, while India's Mars mission was en route to Mars. Bagla, Pallava, and Subhadra Menon. 2014. *Reaching for the Stars: India's Journey to MARS and Beyond.* New Delhi: Bloomsbury Publishing India Private Limited.

871. Lele, Ajey. 2013. *Mission Mars: India's Quest for the Red Planet.* 2014 edition. New Delhi; New York: Springer. P40

872. New Scientist, 9 May 2006, "Indian space agency urges US to lift sanctions". Retrieved from https://www.newscientist.com/article/dn9137-indian-space-agency-urges-us-to-lift-sanctions/

873. After Chandrayaan, It's Mission to Mars: Madhavan Nair. 23 November 2008. *One India.* Retrieved from http://www.oneindia.com/2008/11/23/mission-to-mars-next-ambition-of-isro-madhavan-nair-1227412909.html

874. Saini, Angela. 2016. *Geek Nation.* Hodder Paperbacks. P258. Retrieved from http://www.hodder.co.uk/HodderStoughton/books/detail.page?isbn=97814447101 68.

875. Major Recommendations Emanating from the 97th Indian Science Congress. 2010. http://sciencecongress.nic.in/pdf/97_indian_science.pdf Kalam's recommendations are on P8 and P10 Retrieved from

876. The cause of the failure remains ambiguous. Officially, it was the result of cosmic rays causing corruption of onboard memory chips. But other scenarios have also been proposed. Oberg, James. 16 February 2012. Did Bad Memory Chips Down Russia's Mars Probe? *IEEE Spectrum: Technology, Engineering, and Science News.*

877. Adimurthy, V. 2015. Concept Design and Planning of India's First Interplanetary Mission. *Current Science* 109 (6): 1050–54.

878. Interestingly, whereas Chandrayaan-1 was known as Chandrayaan-1 from the outset, Mangalyaan was not known as Mangalyaan-1. However, ISRO's next mission to Mars is known as Mangalayaan-2. Perhaps, this was also a consequence of the haste under which the Mars mission was executed.

879. Adimurthy, V. 2015. Concept Design and Planning of India's First Interplanetary Mission. *Current Science* 109 (6): 1050–54.

880. Initially, it was a faulty main engine valve that impeded its Trans Mars Injection burn, and later, a powerful solar flare in April 2002 damaged its attitude

control heating system and communications. Despite these problems, the Japanese engineers succeeded in getting Nozomi to Mars in December 2003, but a planned final main engine burn failure prevented a Mars Orbit Insertion on arrival, and Nozomi is now out of contact in a solar orbit.

881. Udupa, S, December 2014, Key Note address - Mars Orbiter Mission, Caltech, USA http://flightsoftware.jhuapl.edu/files/2014/Presentations/Day-2/Session-1/1-MOM-FSW2014_Key_note_address_17122014.pdf P19

882. Some very helpful graphics that illustrate the distances involved and the reasons why spacecraft autonomy is essential for deep space missions. Arya A. S. et al. 2015. Mars Colour Camera: The Payload Characterization/calibration and Data Analysis from Earth Imaging Phase. *Current Science* 109 (6): 1076–86.

883. Arunan, S. and Satish, R. 2015. Mars Orbiter Mission Spacecraft and Its Challenges. *Current Science* 109 (6): 1061–1069. P 1062

884. However, the highly elliptical orbit was not necessarily a disadvantage as was thought at the early planning stage "what initially appeared to be a limitation, can be turned into a unique opportunity by configuring experiments that exploit the highly elliptic orbit."

885. The 4th stage was beyond the view of ISRO's eastern-most ground station Biak in Indonesia. Had the altitude been higher than 100 km, the two large antennae (18 and 32 m) of IDSN could have been used.

886. ISRO leased the two ships for three months to cover launch windows and the unpredictability of the weather given the time of year. Despite the formal separation between civilian ISRO and military DRDO, they are both government organisations and cooperation takes place when required. One of the two terminals for transmission and communication used on one of the ships was borrowed from DRDO. D. S., Madhumathi. 14 September 2013. Scientists on Ships to Track Launch of Mars Orbiter Mission. *The Hindu*. Retrieved from http://icast.org.in/news/2013/sep13/sep14ha.pdf.

887. At about this time, the internal political disagreement had brought to a halt many services provided by the US government. NASA mitigated this impact and confirmed that its DSN would be available to assist the MOM mission as agreed. Retrieved from http://www.isro.gov.in/update/05-oct-2013/nasa-reaffirms-support-mars-orbiter-mission

888. Udupa, Subramanya. 17 December 2014. Mars Orbiter Mission. Presented at the Flight Software Workshop, California Institute of Technology. P37. Retrieved from http://flightsoftware.jhuapl.edu/files/2014/Presentations/Day-2/Session-1/1-MOM-FSW2014_Key_note_address_17122014.pdf

889 Ramachandran, R. 21 September 2014. Mars Orbiter Mission Is on Stable Trajectory: ISRO Chief. *The Hindu*. Retrieved from http://www.thehindu.com/sci-tech/science/interview-with-isro-chief-dr-k-radhakrishnan-isro/article6431827.ece.

890. Adimurthy, V. 2015. Concept Design and Planning of India's First Interplanetary Mission. *Current Science* 109 (6): 1050–54. P1053

891. These objectives were asserted in 2015 after MOM arrived in orbit. February 2015, Dadhwal, V, Mars Orbiter Mission in Orbit http://www.unoosa.org/pdf/pres/stsc2015/tech-03E.pdf

892. This paper authored by the team that built LAP and published a day after MOM's arrival at Mars concluded that "Data downloading and processing are currently under progress". Raja et al. 2015. Lyman Alpha Photometer: A Far-Ultraviolet Sensor for the Study of Hydrogen Isotope Ratio in the Martian Exosphere. *Current Science* 109 (6): 1114–20.

893. The background methane level on Mars is 0.7 ppb (http://www.scientificamerican.com/article/mystery-of-martian-methane-deepens/). A sample of observations is listed below with the year they were made and concentrations in parts per billion. 2013–4: 7 ppb. Curiosity Rover on the Martian surface (http://science.nasa.gov/science-news/science-at-nasa/2014/16dec_methanespike/), 2003: 250 ppb. Earth-based observatories, 2004: 35 ppb. Mars Express (ESA probe in Martian orbit), 1999: 70 ppb. Mars Global Surveyor (NASA probe in Martian orbit) http://www.currentscience.ac.in/Volumes/109/06/1087.pdf For Earth based observations of Martian Methane see Krasnopolsky, V, A et all, 2004, Detection of methane in the Martian atmosphere: evidence for life? Icarus Volume 172, Issue 2, December 2004, Pages 537-547, http://lunar.earth.northwestern.edu/courses/438/krasnopolsky2004.pdf

894. The Curiosity team also reported the detection of organic molecules in the drill samples of mudstones from the floor of Gale Crater. Gronstal, Aaron L. 19 December 2014. Curiosity Detects Methane and Organic Molecules in Gale Crater. *Astrobiology Magazine.* Retrieved from http://www.astrobio.net/news-exclusive/curiosity-detects-methane-organic-molecules-gale-crater/

895. However, MSM may not be able to return data about methane as a result of an inherent design flaw. A recent publication suggests "the reality is we won't be seeing any detections of methane from the Mars methane sensor on MOM". Klotz, Irene. 7 December 2016. India's Mars Orbiter Mission Has a Methane Problem. *Seeker.* Retrieved from http://www.seeker.com/india-mars-orbiter-mission-methane-detector-flaw-red-planet-2133861312.html

896. The data reported in March 2015 was collected in December 2014. See https://www.facebook.com/ISRO/photos/a.1448404935382864.1073741828.144 8364408720250/1607284712828218/?type=1

897. In the following page, the Mars Colour Camera development team discusses the details of the camera. The following sample indicates the distance from Mars and resolution in km, 80000-4.2, 63000-3.3, 43000-2.3 and 372 -0.019. Mathew, K. et al. 2015. Methane Sensor for Mars. *Current Science* 109 (6): 1087–96.

898. Singh, R.P. et all, Thermal Infrared Imaging Spectrometer for Mars Orbiter Mission, Current Science Vol, 10, No. 6 25 September 2015. http://www.i-scholar.in/index.php/CURS/article/viewFile/92259/81779

899. Bhardwaj, A. et al. 16 March 2016. On the Evening Time Exosphere of Mars:

Result from MENCA aboard Mars Orbiter Mission. *Geophysical Research Letters* 43 (5): 1862–1867

900. This "Announcement of Opportunity" is for data from all five of MOM's instruments, not just the Mars Colour Camera. See http://www.isro.gov.in/pslv-c25-mars-orbiter-mission/announcement-of-opportunity-utilising-mars-orbiter-mission-data-mcc

901. A senior NASA scientist, Michale J Mumma who was one of the first to detect Methane n Mars has identified a design error in ISRO Methane sensor for Mars instrument. The signal for Methane is not sufficiently segregated from other data (e.g Carbon Dioxide). The Methane data returned by MOM is not considered to be as reliable as initially designed. Retrieved from https://thewire.in/85859/methane-isro-msm-spectroscopy/

902.   http://www.deccanherald.com/content/595344/isro-mars-orbiter-mission-life.html

903. Zeng, Jianbing. 2009. *Extended Emission Surrounding Nearby Seyfert Galaxies.* Proquest Dissertations and Theses. Maryland: University of Baltimore. P16

904. The instrument in the sounding rocket was designed to make the observation from the Moon. It had a very wide field of view so, although the detection was accurate, the source could not be pinpointed until a year later. Appenzeller, Immo. 2013. *Introduction to Astronomical Spectroscopy.* Cambridge University Press. P22

905. Biswas, S. 1986. Quest for Cosmic Ray Origin: Anuradha Experiment in Spacelab 3. Proceedings of Indian National Science Academy 52A (6):1334–1348. Retrieved from http://insa.nic.in/writereaddata/UpLoadedFiles/PINSA/Vol52A_1986_6_Art04.pdf

906. Astrosat is the product of inputs from numerous research centres. The primary contributors have been Tata Institute of Fundamental Research, Mumbai and Bangalore; Inter-University Centre for Astronomy and Astrophysics, Pune; Raman Research Institute, Bangalore; Physical Research Laboratory, Ahmedabad; Canadian Space Agency, Canada, and Leicester University, the UK.

907. Stars shine by converting hydrogen atoms into helium atoms through nuclear fusion in the core. Energy, including sunlight, is the product. Once all the hydrogen is used up, helium is used for a while instead, but that marks the end of the star's life. How it ends depends on its mass. Low mass stars, like our Sun, will eventually end up first as a white and then a brown dwarf. More massive stars go through a catastrophic explosion called a supernova and end up as a neutron star, pulsar or a black hole.

908. The gas consists of 25% xenon and 75% of something known as P-10. P-10 consists of Argon 90% and Methane 10%. The detectors of the Large Area X-ray Proportional Counters instrument work on a similar principle Chapter 2. Design optimisation of X-ray detectors for SSM. P37 http://shodhganga.inflibnet.ac.in/bitstream/10603/3944/12/12_chapter%202.pdf

909. ISRO. 2016. Astrosat Handbook. Retrieved from http://issdc.gov.in/docs/as1/Astrosat_Handbook.pdf.

910 Department of Space. 'LAXPC Instrument Onboard ASTROSAT Looks at Various X-Ray Sources', 2015. http://www.isro.gov.in/laxpc-instrument-onboard-astrosat-looks-various-x-ray-sources.

911. The actual sensor is a CMOS Fillfactory/ Cypress STAR-250, 512 x 512, 25µm pixels. Kumar, A. et al. 2013. Ultra Violet Imaging Telescope (UVIT) on ASTROSAT. Retrieved from http://arxiv.org/ftp/arxiv/papers/1208/1208.4670.pdf.

912. Kumar, A. et al. 2013. Ultra Violet Imaging Telescope (UVIT) on ASTROSAT. Retrieved from http://arxiv.org/ftp/arxiv/papers/1208/1208.4670.pdf.

913 Singh, K.P. 'Astrosat - India's Multi-Wavelength Astronomy Satellite'. 30 October 2015. http://www.mssl.ucl.ac.uk/www_astro/clusters2015/presentations/kpsingh_astrosat.pdf.

914. Former ISRO Chairman U. R. Rao stated in a newspaper report that MOM was a "great engineering feat" that taught India how to reach the red planet and has sent down good pictures of Mars across millions of kilometres. The MOM-2 spacecraft should ideally have an orbit of 200 km x 2,000 km. It should take better experiments with sharper instruments along and use the bigger GSLV rocket to propel it. Last time, ISRO used the light-lift PSLV." D.S., Madhumathi. 10 August 2016. ISRO Sets the Ball Rolling for Mars Mission-2. *The Hindu.* http://www.thehindu.com/news/national/isro-sets-the-ball-rolling-for-mars-mission2/article8968388.ece.

915. Details of the MOM-2 mission are fluid and will remain so until the proposals have come in and the payload is defined. See http://isro.gov.in/announcement-of-opportunity-ao-future-mars-orbiter-mission-mom-2

916. An announcement Opportunity is a public request for proposals for scientific instruments that should form the package of science instruments for the spacecraft. http://www.isro.gov.in/announcement-of-opportunity-ao-space-based-experiments-to-study-venus.

Chapter 16

917. As a matter of national security, these figures are regarded as top secret, and although a variety of independent sources publish quantitative data, the accuracy cannot be confirmed. These figures come from the Arms Control Association (https://www.armscontrol.org/factsheets/Nuclearweaponswhohaswhat)

918. The report stated that "Kargil highlighted the gross inadequacies in the nation's surveillance capability, particularly through satellite imagery." An executive

summary of the report is available here: http://nuclearweaponarchive.org/India/KargilRCA.html

919. The quality of optical imaging from GEO has been undergoing incremental improvement over the years. TES was the first ISRO spacecraft to exceed 1 m resolution. Cartosat-2A achieved 0.8 m, and Cartosat-2D and Cartosat-3 will be able to achieve 0.65 m and 0.25 m resolution, respectively. Kumar, A. S. Kiran, V. K. Dadhwal and S. K. Shivakumar. March 2013. Indian Remote Sensing Satellite Series and Applications: A Saga of 25 Years. *NNRMS Bulletin*. P23

920. Former ISRO Chairman K. Kasturirangan acknowledges the key role of A. S. Kiran Kumar, ISRO Chairman since January 2015, in developing the technology of step and stare. ISRO. 2016. *From Fishing Hamlet to Red Planet: India's Space Journey*. Noida, Uttar Pradesh: HarperCollins India. P493.

921. See https://directory.eoportal.org/web/eoportal/satellite-missions/t/tes

922. On 2 September 2010, a Bell 430 helicopter crashed in a forest in Andhra Pradesh resulting in the death of the state chief minister and four others onboard. RISAT-2 took 41 images of the forest area in an attempt to locate the helicopter, but the wreckage was not detected. YSR's Death. 6 February 2010. *Mediavigil*. Retrieved from http://mediavigil.blogspot.co.uk/2010/02/ysrs-death.html

923. Satellites in SSPO can enter the Earth's shadow. The exact duration and frequency of eclipses depend on the inclination and altitude of the orbit. To enter the shadow of the Earth, Sun-synchronous orbits must have an inclination of less than 115.36° and an altitude less than 3,316 km. See http://design.ae.utexas.edu/mission_planning/mission_resources/orbital_mechanics/Sun_Synchronous_Orbits.pdf

924. See https://directory.eoportal.org/web/eoportal/satellite-missions/r/risat-1

925. See http://www.aame.in/2012/07/gsat-7-insat-4f-india-military.html

926. IAF's Wait for Own Satellite Gets Longer. 5 January 2017. *Deccan Herald*. Retrieved from http://www.deccanherald.com/content/589900/iafs-wait-own-satellite-gets.html

927. ISRO Gives Eyes in the Sky to Indian Soldiers, Watches Border Day and Night. 2 October 2016. *The Indian Express*. Retrieved from http://indianexpress.com/article/technology/science/isro-gives-eyes-in-the-sky-to-indian-soldiers-watches-border-day-and-night/

928. This information came from a formal announcement by the Defence Minister in the Parliament of India in 2011. See http://www.strategic-affairs.com/details.php? task=other_story&&id=643

929. Rajagopalan, Rajeswari Pillai. January 2014. A New Frontier, Boosting India's Military Presence in Outer Space. *ORF Occasional Paper* #48. Observer Research Foundation. P14

930. In the following paper, Ajey Lele, a former Group Captain with the IAF, documents to a surprising level of detail India's current assets in its Navy, Army

and Air Force. Rajagopalan, Rajeswari Pillai and Narayan Prasad Nagendra. 2017. *Space India 2.0: Commerce, Policy, Security and Governance Perspectives.* Observer Research Foundation. P179. Retrieved from http://www.orfonline.org/research/space-india-2-0-commerce-policy-security-and-governance-perspectives/.

931. In both cases, they were not launched in the numerical order. GSAT-7 was launched before GSAT-6. RISAT-2 was launched before RISAT-1. This is not unusual and can occur as a result of a change in priorities, unavailability of assets or rescheduling due to design or test failures.

932 . These numbers from the Union of Concerned Scientists should be considered as approximate. Apart from the problematic definition of a military satellite, data from satellites not categorised as military can also be used for military purposes. UCS Satellite Database: In-depth Details on the 1,459 Satellites Currently Orbiting Earth. *Union of Concerned Scientists.* Retrieved from http://www.ucsusa.org/nuclear-weapons/space-weapons/satellite-database#.WHowXLaLRE4

933. Many of the spacecraft in orbit have military objectives, and thus details are kept secret. The US's X-37B spaceplane is one such spacecraft that has been on several extended operations in Earth orbit over several years. https://swfound.org/media/205879/swf_x-37b_otv_fact_sheet.pdf

934. In this paper in, Ajey Lele, a former Group Captain with the IAF, documents to a surprising level of detail, India's current assets in its Navy, Army and Air Force. Rajagopalan, Rajeswari Pillai and Narayan Prasad. 2017. *Space India 2.0: Commerce, Policy, Security and Governance Perspectives.* Observer Research Foundation. P179. Retrieved from http://www.orfonline.org/research/space-india-2-0-commerce-policy-security-and-governance-perspectives/

935. Freese, Joan Johson. 2017. Outer Space Treaty and International Relations Theory: For the Benefit of All Mankind. In Lele, Ajey. 2017. *Fifty Years of the Outer Space Treaty. Tracing the Journey.* The Pentagon Press. P20

936. The original 2008 draft is available here: http://www.unog.ch/80256EDD006B8954/(httpAssets)/C4CD83AD4A8B4797C1257CF3003AC425/$file/1319+Russian+Federation+Draft+Updated+PPWT+.pdf

937. Prevention of arms race is a weighty and complex issue, especially when a large number of different national priorities are involved. However, most forum discussions most of the time appear not to be productive. For example, here is an extract from the 2012 conference report "In accordance with the schedule of activities contained in document CD/WP.571/Rev.1, two plenary meetings on agenda item 3 entitled 'Prevention of an arms race in outer space' were held on 5 June and 31 July 2012. There was a lengthy discussion on this issue where delegations reaffirmed their respective positions, which are duly recorded in the plenary records of the sessions (CD/PV.1260 and CD/PV.1265)." See https://documents-dds-

ny.un.org/doc/UNDOC/GEN/G12/626/75/PDF/G1262675.pdf?OpenElement

938. Nuclear Threat Initiative. 25 March 2015. Proposed Prevention of an Arms Race in Space (PAROS) Treaty. See http://www.nti.org/learn/treaties-and-regimes/proposed-prevention-arms-race-space-paros-treaty/.

939. Although 89 nations have signed the Seabed Treaty and 86 have ratified it, only 28 have formally gone through the process of accession. A list of signatories is available here: https://www.state.gov/t/isn/5187.htm#signatory

940. Runoff, Jonathan, and Craig Eisendrath. December 2005. United States, Masters of Space? The US Space Command's "Vision for 2020". Global Security Institute. Retrieved from http://gsinstitute.org/wp-content/uploads/s3/assets/docs/Vision2020_Analysis.pdf.

941. It is unclear how much of the 2020 vision is incorporated into the current defence policy. The analysis is offered in the paper Granoff, Jonathan and Craig Eisendrath. December 2005. United States, Masters of Space? The US Space Command's "Vision for 2020". Global Security Institute. Retrieved from http://gsinstitute.org/wp-content/uploads/s3/assets/docs/Vision2020_Analysis.pdf.

942. Haney, Cecil D. 16 August 2016. Space and Missile Defence Symposium. *USA Strategic Air Command.* Retrieved from http://www.stratcom.mil/Media/Speeches/Article/986484/space-and-missile-defense-symposium/.

943. United Nations. 2016. Conference on Disarmament. Report of the Conference on Disarmament to the General Assembly of the United Nations. P11. Retrieved from http://www.unog.ch/80256EDD006B8954/(httpAssets)/CD32E698D7BD5E29C 125803D0035EFDB/$file/CD2080E.pdf.

944. The complete Ufa 2015 declaration is available here: http://www.brics.utoronto.ca/docs/150709-ufa-declaration_en.html 32 is quoted. "Reaffirming that the exploration and use of outer space shall be for peaceful purposes, we stress that negotiations for the conclusion of an international agreement or agreements to prevent an arms race in outer space are a priority task of the Conference on Disarmament, and support the efforts to start substantive work, inter alia, based on the updated draft treaty on the prevention of the placement of weapons in outer space and of the threat or use of force against outer space objects submitted by China and the Russian Federation...."

945. The Russian Federation addressed to the Secretary-General of the Conference transmitting the comments by China and the Russian Federation regarding the United States of America analysis of the 2014 updated Russian and Chinese texts of the draft treaty on Prevention of the Placement of Weapons in Outer Space and of the Threat or Use of Force Against Outer Space Objects (PPWT)" https://documents-dds-ny.un.org/doc/UNDOC/GEN/G14/050/66/PDF/G1405066.pdf

946. Orlov, Dr Vladimir. 2016. New Threats and Challenges to Global Security: A

View from Russia. In *BRICS 2016 Academic Forum*. Retrieved from http://www.pircenter.org/media/content/files/13/14742852490.pdf

947. Conference on Disarmament Hears from Its President and India on Behalf of the Group of 21.30 June 2015. Retrieved from http://www.unog.ch/80256EDD006B9C2E/(httpNewsByYear_en)/A018DB7463 063189C1257E74005690C8?OpenDocument..

948 The following paper from 2013 discusses the complex picture of the evolving arms race in south Asia. Tandler, Toby Dalton, Jaclyn, and Toby Dalton Tandler Jaclyn. 'Understanding the Arms "Race" in South Asia'. Accessed 13 July 2017. http://carnegieendowment.org/2012/09/13/understanding-arms-race-in-south-asia-pub-49361.

949 Interestingly, the first ASAT test was conducted by the US in 1959. A missile launched from a B47 bomber targeted a US satellite, Explorer 6 in Earth orbit. Bowman, Robert. Star Wars: A Defense Insider's Case Against the Strategic Defense Initiative. 1st Edition. Los Angeles: New York: J P Tarcher, 1986. P14

950. A. P. J. Abdul Kalam, an engineer with deep connections with ISRO and later the President of India, wrote in 1999 "In today's world, technological backwardness leads to subjugation. Can we allow our freedom to be compromised on this account? It is our bounden duty to guarantee the security and integrity of our nation against this threat. Should we not uphold the mandate bequeathed to us by our forefathers who fought for the liberation of our country from imperialism? Only when we are technologically self-reliant will we be able to fulfil their dream" Kalam, A. P. J. Abdul and Arun Tiwari. *Wings of Fire: An Autobiography*. Universities Press, 1999. P91

951. Rai, Cmde Ranjit and Gulshan Luthra. June 2008. Space, China, ASAT, and the Indian Armed Forces. *IndiaStrategic*. Retrieved from http://www.indiastrategic.in/topstories126.htm.

952. Onboard Almaz space station, the USSR tested an R-23 Kartech cannon derived from a 23-millimetre cannon originally designed for a Tupolev-Tu-22 bomber. About 20 rounds were test fired on 24 January 1975 when the Almaz was unmanned and close to re-entry. The rounds, like the Almaz space station, burnt up in the Earth's atmosphere. Zak, Anatoly. 2015. Here Is the USSR's Secret Space Cannon. *Popular Mechanics*. Retrieved from http://www.popularmechanics.com/military/weapons/a18187/here-is-the-soviet-unions-secret-space-cannon/

953. As unusual as it may sound, spacecraft do carry shotguns. The USSR's tradition of equipping re-entry capsules with guns as part of a survival kit should the return to Earth landing occur in an uninhabited area continues to this day. Although not known to have been unpacked, every Soyuz spacecraft to the ISS carries such a gun. Oberg, James. 2001. The Russian Gun at the International Space Station. In Oberg, James. 2001. *Star-Crossed Orbits: Inside the US-Russian Space Alliance*. McGraw-Hill. http://www.jamesoberg.com/russiangun_tec.html.

954. US Statement on Peaceful Use of Outer Space, Thematic Debate of UNGA

First Committee. 19 October 2009. Retrieved from https://geneva.usmission.gov/2009/10/19/outerspace/

955. Rao, Radhakrishna. 6 March 2016. The Strategic Importance of Space Law for India. *Indian Defence Review*. Retrieved from http://www.indiandefencereview.com/news/the-strategic-importance-of-space-law-for-india/.

956. John, Arvind K. August 2012. India and the ASAT Weapon. *ORF Issue Brief #41*. Retrieved from http://www.orfonline.org/wp-content/uploads/2012/08/IssueBrief_41b.pdf.

957. Interview with Dr M. Y. S. Prasad. 12 Jan 2015. See https://www.youtube.com/watch?v=dogXGI55CvM

958. Data breaches have been reported by Indian Railways, Canara Bank and Flipkart. Saha, Preet and Pallavi Reddy. 29 November 2016. Where Are We on Cyber Security in India? *Security Intelligence*. Retrieved from https://securityintelligence.com/where-are-we-on-cybersecurity-in-india/

959. Some reports attribute the loss of ISRO's INSAT-4B to the highly specialised and customised Stuxnet malware. http://www.spamfighter.com/News-15217-Stuxnet-Worm-Responsible-for-Destroying-Indian-Satellite.htm. However, ISRO dismissed the possibility because the specific component, a Programmable Logic Controller, that the Stuxnet worm was designed to attack was not used in INSAT-4B (Find the *Economic Times* article here: http://economictimes.indiatimes.com/tech/internet/cyber-threat-isro-rules-out-stuxnet-attack-on-insat-4-b/articleshow/6733370.cms.) In July 2012, the website of Antrix was the subject of a breach that was initially attributed to Chinese hackers (The news article on *First Post* is here: http://www.firstpost.com/india/website-of-antrix-commercial-arm-of-isro-hacked-2338712.html)

960. Krishnan, M. October 2012. India to Enhance Cyber Defense. *DW*. Retrieved from http://www.dw.com/en/india-to-enhance-cyber-defense/a-16318351

961. The Strategic Importance of Space Law for India' 2016. *Indian Defence Review*. http://www.indiandefencereview.com/news/the-strategic-importance-of-space-law-for-india/.

962. Govt Gets Cracking on Three New Tri-Service Commands. 20 August 2015. *The Times of India*. Retrieved from http://timesofindia.indiatimes.com/india/Govt-gets-cracking-on-three-new-tri-Service-commands/articleshow/48550424.cms.

963. Rajagopalan, Rajeswari Pillai and Narayan Prasad. 2017. *Space India 2.0: Commerce, Policy, Security and Governance Perspectives*. Observer Research Foundation. P209. Retrieved from http://www.orfonline.org/research/space-india-2-0-commerce-policy-security-and-governance-perspectives/.

964. This debris field will evolve over time as collisions between fragments will result in smaller fragments increasing the total number of objects over time. See https://swfound.org/media/9550/chinese_asat_fact_sheet_updated_2012.pdf

965. This was in a statement by Garold Larson, Alternate Representative to the First Committee of the 64th United Nations General Assembly. US Statement on Peaceful Use of Outer Space, Thematic Debate of UNGA First Committee. 19 October 2009. Retrieved from https://geneva.usmission.gov/2009/10/19/outerspace/

966 This is one commitment from a private operator with ambitions to grow space services from LEO. In practice, all new entrants (private or not) will have to enforce de-orbit policies as spacecraft are launched into an increasingly congested area around the Earth. 'OneWeb Vouches for High Reliability of Its Deorbit System - SpaceNews.com'. Accessed 13 July 2017. http://spacenews.com/oneweb-vouches-for-high-reliability-of-its-deorbit-system/

967. Parliamentary Office of Science and Technology. March 2010. Space Debris. Retrieved from http://www.parliament.uk/documents/documents/upload/postpn355.pdf. For more details on the Kessler syndrome, see http://webpages.charter.net/dkessler/files/Kessler%20Syndrome-AAS%20Paper.pdf

968. United Nations. September 2007. Inter-Agency Debris Coordination—Space Debris Mitigation Guidelines.

969. The quote comes from the Bogata declaration. Also in this paper is a reference to an interesting statistic – an upper limit of 2000 satellites in GEO. Probably accurate in the 1980s but current technology can probably exceed that. The current usage of GEO is less than 0.5%. http://scholarlycommons.law.northwestern.edu/cgi/viewcontent.cgi?article=1216&context=njilb

970. Acquiring GEO slots and radio frequencies can be a source of international tension. The COPUOS treaty asserts that these slots are in outer space and thus belong to all nations. In 1976, seven equatorial states, including Colombia, Ecuador and Indonesia, issued the Bogota Declaration, claiming sovereignty over the portion of the geostationary arc above their respective territories. The Bogota signatories argued that the GSO was not, in fact, part of outer space but a distinct region determined by the Earth's gravitational pull. Support for this position by other developing countries has been mixed, partly because ownership by any one nation contradicts the desire of many developing countries for "equitable and/or guaranteed access" to the GSO. However, the equatorial countries obtained the backing of G-77 for a position paper, which declares that "the present regulatory mechanism for assigning orbit positions and radio spectrum does not ensure equitable access to this resource..." and that a new regulatory mechanism is necessary which will take into account "the particular needs of the developing countries including those of the equatorial countries". US Government. 1983. UNISPACE '82: A Context for International Cooperation and Competition. Library of Congress Catalog Card Number 83-600520. P44. Retrieved from http://ota.fas.org/reports/8328.pdf.

971. Lele, Ajey. 2017. *Fifty Years of the Outer Space Treaty: Tracing the Journey.*

Pentagon Press. P23. Retrieved from http://www.idsa.in/book/fifty-years-of-the-outer-space-treaty-tracing-the-journey.

Chapter 17

972. Sarabhai, Vikram. 1970. Atomic Energy and Space Research. A Profile for the Decade 1970-80. Government of India. Retrieved from http://www.iaea.org/inis/collection/NCLCollectionStore/_Public/02/006/2006423.pdf.

973. Prakash, Gyan. 1999. *Another Reason: Science and the Imagination of Modern India*. Princeton: Princeton University Press, P16.

974. The programme was politically motivated, but its ambitions were wide-ranging and noble. In part, its stated aim was "The United States and other free nations have a clear-cut and immediate concern in the material progress of these people. It arises not only from humanitarian impulses but also from the fact that such progress in the underdeveloped areas will advance the cause of freedom and democracy in the world, expand mutually beneficial trade, and help to develop international understanding and good will." Being point four in President Truman's speech, the programme acquired this title. US Department of State. 1950. Point Four: Cooperative Program for Aid in the Development of Economically Underdeveloped Areas. Rev. Washington. P1. Full text is available here: http://pdf.usaid.gov/pdf_docs/Pcaac280.pdf

975. The full text of Nehru's speech at the Bandung Conference in 1955 is available here: http://sourcebooks.fordham.edu/halsall/mod/1955nehru-bandung2.html It was at this conference that the term Third World emerged as an alternative to the capitalist First World and the communist Second World on either side of the Cold War. Initially, intended to represent a collection of countries (mostly former colonial countries) that chose not to align with either, the term Third World today is used interchangeably with the term 'developing nations'.

976. Some key events between 1963 and 2001 that shaped India's international relationships include the war with China in 1962, China's first nuclear test in 1964, war with Pakistan in 1965, launch of China's first satellite in 1970, Bangladesh Liberation War in 1971, India's first nuclear test in 1974, launch of India's first satellite in 1975, Emergency (suspension of democracy) in India from 25 June 1975 to 21 March 1977, the first successful launch of SLV-3 in 1980, launch of the Integrated Guided Missile Development Program in 1983, Pokhran II nuclear tests in 1998 and the terrorist attack in the US on 11 September 2001. The last brought India, as a democracy, closer once again to the US. Except for the three years of Emergency between 1975–77, when Indira Gandhi ruled by decree, democracy has prevailed in India since independence.

977. This delightful term comes from Ajey Lele. Rajagopalan, Rajeswari Pillai and Narayan Prasad. 2017. *Space India 2.0: Commerce, Policy, Security and Governance*

Perspectives. Observer Research Foundation. P180. Retrieved from http://www.orfonline.org/research/space-india-2-0-commerce-policy-security-and-governance-perspectives/

978. Published in 2011, the detailed content and arguments remain relevant even half a decade later. Moltz, J. 2011. *Asia's Space Race: National Motivations, Regional Rivalries, and International Risks*. Columbia University Press. P125.

979. At first glance, this account of ISRO's achievements could come across as exaggerated but is not. Written in an engaging narrative, Touching Lives records first-hand testimony from the people in India's rural communities whose lives ISRO's services were designed to impact. Das, S. K. 2007. *Touching Lives: The Little Known Triumphs of the Indian Space Programme*. Penguin India. P171.

980. The internationally accepted definition is an earnings of $1 (about Rs.67) per day. In India, it is Rs.900 (about $14) per month, which is around 50 US cents per day and fluctuates significantly with exchange rates.

981. Rector, Robert and Rachel Sheffield. 15 September 2014. The War on Poverty After 50 Years. The Heritage Foundation. Retrieved from http://www.heritage.org/research/reports/2014/09/the-war-on-poverty-after-50-years. The US's attempt was not unique. Indian governments, too, have tried similar programmes, for example, Garibi Hatao (Remove Poverty). It was made popular by the Congress Party under Mrs Indira Gandhi, especially during the lead to the 1971 elections. Basu, Raj. 2012. *Understanding the Poverty Amelioration Programmes of the Congress: The Narratives from the Jawaharlal Nehru and Indira Gandhi Years*. Mykolas Romeris University. P372. Retrieved from https://www.mruni.eu/upload/iblock/20c/001_Basu.pdf.

982. Rector, Robert and Rachel Sheffield. 15 September 2014. The War on Poverty After 50 Years. The Heritage Foundation. Retrieved from http://www.heritage.org/research/reports/2014/09/the-war-on-poverty-after-50-years.

983. Panagariya, Arvind. 2008. *India: The Emerging Giant*. Oxford University Press. P137.

984. Tharoor, Shashi. 2015. *India Shastra: Reflections on the Nation in Our Time*. First edition. New Delhi: Aleph Book Company.

985. Bhanumurthy, N. R. and A. Mitra. 2004. Economic Growth, Poverty, and Inequality in the Indian States in the Pre-reform and Reform Periods. *Asian Development Review* 21 (2): 79-99.

986. Chaudhary, Dipanjan Roy. 15 September 2015. India Offers $1 Billion Aid to Afghanistan to Fight Terrorism. *The Economic Times*. Retrieved from http://economictimes.indiatimes.com/news/politics-and-nation/india-offers-1-billion-aid-to-afghanistan-to-fight-terrorism/articleshow/54338150.cms

987. Ibid

988. Report of the Second United Nations Conference on the Exploration and

Peaceful Uses of Outer Space. 1982. P117. Retrieved from http://www.unoosa.org/oosa/documents-and-resolutions/search.jspx?view=&match=a%2Fconf.101

989. Rajagopalan, Rajeswari Pillai and Narayan Prasad. 2017. *Space India 2.0: Commerce, Policy, Security and Governance Perspectives*. Observer Research Foundation. P3. Retrieved from http://www.orfonline.org/research/space-india-2-0-commerce-policy-security-and-governance-perspectives/

990. PricewaterhouseCoopers Private Limited. March 2016. Capacity Crunch Continues: Assessment of Satellite Transponders' Capacity for the Indian Broadcast and Broadband Market. CASBAA. P7. Retrieved from http://edge.casbaa.com/wordpress/wp-content/uploads/2016/04/India-Satellite-Regulation-The-Capacity-Crunch-Continues-March-2016.pdf.

991. Lok Sabha Unstarred Question No. 2412. 11 March 2015. The Government of India, Department of Space. http://www.isro.gov.in/sites/default/files/lu_2412.pdf

992. The payload on GSAT-4 included a GAGAN unit, Hall effect electric propulsion, 6 x C-band and 6 x Ku-band transponders. GSAT-5P payload included 24 x C-band and 12 x Ext C-band transponders.

993. ISRO chairman challenged this number: In total, the country needs 150 additional transponders and not 500 as claimed by private users. ISRO Set to Replace Foreign-leased Transponders in Two Years. 20 May 2015. *The Hindu Business Line*. Retrieved from http://www.thehindubusinessline.com/news/science/isro-set-to-replace-foreignleased-transponders-in-two-years/article7228034.ece

994. Ministry of Science and Technology. 2011. *Working Group Report for the Twelfth Five Year Plan (2012-17)*. The Government of India, P44. Retrieved from http://planningcommission.gov.in/aboutus/committee/wrkgrp12/wg_power1904.pdf.

995. Madhumati D. S. 27 December 2016. Chronic Capacity Shortage Sends ISRO Searching for Lease of Overseas Satellite. *The Hindu*. Retrieved from http://www.thehindu.com/news/national/karnataka/Chronic-capacity-shortage-sends-ISRO-searching-for-lease-of-overseas-satellite/article16946490.ece

996. These figures are from the annual report produced by the Satellite Industry Association. State of the Satellite Industry Report. September 2016. Retrieved from http://www.sia.org/wp-content/uploads/2017/03/SSIR-2016-update.pdf

997. Rajaraman, V. History of Computing in India 1955–2010. 2012. http://www.cbi.umn.edu/hostedpublications/pdf/Rajaraman_HistComputingIndia.pdf. P14

998. Kalam, A. P. J. Abdul and Y. S. Rajan. 2014. *India 2020: A Vision for the New Millennium*. Millennium edition. Penguin. P45.

999 Space India 2.0: Commerce, Policy, Security and Governance Perspectives.

Observer Research Foundation. P215-283. Retrieved from http://www.orfonline.org/research/space-india-2-0-commerce-policy-security-and-governance-perspectives Arup Dasgupta "Unlocking the Potential of Geospatial Data" P56

1000. Foust, Jeff. 2 December 2016. Indian X Prize Team Secures Launch Contract with ISRO. *Spacenews*. http://spacenews.com/indian-x-prize-team-secures-launch-contract-with-isro/

1001. As a member of the Planning Commission, Kasturirangan recalls "But when we discussed the whole thing, they felt that we should enhance the budget of ISRO. In fact, it was the only one instance when the planning commission wanted to increase, whereas they were cutting mercilessly in many other areas. This shows the trust ISRO enjoys in those places". ISRO. 2016. *From Fishing Hamlet to Red Planet: India's Space Journey*. Noida, Uttar Pradesh, India: Harper Collins India, p.494.

1002. A recording of the 31st Professor Brahm Prakash Memorial Lecture delivered by Dr A. S. Kiran Kumar on 21 August 2015 at IISc, Bangalore. In the recording at 22:30, Dr Kumar refers to 40% of the materials used in the Indian cryogenic engine being made in India and the new scheme to engage private industry in risky and innovative R&D. See https://youtu.be/4gfAnfOsjHI

1003. Prashant G. N. 121 Pvt firms help build Mars orbiter, PSLV. 14 October 2014. *Deccan Herald*. Retrieved from http://www.deccanherald.com/content/435822/121-pvt-firms-help-build.html.

1004. See http://www.commercialspaceflight.org/membership/member-organizations/

1005. http://www.deccanherald.com/content/435822/121-pvt-firms-help-build.html.

1006. Details of all spin-offs are listed here: http://isro.gov.in/isro-technology-transfer/space-spin-offs-isro

1007. A short but detailed paper on the capabilities of microprocessor Vikram 1601. Priyanka, R. and S Gnana. 2016. Stage Control Unit for Indigenous Processor-Based System. *IJSTE* 2 (09). Retrieved from http://www.ijste.org/articles/IJSTEV2I9007.pdf.

1008. A list of technologies transferred are available here: http://isro.gov.in/isro-technology-transfer/technologies-transferred

1009. Fok, Evelyn and Malavika Murali. 30 January 2015. Indian Start-ups Are Beginning to Make Their Mark in Space. *The Economic Times*. Retrieved from http://articles.economictimes.indiatimes.com/2015-01-30/news/58625375_1_space-industry-space-sector-isro

1010. In the past, ISRO has expressed its desire to outsource. However, in the absence of companies with mature commercial experience, ISRO has not achieved the level of outsourcing it has sought. The report is still unlikely to have achieved its desired level of engagement with the private sector. In early 2016, ISRO chairman considers the 2020 target date as "tentative" and the scope of the outsourcing as "largely." Laxman, Srinivas. 15 February 2016. Plan to Largely Privatise PSLV Operations by 2020: ISRO Chief. Times of India. Retrieved from http://timesofindia.inia/Plan-to-largely-privatize-PSLV-operations-by-2020-Isro-chief/articleshow/50990145.cms. Also, see http://spacenews.com/india-to-hand-over-pslv-operations-to-private-sector/

1011. Faust, Jeff. The Space Review, Is it time to update the Outer Space Treaty. 5 June 2017. Retrieved from http://www.thespacereview.com/article/3256/1

1012. This short article captures the salient points in the proposed FEAC bill. Sun dahl, Mark J., "The Outer Space Treaty and The Free Enterprise Act: Is International Space Law a Help or a Hindrance? (Op-Ed)" (2017). *Law Faculty Articles and Essays.* 903. http://engagedscholarship.csuohio.edu/fac_articles/903. http://spacenews.com/the-outer-space-treaty-and-the-free-enterprise-act-is-international-space-law-a-help-or-a-hindrance/.

1013. Prasad, Narayan and Prateep Basu. 6 May 2015. Renewing India's Space Vision: A Necessity or Luxury? *NewSpace India.* Retrieved from http://newspaceindia.com/renewing-indias-space-vision/

1014. "Foreign Direct Investment (FDI) up to 100% is allowed in satellites-establishment and operation, subject to the sectoral guidelines of the Department of Space/ISRO, under the government route." See http://www.makeinindia.com/sector/space/

1015. Prokopenkova, Irina. 2015. The Prospects of India and Russia Space Technology Cooperation. *Russian Institute for Strategic Studies.* Retrieved from http://en.riss.ru/analysis/19001/

1016. Rajagopalan, Rajeswari Pillai and Narayan Prasad. 2017. Space India 2.0: *Commerce, Policy, Security and Governance Perspectives.* Observer Research Foundation. P95. Retrieved from http://ww.orfne.org/research/space-india-2-0-commerce-policy-security-and-governance-perspectives/

1017. Joseph, Mathai and Andrew Robinson. 2 April 2014. Policy: Free Indian Science. *Nature* 508 (7494): 36–38. doi:10.1038/508036a.

1018. How the programme worked and an overview of some examples of actual projects undertaken are listed in this presentation: http://www.nrsc.gov.in/uim_2014_proceedings/papers_ppts/UIM2014_US3_Gan eshRaj.pdf. Retrieved on 10 February 2015.

1019. Rao, U. R. 2013. *India's Rise as a Space Power.* Delhi: Cambridge University Press India Pvt Ltd. P191.

1020. Ibid

1021. Patel, V. P. and A. V. Patel. April 2012. Evaluation of Completed RESPOND Projects (April 2010 to December 2011). Retrieved on 10 February

2015   from
http://www.sac.gov.in/respond/Evaluation%20of%20Completed%20Respond%20
Projects.pdf.

1022. RESPOND at SAC Sponsored Research Programme at Academic InstitutesAnnual Report 2014-15. http://www.sac.gov.in/respond/

1023. Coan, Ron. May 2013. Silicon Valley and Route 128: The Camelots of Economic Development. *Journal of Applied Research in Economic Development* 10. Retrieved from http://journal.c2er.org/2013/05/silicon-valley-and-route-128-the-camelots-of-economic-development/.

1024. Indian material scientists offered this explanation when asked why unique alloys were not developed in India. Kalam, A. P. J. Abdul and Y. S. Rajan. 2014. *India 2020: A Vision for the New Millennium.* Millennium Edition. Penguin. P87

1025. Anderson, Robert S. 2010. Nucleus and Nation: Scientists, International Networks, and Power in India. University of Chicago Press, P424.

1026. Ibid

1027. Rajaraman, V. 2012. History of Computing in India 1955–2010. *IEEE Computer   Society.*   Retrieved   from http://www.cbi.umn.edu/hostedpublications/pdf/Rajaraman_HistComputingIndia.pdf. P11

1028. 2017 Nano/Microsatellite Market Forecast published by Spacework Enterprises.   Retrieved   from http://spaceworksforecast.com/docs/SpaceWorks_Nano_Microsatellite_Marke

1029. Federal Aviation Administration. February 2015. The Annual Compendium of   Commercial   Space   Transportation: 2014.   Retrieved   from http://www.globalsecurity.org/space/library/report/2015/2015_compendium.pdf

1030. Perhaps, this was a repeat of the US strategy in 1975. Then, the US provided India free access to their ATS-6 for a year for the SITE experiment. Subsequently, India purchased 4 INSAT-1(A-D) from the US. Perhaps, with this free offer of a transponder, India hopes to commercially engage SAARC nations in future space services.

1031. Sinha, Amitabh. 26 March 2015. SAARC Satellite India's Counter to Other Players Entering Region: ISRO Chief A S Kiran Kumar. *The Indian Express.* Retrieved from http://indianexpress.com/article/india/india-others/saarc-satellite-indias-counter-to-other-players-entering-region-isro-chief-a-s-kiran-kumar/.

1032. In his piece on exploring space as an instrument of India's foreign policy, Video Sagar Reddy goes further and suggests that India should, at least in part, help build the ground stations and provide training, too. Rajagopalan, Rajeswari Pillai and Narayan Prasad. 2017. *Space India 2.0: Commerce, Policy, Security and Governance Perspectives.* Observer Research Foundation. P173. Retrieved from http://www.orfonline.org/research/space-india-2-0-commerce-policy-security-and-governance-perspectives/

1033. Ajey Lele explains the deeper long-lasting reasons why this venture was probably doomed from the outset. "Almost for the last three decades, the Pakistani military leadership/Inter-Services Intelligence agency (ISI) has directly or indirectly ensured that civilian governments would not be able to transform peace talks into a sustainable process." Lele, Ajey. 19 April 2016. Satellite for SAARC: Pakistan's Missed Opportunity. Institute for Defence Studies and Analyses. Retrieved from http://www.idsa.in/idsacomments/satellite-for-SAARC_alele_190416

1034. Section IV International Cooperation has six interesting articles reflecting how France, the US, Russia, Israel, Japan and Australia have collaborated in the past or can do so in the future with India. Rajagopalan, Rajeswari Pillai and Narayan Prasad (Eds). 2017. *Space India 2.0: Commerce, Policy, Security and Governance Perspectives*. Observer Research Foundation. P215-283. Retrieved from http://www.orfonline.org/research/space-india-2-0-commerce-policy-security-and-governance-perspectives/

1035. Deshmukh, Chintamani. 2010. *Homi Jahangir Bhabha*. New Delhi: National Book Trust. P5

1036. The International Lunar Decade's roadmap starts with a series conferences in 2017, endorsement by G20, UNISPACE +50 and the UN GA in 2018 and the official launch in 2019. Retrieved from https://ildwg.wordpress.com

1037. The IIS is the product of a 1998 long-term Inter-Governmental Agreement between four countries (Canada, the US, Russian Federation and Japan) and the 11 members of ESA (Belgium, Denmark, France, Germany, Italy, The Netherlands, Norway, Spain, Sweden, Switzerland and the UK). The agreement shares between its members the management and costs of operation of a permanently inhabited civil space station for peaceful purposes under international law. The complete framework is described here: http://www.esa.int/Our_Activities/Human_Spaceflight/International_Space_Station/International_Space_Station_legal_framework.

1038. Pallava, Bagla. 10 April 2016. India, China Set Eyes on Joint Satellite Constellation for BRICS. *The Economic Times*. Retrieved from http://economictimes.indiatimes.com/news/science/india-china-set-eyes-on-joint-satellite-constellation-for-brics/articleshow/51763274.cms. Also, the director of ICAO's Air Navigation Bureau cited COSPAS-SARSAT as "an interesting and potentially useful precedent for international cooperation on funding, ownership, management, and operation of space systems." See https://www.cospassarsat.int/images/stories/media/Documents/CospasSarsat_quiets uccessstory.pdf

1039. This is a quote from the inaugural speech of President Truman. The full speech is available here: http://www.thisnation.com/library/inaugural/truman.html

1040. This was at a time when the relations between the US and Russia were particularly low as a consequence of the geopolitical relations, especially in connection with Ukraine, Crimea, Syria, Iraq, Iran and Turkey. See http://rbth.com/news/2016/04/12/putin-notes-importance-of-space-cooperation-with-other-nations_584201

1041. Telephone interview with Al Worden. 3 September 2014.

1042. Interview with the Rakesh Sharma. 13 August 2013. See https://youtu.be/up_ANSNTB-U

1043. Tal, Alon. 2006. Speaking of Earth: Environmental Speeches That Moved the World. Rutgers University Press. P126.

1044. An interview recorded with Bill Anders on the 40th anniversary of the mission. See https://www.theguardian.com/science/2008/dec/20/space-exploration-usa-earth-moon

1045. ISRO. 2016. From Fishing Hamlet to Red Planet: India's Space Journey. Noida, Uttar Pradesh: HarperCollins India. P62.

1046. Kazuto Suzuki concluded that "even when Japanese technology became more mature, the industry still considered that space activities were only a 'jacket for entering a major industrialised countries' club'. It has never considered seriously taking the risks of 'industrialisation' and 'commercialisation' of space, and instead continued space R&D for their sheer prestige". Rajagopalan, Rajeswari Pillai and Narayan Prasad. 2017. Space India 2.0: Commerce, Policy, Security and Governance Perspectives. Observer Research Foundation. P276. Retrieved from http://www.orfonline.org/research/space-india-2-0-commerce-policy-security-and-governance-perspectives/

1047. He was working at the Department or Atomic Energy at the time that INCOSPAR was established and was one of initial engineers to join the space programme that Sarabhai interviewed personally. Aravamudan, R. and Gita Aravamudan. ISRO: A Personal History. Noida, Uttar Pradesh, India: HarperCollins India, 2017.

1048. International Science and Technology. October 1963. Homi Bhabha Interview on India's Development Strategy. P93. This reference is also available in TIFR. 2009. Homi Jahangir Bhabha on Indian Science and the Atomic Energy Programme: A Selection. TIFR. P143.

1049. Rai, Saritha. 6 January 2016. India Just Crossed 1 Billion Mobile Subscribers Milestone and the Excitement's Just Beginning. Forbes. Retrieved from https://www.forbes.com/sites/saritharai/2016/01/06/india-just-crossed-1-billion-mobile-subscribers-milestone-and-the-excitements-just-beginning/#261e4be87db0

1050. For example, SpaceX is planning for high-speed, global Internet operation in the early 2020s. It will be provided by 4,425 satellites operating in 83 orbital planes at an altitude ranging from 1,110 km to 1,325 km. See https://www.theguardian.com/technology/2016/nov/17/elon-musk-satellites-internet-spacex. Plans of ViaSat and aerospace firm Boeing include just three satellites providing high speed, high capacity, global coverage Internet access by 2020. See http://www.zdnet.com/article/the-internet-from-space-how-satellites-could-soon-play-a-bigger-role-in-broadband/. Bengaluru-based Astrome Technologies plans to launch 150 satellites weighing 120 kg to provide 50–400

Mbps Internet to a limited number of developing nations by 2020. See http://tech.economictimes.indiatimes.com/news/technology/bengaluru-based-space-startup-astrome-aims-to-provide-superfast-internet-via-satellites/55023828

1051. Specifically, he listed (a) Manned missions to the Moon and Mars and the establishment of space industries. (b) Cost effective space transport systems using hypersonic reusable vehicles (SSTO). (c) Harnessing space energy for generating power and drinking water (d)Developing solar seal for interplanetary missions. (e)Integrating disaster management system utilising space technology. (f) Operational Indian navigation satellites Kalam, A.P.J. and Rajan, Y.S, The Scientific Indian by APJ Abdul Kalam and YS Rajan (Penguin Books India)2010.

1052. Rao, U.R. 2001. *Space Technology for Sustainable Development*. New Delhi: McGraw-Hill Education.

1053. Ibid

1054. Kasturirangan, K. 20 June 2006. India's Space Enterprise—A Case Study in Strategic Thinking and Planning. Dr K. Narayanan Oration. Australian National University. P197. Retrieved from https://crawford.anu.edu.au/acde/asarc/pdf/narayanan/2006_powerpoint.pdf.

www.ingramcontent.com/pod-product-compliance
Lightning Source LLC
Chambersburg PA
CBHW021840020426
42334CB00013B/135